NEMATOLOGY

NEMATOLOGY

Editor

Wendel Garver

scitus
academics

Nematology

Edited by **Wendel Garver**

This Book is Printed on Acid-Free Paper. ♾

ISBN: 9781681179155
Library of Congress Control Number: 2017948944

SCITUS Academics LLC,
616, Corporate Way, Suite 2, 4766,
Valley Cottage, NY 10989

www.scitusacademics.com

Notice

Printed in United States of America.

Preface

Plant parasitic nematodes are so wide spread and universal in distribution that no living plant is likely to be free from their attack. All cultivated crops are usually attacked by one or more species of nematodes. Crop losses by pests (insects, diseases and weeds) are as old as plant themselves but as agriculture are intensified and cropping patterns including the cultivation of high yielding varieties and hybrids are changing over time the impact of the pests becoming increasingly important. It will indeed be a wonder if any crop is free from plant parasitic nematodes [PPN]. Many a times in olden days, nematodes have caused people to migrate due to soil sickness. It has been estimated by the International Meloidogyne Project that nematodes cause annual looses of 78 billion US dollars in developed countries and more than 100 billion in the developing countries. Nematode problems are more severe and complicated in warmer than cooler areas, in horticultural than field crops, mono-culture than multi-culture, plantation crops than natural forests and vegetation. Horticultural crops are more efficient producers of biomass and harvestable produce than the agronomical crops. Nematodes pose a constraint to horticultural development and intensive cultivation. It has been estimated that annually an average 6% loss in field crops, 12% in fruit and nut crops and 11% in vegetables and 10% in ornamental crops is due to nematode infections. Besides causing quantitative losses, nematodes are known to reduce vitamins and minerals in edible plant parts. Nematode damage is less obvious and many a times goes unnoticed. It causes gradual decline in yield. Nematodes cause complex diseases in association with other soil-borne pathogens. Management of plant-parasitic nematodes is complicated by the complexity of the soil environment. Chemical, biological, and cultural methods along with the use of host plant resistance comprise management strategies that have decreased the risk of damage by many nematode species.

This book 'Nematology' covers the nematodes specificities and their interactions with plants are reviewed, considering host plants quality and compatibility. The potentials of nematode resistance and diversity of antagonists and parasitism are discussed, in relation to host specificity and obligate multitrophic relationships. The ecology and management of nematode communities are also discussed, focusing on soil health approaches and new paradigms for plant protection.

Table of Contents

CHAPTER 1

Nematodes: The Worm and Its Relatives

Mark Blaxter

Institute of Evolutionary Biology, The University of Edinburgh, Ashworth Laboratories, Edinburgh, United Kingdom

Browse recently published articles in most issues of leading journals, and there will be mention of "the worm". What is this worm, why is it so keenly studied by so many, and what has it told us about the diversity of life? And why this worm, and not one of the many other worms?

Caenorhabditis elegans Is *the* Worm

The worm is *Caenorhabditis elegans*, a small, bacteriovorous nematode (or roundworm) first described by Emile Maupas in 1900 [1]. While *C. elegans* had been known and studied in the laboratories of nematologists for many years, it was not until Sydney Brenner in Cambridge, United Kingdom, selected this species for his new programme in genetic research [2],[3] that it became a global phenomenon. He wanted a species that was easy to keep, that had tractable genetics (so that mutants could be isolated and crosses made), and that was easy to observe. Brenner attracted a remarkable team of geneticists to join him, and *C. elegans* researchers have won three Nobel prizes for discoveries made using his new model organism.

So, why *C. elegans*? One key feature of this nematode is how easy it has turned out to be to grow, observe, analyse, and manipulate (See Box 1). It thrives in simple petri-dish culture, and has a simple life cycle (Figure 1). It is small, but easy to visualise under the microscope. It is see-through at all stages of development, facilitating the analysis of changes in development, or following experimental manipulation. *C. elegans* is an animal, and so has, like other animals, muscles, a nervous system, a digestive system, skin, and so on. Remarkably, and attractively, in *C. elegans* all these organs and tissues are built with very few cells: Brenner's postdoc John Sulston counted 558 nuclei in a hatching larva, and 959 in an adult hermaphrodite (excluding the germline) [4]–[6]. Sulston and colleagues mapped the origins and fates of all these nuclei during development in the beautifully transparent embryos. *C. elegans*embryos undergo a stereotypical pattern of cleavage from the just-fertilised zygote to the emerging first stage larva, such that (with a few important exceptions) the cell lineage is invariant [4]–[6]. For each cell in any embryo, it is possible to say with certainty where it came from (which cells in earlier embryos were its progenitors) and which cells (and tissues) the cell would contribute to the mature animal.

BOX 1. Setting Up to Study the Worm

There are many small animal species, yet *C. elegans* is the pre-eminent model. This is in part due to the ease of culture, manipulation, and observation of this nematode. Starting a lab to work on the worm requires, initially, only a few key tools: an incubator that maintains a ~20°C environment, a good dissection microscope, and a good Internet connection. To observe developing embryos, an inverted Nomarski (differential interference contrast) compound microscope is sufficient.

- *C. elegans* does not need complex rearing conditions: it feeds on bacteria, and in the lab can be maintained at room temperature on agar plates covered with a lawn of the standard molecular biology bacterium *Escherichicia coli*. No bio-containment is required.
- It is small (adults are ~1 mm in length), and thus millions of nematodes can be housed in a small space.
- It is transparent throughout the life cycle, making it easy to directly observe changes at the cellular level using standard live microscopy. This includes following the development of the embryo from fertilisation to hatching.
- It has a short life cycle, taking only 3 days to proceed from a fertilised egg to a sexual adult (Figure 1). Thus, genetic experiments involving multiple generations can be completed in only a few days.
- Propagation is simple, as the standard sexual morph is the self-fertilising hermaphrodite. Because of this mode of reproduction, issues of inbreeding depression (where inbreeding results in lowered reproductive fitness of lines because of homozygous deleterious mutations) are largely absent. Matrilineal stocks can be propagated for decades.
- Genetic crossing is still possible, as *C. elegans* can also exist as fertile males that successfully mate with hermaphrodites to produce outcross offspring.

- *C. elegans* can be cryopreserved at −80°C, allowing strains to be archived securely.
- The *C. elegans* community has sponsored strain and genetic resources collections, and these are searchable online. Mutant strains can be ordered online, and delivered in days through standard mail.
- The genome sequence, and resources of transgenic strains and of RNA interference reagents targeting all the genes in the genome, make the process of identifying and detailing the genetic underpinnings of traits streamlined.

Many successful researchers have started their independent *C. elegans* labs by using these basic resources to perform imaginative screens for mutations affecting particular phenotypes of interest, and thus identifying new genes controlling key biological systems.

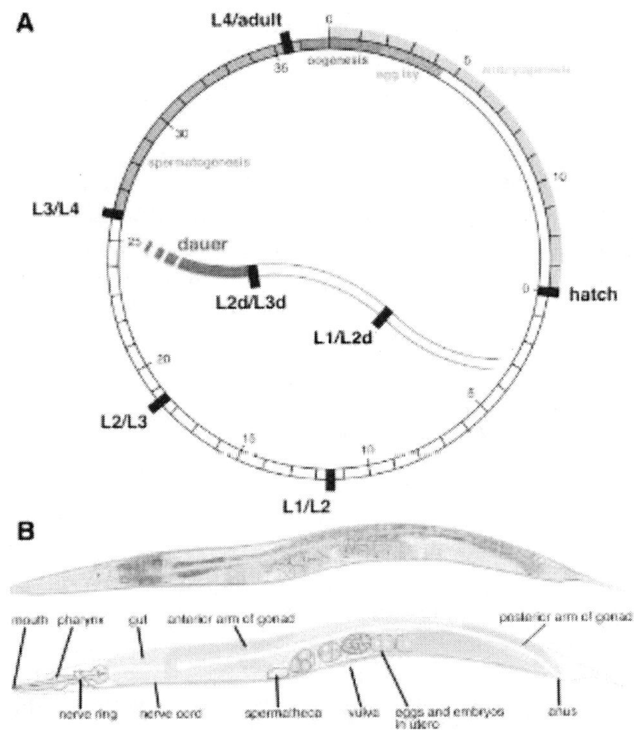

Figure 1. The simple life cycle and anatomy of *C. elegans*.

(A) *C. elegans* has a direct life cycle, with eggs developing through four larval stages into sexual adults. The larvae resemble the adults except in the lack of fully developed gonads, and their smaller size. The illustration shows the timing of developmental events at 25°C, with hours since fertilisation on the outside of the circle, and hours since hatching on the inside. Moults are indicated by solid black bars. In the hermaphrodite, the first ~250 germ cells develop as sperm (after the L3 to L4 moult); later germ cells develop as oocytes. In conditions of overcrowding, starvation, or high temperature, *C. elegans* L1 commit to enter an alternate developmental pathway (via a lipid-storing alternate L2d) that results in the production of a diapausal dauer ("enduring") L3d. The L3d is non-feeding, resistant to environmental insult, and displays arrested ageing. The L3d resumes development when exposed to sufficient food resources. Other nematodes also have a five-stage life cycle, punctuated by four moults, and many species, including parasites, also have a dauer-like L3 stage. (B) The adult hermaphrodite anatomy is simply observed under light microscopy. Above is an adult animal (length ~1 mm). In the cartoon below the major organ systems are indicated.

C. elegans "behaves" much as other animals do—finding food, finding mates, and avoiding danger. However, these behaviours are achieved with a tiny number of neurons: only 302 cell nuclei are present in the adult hermaphrodite nervous system. John White, Sydney Brenner, and colleagues used serial transmission electron microscopy to reconstruct the anatomy and, more importantly, connectivity of this simple nervous system in individual animals [7]. The neurons could be grouped into 118 classes, and their interactions through 7,600 synapses were identified. It remains the only animal nervous system with such a complete wiring diagram, but, frustratingly, it proved impossible to "compute" *C. elegans* behaviour from this, and thus the dynamic field of *C. elegans* neurobiology was founded.

FROM LOCUS TO GENE TO GENOME

Brenner's first paper [3] described 619 visibly mutant strains picked from spontaneously arising variants and from cultures treated with the mutagen ethyl methanesulphonate. These were mapped and used to define six linkage groups, confirming the karyotype (2n=12) and mode of sex determination (males have 2n=11, and sex is determined by the number of X chromosomes). Importantly, these mutants include several that affect development, changing or deleting the fates of cells in the lineage. From these small, promising beginnings, a worldwide community of *C. elegans* researchers grew, using mutagenesis and careful developmental and biological analyses to reveal the genetic underpinnings of development, neurosensation, ageing, and many other phenotypes. The *C. elegans* research field has been openly collaborative from the beginning, with *The Worm Breeder's Gazette* an early example of open-access publishing of research findings by and to a self-defined community (see Table 1). One of the key products of this collaboration was the development of a genetic map, placing all the loci identified across the world on a common framework [8].

Table 1. Resources for *C. elegans* and other nematodes.

Name	URL	Content
WormBase	http://www.wormbase.org/	The *C. elegans* genome database, including the genome sequence, expression pattern data, and genetic mapping information. Also includes comparative analyses of the genomes of other nematodes.
WormBook	http://www.wormbook.org/	The online, peer-reviewed, open-access textbook on *C. elegans* biology, genetics, development, and evolution. Includes the archive of *Worm Breeder's Gazette*. Freely searchable and downloadable.
WormAtlas	http://www.wormatlas.org/	The online virtual worm. This extraordinarily detailed atlas is built from high resolution electron micrographs and includes gene expression patterns and neural connectivity reconstructions.
CGC	http://www.cbs.umn.edu/CGC/	The *Caenorhabditis* Genetics Center, which distributes stocks of *C. elegans* strains and mutants, and many additional species as well.
NEMBASE	http://www.nematodes.org/nembase4/	The Edinburgh comparative nematode transcriptome database
Nematode.net	http://www.nematode.net/	The Washington University in St. Louis nematode genomics server
959 Nematode Genomes	http://www.nematodegenomes.org/	A collaborative wiki collating information on the many nematode genome projects underway or planned round the world and across the phylum.
SON	http://www.nematologists.org/	The US-based Society of Nematologists is the key professional organisation for nematologists worldwide.
ESN	http://www.esn-online.org/	The European Society of Nematologists
RhabditinaDB and WSRN	http://wormtails.bio.nyu.edu/Home.html	David Fitch's reference Web site including Rhabditina evolution, the Worm Systematics Resource Network, and the NYU collection of wild nematode species.

Understanding the action of genes through their mutant phenotypes is revealing, but deeper insight can be won from the molecular nature of their gene products and the details of the lesions induced by mutation. To this end, research teams started using molecular biological tools to isolate the DNA for their genes and describing the biochemical and physiological functions. This process was aided by another community project, undertaken by John Sulston, Alan Coulson, and colleagues, of the generation of a physical map of the *C. elegans* genome [9],[10]. Using a DNA fingerprinting technique, long, contiguous stretches of the chromosomes were assembled from overlapping cosmid clones. As these clones were further analysed, and the marker loci used in genetic mapping were cloned and placed on the physical map, it became ever easier to "clone your gene" from these mapped cosmids.

In the late 1980s, the nascent human genome sequencing program was looking for test beds for technologies to tackle the 3-gigabase human genome. The *C. elegans* genome had been sized at 100 megabases (Mb) [11], and the physical map of overlapping cosmids was ideally suited to the DNA sequencing technologies available. Thus the *C. elegans* genome project was born. In a few short years, the high-quality genome sequence emerging from teams in Cambridge, UK (later at the Sanger Institute), and St. Louis, United States, revolutionised the way *C. elegans* researchers did their science [12]. The publication of the near-complete sequence in 1998 [13] meant that *C. elegans* was the first animal for which the genome was known. The availability of this sequence changed the ways in which the worm could be approached experimentally, and large-scale projects examining gene expression, gene knockout phenotypes, and genetic interactions joined the roster of single-gene, focussed projects. For the human genome project, the *C. elegans* genome consortium proved that dedicated teams, using a clone-by-clone sequencing strategy and the new assembly and analysis tools they developed, could indeed tackle large genomes. Many technologies first developed and used for the *C. elegans* genome, such as fingerprint mapping of large insert clones, using yeast artificial chromosome cloning systems, and the first generation of automated gene finders, have subsequently been used widely.

THE *C. ELEGANS* TOOLKIT

C. elegans has proved to be an excellent model research organism. It is not only easy to grow and study under the microscope, but it also is uniquely amenable to many genetic and other manipulations. Its transparency enables direct screening for defects and changes under the microscope, and technologies such as laser ablation (where individual nuclei are killed by the action of a laser directed through the objective of a microscope), and cell-specific optogenetic manipulation (where light-responsive ion channels and enzymes can be specifically induced in a single or a few cells) are key tools for cell-level investigation of neural and developmental systems. *C. elegans* can be genetically transformed by microinjection of foreign DNA, allowing transgenic analysis of gene function [14],[15]. The use of green fluorescent protein as a transgenic marker was pioneered in *C. elegans* [16]. The phenomenon of RNA interference (RNAi; where double-stranded RNA applied to the organism specifically knocks down expression of the targeted gene) was first discovered and applied in *C. elegans* [17]. *C. elegans* has proved to be uniquely susceptible to RNAi: genes can robustly be knocked down by feeding nematode cultures on *Escherichicia coli* that express double-stranded RNA from the gene of interest. The simplicity of this method means that RNAi "feeding" libraries targeting all of the genes in the genome are available for use in screening [18]. *C. elegans* can be grown in bulk liquid culture and phenotyped, sorted, and counted automatically for high-throughput screening of drugs and other treatments.

"Four-dimensional" microscopy, tracking cells in space and time through development, can be used to define the effects of developmental mutants in a tiny fraction of the time taken by Sulston and colleagues to determine the wild-type lineage [19],[20]. The small genome size and high quality of the sequence (it remains to this day the only absolutely complete animal genome) has in turn enabled all sorts of whole-genome assays. Thus, the model organism Encyclopaedia of DNA Elements (modENCODE) teams have used the full battery of next generation analysis tools (microarrays, DNA methylation analyses, deep sequencing transcriptomics, immunoprecipitation of chromatin bound to transcription factors) to define the regulation of the *C. elegans* genome through development [21],[22]. All of these global surveys, and the many thousands of single-gene and single-system analyses, are collated and cross-referenced in the openly accessible online database WormBase [23] (see Table 1 for *C. elegans* and other data resources).

The simple and accessible nervous system has permitted analysis of many aspects of nervous system development and function of wide importance, including issues such as how neural cells take on specific fates [24], how growing axons find their way and make the correct connections, and how individual neurons integrate the many inputs they experience. While *C. elegans* has very few sensory neurons (the sensory nervous system includes only 39 sensory neurons, most concentrated in the anterior amphids and labial sensillae [7]), the genome sequence surprisingly revealed over 1,200 putative G-protein-coupled transmembrane receptors likely to be involved in sensing the environment. Multiple receptors are expressed in a single neuron, and generation of appropriate responses involves intra- and inter-cellular regulation. The nervous system in *C. elegans*, as in other organisms, is closely integrated with hormonal control of physiology, including the regulation of dauer entry and exit, fat storage, body size, and longevity [25].

C. ELEGANS IS A MODEL ANIMAL

The pattern of development observed in *C. elegans* is markedly different from that seen in other well-studied organisms such as fruit flies or mammals. In flies and mammals, deleting one or a few cells from an embryo usually has no effect on subsequent development: the embryos regulate to replace the structures that would have been produced by the missing cells. In *C. elegans*, however, removal of cells from the embryo is like removing tiles from a mosaic: the other cells cannot change fates to replace the missing parts. Does this mean that work on *C. elegans* is merely the study of a curiosity of little wider relevance? Mosaic development is actually common in small non-vertebrates, and may be an adaptation to the need for rapid, reliable embryogenesis [26], so *C. elegans'* developmental mechanisms are derived from regulative ancestors. Indeed, in the *C. elegans* embryo, the near-invariant pattern of the cell lineage is in fact set up by a series of complex cell–cell interactions. Importantly, this means that the processes and genetic circuits underpinning *C. elegans* development are likely to be common to all animals, and thus work on this simpler model has informed human and other research, and has had a huge impact on medical science.

The importance of *C. elegans* for the study of human biology has two facets. One is the startling finding that many of the genes in the *C. elegans* genome have close homologues in the human, and that many human disease genes are present in the worm. The simplicity of the nematode system makes it a favoured test bed for investigation of the function and interactions of these genes in biological systems affected in disease, including syndromes such as ageing and obesity. The second is the ability to ask simple, direct questions of the *C. elegans* system and thus get simple, direct answers of universal significance.

For example, Robert Horvitz, Paul Sternberg, and colleagues showed that the cell–cell and intracellular signalling pathways involved in the production of the hermaphrodite vulva (a process that takes place in the L3 and L4 stages) are common to all animals, and are also involved in embryogenesis and cancer in humans [27]. Horvitz and colleagues also were the first to define the pathway that controlled the programmed death (apoptosis) of specific cells during *C. elegans* embryogenesis [28]: this pathway is also found in humans, where it is an important regulator of cancerous growth.

As outlined above, RNAi was defined in *C. elegans*, and the phenomenon of RNAi is now known to use systems that are involved in innate immunity to viruses in humans and other organisms. Excitingly, genes encoding endogenous small RNAs, similar to the effector RNAs active in RNAi, were found in *C. elegans* through standard genetic screens investigating developmental mutants [29]. These defined the now burgeoning field of microRNAs (miRNAs), regulatory effectors critical in development and disease in humans, other animals, and plants.

Lastly, the dauer L3 is a non-ageing stage, and the genes that control entry and exit from the dauer were shown to affect the life span of *C. elegans*, even when they did not passage through dauer [30]. This ageing pathway is also effective in other animals, and analysis of Methuselah-like *C. elegans* mutants that live twice as long as wild type has implicated other deeply conserved pathways such as those of insulin signalling. These pathways are also implicated in ageing in other species, including humans.

C. ELEGANS IN THE WILD

In the laboratory, *C. elegans* grows and thrives in a two-dimensional world of agar plates, and copious food in the form of *E. coli*. Obviously, this is an artificial environment. *C. elegans* is often introduced as a "soil nematode" but it is very rarely isolated from soils. The reference strain used since Brenner's pioneer experiments is "N2", established from spent mushroom compost [31], and most isolations have been from organic-rich environments such as urban compost heaps. However, while compost heaps are wilder than agar plates, they are still artificial environments constructed by humans. Where do *C. elegans* live when not living on human-concentrated rotting vegetation, or being cosseted on agar plates? A worldwide search for *C. elegans* by Marie-Anne Félix, Asher Cutter, and their colleagues has identified rotting fruits in temperate regions as a likely true wild habitat for this species [32]–[36].

This discovery has made the task of collecting wild *C. elegans* a much more reliable pursuit, but raises new questions. How does *C. elegans* get to rotting fruit? What does the species do outside the fruiting season? The answers to these questions are still being worked out, but it is likely that the dauer L3 plays a key role. The dauer is an arrested form, and dauers can be harvested from the soils around rotting fruits: it is likely that they persist in the environment until the next food source drops from the tree. Dauers of *Caenorhabditis* species are also often found attached to the outsides of insects, woodlice, and millipedes. These arthropod species probably act as transport hosts for the nematodes, carrying them from one food source to another. *C. elegans* has been isolated from temperate sites worldwide, from Australia to Africa, and Canada to Asia [32],[37]. The isolates have usually been from locations constructed by human action (e.g., compost heaps), and it is thus likely that the nematodes have been spread also by human action. Global transport of rooted plants and fruit, and wholesale transfer of soils, will also have efficiently carried *C. elegans*. As would be expected from this model, there is little global differentiation across *C. elegans* populations. Using highly variable microsatellite genetic markers, no evidence of isolation by distance was found, and small local areas contained as much genetic diversity as different continents. In this, *C. elegans* resembles the other key non-vertebrate model organism, the fruit fly *Drosophila melanogaster*. *D. melanogaster*, another lover of rotting fruit, has also been recently dispersed by human action from its origins in West Africa, and these diaspora populations show low levels of genetic distinction.

Interestingly, the "wild type" reference *C. elegans*, Brenner's N2 strain, is actually a multiple mutant, selected for growth in artificial lab conditions, and it may not be representative of most truly "wild" *C. elegans*. Wild males secrete a mucus plug over the hermaphrodite vulva during mating [38], but

N2 does not plug, due to a recent loss-of-function mutation [39]. N2 nematodes range widely on the agar plates seeded with *E. coli*, leaving the bacterial lawn frequently, but most wild strains do not leave the bacterial lawns, clumping wherever the bacterial growth is thickest. This difference is due to another recent reduction-in-function mutation in N2 in a neuropeptide receptor gene [40],[41].

NOT ALL NEMATODES ARE *C. ELEGANS*

When "traps" are laid to catch *C. elegans*, most of the nematodes that are caught are not the chosen worm. There are many bacteriovorous and fungivorous nematodes in soil and compost attracted to the rotting baits. Some of these are other *Caenorhabditis* species, such as the *C. briggsae* that Brenner initially worked on [34]. There are now about 25 known species in the genus *Caenorhabditis* [37],[42],[43] and many of these have been developed as satellite models to the main project. Using these species, it is possible to examine how the specific traits and genomic architectures of *C. elegans* came to be as they are, and thus develop predictive models of evolution. Species from other relatively closely related genera such as*Pristionchus* [44],[45] and *Oscheius* [46] have also been used as alternate models.

Caenorhabditis is part of a diverse radiation of terrestrial nematodes, the Rhabditina. The Rhabditina includes not only free-living species such as *C. elegans*, but also nematodes that associate with insects and other arthropods, and species that are important animal parasites. The free-living rhabditids are important members of terrestrial ecosystems, part of the ecological webs that drive soil productivity. The arthropod-associated species include those that just use their hosts for transport, and several that are pathogens or parasites of insects. Some of the insect-pathogenic nematodes have been developed as safe biocontrol agents for crop pests, and can be purchased (as arrested dauer stages) from garden stores. The Rhabditina also includes a very important group of vertebrate parasites, the Strongyloidea. Strongyloids such as the human hookworm *Necator americanus* are important determinants of human health in tropical countries [47],[48], and major efforts are underway to develop new drugs and vaccines for the devastating diseases they cause. In these efforts, *C. elegans*research plays a major role, acting as a test bed for drugs, and an archetype onto which the specific details of parasite biology can be mapped. For example, the infective stage in Strongyloids is a dauer-like L3, and discovery of drugs that prevent dauer exit, or mis-specify post-dauer development, may have important roles in community control programmes. Many agricultural animals are also susceptible to infection by a range of strongyloid species, and again *C. elegans* is used in preliminary studies for veterinary drug development.

THE PHYLUM NEMATODA

Rhabditina is only one small part of the diversity of the phylum Nematoda. Nematodes are very diverse, not only in morphology (despite a general perception that nematodes are boring, they in fact have lots of morphological diversity), but also in size (adults from less than a millimetre to over 6 metres), life cycles (from parthenogens to complex cycles of alternating sexual strategies), and ecology (including parasites of almost all other large multicellular organisms, plant and animal). While only about 23,000 species have been described, current estimates suggest that there may be over a million nematode species on Earth [49]. Most species are members of the meiofauna that lives in marine sediments, where nematodes outnumber all other animals many fold [50]. Nathan Cobb, a pioneer nematologist, asked his readers to imagine a world where everything except the nematodes had been magically taken away: "our world would still be dimly recognizable...we should find its mountains, hills, vales, rivers, lakes, and oceans represented by a film of nematodes" [51].

Understanding of the phylogenetic relationships of nematodes has been changed by the use of DNA sequence data [52]–[54]. The new view of phylum Nematoda (Figure 2) [55],[56] shows three major branches, the Enoplia, Dorylaimia, and Chromadoria. *C. elegans* is placed in the Chromadoria, along with the Tylenchina (a group that includes important plant parasites, including many that devastate crops worldwide, such as *Meloidogyne incognita*, a species that can parasitise a surprisingly wide range of hosts, as well as free-living and animal parasitic species), Spirurina (which are all animal parasites, including those causing human filariases—river blindness [*Onchocerca volvulus*] and elephantiasis [*Brugia malayi*]), and other Rhabditina. In the Dorylaimia are terrestrial predatory species that play key roles in food webs, and insect and animal parasites. One of these dorylaim parasites is *Trichinella spiralis*, the trichina worm, a fascinating species that can infect many vertebrates and non-vertebrates, and causes a nasty disease in humans when diapausing larvae (the L1 stage in this case) are ingested in uncooked meats, usually pig or wild meats such as bear. The Enoplia are mainly marine, and include microbivores, predators, and a group of terrestrial herbivores (or plant parasites), the Trichodoridae. Trichodorids such as *Xiphinema index* affect their plant hosts by both feeding on the roots, and through specific transmission of devastating viruses. Parasitism of animals and plants has arisen multiple times in the Nematoda, and convergent evolution in other traits is also common [56]–[58].

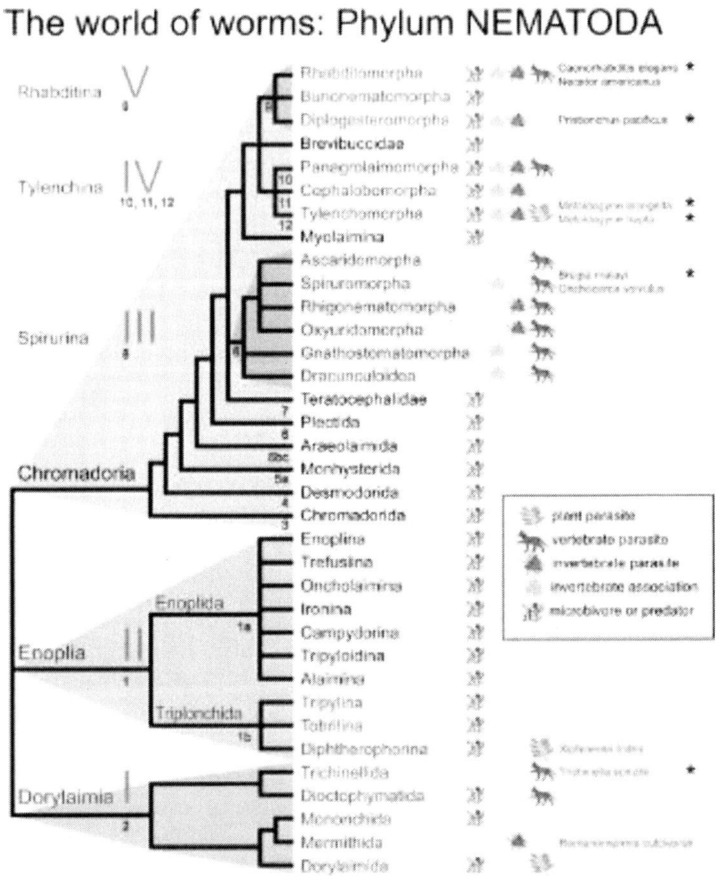

Figure 2. The relationships of the Nematoda.

This phylogeny is based on molecular phylogenetic analyses utilising the small subunit ribosomal RNA gene. The systematic names given by De Ley and Blaxter [55],[56] are given, as is the "clade" naming convention introduced by Blaxter et al. in 1998 [52]. More recently, Helder and colleagues [53],[77] have introduced a numerical clade name scheme: this is given in outlined letters below the relevant branches. Feeding mode, and animal and plant parasitic and vector associations, are indicated by small icons, and representative species are named for some groups. Species with a sequenced genome are indicated by an asterisk.

One of the important results to emerge from the comparison to other nematodes is that the extreme mosaicism seen in *C. elegans* development is not found in all species [59]–[62]. Mosaic development in *C. elegans*, and related nematodes in the Chromadoria, is a derived trait. These and other comparisons are contextualising the details of the *C. elegans* project, as well as pointing out where this model nematode has followed a very idiosyncratic evolutionary path.

NEMATODE GENOME PROJECTS

Research on the huge number of other nematode species does not approach that on *C. elegans* in its depth or detail, but there are especially large literatures on the human parasites and the diseases they cause. One way in which the diversity of nematodes has been approached is through comparative genomics. Initially, this was achieved through directed sequencing of the expressed genes of the target species (the transcriptome approach). Over 60 transcriptome datasets have been generated for free-living, animal-parasitic, and plant-parasitic species [63]. Furthermore, using the *C. elegans* genome project as a methodological and biological guide, teams have developed complete genome sequences for plant parasites (*M. incognita* [64] and *Meloidogyne hapla* [65]) and animal parasites (*B. malayi* [66] and *T. spiralis* [67],[68]), as well as additional free-living species (*Pristionchus pacificus* [69],[70] and additional *Caenorhabditis* species [71]). The *C. elegans* genome, at 100 Mb, is small compared to that of humans (which is 30 times bigger), but appears to be about standard for nematodes (the other sequenced species genomes range from 50 Mb to 120 Mb). The advent of new sequencing technologies has spurred a major increase in the scale of nematode genomics, and nearly a hundred genome projects are under way or planned [72]. These new genomes will reveal not only the special biology of the individual species they represent, but also expand the reach and universality of the ongoing *C. elegans* programme.

PUTTING *THE* WORM ON THE TREE OF LIFE

Molecular data have also clarified the position of Nematoda in relation to other animals. Before the late 1990s, nematodes, along with a rag-bag of other soft-bodied, "wormy" phyla, had been placed in a group termed the Pseudocoelomata (describing the nature of the body cavity in these taxa). However, the morphological arguments supporting this superphylum were never strong, and despite the absolute certainty expressed in textbook treatments of the phylogeny of the animals, leaders in the field, such as

Libby Hyman, always expressed grave doubts as to the biological reality of this grouping [73]. Analysis of ribosomal RNA sequence data from a range of nematodes, however, suggested instead a radical rearrangement of the animal part of the tree of life [74]. In this new model, which has strong support from several genes and some support from morphological data, Nematoda is part of a superphylum of moulting animals, the Ecdysozoa [74], that includes Arthropods (and thus *D. melanogaster*, the other major non-vertebrate model), Nematomorpha (horsehair *worms*), Onychophora (velvet *worms*), Tardigrada (water bears), Priapulida (penis *worms*), and other minor phyla. The rest of the "pseudocoelomates" are now placed in the Lophotrochozoa [75],[76], a group that includes Mollusca (snails and clams), Annelida (rag*worms* and earth*worms*), and Platyhelminthes (flat*worms*), amongst others.

Thus, *the* worm is only one nematode of many, and nematodes are only one sort of worm. Despite this, *C. elegans* is still a model organism *par excellence*: it is a good model nematode, and a good model animal, and a good model for the basic biology that underpins all life.

REFERENCES

1. Maupas E (1900) Modes et formes de reproduction des nematodes. Archives de Zoologie Experimentale et Generale 8: 463–624.E. Maupas1900Modes et formes de reproduction des nematodes.Archives de Zoologie Experimentale et Generale8463624
2. Brenner S (2010) Sydney Brenner and Caenorhabditis elegans. S. Brenner2010Sydney Brenner and Caenorhabditis elegans.Caenorhabditis elegans WWW Server. Available:http://elegans.swmed.edu/Sydney.html. Accessed 14 March 2011. Caenorhabditis elegans WWW Server. Available: http://elegans. swmed.edu /Sydney.html. Accessed 14 March 2011.
3. Brenner S (1974) The genetics of Caenorhabditis elegans. Genetics 77: 71–94.S. Brenner1974The genetics of Caenorhabditis elegans.Genetics777194
4. Sulston J. E (1985) Neuronal cell lineages of the nematode Caenorhabditis elegans. Cold Spring Harb Symp Quant Biol 50: 443–452.J. E. Sulston1985Neuronal cell lineages of the nematode Caenorhabditis elegans.Cold Spring Harb Symp Quant Biol50443452
5. Sulston J. E, Schierenberg E, White J. G, Thompson J. N (1983) The embryonic cell lineage of the nematode Caenorhabditis elegans. Dev Biol 100: 64–119.J. E. SulstonE. SchierenbergJ. G. WhiteJ. N. Thompson1983The embryonic cell lineage of the nematode Caenorhabditis elegans.Dev Biol10064119
6. Sulston J, Horvitz H. R (1977) Post-embryonic cell lineages of the nematodeCaenorhabditis elegans. Dev Biol 56: 110–156.J. SulstonH. R. Horvitz1977Post-embryonic cell lineages of the nematode Caenorhabditis elegans.Dev Biol56110156
7. White J. G, Southgate E, Thomson J. N, Brenner S (1986) The structure of the nervous system of the nematode C. elegans. Philos Trans R Soc Lond Series B Biol Sci 314: 1–340.J. G. WhiteE. SouthgateJ. N. ThomsonS. Brenner1986The structure of the nervous system of the nematode C. elegans.Philos Trans R Soc Lond Series B Biol Sci3141340

8. Hodgkin J (1997) Appendix 1: genetics. In: Riddle D, Blumenthal T, Meyer B, Priess J, editors. C. elegans II. Cold Spring Harbor: Cold Spring Harbor Press. pp. 881–1047.J. Hodgkin1997Appendix 1: genetics.D. RiddleT. BlumenthalB. MeyerJ. PriessC. elegans IICold Spring HarborCold Spring Harbor Press8811047
9. Coulson A. R, Sulston J. E, Brenner S, Karn J (1986) Towards a physical map of the genome of the nematode C. elegans. Proc Natl Acad Sci U S A 83: 7821–7825.A. R. CoulsonJ. E. SulstonS. BrennerJ. Karn1986Towards a physical map of the genome of the nematode C. elegans.Proc Natl Acad Sci U S A8378217825
10. Coulson A, Huynh C, Kozono Y, Shownkeen R (1995) The physical map of the Caenorhabditis elegans genome. Methods Cell Biol 48: 533–550.A. CoulsonC. HuynhY. KozonoR. Shownkeen1995The physical map of the Caenorhabditis elegans genome.Methods Cell Biol48533550
11. Sulston J. E, Brenner S (1974) The DNA of Caenorhabditis elegans. Genetics 77: 95–104.J. E. SulstonS. Brenner1974The DNA of Caenorhabditis elegans. Genetics 7795104
12. Wilson R, Ainscough R, Anderson K, Baynes C, Berks M, et al. (1994) 2.2 Mb of contiguous nucleotide sequence from chromosome III of C. elegans. Nature 368: 32–38.R. WilsonR. AinscoughK. AndersonC. BaynesM. Berks19942.2 Mb of contiguous nucleotide sequence from chromosome III of C. elegans.Nature3683238
13. The C. elegans Genome Sequencing Consortium (1998) Genome sequence of the nematode C. elegans: a platform for investigating biology. Science 282: 2012–2018.The C. elegans Genome Sequencing Consortium1998Genome sequence of the nematode C. elegans: a platform for investigating biology.Science28220122018
14. Fire A (1986) Integrative transformation of Caenorhabditis elegans. EMBO J 5: 2673–2680.A. Fire1986Integrative transformation of Caenorhabditis elegans.EMBO J526732680
15. Stinchcomb D. T, Shaw J. E, Carr S. H, Hirsh D (1985) Extrachromosomal DNA transformation of Caenorhabditis elegans. Mol Cell Biol 5: 3484–3496.D. T. StinchcombJ. E. ShawS. H. CarrD. Hirsh1985Extrachromosomal DNA transformation of Caenorhabditis elegans.Mol Cell Biol534843496
16. Chalfie M, Tu Y, Euskirchen G, Ward W. W, Prasher D. C (1994) Green fluorescent protein as a marker for gene expression. Science 263: 802–805.M. ChalfieY. TuG. EuskirchenW. W. WardD. C. Prasher1994Green fluorescent protein as a marker for gene expression.Science263802805
17. Fire A, Xu S, Montgomery M. K, Kostas S. A, Driver S. E, et al. (1998) Potent and specific genetic interference by double-stranded RNA in Caenorhabditis elegans. Nature 391: 806–811.A. FireS. XuM. K. MontgomeryS. A. KostasS. E. Driver1998Potent and specific genetic interference by double-stranded RNA in Caenorhabditis elegans.Nature391806811
18. Kamath R. S, Fraser A. G, Dong Y, Poulin G, Durbin R, et al. (2003) Systematic functional analysis of the Caenorhabditis elegans genome using RNAi. Nature 421: 231–237.R. S. KamathA. G. FraserY. DongG. PoulinR. Durbin2003Systematic functional analysis of the Caenorhabditis elegans genome using RNAi. Nature421231237
19. Fire A (1994) A four-dimensional digital image archiving system for cell lineage tracing and retrospective embryology. Comput Appl Biosci 10: 443–447.A. Fire1994A four-dimensional digital image archiving system for cell lineage tracing and retrospective embryology.Comput Appl Biosci10443447
20. Schnabel R, Hutter H, Moerman D, Schnabel H (1997) Assessing normal embryogenesis in Caenorhabditis elegans using a 4D microscope: variability of development and regional specification. Dev Biol 184: 234–265.R. SchnabelH. HutterD. MoermanH. Schnabel1997Assessing normal embryogenesis in Caenorhabditis elegans using a 4D microscope: variability of development and regional specification.Dev Biol184234265
21. Niu W, Lu Z. J, Zhong M, Sarov M, Murray J. I, et al. (2011) Diverse transcription factor binding features revealed by genome-wide ChIP-seq in C. elegans. Genome

Res 21: 245–254.W. NiuZ. J. LuM. ZhongM. SarovJ. I. Murray2011Diverse transcription factor binding features revealed by genome-wide ChIP-seq in C. elegans.Genome Res21245254

22. Gerstein M. B, Lu Z. J, Van Nostrand E. L, Cheng C, Arshinoff B. I, et al. (2010) Integrative analysis of the Caenorhabditis elegans genome by the modENCODE project. Science 330: 1775–1787.M. B. GersteinZ. J. LuE. L. Van NostrandC. ChengB. I. Arshinoff2010Integrative analysis of the Caenorhabditis elegans genome by the modENCODE project.Science33017751787
23. Harris T. W, Antoshechkin I, Bieri T, Blasiar D, Chan J, et al. (2010) WormBase: a comprehensive resource for nematode research. Nucleic Acids Res 38: D463–D467.T. W. HarrisI. AntoshechkinT. BieriD. BlasiarJ. Chan2010WormBase: a comprehensive resource for nematode research.Nucleic Acids Res38D463D467
24. Hobert O, The C. elegans Research Community, editor (2010) Neurogenesis in the nematode Caenorhabditis elegans. WormBook. O. HobertThe C. elegans Research Community, editor2010Neurogenesis in the nematode Caenorhabditis elegans.WormBookAvailable: http://www.wormbook.org/chapters/www_specnervsys.2/neurogenesis.html. Accessed 14 March 2011.
25. Bargmann C. I, The C. elegans Research Community, editor (2006) Chemosensation in C. elegans. WormBook. C. I. BargmannThe C. elegans Research Community, editor2006Chemosensation in C. elegans.WormBookAvailable: http://www.wormbook.org/chapters/www_chemosensation/chemosensation.html. Accessed 14 March 2011.
26. Gabriel W. N, McNuff R, Patel S. K, Gregory T. R, Jeck W. R, et al. (2007) The tardigrade Hypsibius dujardini, a new model for studying the evolution of development. Dev Biol 312: 545–559.W. N. GabrielR. McNuffS. K. PatelT. R. GregoryW. R. Jeck2007The tardigrade Hypsibius dujardini, a new model for studying the evolution of development.Dev Biol312545559
27. Sternberg P. W, Horvitz H. R (1989) The combined action of two intercellular signalling pathways specifies three cell fates during vulval induction in C. elegans. Cell 58: 679–693.P. W. SternbergH. R. Horvitz1989The combined action of two intercellular signalling pathways specifies three cell fates during vulval induction in C. elegans.Cell58679693
28. Ellis H. M, Horvitz H. R (1986) Genetic control of programmed cell death in the nematode C. elegans. Cell 44: 817–829.H. M. EllisH. R. Horvitz1986Genetic control of programmed cell death in the nematode C. elegans.Cell44817829
29. Reinhart B. J, Slack F. J, Basson M, Pasquinelli A. E, Bettinger J. C, et al. (2000) The 21-nucleotide let-7 RNA regulates developmental timing in Caenorhabditis elegans. Nature 403: 901–906.B. J. ReinhartF. J. SlackM. BassonA. E. PasquinelliJ. C. Bettinger2000The 21-nucleotide let-7 RNA regulates developmental timing in Caenorhabditis elegans.Nature403901906
30. Kimura K. D, Tissenbaum H. A, Liu Y, Ruvkun G (1997) daf-2, an insulin receptor-like gene that regulates longevity and diapause in Caenorhabditis elegans. Science 277: 942–946.K. D. KimuraH. A. TissenbaumY. LiuG. Ruvkun1997daf-2, an insulin receptor-like gene that regulates longevity and diapause in Caenorhabditis elegans.Science277942946
31. Nigon V, Dougherty E. C (1949) Reproductive patterns and attempts at reciprocal crossing of Rhabditis elegans Maupas, 1900, and Rhabditis briggsae Dougherty & Nigon, 1949 (Nematoda, Rhabditida). J Exp Zool 112: 485–503.V. NigonE. C. Dougherty1949Reproductive patterns and attempts at reciprocal crossing of Rhabditis elegans Maupas, 1900, and Rhabditis briggsae Dougherty & Nigon, 1949 (Nematoda, Rhabditida).J Exp Zool112485503
32. Dolgin E. S, Félix M. A, Cutter A. D (2008) Hakuna Nematoda: genetic and phenotypic diversity in African isolates of Caenorhabditis elegans and C. briggsae. Heredity 100: 304–315.E. S. DolginM. A. FélixA. D. Cutter2008Hakuna Nematoda: genetic and phenotypic diversity in African isolates of Caenorhabditis elegans and C. briggsae.Heredity100304315

33. Barriere A, Félix M. A (2007) Temporal dynamics and linkage disequilibrium in natural Caenorhabditis elegans populations. Genetics 176: 999–1011.A. BarriereM. A. Félix2007Temporal dynamics and linkage disequilibrium in natural Caenorhabditis elegans populations.Genetics1769991011
34. Cutter A. D, Félix M. A, Barriere A, Charlesworth D (2006) Patterns of nucleotide polymorphism distinguish temperate and tropical wild isolates of Caenorhabditis briggsae. Genetics 173: 2021–2031.A. D. CutterM. A. FélixA. BarriereD. Charlesworth2006Patterns of nucleotide polymorphism distinguish temperate and tropical wild isolates of Caenorhabditis briggsae.Genetics17320212031
35. Barriere A, Félix M. A (2005) High local genetic diversity and low outcrossing rate in Caenorhabditis elegans natural populations. Curr Biol 15: 1176–1184.A. BarriereM. A. Félix2005High local genetic diversity and low outcrossing rate in Caenorhabditis elegans natural populations.Curr Biol1511761184
36. Barriere A, Félix M. A, The C. elegans Research Community, editor (2006) Isolation of C. elegans and related nematodes. WormBook. A. BarriereM. A. FélixThe C. elegans Research Community, editor2006Isolation of C. elegans and related nematodes.WormBook
37. Kiontke K, Sudhaus W, The C. elegans Research Community, editor (2006) Ecology of Caenorhabditis species. WormBook. K. KiontkeW. SudhausThe C. elegans Research Community, editor2006Ecology of Caenorhabditis species.WormBook
38. Hodgkin J, Doniach T (1997) Natural variation and copulatory plug formation in Caenorhabditis elegans. Genetics 146: 149–164.J. HodgkinT. Doniach1997Natural variation and copulatory plug formation in Caenorhabditis elegans. Genetics146149164
39. Palopoli M. F, Rockman M. V, TinMaung A, Ramsay C, Curwen S, et al. (2008) Molecular basis of the copulatory plug polymorphism in Caenorhabditis elegans. Nature 454: 1019–1022.M. F. PalopoliM. V. RockmanA. TinMaungC. RamsayS. Curwen2008Molecular basis of the copulatory plug polymorphism in Caenorhabditis elegans.Nature45410191022
40. Gloria-Soria A, Azevedo R. B (2008) npr-1 Regulates foraging and dispersal strategies in Caenorhabditis elegans. Curr Biol 18: 1694–1699.A. Gloria-SoriaR. B. Azevedo2008npr-1 Regulates foraging and dispersal strategies in Caenorhabditis elegans.Curr Biol1816941699
41. Rogers C, Reale V, Kim K, Chatwin H, Li C, et al. (2003) Inhibition of Caenorhabditis elegans social feeding by FMRFamide-related peptide activation of NPR-1. Nat Neurosci 6: 1178–1185.C. RogersV. RealeK. KimH. ChatwinC. Li2003Inhibition of Caenorhabditis elegans social feeding by FMRFamide-related peptide activation of NPR-1.Nat Neurosci611781185
42. Kiontke K, Fitch D. H, The C. elegans Research Community, editor (2005) The phylogenetic relationships of Caenorhabditis and other rhabditids. WormBook. K. KiontkeD. H. FitchThe C. elegans Research Community, editor2005The phylogenetic relationships of Caenorhabditis and other rhabditids.WormBook
43. Sudhaus W, Kiontke K (1996) Phylogeny of Rhabditis subgenus Caenorhabditis (Rhabditidae, Nematoda). J Zoo Syst Evol Research 34: 217–233.W. SudhausK. Kiontke1996Phylogeny of Rhabditis subgenus Caenorhabditis (Rhabditidae, Nematoda).J Zoo Syst Evol Research34217233
44. Mayer W. E, Herrmann M, Sommer R. J (2007) Phylogeny of the nematode genus Pristionchus and implications for biodiversity, biogeography and the evolution of hermaphroditism. BMC Evol Biol 7: 104.W. E. MayerM. HerrmannR. J. Sommer2007Phylogeny of the nematode genus Pristionchus and implications for biodiversity, biogeography and the evolution of hermaphroditism.BMC Evol Biol7104
45. Hong R. L, Sommer R. J (2006) Pristionchus pacificus: a well-rounded nematode. Bioessays 28: 651–659.R. L. HongR. J. Sommer2006Pristionchus pacificus: a well-rounded nematode.Bioessays28651659

46. Baille D, Barriere A, Félix M. A (2008) Oscheius tipulae, a widespread hermaphroditic soil nematode, displays a higher genetic diversity and geographical structure than Caenorhabditis elegans. Mol Ecol 17: 1523–1534.D. BailleA. BarriereM. A. Félix2008Oscheius tipulae, a widespread hermaphroditic soil nematode, displays a higher genetic diversity and geographical structure than Caenorhabditis elegans.Mol Ecol1715231534
47. Cantacessi C, Mitreva M, Jex A. R, Young N. D, Campbell B. E, et al. (2010) Massively parallel sequencing and analysis of the Necator americanus transcriptome. PLoS Negl Trop Dis 4: e684.C. CantacessiM. MitrevaA. R. JexN. D. YoungB. E. Campbell2010Massively parallel sequencing and analysis of the Necator americanus transcriptome.PLoS Negl Trop Dis4e684
48. Blaxter M. L (2000) Genes and genomes of Necator americanus and related hookworms. Int J Parasitol 30: 347–355.M. L. Blaxter2000Genes and genomes of Necator americanus and related hookworms.Int J Parasitol30347355
49. Lambshead P. J. D, Boucher G (2003) Marine nematode deep-sea biodiversity - hyperdiverse or hype? J Biogeogr 30: 475–485.P. J. D. LambsheadG. Boucher2003Marine nematode deep-sea biodiversity - hyperdiverse or hype?J Biogeogr30475485
50. Creer S, Fonseca V. G, Porazinska D. L, Giblin-Davis R. M, Sung W, et al. (2010) Ultrasequencing of the meiofaunal biosphere: practice, pitfalls and promises. Mol Ecol 19: 4–20.S. CreerV. G. FonsecaD. L. PorazinskaR. M. Giblin-DavisW. Sung2010Ultrasequencing of the meiofaunal biosphere: practice, pitfalls and promises.Mol Ecol19420
51. Cobb N. A (1915) Nematodes and their relationships. Year Book Dept Agric 1914. Washington (D.C.): Department of Agriculture. pp. 457–490.N. A. Cobb1915Nematodes and their relationships. Year Book Dept Agric 1914Washington (D.C.)Department of Agriculture457490
52. Blaxter M. L, De Ley P, Garey J. R, Liu L. X, Scheldeman P, et al. (1998) A molecular evolutionary framework for the phylum Nematoda. Nature 392: 71–75.M. L. BlaxterP. De LeyJ. R. GareyL. X. LiuP. Scheldeman1998A molecular evolutionary framework for the phylum Nematoda.Nature3927175
53. Holterman M, van der Wurff A, van den Elsen S, van Megen H, Bongers T, et al. (2006) Phylum-wide analysis of SSU rDNA reveals deep phylogenetic relationships among nematodes and accelerated evolution toward crown Clades. Mol Biol Evol 23: 1792–1800.M. HoltermanA. van der WurffS. van den ElsenH. van MegenT. Bongers2006Phylum-wide analysis of SSU rDNA reveals deep phylogenetic relationships among nematodes and accelerated evolution toward crown Clades.Mol Biol Evol2317921800
54. Meldal B. H, Debenham N. J, De Ley P, De Ley I. T, Vanfleteren J. R, et al. (2007) An improved molecular phylogeny of the Nematoda with special emphasis on marine taxa. Mol Phylogenet Evol 42: 622–636.B. H. MeldalN. J. DebenhamP. De LeyI. T. De LeyJ. R. Vanfleteren2007An improved molecular phylogeny of the Nematoda with special emphasis on marine taxa.Mol Phylogenet Evol42622636
55. De Ley P, Blaxter M (2004) A new system for Nematoda: combining morphological characters with molecular trees, and translating clades into ranks and taxa. In: Cook R, Hunt D. J, editors. Nematology monographs and perspectives. Leiden: E.J. Brill. pp. 633–653.P. De LeyM. Blaxter2004A new system for Nematoda: combining morphological characters with molecular trees, and translating clades into ranks and taxa.R. CookD. J. HuntNematology monographs and perspectivesLeidenE.J. Brill633653
56. De Ley P, Blaxter M. L (2002) Systematic position and phylogeny. In: Lee D, editor. The biology of nematodes. London: Taylor & Francis. pp. 1–30.P. De LeyM. L. Blaxter2002Systematic position and phylogeny.D. LeeThe biology of nematodesLondonTaylor & Francis130
57. Blaxter M, Floyd R, Dorris M, Eyualem A, De Ley P (2004) Utilising the new nematode phylogeny for studies of parasitism and diversity. In: Cook R, Hunt D. J,

editors. Nematology monographs and perspectives. Leiden: E.J. Brill. pp. 615–632.M. BlaxterR. FloydM. DorrisA. EyualemP. De Ley2004Utilising the new nematode phylogeny for studies of parasitism and diversity.R. CookD. J. HuntNematology monographs and perspectivesLeidenE.J. Brill615632
58. Dorris M, De Ley P, Blaxter M. L (1999) Molecular analysis of nematode diversity and the evolution of parasitism. Parasitol Today 15: 188–193.M. DorrisP. De LeyM. L. Blaxter1999Molecular analysis of nematode diversity and the evolution of parasitism.Parasitol Today15188193
59. Schulze J, Schierenberg E (2009) Embryogenesis of Romanomermis culicivorax: an alternative way to construct a nematode. Dev Biol 334: 10–21.J. SchulzeE. Schierenberg2009Embryogenesis of Romanomermis culicivorax: an alternative way to construct a nematode.Dev Biol3341021
60. Schierenberg E, The C. elegans Research Community, editor (2006) Embryological variation during nematode development. WormBook. E. SchierenbergThe C. elegansResearch Community, editor2006Embryological variation during nematode development.WormBook
61. Voronov D. A (1999) The embryonic development of Pontonema vulgare (Enoplida: Oncholaimidae) with a discussion of nematode phylogeny. Russ J Nematol 7: 105–114.D. A. Voronov1999The embryonic development of Pontonema vulgare (Enoplida: Oncholaimidae) with a discussion of nematode phylogeny.Russ J Nematol7105114
62. Voronov D. A, Panchin Y. V (1998) Cell lineage in marine nematode Enoplus brevis. Development 125: 143–150.D. A. VoronovY. V. Panchin1998Cell lineage in marine nematode Enoplus brevis.Development125143150
63. Wasmuth J, Schmid R, Hedley A, Blaxter M (2008) On the extent and origins of genic novelty in the phylum nematoda. PLoS Negl Trop Dis 2: e258.J. WasmuthR. SchmidA. HedleyM. Blaxter2008On the extent and origins of genic novelty in the phylum nematoda.PLoS Negl Trop Dis2e258
64. Abad P, Gouzy J, Aury J. M, Castagnone-Sereno P, Danchin E. G, et al. (2008) Genome sequence of the metazoan plant-parasitic nematode Meloidogyne incognita. Nat Biotechnol 26: 882–884.P. AbadJ. GouzyJ. M. AuryP. Castagnone-SerenoE. G. Danchin2008Genome sequence of the metazoan plant-parasitic nematode Meloidogyne incognita.Nat Biotechnol26882884
65. Opperman C. H, Bird D. M, Williamson V. M, Rokhsar D. S, Burke M, et al. (2008) Sequence and genetic map of Meloidogyne hapla: a compact nematode genome for plant parasitism. Proc Natl Acad Sci U S A 105: 14802–14807.C. H. OppermanD. M. BirdV. M. WilliamsonD. S. RokhsarM. Burke2008Sequence and genetic map of Meloidogyne hapla: a compact nematode genome for plant parasitism.Proc Natl Acad Sci U S A1051480214807
66. Ghedin E, Wang S, Spiro D, Caler E, Zhao Q, et al. (2007) Draft genome of the filarial nematode parasite Brugia malayi. Science 317: 1756–1760.E. GhedinS. WangD. SpiroE. CalerQ. Zhao2007Draft genome of the filarial nematode parasite Brugia malayi.Science31717561760
67. Mitreva M, Jasmer D. P, The C. elegans Research Community, editor (2006) Biology and genome of Trichinella spiralis. WormBook:. WormBook. M. MitrevaD. P. JasmerThe C. elegans Research Community, editor2006Biology and genome of Trichinella spiralis. WormBook:WormBook
68. Mitreva M, Jasmer D. P (2008) Advances in the sequencing of the genome of the adenophorean nematode Trichinella spiralis. Parasitology 135: 869–880.M. MitrevaD. P. Jasmer2008Advances in the sequencing of the genome of the adenophorean nematode Trichinella spiralis.Parasitology135869880
69. Dieterich C, Clifton S. W, Schuster L. N, Chinwalla A, Delehaunty K, et al. (2008) The Pristionchus pacificus genome provides a unique perspective on nematode lifestyle and parasitism. Nat Genet 40: 1193–1198.C. DieterichS. W. CliftonL. N. SchusterA. ChinwallaK. Delehaunty2008The Pristionchus pacificus genome provides a unique perspective on nematode lifestyle and parasitism.Nat Genet4011931198

70. Rae R, Riebesell M, Dinkelacker I, Wang Q, Herrmann M, et al. (2008) Isolation of naturally associated bacteria of necromenic Pristionchus nematodes and fitness consequences. J Exp Biol 211: 1927–1936.R. RaeM. RiebesellI. DinkelackerQ. WangM. Herrmann2008Isolation of naturally associated bacteria of necromenic Pristionchus nematodes and fitness consequences.J Exp Biol21119271936
71. Stein L. D, Bao Z, Blasiar D, Blumenthal T, Brent M. R, et al. (2003) The genome sequence of Caenorhabditis briggsae: a platform for comparative genomics. PLoS Biol 1: e45.L. D. SteinZ. BaoD. BlasiarT. BlumenthalM. R. Brent2003The genome sequence of Caenorhabditis briggsae: a platform for comparative genomics.PLoS Biol1e45
72. Kumar S, Blaxter M (2010) 959 nematode genomes initiative. S. KumarM. Blaxter2010959 nematode genomes initiative.Available: http://www.nem atodegenomes.org/. Accessed 14 March 2011. Available: http://www.nemato degenomes.org/. Accessed 14 March 2011.
73. Hyman L. H (1951) The invertebrates: Acanthocephala, Aschelminthes, and Entoprocta - the pseudocoelomate Bilateria. In: Boell E. J, editor. New York: Mc Graw-Hill Book Company, Inc. 572 p.L. H. Hyman1951The invertebrates: Acanthocephala, Aschelminthes, and Entoprocta - the pseudocoelomate Bilateria.E. J. BoellNew YorkMc Graw-Hill Book Company, Inc572
74. Aguinaldo A. M. A, Turbeville J. M, Linford L. S, Rivera M. C, Garey J. R, et al. (1997) Evidence for a clade of nematodes, arthropods and other moulting animals. Nature 387: 489–493.A. M. A. AguinaldoJ. M. TurbevilleL. S. LinfordM. C. RiveraJ. R. Garey1997Evidence for a clade of nematodes, arthropods and other moulting animals.Nature387489493
75. Halanych K. M, Bacheller J. D, Aguinaldo A. M. A, Liva S. M, Mills D. H, et al. (1995) Evidence from 18S ribosomal DNA that the lophophorates are protostome animals. Science 267: 1641–1643.K. M. HalanychJ. D. BachellerA. M. A. AguinaldoS. M. LivaD. H. Mills1995Evidence from 18S ribosomal DNA that the lophophorates are protostome animals.Science26716411643
76. Philippe H, Lartillot N, Brinkmann H (2005) Multigene analyses of bilaterian animals corroborate the monophyly of Ecdysozoa, Lophotrochozoa and Protostomia. Mol Biol Evol 1246–1253.H. PhilippeN. LartillotH. Brinkmann2005Multigene analyses of bilaterian animals corroborate the monophyly of Ecdysozoa, Lophotrochozoa and Protostomia.Mol Biol Evol12461253
77. van Megen H, van den Elsen S, Holterman M, Karssen G, Mooyman P, et al. (2009) A phylogenetic tree of nematodes based on about 1200 full-length small subunit ribosomal DNA sequences. Nematology 11: 927–950.H. van MegenS. van den ElsenM. HoltermanG. KarssenP. Mooyman2009A phylogenetic tree of nematodes based on about 1200 full-length small subunit ribosomal DNA sequences. Nematology11927950

CHAPTER 2

The Evolutionary Position of Nematodes

Jaime E Blair[1], Kazuho Ikeo[2], Takashi Gojobori[2] and S Blair Hedges[1]

[1]*Astrobiology Research Center and Department of Biology, The Pennsylvania State University, University Park, PA 16802, USA and*

[2]*Center for Information Biology, National Institute of Genetics, Mishima, Shizuoka-ken 411-8540, Japan*

ABSTRACT

Background

The complete genomes of three animals have been sequenced by global research efforts: a nematode worm (*Caenorhabditis elegans*), an insect (*Drosophila melanogaster*), and a vertebrate (*Homo sapiens*). Remarkably, their relationships have yet to be clarified. The confusion concerns the enigmatic position of nematodes. Traditionally, nematodes have occupied a basal position, in part because they lack a true body cavity. However, the leading hypothesis now joins nematodes with arthropods in a molting clade, Ecdysozoa, based on data from several genes.

Results

We tested the Ecdysozoa hypothesis with analyses of more than 100 nuclear protein alignments, under conditions that would expose biases, and found that it was not supported. Instead, we found significant support for the traditional hypothesis, Coelomata. Our result is robust to different rates of sequence change among genes and lineages, different numbers of taxa, and different species of nematodes.

Conclusion

We conclude that insects (arthropods) are genetically and evolutionarily closer to humans than to nematode worms.

BACKGROUND

Traditionally, the animal body cavity (coelom) has played a major role in interpretations of metazoan evolution, from groups (e.g., flatworms) lacking a coelom to those (e.g., nematodes) with a false coelom and finally to the bulk of animal phyla having a true coelom (Coelomata) [1, 2]. There has never been complete agreement on animal phylogeny and classification, but most researchers have divided living coelomate animals into deuterostomes (echinoderms, hemichordates, urochordates, cephalochordates, and vertebrates) and protostomes (arthropods, annelids, mollusks, and other phyla) based on differences in early embryonic development. An analysis of small subunit ribosomal RNA (18S rRNA) sequences challenged this arrangement by placing acoelomate and pseudocoelomate phyla in more derived positions among the protostomes, and in further defining a clade (Ecdysozoa) of molting animals that includes arthropods and nematodes [3]. This "Ecdysozoa" hypothesis has influenced diverse fields [4] and interpretations of developmental evolution in animals [5, 6, 7]. Since its publication, evidence has appeared both for and against this hypothesis [8, 9, 10, 11, 12, 13, 14, 15]. Knowing the branching order of the major animal lineages, especially those three with fully sequenced genomes, is of importance to diverse fields such as medical genetics, physiology, neurobiology, paleontology, and astrobiology. With a genealogy of animals, it will be easier to determine the origins and inheritance of mutations, genes, gene functions, and structures.

The three possible relationships of these animal phyla are: (I) arthropods + vertebrates, (II) arthropods + nematodes, and (III) nematodes + vertebrates. The first hypothesis corresponds to the traditional grouping Coelomata and the second corresponds to Ecdysozoa [3]. For convenience, we will use these names in reference to the two hypotheses while recognizing that this study, by necessity, involves only a subset of all animal phyla. The third hypothesis will be referred to as "hypothesis III" (Figure 1). To test each hypothesis, sequence alignments of more than 100 nuclear proteins were assembled and subjected to a series of analyses designed to reveal biases that could result in an incorrect phylogeny.

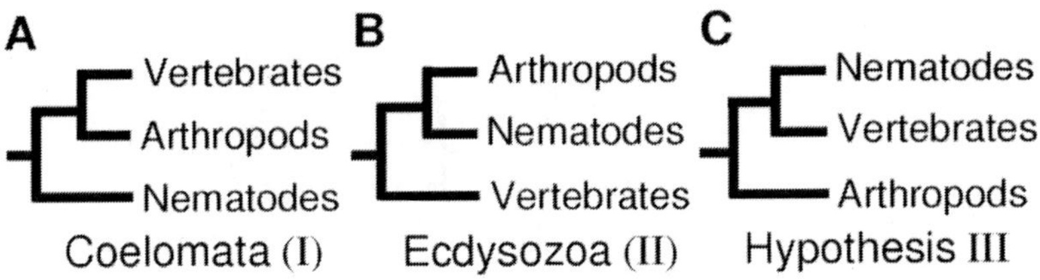

Figure 1. The three possible relationships of vertebrates, arthropods, and nematodes.

RESULTS AND DISCUSSION

Analyses of the individual protein datasets using neighbor-joining [16] show that most (62%) support Coelomata while 25% support Ecdysozoa and 13% support hypothesis III (Supplemental Table 1, four-taxon analysis). Of the 25 proteins in which support for one of the three hypotheses is significant (≥ 95%), the results are 84% (21 proteins), 16% (4), and 0%, respectively; for those ten proteins with a highly significant (≥ 99%) topology, the results are 90% (9 proteins), 10% (1), and 0%, respectively. The four proteins showing significant support (>95%) for a hypothesis other than Coelomata were reanalyzed using maximum parsimony and maximum likelihood; bootstrap values were not significant using other methods. Such divided results are typical of single-gene analyses because of limited information (~400 amino acids), necessitating combined analysis. Coelomata was supported significantly (100% bootstrap confidence, posterior probability = 1.0) when the 100 four-taxon protein alignments were concatenated and analyzed using neighbor-joining, maximum parsimony, maximum likelihood, and Bayesian inference (Figure 2A). Using the Shimodaira-Hasegawa (SH) test [17], this maximum likelihood topology was significantly different from the alternative hypotheses of Ecdysozoa ($P < 0.001$) and Hypothesis III ($P < 0.001$). These results agree with earlier studies involving 10–36 nuclear proteins [8, 9, 11].

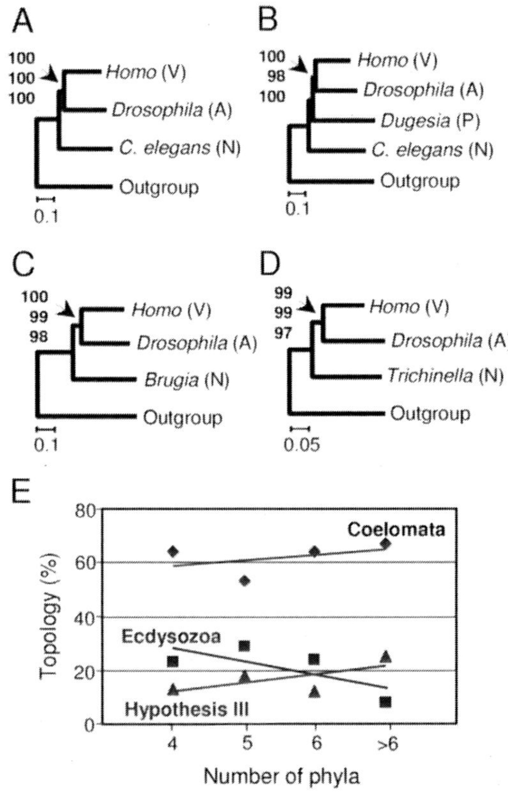

Figure 2. Phylogenetic analyses of individual and combined (concatenated) sequence alignments bearing on the position of nematodes. V = vertebrate, A = arthropod, N = nematode, P = platyhelminth. Bootstrap values (>95%) are shown for neighbor-joining, maximum parsimony, and maximum likelihood, respectively; all are indicated for the node joining *Homo* and *Drosophila* (=Coelomata). Posterior probabilities are not shown (all highlighted nodes = 1.0). (*A*) Four-taxon analysis of 100 combined protein alignments (44,214 amino acids), using nematode *Caenorhabditis elegans* (Chromadorea, Rhabditida, Rhabditoidea, Rhabditidae); the nematode branch is approximately 16% longer than the vertebrate and arthropod branches. (*B*) Five-taxon analysis of 100 combined proteins includes planarian EST sequences (14,041 amino acids); the nematode branch is approximately 23% longer. Other trees show different representative nematodes. (C) *Brugia* (Chromadorea, Spirurida, Filarioidea, Onchocercidae), based on 18 combined proteins (4598 amino acids); nematode branch= 15% longer. (D) *Trichinella* (Enoplea, Trichocephalida, Trichinellidae), based on 6 combined proteins (2261 amino acids); nematode branch = 24% longer than the vertebrate branch and 5% shorter than the arthropod branch. (E) Proportion of individual protein analyses supporting each of the three possible topologies with differing numbers of phyla included (4 taxa = 124 proteins, 5 taxa= 107 proteins, 6 taxa= 66 proteins, >6 taxa = 12 proteins).

To test the stability of Coelomata to taxon sampling, we included new sequences of the planarian *Dugesia japonica* (Phylum Platyhelminthes) in 100 five-taxon protein alignments (Supplemental Table 1, five-taxon analysis). The results, upon concatenation with this additional taxon, were unchanged (Figure 2B): Coelomata continued to be significantly supported (≥ 98% bootstrap support, posterior probability = 1.0) and the alternative hypotheses were both rejected using the SH test ($P < 0.001$), although the relationships among the basal phyla (Nematoda and Platyhelminthes) could not be resolved.

Another potential bias is the specific taxa included in an analysis. For example, the original support for Ecdysozoa was obtained only with a particular genus of nematode, *Trichinella,* that had a short branch in the 18S rRNA tree [3]. To test this, phylogenies were constructed using different species of nematodes. The Coelomata hypothesis was significantly supported (≥ 97% bootstrap support, posterior probabilities = 1.0; Ecdysozoa rejected by SH test, $P < 0.006$; Hypothesis III rejected by SH test, $P < 0.027$) using either a genus in a different order, *Brugia* (18 proteins), or *Trichinella* (six proteins) (Figure 2C, 2D). To further address the possibility that these results could be biased by taxon sampling, we included representatives from all available phyla for each protein. The results indicate that an increase in the number of taxa does not decrease single-protein support for Coelomata; in fact, the trend is the reverse (Figure 2E). Simulation studies have shown that incomplete taxon sampling does not increase topological errors, and that most error is caused by limited sequence data [18].

In the initial study defining Ecdysozoa [3], rate variation was considered to be the major bias affecting the phylogenetic position of nematodes. In the 18S rRNA gene, nematodes typically have long branches indicating an increased rate of sequence change. Other nuclear genes also show this pattern, but to a lesser degree [8, 9]. Phylogenetic methods can accommodate moderate amounts of rate variation among lineages without producing an incorrect phylogeny [19]. However, if the rate of change is sufficiently large, longer branches in a phylogeny will sometimes attract one another [20]. If that happens, an ingroup species with a long branch may move to a more basal position in the tree. In analyses of the 18S rRNA gene, nematodes typically appear basal to arthropods + vertebrates. Because the use of a short branch nematode (*Trichinella*) resulted in a tree whereby nematodes clustered with arthropods, the basal position of nematodes in typical 18S analyses has been interpreted as long-branch attraction [3].

If nematodes cluster basally because of long-branch attraction, then the strongest support for Ecdysozoa should be obtained with the slowest evolving proteins. This was tested in an analysis of 36 nuclear proteins [8], but the results were equivocal. Therefore, we tested this suggestion with our four-taxon data set of 100 proteins, ordered by rate of evolution. Rate orders were determined in two ways: (i) nematode branch length and (ii) vertebrate-arthropod pairwise distance. The 100 proteins were grouped into concatenations of 10 proteins and 20 proteins to increase statistical resolution. The results show support for Coelomata at all rate orders, but the support is significant with the slowest evolving proteins, regardless of rate measure or number of proteins combined (Figure 3). Concatenations of slow evolving proteins also show compositional homogeneity (pairwise disparity index test, $P < 0.05$) [21], suggesting the basal position of the nematode results from true phylogenetic signal and not compositional bias. Support for Coelomata was weakest with the fastest evolving proteins (which also showed compositional heterogeneity), indicating that Ecdysozoa, not Coelomata, may be the result of a rate bias, compositional bias, or other artifact.

To ensure that this result was not affected by mutational saturation in our data set, the mean number of variants per variable site was determined for each protein and averaged for ten groups of proteins ordered by evolutionary rate (Fig 4A). As predicted, variable sites in the faster evolving proteins showed a higher number of variants than those in slow-evolving proteins. We also examined the minimum number of nucleotide changes required for sites where only the nematode sequence varied (Fig 4B). Slow-evolving proteins showed a smaller number of nucleotide changes required to alter amino acid identity, while faster evolving proteins required more changes. Thus, the nematode sequences in the slow-evolving proteins do not appear to be mutationally saturated.

Finally, the affect of lineage-specific rate variation on support for Coelomata was tested with the use of relative rate tests. Presumably, the selective elimination of genes with long branches will increase statistical support for the correct topology. Individual proteins from the four-taxon data set were each subjected to two different relative rate tests [22, 23]. Proteins determined to be rate-constant at the typically applied stringency level (5% significance), and at two greater stringency levels (10% and 40% significance) were concatenated and bootstrap support was determined using neighbor joining. The results of the two tests were similar. As stringency increased, 40–83% of proteins were rejected, and the relative nematode branch length (to the arthropod and vertebrate branches) dropped from 16% to 0%. However, in all cases, Coelomata remained highly significant (Figure 5). Thus, the suggestion that a basal position of nematodes is the result of long-branch attraction [3] can be rejected.

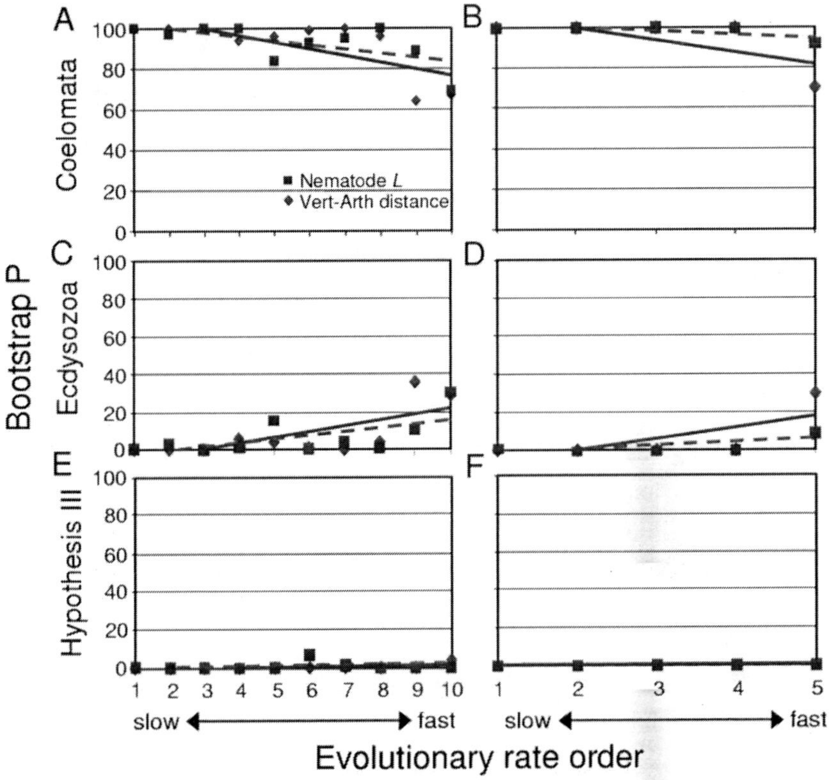

Figure 3. Effect of genetic distance on bootstrap support for the three hypotheses from analysis of 100 nuclear proteins with four taxa. (*A, B*) show bootstrap support for Coelomata; (*C, D*) for Ecdysozoa; (*E, F*) for Hypothesis III. Proteins were ordered from slowest evolving to fastest evolving based on two criteria: vertebrate-arthropod pairwise distance (diamonds) and nematode branch length (squares). Proteins were concatenated into ten groups often (*A, C, E*) and five groups of twenty (*B, D, F*). Graphs show rate from slowest to fastest evolving (left to right). Trend lines are indicated (solid for vertebrate-arthropod distance, dashed for nematode branch length).

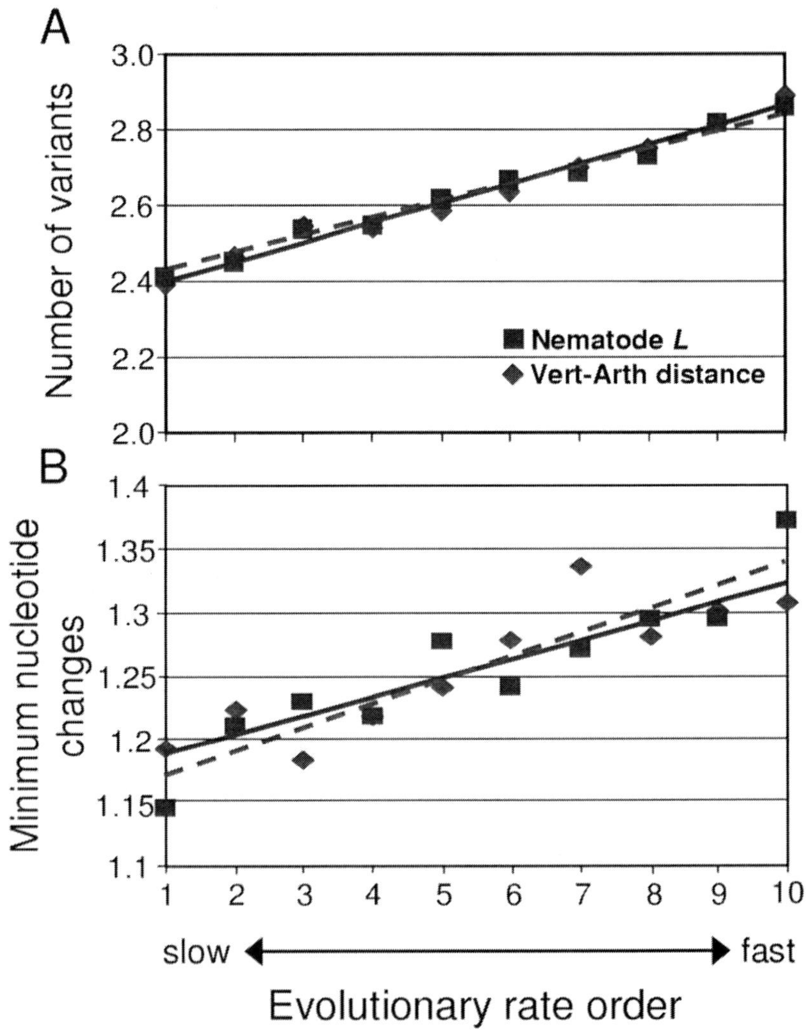

Figure 4. Test of mutational saturation in the four-taxon data set. (*A*) The mean number of variants per variable site was averaged for ten groups often according to evolutionary rate (vertebrate-arthropod distance = diamonds, nematode branch length = squares). (*B*) The minimum number of nucleotide changes required for unique nematode variants were also averaged according to evolutionary rate. Trend lines are indicated (solid for vertebrate-arthropod distance, dashed for nematode branch length).

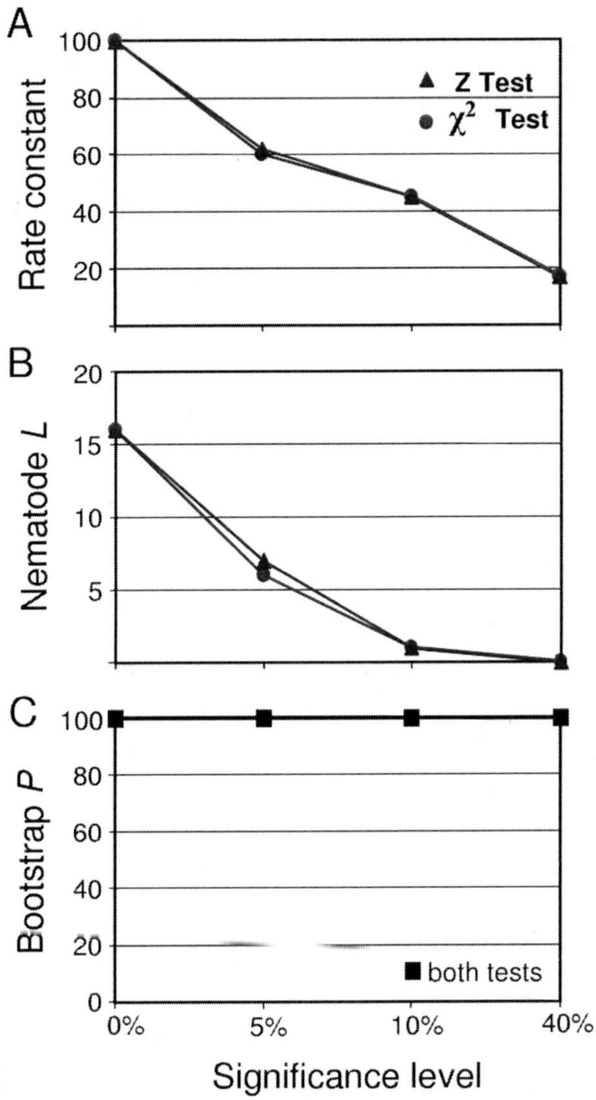

Figure 5. Effect of rate constancy on bootstrap support for Coelomata in four-taxon analysis. Graphs show results before application of tests (left, 0-level) followed by increasing stringency (5, 10, 40% significance) of the chi-square test [22] (circles) and Z-test [23] (triangles); the 5% level is normally used. (*A*) number of proteins passing rate constancy at each cutoff level. (*B*) relative nematode branch length upon concatenation of all rate constant proteins at each level. (*C*) bootstrap support for Coelomata for each rate-constant concatenation.

The importance of knowing the branching order of these species is illustrated by the immediate and wide acceptance of the Ecdysozoa hypothesis and its use in tracing patterns of developmental evolution [5, 6, 7, 10]. However, in the initial analysis of 18S rRNA sequences [3], Ecdysozoa was statistically significant only when a paralinear distance method was used; three other methods did not yield significant bootstrap support. In that study, Ecdysozoa also was not significant, using any method, when the flatworm sequence was included [3]. Subsequent analyses of the 18S rRNA gene have been interpreted differently [24, 25], but none has yielded statistically significant results supporting Ecdysozoa. Moreover, the molting cuticles of arthropods (chitin) and nematodes (collagen) are not homologous [4]. The significance of other morphological characteristics bearing on the position of nematodes continues to be debated [26].

Besides the 18S rRNA evidence, other genetic evidence for the grouping of nematodes and arthropods has come from qualitative interpretations of *Hox* gene [10] and β-thymosin [12] evolution. In the case of *Hox* genes, support comes from a single posterior gene sequence (Y75B8A.1) of the nematode *Caenorhabditis elegans* argued to have greater amino acid similarity with a posterior *Hox* genes of *Drosophila* and *Priapulus*[10]. Unfortunately, the *Hox* homeodomain is a short (60 amino acid) region with many sequence differences between these taxa. Definition of "sequence signatures" is qualitative and has not been tested statistically. In a subsequent study of nematode posterior *Hox* genes, other researchers were unable to determine if the simple nematode *Hox* cluster of six genes is an ancestral or a derived condition [13].

In the case of β-thymosin, a sequence signature also has been argued to support a grouping of *Drosophila* and *Caenorhabditis*[12]. However, it is a gene family known to have paralogs within animals, the position of introns differs between sequences from the two species, and only four other metazoan taxa were surveyed. In addition, knowing the presence or absence of a gene can be problematic without the complete genome sequence of an organism (in this case, genomes were known only in *Drosophila* and *Caenorhabditis*). Thus, although suggestive, it is too soon to judge the significance of this sequence signature. One difficulty with interpreting such qualitative evidence, including *Hox* gene orthology, is that almost any pattern can be found in nature if one looks. In other words, sequence signatures have not yet been surveyed systematically and objectively. In contrast, sequence evidence from randomly selected genes, analyzed phylogenetically, provides a more unbiased database amenable to statistical analysis.

CONCLUSIONS

Although it is possible that a basal position of nematodes is the result of some unknown and widespread bias not yet identified, a simpler explanation is that the grouping of nematodes with arthropods is an artifact that arose from the analysis of a single gene, 18S rRNA. The results presented here suggest caution in revising animal phylogeny from analyses of one or a few genes or sequence signatures. Although many other aspects of animal phylogeny remain unresolved, our results indicate that insects (arthropods) are genetically and evolutionarily closer to humans (vertebrates) than to nematodes.

MATERIALS AND METHODS

DNA sequences from *Dugesia japonica*[27] were used to search the public protein database (Entrez) for orthologous counterparts in *Drosophila melanogaster, Caenorhabditis elegans, Homo sapiens, Arabidopsis thaliana, Saccharomyces cerevisiae,* and *Schizosaccharomyces pombe.* When available, sequences from other major animal phyla (e.g., Mollusca, Echinodermata, Annelida) also were obtained. In addition, the database was searched for all proteins from two other nematodes, *Brugia* and *Trichinella,* with orthologs in *Drosophila, Homo,* and *Arabidopsis, Saccharomyces,* or *Schizosaccharomyces. Arabidopsis* was used as the primary outgroup (95 out of 100 proteins in the four-taxon analysis, 94 out of 100 proteins for the five-taxon analysis, all proteins in *Brugia* and *Trichinella* analyses); yeast was used as the outgroup when *Arabidopsis* sequences were unavailable or paralogous. This was because many more genes were available for rooting with the plant than with the fungus. All three kingdoms are about equidistant from each other in terms of branch lengths [9] and therefore a plant serves about equally well for rooting an animal phylogeny as does a fungus. Orthology was assessed using reciprocal BLAST searches of the public protein database; those sequences receiving high scores in each search were also analyzed phylogenetically to ensure orthology. Short (<100 amino acids) sequences were omitted.

Sequences were aligned using Clustal X [28] and each alignment was visually inspected. Primary analyses of aligned protein data sets were conducted in MEGA2 [29]. Phylogenies were reconstructed using neighbor-joining [16] under a Poisson correction and a gamma distance (α = 2, or estimated from the data for combined analyses), with bootstrapping (2000 replications) for all analyses. Gamma parameters were estimated from the combined data using maximum likelihood under a Poisson correction [30]

(4-taxon, α = 1.62; 5-taxon, α = 0.94, *Brugia*, α = 0.87; *Trichinella*, α = 0.66). In addition, phylogenetic analyses were conducted with maximum likelihood (JTT-F option) [31] and maximum parsimony (Max-Mini Branch & Bound option) [29] on combined data sets; in all cases they resulted in similar results (topology and significance) to the neighbor-joining analyses. Posterior probabilities of concatenated files were computed using Bayesian inference [32] (Jones model with gamma estimated from data; 10,000 generations; 4 chains with temp = 0.2). Shimodaira-Hasegawa tests [17] were performed in PAML [30] (JTT-F option, fixed gamma); p-values for each topology were recorded.

Rate constancy was assessed using a chi-square test [22] under increasing stringency (5, 10, 40% significance levels); p-values were recorded for each protein. A Z-test [23] was also used under increasing stringency; z-values were recorded for each protein. Proteins determined to be rate constant at different significance levels were concatenated and analyzed in MEGA2 [29]. Nematode position and evolutionary distance were determined for each concatenation. New sequences, accession numbers of sequences, and sequence alignments may be found at the Evogenomics website http://www.evogenomics.org/publications/data/nematode/.

ACKNOWLEDGEMENTS

This work was supported by grants from the NASA Astrobiology Institute and National Science Foundation (to S.B.H.).

AUTHORS' CONTRIBUTIONS

JEB carried out the bioinformatics research and drafted the manuscript. KI and TG supplied the unpublished EST sequences of the platyhelminth *Dugesia japonica*. SBH directed the research and assisted with drafting of the manuscript.

REFERENCES

1. Hyman LH: The Invertebrates. 1940, New York: McGraw-Hill
2. Jenner RA: Evolution of animal body plans: the role of metazoan phylogeny at the interface between pattern and process. Evol Dev. 2000, 2: 208-221.
3. Aguinaldo AM, Turbeville JM, Linford LS, Rivera MC, Garey JR, Raff RA, Lake JA: Evidence for a clade of nematodes, arthropods and other moulting animals. Nature. 1997, 387: 489-93.

4. Adoutte A, Balavoine G, Lartillot N, Lespinet O, Prud'homme B, de Rosa R: The new animal phylogeny: reliability and implications. Proc Natl Acad Sci USA. 2000, 97: 4453-6.
5. Carroll SB, Grenier JK, Weatherbee SD: From DNA to Diversity. Edited by: Malden MA. 2001, Blackwell Science
6. Davidson EH: Genomic Regulatory Systems. 2001, San Diego: Academic Pres
7. Valentine JW, Collins AG: The significance of moulting in Ecdysozoan evolution. Evol Dev. 2000, 2: 152-6. 10.
8. Mushegian AR, Garey JR, Martin J, Liu LX: Large-scale taxonomic profiling of eukaryotic model organisms: a comparison of orthologous proteins encoded by human, fly, nematode, and yeast genomes. Genome Research. 1998, 8: 590-598.
9. Wang DY, Kumar S, Hedges SB: Divergence time estimates for the early history of animal phyla and the origin of plants, animals and fungi. Proc R Soc Lond B Biol Sci. 1999, 266: 163-71.
10. de Rosa R, Grenier JK, Andreeva T, Cook CE, Adoutte A, Akam M, Carroll SB, Balavoine G: Hox genes in brachiopods and priapulids and protostome evolution. Nature. 1999, 399: 772-6.
11. Hausdorf B: Early evolution of the Bilateria. Syst. Biol. 2000, 49: 130-142.
12. Manuel M, Kruse M, Muller WE, Le Parco Y: The comparison of beta-thymosin homologues among metazoa supports an arthropod-nematode clade. J Mol Evol. 2000, 51: 378-81.
13. Van Auken K, Weaver DC, Edgar LG, Wood WB: Caenorhabditis elegans embryonic axial patterning requires two recently discovered posterior-group Hox genes. Proc Natl Acad Sci USA. 2000, 97: 4499-4503.
14. Haase A, Stern M, Wachtler K, Bicker G: A tissue-specific marker of Ecdysozoa. Dev Genes Evol. 2001, 211: 428-33.
15. Peterson KJ, Eernisse DJ: Animal phylogeny and the ancestry of bilaterians: inferences from morphology and 18S rDNA gene sequences. Evol Dev. 2001, 3: 170-205.
16. Saitou N, Nei M: The neighbor-joining method: a new method for reconstructing phylogenetic trees. Mol Biol Evol. 1987, 4: 406-25.
17. Shimodaira H, Hasegawa M: Multiple comparisons of log-likelihoods with applications to phylogenetic inference. Mol Biol Evol. 1999, 16: 1114-1116.
18. Rosenberg MS, Kumar S: Incomplete taxon sampling is not a problem for phylogenetic inference. Proc Natl Acad Sci USA. 2001, 98: 10751-6.
19. Nei M, Kumar S: Molecular Evolution and Phylogenetics. 2000, New York: Oxford University Press
20. Felsenstein J: Cases in which parsimony or compatibility methods will be positively misleading. Syst. Zool. 1978, 27: 401-410.
21. Kumar S, Gadagkar SR: Disparity index: a simple statistic to measure and test the homogeneity of substitution patterns between molecular sequences. Genetics. 2001, 158: 1321-7.
22. Tajima F: Simple methods for testing the molecular evolutionary clock hypothesis. Genetics. 1993, 135: 599-607.
23. Takezaki N, Rzhetsky A, Nei M: Phylogenetic test of the molecular clock and linearized trees. Mol Biol Evol. 1995, 12: 823-33.
24. Wagele JW, Erikson T, Lockhart P, Misof B: The Ecdysozoa: artifact or monophylum. J Zool Syst Evol Research. 1999, 37: 211-223.
25. Giribet G, Wheeler WC: The position of arthropods in the animal kingdom: Ecdysozoa, islands, trees, and the "Parsimony ratchet". Mol Phylogenet Evol. 1999, 13: 619-23.
26. Rouse GW, Fauchald K: Cladistics and polychaetes. Zoologica Scripta. 1997, 26: 139-204.
27. Tazaki A, Gaudieri S, Ikeo K, Gojobori T, Watanabe K, Agata K: Neural network in planarian revealed by an antibody against planarian synaptotagmin homologue. Biochem Biophys Res Commun. 1999, 260: 426-32.

28. Thompson JD, Gibson TJ, Plewniak F, Jeanmougin F, Higgins DG: The CLUSTAL_X windows interface: flexible strategies for multiple sequence alignment aided by quality analysis tools. Nucleic Acids Res. 1997, 25: 4876-82.
29. Kumar S, Tamura K, Jakobsen I, Nei M: MEGA2: molecular evolutionary genetics analysis software. Bioinformatics. 2001, 17: 1244-1245.
30. Yang Z: PAML: a program package for phylogenetic analysis by maximum likelihood. Comput Appl Biosci. 1997, 13: 555-6.
31. Adachi J, Hasegawa M: MOLPHY version 2.3: Programs for Molecular Phylogenetics based on Maximum Likelihood. 1996, Tokyo: Institute of Statistical Mathematics
32. Huelsenbeck JP, Ronquist F: MRBAYES: Bayesian inference of phylogenetic trees. Bioinformatics. 2001, 17: 754-5.

CHAPTER 3

Taxonomy of Plant Parasitic Nematodes in Global Context vis-ÃƒÂ -vis India

Matiyar Rahaman Khan*

Department of Agricultural Entomology, Bidhan Chandra Krishi Viswavidyalaya, Kalyani, PIN-741235, West Bengal, India.

Keywords: Nematodes, Aphelenchoides besseyi, *M. javanica*

EDITORIAL

NEMATODES are most abundant multicellular animals on earth, almost similar in number to that of insects. The phylum Nematoda comprises of over 26650 described species [1] and many of them are parasitizing **plants** or animals. Nearly 4100 phytoparasitic nematode species belonging to 197 genera have been described all over the world; they represent about 10% of total number of nematodes known so far. Plant parasitic nematodes deserve adequate attention for damaging crops and are often considered as a limiting factor in crop cultivation. At least 10 nematode species have been recognized for their economic significance in **AGRICULTURE** [2]. Identification of plant parasitic nematode species is relatively difficult for its small size and limited number of differentiating characters which requires trained specialist for dealing with the population. Nematode parasitizing on animals are known since ancient times but knowledge on plant feeding nematode are rather recent. Prior to 1950, a few nematode species were known attacking plants and posing threat to **crop cultivation**. The taxonomic contributions of N.A. Cobb, H.C. Bastian, H. Micoletzky, I.N. Filipjev, J.B. Goodey, T. Goodey, A.A. Paramonov, M.W. Allen, G. Thorne, B.G. Chitwood, A.L. Taylor, A. Coomans, E. Geraert etc. are remembered in history of Nematology. In India, some economically important nematodes of crops were reported from time to time; root knot nematode on tea from south India by Barber [3], Ditylenchus angustus inducing ufra disease on rice by Butler [4], wheat seed gall nematode (Anguina tritici) on wheat by Hutchinson [5], Aphelenchoides besseyi inducing white tip disease of rice in the Central Province (now Madhya Pradesh) by Dastur [6], M. incognita on jute by Chattopadhyay and Sengupta [7] and M. javanica on vegetables by Sen [8], Tylenchulus semipenetrans on citrus from Uttar Pradesh by Siddiqi [9] and Radopholus similis on banana from Kerala by Nair et al. [10]. A strong foundation to Nematology in the country was built with the devotion of nematologists worked at Aligarh Muslim University, Aligarh, and Indian Agricultural Research Institute, New Delhi and some other universities like Osmania University, CCS-Haryana Agricultural University, Tamil Nadu Agricultural University, Orissa University of Agriculture and Technology, Assam Agricultural University to name a few. About 631 phytoparasitic species belonging to 56 genera under Tylenchida have been reported from India; it is nearly one-third of the total known plant **PARASITIC** nematodes (1900 species of 116 genera) around the world. During 1960s and 1970s, more than 500 species was described from India; this was within a span of less than 13 years [11]. Later the taxonomists shifted their attention for developing sound classification and studying intra-specific variations in major nematode species. A quick overview of the published works in the three major journals viz., Nematologica (now Nematology), Indian Journal of Nematology (IJN), Journal of Nematology (JON) was scanned for taxonomic contributions from India and by the Indian taxonomists. It was evident from the IJN (1971-2014) that number of taxonomic publications varied with the years and the maximum taxonomic papers appeared (14-32 per volume) during 1980s and a marked decline was noted since 1990 onwards. **Taxonomic** contributions of Indian nematologists in the Nematologica during 1956 to 1998 were notably high; 38 during 1956-1966, 59 during 1967-1977, 41 during 1978-1988, and 20 during 1989-1998. The contributions of Indian nematologists in the JON were very meager, and in Nematology it was only 8 during 1998-2014. The approaches of taxonomy have shifted from conventional description of species primarily based on morphological characters to biochemical and molecular parameters. Most of the reputed journals accept taxonomic papers provided with traditional descriptions of species supplemented with biological, **BIOCHEMICAL** and molecular information. Indian nematode taxonomists by and large are not yet equipped with modern identification tools (**DNA barcoding**, 4D microscopy, video vouchering) for description and discrimination of species as evidenced by a conspicuous lack of such publications. Majority of the newly described species provide adequate illustration, diagnostic relationship with closely related species and supplementary information. This led to sizeable number of new species described that find place in the list of synonymy of species. This has become a major impediment for Indian nematode taxonomists to meet the current criteria for accessing high impact factor journals for publishing their works. Modern approaches of taxonomy attract **molecular biologists** to take the stage of traditional taxonomists. In fact, the taxonomy field is waning and remains unattractive for the younger generation of nematologists. Currently, nematode taxonomy is handicapped through a shortfall of nematode taxonomists with the superannuation of senior taxonomists from most reputed institutions. However, taxonomy is not only necessary but also imperative for conserving known and exploring unknown biological diversity on earth. There is an urgent need to take initiative for global and regional capacity building on taxonomy and systematics, and utilize Information and Communication Technology (IC&T) for taxonomic research [12]. Therefore, Global Taxonomic Initiative (GTI) was launched in 2002 Smith and Klopper [13] to reduce taxonomy impediment but its progress is not encouraging. Much of molecular works have been focused on plant nematodes of quarantine significance. Further attention is required for developing **GENOMICS** resources and tools for multiple plant parasitic nematode species, and designing molecular tools for identification and surveillance. However, there is pressing need for a synergy between traditional and molecular maker based identification in nematode taxonomy to pinpoint the species identity at both levels.

REFERENCES

1. Hugot JP et al. Biodiversity in helminthes and nematodes as a field of study: An overview. Nematology. 2001; 3: 199-208.
2. Jones JT et al. Top 10 plant-parasitic nematodes in molecular plant pathology. Mol. Pl. Pathol2013. DOI: 10.1111/mpp.12057.
3. Barbar CA, Department land records and agriculture. Madras Agric.1901; 45: 227-234.
4. Butler EJ, An eelworm disease of rice. Agr. Res. Inst., PusaBull.1913; 34.
5. Hutchinson CM, A bacterial disease of wheat in Punjab. Memoirs of Department of Agriculture, India,1917; Bacteriological Series 1.
6. Dastur JF, A nematode disease of rice in Central Provinces. Proc. Indian Acad. Sci.1936; 4: 108-122.
7. Chattopadhyayand SB Sengupta SK, Root-knot disease of jute in West Bengal. Curr. Sci.1955; 24: 276-77.
8. Sen AC, In: Proceedings of the 46th Indian Science Congress III.1959; p.506.
9. Siddiqi MR, Occurrence of the citrus nematode, Tylenchulussemipenetrans Cobb, 1913 and the reniform nematode, Rotylenchulusreniformis in India.Proc. 48 Indian Sci. Congr.1961; 3: 504.

10. Nair MRGK, et al. On the occurrence of the burrowing nematode, Radopholussimilis (Cobb, 1989) Thorne, 1949 on banana in Kerala. Indian. J. Entomol. 1966; 28: 553-554.
11. Sitaramaih K, et al. Plant Parasitic nematodes of India. Uttar Pradesh Agriculture University, Panthnagar and United States Department of Agriculture, Expt. Stn. Bull. 1971; 3:70.
12. Khan MR andDasgupta MK, Innovations in documentation and Information and communication Technology in Nematology. Science International.2013; 1: 271-281. DOI:10.5567/sciintl.2013.271.280.
13. Smith GF and Kopper RR, A southern perspective on the All Species Project and the global taxonomic imperative. Taxon.2002; 51: 359-361.

CHAPTER 4

Nutritional Requirements of Soybean Cyst Nematodes

Steven C. Goheen[1], James A. Campbell[1] and Patricia Donald[2]

[1] *Pacific Northwest National Laboratory, Richland, Washington, USA*

[2] *U. S. Department of Agriculture/ARS, Jacksonville, Tennessee, USA*

1. INTRODUCTION

Soybeans [*Glycine max*] are the second largest cash crop in US Agriculture, but the soybean yield is compromised by infections from *Heterodera glycines*, also known as Soybean Cyst Nematodes [SCN]. SCN are the most devastating pathogen or plant disease soybean farmers confront. This obligate pathogen requires nutrients from the plant to complete its life cycle. To date, SCN nutritional requirements are not clearly defined. Growth media supporting SCN still contain soy products. Understanding the SCN nutritional requirements and how host plants meet those requirements should lead to the control of SCN infestations. The nutritional requirements of SCN are reviewed in this chapter and those requirements are compared to those of other nematodes. Carbohydrates, vitamins, amino acids, lipids, and other nutritional requirements are discussed.

The survival of parasitic nematodes requires adequate nutrition. These essential nutrients are at least partially supplied by the host. But, availability of nutrients may not alone be sufficient for survival and reproduction. The parasite must also be able to establish a feeding site. Both the establishment of the feeding site and the presence of adequate nutrients for the soybean cyst nematode [SCN] are discussed below.

1.1. Feeding Site Establishment

Nematodes have differing mouth part structures which are adapted to their food source [1]. In the case of plant-parasitic nematodes, a stylet [analogous to a hypodermic needle], is used to puncture plant cells and a pump mechanism located in the nematode esophagus allows for exchange of fluids between the nematode and plant [1]. Most studies of the economically important root-knot and cyst-forming plant-parasitic nematodes have focused on what fluids are secreted by the nematode and how this facilitates establishment of a feeding site [2-4]. Specific information on the essential nutrients provided by the plant is lacking. In this chapter we focus on what is known about nutrient requirements for soybean cyst nematode, SCN.

The SCN is an obligate parasite requiring a host plant to complete its life cycle (see Figure 1). The cysts are found in the soil and contain eggs and first stage juveniles. The second stage juvenile hatches from the egg and penetrates plant roots. If the roots are a plant that is a host for SCN, the third and fourth stage juveniles molt into an enlarged shape called a sausage once a feeding site is successfully established where the primary goal is removing nutrients from the plant for use by the nematode. After enough nutrients have been obtained by the nematodes, those destined to become males molt into a worm-shape again and migrate out of the roots in search of a female. As the females mature, their size increases breaking root epidermal cells and the nematode is exposed to the soil where she emits pheromones to attract the males already in the soil. Once fertilization of the eggs has occurred, the female dies and her hardened body becomes the cyst which protects the eggs from environmental extremes and organisms which can kill the eggs. Some eggs are extruded into the soil in a gelatinous matrix and these eggs are thought to hatch once conditions favor hatch. The eggs within the cyst go through diapause and can survive within the cyst for more than a dozen years under the right conditions. Juveniles which enter nonhost plant roots may molt into a third stage juvenile but a successful feeding site will not be established and the plant will recognize the nematode as an invader and form necrotic cells surrounding the nematode effectively killing the nematode. Alternatively, some plants are slower to recognize the nematode as an invader and a molt to the third stage may occur but no further development of the nematode will occur. Once the nematode reaches the sausage stage, it lacks the muscles to leave the root and it dies.

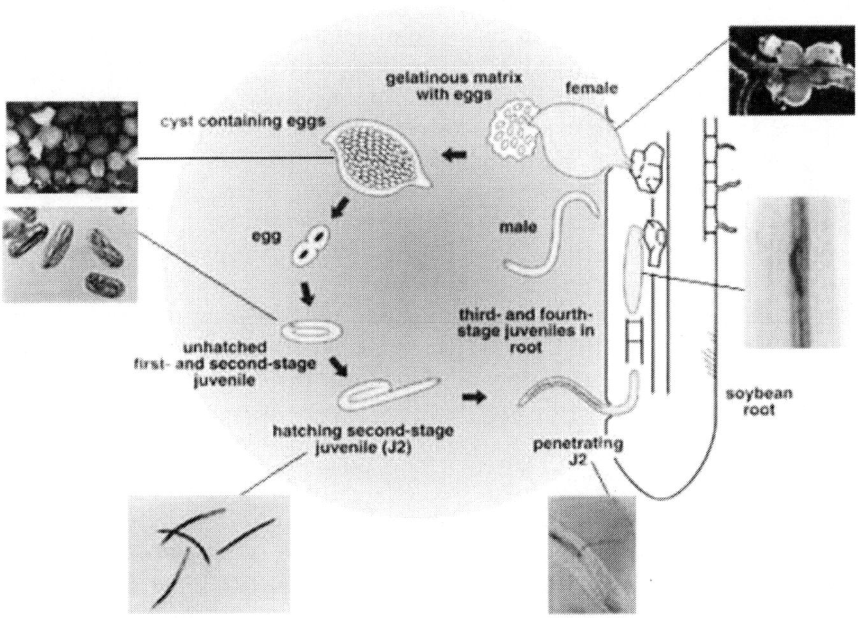

Figure 1. The life cycle of the soybean cyst nematode (SCN) is shown. Soil contains cysts with eggs as well as first stage juveniles. Second stage juveniles hatch from eggs and can then penetrate plant roots. The third and fourth stage juveniles feed off the plant. Males migrate out of the roots in search of a female. Maturing females rupture the root, releasing pheromones to attract males from soil. Females die after egg fertilization and her body becomes the cyst. This figure was obtained with permission from www.extension.umn.edu.

As an important crop in the United States [5], there are over 120 soybean lines which have some level of resistance to SCN [6]. Commercial soybean varieties primarily contain one or more different sources of resistance but 95% of all resistance is found from one source, PI 88788. Peking [PI 548402] and Hartwig [PI 437654] are also found in a few commercial varieties. Genetics of resistance is complex with multiple genes involved and interaction of minor genes or nongenetic sources complicates understanding of the process. In a resistant reaction, cytological changes occur and these have been documented [7-19]. Initial reaction to the nematode during the formation of the syncytium in both susceptible and certain resistant lines is identical for the first 4 days after infection [7. 9. 11]. Resistant reactions can be seen about day 4-5 [7, 9-11].

Cyst nematode juveniles hatch from eggs within the cyst or in the soil and enter plant roots typically in the zone of root elongation. They migrate to the pericycle and establish a feeding site [20]. Cellulases break polysaccharide chains and associated proteins in the plant cell walls. Other enzymes have been shown to be secreted by the nematodes as they move through plant tissue [21]. Rapid response by the plant to the nematode inhibits formation of a successful feeding site. A successful feeding site initiation results when the plant fails to respond or responds slowly to the presence of the nematode. One of the ways plant-parasitic nematodes protect themselves from plant responses to the nematodes is through secretion of peroxiredoxin, glutathione periosidase, and secreted lipid binding proteins within the surface coat of the nematode [22]. Although considerable knowledge is now available on the morphological changes in the plant cells due to the presence of the nematode feeding site and molecular studies have advanced our understanding of the interactions on a molecular level, the details of host specificity are unknown [23].

Information is available on the changes that occur within soybean plants when a compatible interaction between SCN and the plant occur. Information is also present on incompatible reactions when plant resistance inhibits SCN reproduction through either a hypersensitive response or formation of small syncytia which limit SCN reproduction. Infection of plant-parasitic nematodes is thought to alter plant products from the shikimic pathway. Infection by SCN increases the concentration of glucose, K, Ca and Mg in the roots but information is not available on whether these increases are products SCN then extracts from plant cells or whether these are responses by the plant to the presence of the nematode.

1.2. Nutritional Requirements

Heterodera glycines is considered to have a wide host range. Riggs and Hamblen tested 1152 entries from the Leguminosae family and found that 399 of these entries from 23 genera were susceptible. Poor hosts included 270 entries in 12 other genera [24]. Additional host studies have been conducted by Riggs and Hamblen [25-26], Miller and Gray [27-28], Venkatesh et al, [29], and Venkatesh et al [30]. Variability in host status within a plant species potentially makes identification of necessary nutrients required for establishment of the obligate feeding site easier but to date the specifics have eluded scientists.

A summary of the plants invaded by SCN are shown in Table 1. Most hosts of SCN are legumes and are limited to three subfamilies of the Leguminosae; however, approximately 50 genera in 22 families including nonlegumes are also hosts [31-32]. Some plants allow SCN to penetrate plant roots but limit reproduction of SCN [33]. The reason for this could be nutritional, or it could be due to other barriers within the plant. To determine which of those two possibilities are controlling virulence of SCN, nutritional requirements should be investigated more fully.

Table 1. Common names for plants that have been identified as good hosts for soybean cyst nematode [24-31].

Host Common Name	Host Scientific Name	Use
azuki bean	*Vigna angularis*	edible
. bean tree	*Laburnum sp*	ornamental
beans, green, dry	*Phaseolus vulgaris*	edible
beard tongue	*Penstemon digitalis*	ornamental
begger tick	*Desmodium ovalifolium*	weed
bells of Ireland	*Mollucella laevis*	ornamental
bitter cress	*Barbarea vulgaris*	spice
bladder senne	*Colutea arborescens*	shrub -ornamental
bush clover	*Lespeza capitata*	prairie plant
California burclover	*Medicago hispida*	weed
common chickweed	*Stellaria media*	weed
common lespedeza	*Lespedeza striata*	weed
coral bells	*Heuchera sanglunea*	ornamental
cranesbill	*Geranium maculatum*	weed
largeflowered beardtongue	*Penstemon gradiflorus*	wildflower
field pea tuberous vetch	*Lathyrus tuberosus*	edible/weed
fennugreek	*Trigonella goenum-gracum*	spice
foxglove	*Digitalis sp.*	weed
. geranium	*Pelargonium sp*	ornamental
gold apple	*Lycopersicon esulentum*	weed
golden chain	*Laburnum anagyroides*	ornamental

grass pea vine	*Lathyrus sativa*	edible/ornamental
green pea	*Pisum sativum*	edible
hairy vetch	*Vicia villosavillosa*	forage /cover crop
hemp sesbania	*Sesbania exaltata*	weed
henbit	*Lamium amplexicaule*	weed
hog peanut	*Amphicarpa bracteata*	weed
Indian joint vetch	*Aeschynomene virginica*	weed
indigo	*Indigofera parodiana*	shrub/herbaceous/small tree
clover Kenyan clover	*Trifolium*	ornamental
Korean lespedeza	*Lespedeza stiulacea*	forage
lance leaf rattlebox	*Crotalaria lanceolata*	weed
large flowered beard tongue	*Penstemon grandiflorus*	wild flower
large leaf lupine	*Lupinus polyphyllus*	wild flower
licorice milk vetch	*Astragalus glaucophyllus*	forage
little bur clover	*Medicago minima*	weed
milk vetch	*Astragalus canadensis*	forage
milky purslane	*Euphorbia supine*	weed
mouse ear chickweed	*Cerastium vulgatum*	weed
Common mullein	*Verbascum thapsus*	weed
nasturtium	*Tropaelum pergrinum*	ornamental
old field toadflax	*Linaria canadensis*	weed
pigeon pea	*Cajanus cajan*	edible
Americana pokeweed	*Phytolacca*	weed
purple deadnettle	*Lamium purpureum*	weed
purslane	*Portulaca oleracea*	weed
rainbow pink	*Dianthus chinensis*	ornamental
river bank lupine	*Lupinus rivularis*	edible
Rusian sickle milk vetch	*Astragalus falcate*	weed
service lespedeza	*Lespedeza cuneata*	weed
shrub lespedeza	*Lespedeza bicolor*	ornamental

Siberian pea tree	*Caragana arborescens*	ornamental
sicklepod	*Cassia tora*	weed
small flowered buttercress	*Cardamine parviflora*	weed
soybean	*Glycine max*	edible
Spanish broom	*Spartium junceum*	ornamental
speedwell	*Veronica peregrine*	weed
spider flower	*Cleome spinosa*	ornamental
spotted burclover	*Medicago arabica*	forage
stinking clover	*Cleome serrulata*	weed
sweet clover	*Melilotus taurica*	weed
sweet pearl lupine	*Lupinus mutabilis*	edible
tiny vetch	*Vicia hirsute*	ornamental vine
white horsehound	*Marrubium vulgare*	medicinal plant
white lupine	*Lupinus albus*	livestock feed
white pea	*Lathyrus ochrus*	wild flower
Wilcox penstemon	*Penstemon wilcoxi*	wilflower
winged pigweed	*Cycloloma atriplicifolia*	weed
yellow lupine	*Lupinus lateus*	wild flower

In many ways, it is inappropriate to compare humans to nematodes. But, from a nutritional perspective, much more is known about human nutrition than what is known about nutritional requirements of nematodes. For humans, numerous biochemical and mineral components are essential nutrients. But, for nematodes, only a few are known. Yet, nematodes have a comparatively simple digestive system. So, it would be reasonable to predict that nutritional requirements for these organisms are more extensive than what is currently known.

It is also inappropriate to generalize nutritional needs from studies on one nematode to all the nematodes within the various trophic categories. Certainly there should be similarities, but it is clear from the literature that animal parasitic nematodes have different needs from the plant parasites. And, it may also be that those plant parasites infecting specific organisms, such as SCN might have nutritional needs that synergize with the contents of the host soybean plant.

Survival is best understood when chemically defined culture media can be shown to not only sustain life, but also to promote reproduction. Chemically defined media have been identified for the survival of some nematodes and this work has recently been reviewed [34]. The successful media originally included all the amino acids in *Escherichia coli*, and in the amino acid ratios found in *E. coli*. Nematode growth media has been since modified to include a greater number of constituents including glucose, minerals, growth factors, nucleic acid precursors, vitamins, a sterol and heme source. However, SCN has not yet been shown to survive or reproduce on these media. Currently, the only growth media known to sustain SCN includes soy products [35].

Articles published on the nutritional requirements of a wide range of nematodes, generally do not specify SCN [1. 36-37]. While a few nutritional requirements for individual nematode species have been studied, these requirements are limited and their applicability to SCN is unknown. It is assumed that plant- and animal-parasitic nematodes may have different nutritional requirements from entomopathogenic, and microbivorous nematodes.

2. LIPIDS

Lipids consist of many non-water soluble components including free fatty acids, phospholipids, triglycerides, sterols, and other species. Many of these classes have been studied at least in one host-nematode relationship and are the most studied with the exception of nucleic acids due to their great structural variety and importance as food reserves. For example, Krusberg [38] reported the total lipids and fatty acids from 5 species of plant parasitic nematodes, and their common hosts. They found that the nematodes had the same fatty acids as the hosts, with the exception of the polyunsaturated fatty acids. These appeared to be synthesized by the nematodes. There was also some speculation that nematode fatty acid synthesis resembled that of bacterial pathways rather than that of higher animals. It was not clear from the study whether intestinal flora of the nematode could have been at least partially responsible for this difference, or whether the nematode itself synthesized the fatty acids. Some nematodes are clearly capable of synthesizing longer chain fatty acids from shorter chain precursors. They are also capable of desaturating the fatty acids [39].

Entomopathogenic nematodes infecting locusts consume host fat and protein [40]. A decrease in lipid reserves has been seen in starved nematodes which can be related to decreased infectivity [41]. Lipid content is also known to decrease when nematodes come out of anhydrobiosis [42]. Lipids associated with the nematode surface [cuticle] are triacylglycerols, sterols, specific phospholipids, and other glycolipids [43-45].

The most widely known class of essential nutrients for nematodes is sterol [36,46]. This nutritional requirement was first discovered by Dutky et al.

[47] and thought to be potentially a means for control of plant parasitic nematodes. A recent review further confirms this nutritional sterol requirement for the nematode *C. elegans* [48]. Nematode parasites of animals also require sterol for larval development [49]. The biochemical mechanism which converts sitosterol to cholesterol appears to be lacking in nematodes [50]. Nematodes are capable of modifying sterols obtained from their diet [46] but degradation of sterols to CO_2 by nematodes is not clear [51]. More than 63 sterols have been identified from free-living and plant-parasitic nematodes. Characteristics of sterols which can be used by nematodes include those which have a hydroxyl group at C-3, a trans-A/B ring system and an intact nonhydroxylated side chain but lack methyl groups at C-4 [52]. Plant sterols are different than animal sterols with plants being unique in methyl, ethyl or related alkyl groups at the C-24 position of the sterol side chain [52]. There are also differences between plant sterols and plant-parasitic nematode sterols. These findings suggest that nematodes ingest plant sterols and remove the C-24 side chain. In addition, the nematode saturates the double bonds in the four-membered ring system to produce stannols [52]. Steroid hormones are important in development processes and in transition to different life stages [53]. Most likely genetic and biochemical methods will be needed to determine the function of hormones found in nematodes [54]. Novel genes involved in the production of 17β-hydroxysteroid dehydrogenase in the soybean cyst nematode have been reported [55].

Sterols were first reported in soy oil by Kraybill et al. [56]. Formononetin is an o-methyl-isoflavone mainly produced in legumes, including soybean plants [57]. It helps stimulate the production of steroids in mammals, and possibly also in nematodes. Research in this area by the USDA was reviewed by Chitwood [58].

3. AMINO ACIDS AND PROTEINS

There are no clearly defined requirements for proteins, amino acids, or peptides for SCN. However, it is unlikely that nematodes synthesize all the amino acids. For humans, there are 9 essential amino acids [phenylalanine, valine, threonine, tryptophan, isoleucine, methionine, leucine, lysine, and histidine]. Some others are required under special circumstances [arginine, cysteine, glutamine, proline, serine, tyrosine, and asparagiene]. Cysteine, tyrosine, and arginine are required during rapid growth, such as in infancy. And, arginine, cysteine, glycine, glutamine, histidine, proline, serine and tyrosine are required by some individuals because these amino acids are not adequately synthesized by these individuals. These are essential components for the synthesis of many essential enzymes and structural proteins ; it is anticipated there are similar needs in the nematode diet.

Protein consumed by parasitic nematodes can severely damage the host. Juveniles have high protein requirements and consuming the host protein can severely weaken the plant [46].

There have been efforts to identify the essential amino acids of nematodes [59-61], but so far common requirements have not been identified. However, protein synthesis in cotton roots is modified when the root-knot nematode [RKN] infects susceptible plants. These plant-parasitic nematodes influence the distribution of amino acids in cotton root galls [61]. Also, there is one genetic modification of the cotton plant which makes them less susceptible to infection by the RKN. This modification is responsible for the synthesis of a 14 kDa protein [60].

For the snail parasitic nematode, *Rhabditis maupasi*, five essential amino acids have been identified. These include lysine, methionine, phenylalanine, tryptophane, and valine [62]. In the entomophilic locust parasite, *M. migrescens*, essential nutrients include protein nitrogen [63]. Essential amino acids have also been identified for the nematode *C. briggsae* [64].

4. VITAMINS

There are 13 essential vitamins required by humans. These include Vitamin A [Retinol] Vitamin B_1[Thiamine] Vitamin C [Ascorbic acid] Vitamin D [Calciferol] Vitamin B_2 [Riboflavin] Vitamin E [Tocopherol] Vitamin B_{12} [Cobalamins] Vitamin K_1 [Phylloquinone] Vitamin B_5 [Pantothenic acid] Vitamin B_7 [Biotin] Vitamin B_6 [Pyridoxine] Vitamin B_3 [Niacin] Vitamin B_9 [Folic acid]. Of these, vitamin E is known to be a nutritional requirement for the gastrointestinal parasite, *Heligmosomoides bakeri* [65], and several of the B vitamins are known to be essential nutrients of *C. elegans* [66-68].

For SCN, DNA sequences responsible for the biosynthesis of enzymes that can produce some of the B vitamins *de novo* have been discovered [69]. Therefore, SCN may not need the same B vitamins as *H. bakeri*, for example. And, it is likely that there are other differences in vitamin and supplement requirements across all nematodes.

5. MINERALS

Considerable research on mineral requirements for nematodes has been reported in mammalian parasites. For example, the gastrointestinal nematode, *H. bakeri*, requires boron [70], zinc [71], and selenium [65] for survival. And, other nematodes have similar mineral requirements [72-74]. For example, magnesium, sodium, potassium, manganese, calcium and copper are required nutrients of *C. elegans* [5]. However, SCN mineral requirements remain unclear.

Whether minerals, influence nematode survival may not help in their control if necessary minerals are readily available in soil, and essential to the host organisms. But, elements not essential to survival of the host could be controlled in soils to help control SCN survival.

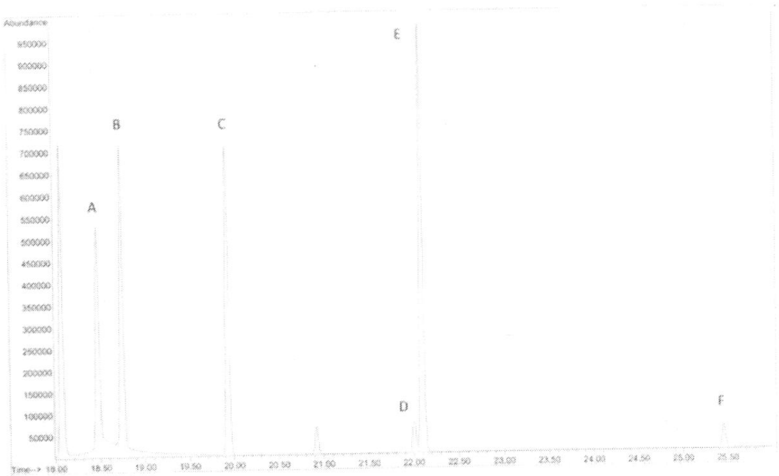

FIGURE 2. A total ion chromatogram of derivatized soybean root extract is shown. A = D-(-)-Fructose, B = D-Pinitol, C = D-(+)-Glucose, D = D-*chiro*-inositol, E = β-D-(+)-Glucose, F = *Myo*-inositol. Reproduced with permission from [79].

6. CARBOHYDRATES

Nematodes require carbohydrates for energy, usually in the form of glycogen. One study showed that several different carbohydrates were sufficient to provide a carbon, or energy source for *C. elegans*, and that glucose was more effective than fructose or sucrose [76]. For *C. elegans*, glucose along with cytochrome c and β-sitosterol were sufficient to sustain a healthy population.

One of the most striking features of soybean chemistry is the abundance of pinitol [77-79]. Pinitol is a carbohydrate with unusual nutritional properties [77]. Figure 2 shows a total ion chromatogram of a derivatized extract of soybean roots. It is unusual for a plant to have so much pinitol. The levels shown in this study indicate pinitol is present at a concentration of 26 mg/g (dry weight) compared to peanuts with only 4.7 mg/g or clover with 14 mg/g [79]. However, there is no evidence that pinitol, or any of the related inositols are needed for SCN survival [79].

7. OTHER NUTRIENTS OR FEEDING REQUIREMENTS

The nematode *Rhabditis maupasi* requires hemin or another iron porphyrin for survival [62]. Similarly, *C elegans* also requires a heme source for survival [34]. It is likely that many other nematodes require heme, or a closely related hemin. There is also good evidence that SCN requires a heme source [80].

8. DISCUSSION

In comparison to our knowledge of human nutrition, our understanding of nutritional requirements of SCN is in its infancy. Limited information is available for members of the Nematoda Phyllum, but such a small amount of information is available that extrapolation across trophic groups and even within genera may be misleading. Finding a successful artificial diet would be a reasonable first step in defining the nutritional needs of SCN. But, this data needs to be coupled with a good understanding of feeding site establishment and plant responses to SCN infections.

Studying biochemical pathways would be a valuable approach, and could also help identify pathways that could be blocked to help minimize SCN survival. Our laboratory began by examining the chemistry of the plant to identify unique nutrients necessary for SCN survival, but that approach was not immediately successful. Another approach is to continue to use DNA mapping to better understand potential plant and parasite pathways. While this approach is less direct, it is currently a very active area of investigation, and can reveal more information than simply nutritional requirements.

Details of the SCN host-parasite responses during infection and feeding site establishment have been more extensively investigated than nutritional requirements. Relationships between the available nutrients from host plants compared to non-hosts could provide valuable clues on these requirements. And, once an adequate media for SCN survival has been well defined, methods to control this pest should follow.

ACKNOWLEDGEMENTS

The authors acknowledge support from the USDA and Battelle.

REFERENCES

1. Munn, E.A. Munn, P. D. Feeding and digestion in Lee, D. L. (ed). The Biology of Nematodes. Taylor and Frances. New York; 2002. p211-232.
2. Davis, E. L., Hussey, R. S., Baum, T. J. Getting to the roots of parasitism by nematodes. *Trends in Parasitology*. 2004; 20:134-141.
3. Hussey, R. S., Mims, C. W. Ultrastructure of esophageal glands and their secretory granules in the root-knot nematode, *Meloidogyne incognita*. *Protoplasma*. 1990; 156: 9-18.
4. Wyss, U. Feeding behavior of plant-parasitic nematodes.. In: Lee, D. L. (ed.). The Biology of Nematodes. Taylor and Francis: New York; 2002. p233-259.
5. Wrather, A., Koenning, S. Effects of Diseases on Soybean Yields in the United States 1996 to 2007. *Plant Management Network*. 2009; (http://www. plantmanagementnetwork.org/pub/php/research/2009/yields/).
6. Shannon, J.G., Arelli, P. R., Young, *L. D. Breeding for resistance and tolerance. In Schmitt, D. P., Wrather, J. A., Riggs, R. D. eds, Biology and Management of Soybean Cyst Nematode, Ed 2. Schmitt & Associates of Marceline, Marceline, MO,* 2004: p155–180.
7. Endo, B.Y. Histological responses of resistant and susceptible soybean varieties, and backcross progeny to entry development of *Heterodera glycines*. *Phytopathology*. 1965; 55:3.75-381.
8. Endo, B.Y. Ultrastructure of initial responses of resistant and susceptible soybean roots to infection by *Heterodera glycines*. *Revue of Nematology*. 1991; 14:73-94.
9. Riggs, R.D., Kim, K.S., Gipson, I. Ultrastructural changes in Peking soybeans infected with *Heterodera glycines*. *Phytopathology*. 1973; 63:76-84.
10. Acido, J. R., Dropkin, V. H., Luedders, V. D.. Nematode population attrition and histopathology of *Heterodera glycines*-soybean associations. *Journal of Nematology*. 1984; 16:48-57.
11. Kim, Y. H., Riggs, R. D., Kim, K. S. Structural changes associated with resistance of soybean to *Heterodera glycines*. *Journal of Nematology*. 1987; 19:177-187.
12. Halbrendt, J.M., Lewis, S.A., Shipe, E. R.. A technique for evaluating *Heterodera glycines* development in susceptible and resistant soybean. *Journal of Nematology*. 1992; 24:84-91.
13. Kim, Y. H., Riggs, R. D. Cyclopathological reactions of resistant soybean plants to nematode invasion. in Wrather, J.A., Riggs, R. D. (eds). Biology and management of the soybean cyst nematode. APS Press. St. Paul. 1992; p157-168.

14. Mahalingam, R., Skorupska, H.T.. Cytological expression of early response to infection by *Heterodera glycines* Ichinohe in resistant PI 437654 soybean. *Genome*. 1996; 39:986-998.
15. Klink, V.P. Overall, C. C., Alkharouf, N., MacDonald, M. H., Matthews, B. F. Laser capture microdissection (LCM) and comparative microarray expression analysis of synctial cells isolated from incompatible and compatible soybean roots infected by soybean cyst nematode (*Heterodera glycines*). *Planta*. 2007; 226:1389-1409.
16. Klink, V.P., Overall, C. C., Alkharouf, N., MacDonald, M. H., Matthews, B. F. A comparative microarray analysis of an incompatible and compatible disease response by soybean (*Glycine max*) to soybean cyst nematode (*Heterodera glycines*) infection. *Planta*. 2007; 226:1423-1447.
17. Klink, V.P., Hosseini, P., Matsye, P., Alkharouf, N., Matthews, B. F. A gene expression analysis of syncytia laser microdissected from the roots of the *Glycine max* (soybean) genotype PI 548402 (Peking) undergoing a resistant reaction after infection by *Heterodera glycines* (soybean cyst nematode). *Plant Molecular Biology*. 2009; 71:525-567.
18. Klink, V.P., Hosseini, P., Matsye, P., Alkharouf, N. W. Matthews, B. F. Syncytium gene expression in *Glycine max* (PI 88788) roots undergoing a resistant reaction to the parasitic nematode *Heterodera glycines*. *Plant Physiological and Biochemistry*. 2010; 48:176-193.
19. Klink, V.P., Overall, C. C., Alkharouf, N., MacDonald, M. H., Matthews, B. F. Microarray detection calls as a means to compare transcripts expressed within synctial cells isolated from incompatible and compatible soybean (*Glycine max*) roots infected by the soybean cyst nematode (*Heterodera glycines*). *Journal of Biomedicine and Biotechnology*. 2010:491217 (1-30).
20. Wyss, U., Stender, C., Lehmann, H.. Ultrastructure of feeding sites of the cyst nematode *Heterodera schachtii* Schmidt in roots of susceptible and resistant *Raphanus sativus* L. var. *oleiformis*. Pers. cultivars. *Physiological Plant Pathology*. 1984; 25: 21-37.
21. Davis, E. L., Hussey, R. S. Baum T. J., Bakker, J., Schots, A. Nematode parasitism genes. *Annual Review of Phytopathology*. 2000; 38:365-396.
22. Gheysen, G., Jones, J. T. Molecular aspects of plant-nematode interactions. In Perry, R. N., Moens, M. (eds). *Plant Nematology*. CABI. 2006; p234-254.
23. Vanholme, B., De Meutter, J., Tytgat, T., Van Montagu, M., Coomans, A. Gheysen, G. Secretions of plant-parasitic nematodes: a molecular update. *Gene*. 2004; 332: 13-27.
24. Riggs, R.D., Hamblen, M.L. Soybean cyst nematode host studies in the family Leguminosae. Arkansas Agricultural Experiment Station Report Series 110. Fayetteville, AR 1962; 17p.
25. Riggs, R.D., Hamblen, M.L. Additional weed hosts of *Heterodera glycines*. *Plant Disease Reporter*. 1966; 50:15-16.
26. Riggs, R.D., Hamblen, M.L. Further studies on the host range of the soybean cyst nematode. Arkansas Agricultural Experiment Station Bulletin 718. Fayetteville, AR 1966; 19p.
27. Miller, L.I., Gray, B. J. Reaction of lambsquarter, Swiss chard, and spinach to eleven isolates of the soybean cyst nematode. *Virginia Journal of Science*. 1966; 17: 246.
28. Miller, L.I., Gray, B. J. Development of eleven isolates of *Heterodera glycines* on six legumes. *Phytopathology*. 1967; 57:647.
29. Venkatesh, R., Harrison, S. K., Riedel, R. M. Weed hosts of soybean cyst nematode (*Heterodera glycines*) in Ohio. *Weed Technology*. 2000; 14:156-160.
30. Venkatesh, R., Harrison, S. K., Regnier, E. E., Riedel, R. M. Purple deadnettle effects on soybean cyst nematode populations in no-till soybean. North Central Weed Science. Society. 2004; 59:56.
31. Riggs, R.D., Host Range. In Riggs, R. D., Wrather, J. A. (eds). Biology and management of the soybean cyst nematode. APS Press, St Paul, MN. 1992; p107-114.

32. Mock, V.A., Creech, J.E., Davis, V.M., Johnson. W. G. Plant growth and soybean cyst nematode response to purple deadnettle (*Lamium pupureum*), annual ryegrass, and soybean combinations. *Weed Science*. 2009; 57:489-493.
33. Riggs, R.D., Nonhost root penetrations by soybean cyst nematode. *Journal of Nematology*. 1987; 19:251-254.
34. Braeckman, B. P., Houthoofd, K., Vanfleteren, J. R. Intermediary metabolism, WormBook, (ed. The C. elegans Research Community), WormBook, doi/10.1895/wormbook. 2009; 1.146.1, http://www.wormbook.org.
35. Krusberg, L.R. Studies on the culturing and parasitism of plant-parasitic nematodes, in particular *Ditylenchus dipsaci* and *Aphelenchoidesrizemabosi* on alfalfa tissues. *Nematologica*. 1961; 6:181-200.
36. Chitwood, D. J. Biochemistry and function of nematode steroids. *Critical Reviews in Biochemistry and Molecular Biology*. 1999; 34(4) 273-284.
37. DeLey, P., Mundo-Ocampo, M. Cultivation of nematodes.. In Nematology: Advances and Perspectives Vol 1. Nematode morphology, physiology and ecology. CABI Publishing, Cambridge, Mass. 2004; p541-608.
38. Krusberg, L. R. Analyses of total lipids and fatty acids of plant-parasitic nematodes and host tissues. *Comparative Biochemistry and Physiology*. 1967; 21: 83-90.
39. Ruess, L., Haggblom, M. H., Zapta, E. J. G., Dighton, J. Fatty acids of fungi and nematodes – possible biomarkers in the soil food chain?" *Soil Biology & Biochemistry*. 2002; 34: 745-756.
40. Gordon, R, Webster, J. M., Hislop, T. G. Mermithid parasitism, protein turnover and vetillogenesis in the desert locust, *Schistocerca gregaria* Forskal. *Comparative Biochemistry and Physiology*. 1973; 46B: 575-593.
41. Barrett, J., Wright, D. J. Intermediary metabolism. In Perry, R. N., Wright, D. J. (eds). The physiology and biochemistry of free-living and plant-parasitic nematodes. CABI. London. 1998; p331-353.
42. Crowe, J. H., Madin, K. A. C., Loomis, S. H. Anhydrobiosis in nematodes: metabolism during resumption of activity. *Journal of Experimental Zoology*. 1977; 201:57-63.
43. Blaxter, M. L. Cuticle surface proteins of wild type and mutant *Caenorhabditis elegans*. *Journal of Biological Chemistry*. 1993; 2638: 6600-6609.
44. Bird, A.F., Bird, J. The Structure of Nematodes. 2nd edition. Academic Press, San Diego. 1991.
45. Spiegel, Y. and M.A. McClure.. The surface coat of plant-parasitic nematodes: chemical composition, origin, and biological role- a review. *Journal of Nematology*. 1995; 27:127-134.
46. Chitwood, D. J., Lusby, W. R. Metabolism of plant sterols by nematodes. *Lipids*. 1991; 26 (8): 619-627.
47. Dutky, S. R., Robbins, W. E., Thompson, J. V. The demonstration of sterols as requirements for the growth, development, and reproduction of the DD-136 nematode. *Nematologica*. 1967; 13: 140.
48. Entchev, E., Kurzchalia, T. V. Requirement of sterols in the life cycle of the nematode *Caenorhabditis elegans" Seminars in Cell & Developmental Biology*. 2005; 16: 175-182.
49. Bolla, R.I, Weinstein, P.P., Cain, G.D. Fine structure of the coelomocyte of adult *Ascaris suum*. *Journal of Parasitology*. 1972; 58: 1025-1036.
50. Cheong, C. M., Na, K., Kim, H., Jeong, S., Joo, H., Chitwood, D. J., Paik, Y. A potential biochemical mechanism underlying the influence of sterol deprivation stress on *Caenorhabditis elegans* longevity. *Journal of Biological Chemistry*. 2011; 286: 7248-7256.
51. Rothstein, M. Nematode biochemistry. IX. Lack of sterol biosynthesis in free-living nematodes. *Comparative Biochemistry and Physiology*. 1968; 27: 309-317.
52. Chitwood, D. J. Biosynthesis in Perry, R. N., Wright, D. J. (eds). The physiology and biochemistry of free-living and plant-parasitic nematodes. CABI. London. 1998; p303-330.

53. Riddle, D.L. and P.S. Albert. Genetic and environmental regulation of dauer larva development. In *C. elegans* II. Riddle, D., Meyer, B., Priess, J., and Blumenthal, T. (ed). Cold Spring Harbor: Cold Spring Harbor Press. 1997; p739-768.
54. Chervitz, S.A., Aravind, L., Herloc, G. S., Ball, C. A., Koonin, E. V., Dwight, S. S.. Comparison of the complete protein sets of worm and yeast: orthology and divergence. *Science*. 1998; 282: 2022-2028.
55. Skantar, A. M., Guimond, N. A., Chitwood, D. J. Molecular characterization of two novel 17β-hydroxysteroid dehydrogenase genes from the soybean cyst nematode *Heterodera glycines*. *Nematology*. 2006; 8(3) 321-333.
56. Kraybill, H. R., Thornton, M. H., Eldridge, K. E. Sterols from crude soybean oil. *Industrial and Engineering Chemistry*. August, 1940:1138-1139.
57. Medjakovic, S. and Jungbauer, A. "Red Clover Isoflavones Biochanin A and Formononetin are Potent Ligands of the Human Aryl Hydrocarbon Receptor". *The Journal of Steroid Biochemistry and Molecular Biology*. 2008; 108 (1–2) 171–177.
58. Chitwood, D. J. Research on plant-parasitic nematode biology conducted by the United States Department of Agriculture-Agricultural Research Service." *Pest Management Science*. 2003; 59: 748-753.
59. Young, J. R., Riggs, R. D. Identification of the free amino acids of nematode resistant and susceptible soybeans. *Arkansas Academy of Science Proceedings*. 1964; 18: 46-49.
60. Hedin, P. A., Creech, R. G. Altered amino acid metabolism in root-knot nematode inoculated cotton plants. *Journal of Agriculture and Food Chemistry*. 1998; 46: 4413-4415.
61. Hedin, P. A., Creech, R. G. Effects of root-knot nematodes on distribution of amino acids in cotton root galls. Mississippi State University Experiment Station Bulletin 1103 Office of Agricultural Communications, Mississippi. 2001; p1-6.
62. Brockelman, C. R., Jackson, G. J. Amino acid, heme, and sterol requirements of the nematode, *Rhabditis maupasi Journal of Parasitology*. 1978; 64(5): 803-809.
63. Baylis, H. A. Observations on the nematode *Mermis migrescens* and related species. *Parasitology*. 1944; 36: 122-132.
64. Vanfleteren, J. R. Amino acid requirements of the free-living nematode *Caenorhabditis briggsae*. *Nematologica*. 1973; 19: 93-99.
65. Smith, A., Madden, K. B., Yeung, K. J. A., Zhao A., Elfrey, Finkelman, , F. J., Lavender, O., Shea-Donohue, T., Uban, J. F. Deficiencies in selenium and/or vitamin E lower the resistance of mice to *Heligmosomides polygyrus* infections. *Journal of Nutrition*. 2005; 135: 830-836.
66. Liu , C., Lu, N. Biotin requirement and its biosynthesis blockage in the free-living nematode, *Caenorhabditis elegans*. *Federation of American Societies for Experimental Biology*. 2008; 22: 1102-1104.
67. Nicholas, W. L. Hansen, E., Dougherty, E. C. The B-vitamins required by *Caenorhabditis briggsae (Rhabditidae)*. *Nematologicia*. 1962; 8: 129-135.
68. Szewczyk, N. J. , Kozak, E., Conley, C. A. Chemically defined medium and *Caenorhabditis elegans*. *BMC Biotechnology*. 2003; 3: 19.
69. Craig, J. P., Bekal, S. , Niblack, T. , Domier, L., Lambert, K. N. Evidence for horizontally transferred genes involved in the biosynthesis of vitamin b1, b5, and b7 in *Heterodera glycines*. *Journal of Nematology*. 2009; 41(4) 281–290.
70. Bourgeois, A. C., Scott, M. E., Sabally, K., Koski, K. G. Low dietary boron reduces parasite (nematoda) survival and alters cytokine profiles but the infection modifies liver minerals in mice. *Journal of Nutrition*. 2007; 137(9): 2080-2086.
71. Shi, H. N., Scott, M. E., Koski, K. G., Boulay, M., Stevenson, M. Energy restriction and severe zinc deficiency influence growth, survival and reproduction of *Heligmosomoides polygyrus* (Nematoda) during primary and challenge infections in mice. *Parasitology*. 1995; 110: 599-609.
72. Coop, R. L., Kyriazakis, I. Influence of host nutrition on the development and consequences of nematode parasitism in ruminants. *Trends in Parasitology*. 2001; 17(7) 325-330.

73. Koski, K. G., Scott, M. E. Gastrointestinal nematodes, trace elements, and immunity. *Journal of Trace Elements in Experimental Medicine.* 2003; 16: 237-51.
74. McClure, S. J., McClure, T. J., Emery, D. I. Effects of molybdenum intake on primary infection and subsequent challenge by the nematode parasite *Trichostromgylus colubriformis* in weaned Merino lambs. *Research in Veterinary Science.* 1999; 67: 17-22.
75. Lu, N. C., Cheng A. C., Briggs, G. M. A Study of Mineral Requirements in *Caenorhabditis elegans. Nematologica.* 1983; 29: 425-434.
76. Lu, N. C., Goetsch, K. M. Carbohydrate Requirement of *Caenorhabditis elegans* and the Final Development of a Chemically Defined Medium. *Nematologica.* 1993; 39: 303-311.
77. Campbell, J.A., Goheen, S.C., Donald, P. Extraction and Analysis of Inositols and Other Carbohydrates from Soybean Plant Tissues. in Recent Trends for Enhancing the Diversity and Quality of Soybean Products, Krezhova, D. ed. InTech publisher http://www.intechopen.com/articles/show/title/extraction-and-analysis-of-inositols-and-other-carbohydrates-from-soybean-plant-tissues. 2011; p421 – 446.
78. Garland, S., Goheen, S., Donald, P.A., Campbell, J. Application of derivatization gas chromatography/mass spectrometry for the identification and quantitation of pinitol in plant roots. *Analytical Letters.* 2009; 42: 2096-2105.
79. McDonald IV, L. W., Goheen, S. C., Donald, P. A., Campbell J. A. Identification and quantitation of various inositols and o-methylinositols present in plant roots related to soybean cyst nematode host status. *Nematropica.* 2012; 42: 1-8.
80. Ko, M. P., Huang, P., Huang, J., Barker, K. R. The occurrence of phytoferritin and its relationship to effectiveness of soybean nodules. *Plant Physiology.* 1987; 83: 299-305.

CHAPTER 5

Hsp-90 and the Biology of Nematodes

Nik AIIN Him†[1], Victoria Gillan†[1], Richard D Emes†[2], Kirsty Maitland[1] and Eileen Devaney*[1]

[1]Parasitology Group, Institute of Comparative Medicine, School of Veterinary Medicine, University of Glasgow, Bearsden Road, Glasgow G61 1QH, UK and

[2]The School of Veterinary Medicine and Science, The University of Nottingham, Sutton Bonington Campus, Sutton Bonington, Leicestershire, LE12 5RD, UK

ABSTRACT

Background

Hsp-90 from the free-living nematode *Caenorhabditis elegans* is unique in that it fails to bind to the specific Hsp-90 inhibitor, geldanamycin (GA). Here we surveyed 24 different free-living or parasitic nematodes with the aim of determining whether *C. elegans* Hsp-90 was the exception or the norm amongst the nematodes. We combined these data with codon evolution models in an attempt to identify whether *hsp-90* from GA-binding and non-binding species has evolved under different evolutionary constraints.

Results

We show that GA-binding is associated with life history: free-living nematodes and those parasitic species with free-living larval stages failed to bind GA. In contrast, obligate parasites and those worms in which the free-living stage in the environment is enclosed within a resistant egg, possess a

GA-binding Hsp-90. We analysed Hsp-90 sequences from fifteen nematode species to determine whether nematode *hsp-90*s have undergone adaptive evolution that influences GA-binding. Our data provide evidence of rapid diversifying selection in the evolution of the *hsp-90* gene along three separate lineages, and identified a number of residues showing significant evidence of adaptive evolution. However, we were unable to prove that the selection observed is correlated with the ability to bind geldanamycin or not.

Conclusion

Hsp-90 is a multi-functional protein and the rapid evolution of the *hsp-90* gene presumably correlates with other key cellular functions. Factors other than primary amino acid sequence may influence the ability of Hsp-90 to bind to geldanamycin.

BACKGROUND

Heat shock protein 90 (Hsp-90) is essential in all eukaryotes because of its role in chaperoning proteins such as kinases, growth factors and other signal transduction proteins [1]. As knock out of *hsp-90* is generally lethal, specific inhibitors have been important for defining the cellular functions of Hsp-90. Of these, the best characterised is geldanamycin (GA) a naturally occurring ansamycin compound produced by *Streptomyces hygroscopicus* [2]. The mode of action of GA is well documented and the crystal structure of yeast Hsp-90 complexed to GA has been solved [3]. Most eukaryotic Hsp-90s are susceptible to inhibition with GA and will bind the drug in pull-down assays using GA linked to a solid support [4]. The exception to this rule is Hsp-90 (DAF-21) from the free-living nematode *Caenorhabditis elegans* (*Ce*), which fails to bind to GA in pull-down assays [5] and which shows no apparent ill-effect when maintained on a lawn of *S. hygroscopicus*. In contrast GA binds to Hsp-90 from the parasitic nematode *Brugia pahangi* (*Bp*) and is lethal to all life cycle stages [6]. The inability of *Ce*-Hsp-90 to bind GA was proposed to represent an example of adaptive evolution, as both *C. elegans* and the *Streptomyces* species that synthesizes GA are soil dwelling and so may share the same ecological niche [5]. In this study we adopted an integrated approach to investigating the possibility that nematode *hsp-90*s may have undergone adaptive evolution that influences GA-binding. We surveyed Hsp-90 from a range of free-living or parasitic nematodes for their ability to bind GA, and combined these data with statistical methods to compare substitution rates using maximum likelihood (ML) models of codon evolution [7, 8], in an attempt to identify if significant evidence for adaptive diversifying evolution ($d_N/d_S > 1$) relating to GA-binding exists. Specifically we investigated whether *hsp-90* from GA-binding nematodes has evolved under differing evolutionary constraints compared to non-GA binding nematode species

Nematodes are an extensive group of animals that have evolved to colonise most ecological niches. Many nematodes are important parasites of humans, domestic animals and plants and are responsible for considerable economic losses worldwide [9]. Since the sequencing of the *C. elegans* genome, the genomes of several additional free-living nematodes have been sequenced, including two other species of *Caenorhabditis* (*briggsae* and *remanei*) and *Pristionchus pacificus* [10], a species that associates with beetles. Draft genome sequences are available for *C. brenneri* and *japonica*, and for the parasitic species *Trichinella spiralis* [11], *Brugia malayi* [12], *Haemonchus contortus* http://www. sanger.ac.uk/Projects/H_contortus/ and *Meloidogyne hapla* [13] and *incognita* [14], plant parasitic nematodes. Furthermore, sequencing projects are planned for several other important parasitic species http://www. sanger.ac.uk/Projects/Helminths/. In addition EST projects have provided an abundance of sequence information on the phylum [15]. A phylogeny of the nematodes based on the small subunit ribosomal DNA sequences has been proposed [16] in which the nematodes are divided into five clades. Each of these clades contain parasitic species, suggesting that parasitism has arisen on multiple occasions during the evolution of the phylum. *C. elegans* is grouped in clade V, along with many important parasites of veterinary significance, while the filarial nematode *Brugia malayi* lies in clade III, as do some important ascarid parasites of humans and animals.

While free-living nematodes, such as Caenorhabditis species, often exhibit relatively simple life histories developing from egg to adult in one environment, parasitic species display a huge diversity in life cycles. Some parasitic species are transmitted direct from host to host and many of these have a free-living stage within the environment (e.g. the Trichostrongyles), while others, such as the filarial nematodes are obligate parasites. These worms have no free-living stages and are transmitted from host to host by the bite of an infected intermediate host, such as a mosquito. Likewise, the clade I species Trichinella spiralis is permanently parasitic and transmitted by ingestion of infected tissue. In this study, we set out to investigate whether the life history of a species was correlated with the ability of Hsp-90 to bind GA by analysing an additional twenty-four species of free-living or parasitic nematode, belonging to four of the five clades. These data were combined with an analysis of hsp-90 DNA sequences from a range of nematode species, in an attempt to determine whether the hsp-90 genes of

GA-binding nematodes have significantly different ratios of non-synonomous (amino acid-changing) to synonomous (amino acid-retaining) nucleotide substitutions compared to non-binding species.

RESULTS

GA pull-down assays

The ability of nematode Hsp-90 to bind to GA was assessed using a standard pull-down procedure in which GA bound to a solid support was used to select Hsp-90 from worm extract. The presence of Hsp-90 in the pull-down was assessed by immuno-blotting. Prior to use in pull-down experiments, extracts of each worm were probed with an anti-*B. pahangi* Hsp-90 antibody or with AC88, a monoclonal antibody to Hsp-90, to confirm that Hsp-90 could be detected on a blot. Extracts of three species failed to react with either antibody over a range of antibody concentrations and exposure times: *Trichinella spiralis*, *Trichuris muris* and *Anguillicolla crassus*. In addition, extracts of adult worms of *Haemonchus contortus* did not react with either antibody, despite the inclusion of *hsp-90* sequences in the *Haemonchus* EST database with reasonable overlap to the 3' end of *Bp*-Hsp-90 that was used to make the antibody [6]. Extracts of adult *Haemonchus* were prepared in the presence of an additional protease inhibitor (E44) or directly by boiling in SDS-PAGE sample cocktail. Only when worms were extracted directly into sample cocktail could Hsp-90 be detected, suggesting that the aqueous extract may be subject to proteolytic degradation. In contrast, a soluble extract of the L3 stage reacted well with the *Brugia* antibody and was used in subsequent pull-down assays.

Nine clade V nematodes were tested in the GA pull-down assay, five of which were free-living. These included *C. elegans* N2, *C. elegans* JT6130, a *daf-21* mutant which carries a point mutation in *hsp-90* (E292 to K), conferring a temperature-sensitive dauer-constitutive phenotype [17], *Caenorhabditis briggsae* (result not shown), *Oscheia tipulae* and *Pristionchus pacificus*. In addition four parasitic species were tested: *Teladorsagia circumcincta*, *Nippostrongylus brasilensis*, *Heligmosomoides polygyrus* and *H. contortus*. Hsp-90 from all clade V species reacted positively with the anti-*Brugia* Hsp-90 antibody (Figure 1, panels a and c), but none bound to GA in the pull-down assay, whether from free-living or parasitic species (see Figure 1, panels b and d). Blots were exposed for extended periods of time but no positive reaction was observed, while the control *B. pahangi* lysate always yielded a strong positive result. A similar picture was observed with clade IV species, where Hsp-90 from *Strongyloides ratti*, two species of the plant parasite *Globodera* and the free-living *Panagrellus redivivus* all failed to bind GA (data not shown). For clade III nematodes, the situation was more diverse with Hsp-90 from all five filarial worms tested *B. pahangi, Acanthocheilonema viteae, Dirofilaria immitis, Onchocerca ochengi* and *Litomosoides sigmondontis* (result not shown) binding to GA (Figure 2, panel c). Hsp-90 from three ascarid species (*Toxocara cati, Ascaris suum* and *Parascaris equorum*) also bound to GA, but Hsp-90 from the related anisakids (*Anisakis simplex* and *Pseudoterranova decipens*) did not (Figure 2, panel d, lanes 9 & 10) while *A. crassus* failed to react with either antibody (Figure 2 panel c, lane 11). Where a positive signal was obtained in the pull-down assay, experiments were repeated in the presence of excess GA to block free binding sites. In all cases free GA blocked binding of Hsp-90 to the GA beads demonstrating the specificity of the interaction (data not shown).

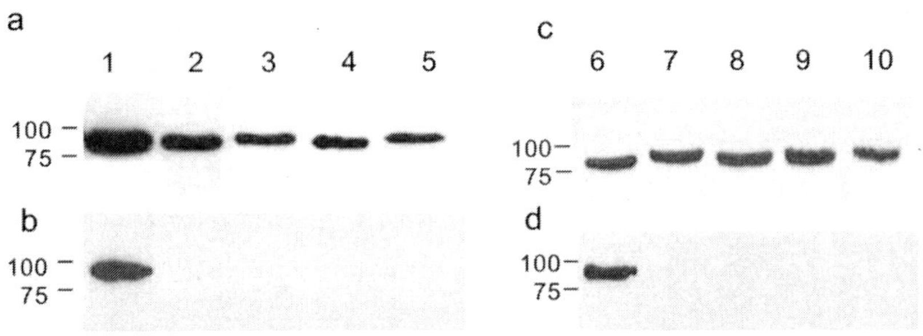

Figure 1. HSP-90 FROM CLADE V NEMATODES DOES NOT BIND GA IN A SOLID PHASE PULL-DOWN ASSAY. Extracts of various nematodes were analysed by immuno-blots (panels a and c) and by GA pull-down assay (panels b and d). Equivalent amounts of protein (determined by Bio-Rad protein assay) were analysed by SDS-PAGE on 10% gels and transferred to NCP. Blots were probed with 1:5000 rabbit anti-*Brugia* Hsp-90 antibody followed by 1:10,000 HRP-labelled anti-rabbit IgG. Bound antibody was detected by chemiluminesence. GA pull-down assays (panels b and d) were carried out with 500 μg of worm lysate using control or GA beads. Control beads were negative in all cases and are not shown. GA pull-down assays were analysed by SDS-PAGE and immuno-blotting as described above. Lane 1, *B. pahangi*, lane 2, *C. elegans* N2, lane 3, *C. elegans daf-21* mutant, lane 4, *O. tipulae*, lane 5, *P. pacificus*, lane 6, *B. pahangi*, lane 7, *T. circumcincta*, lane 8, *N. brasiliensis*, lane 9, *H. polygyrus*, lane 10, *H. contortus*.

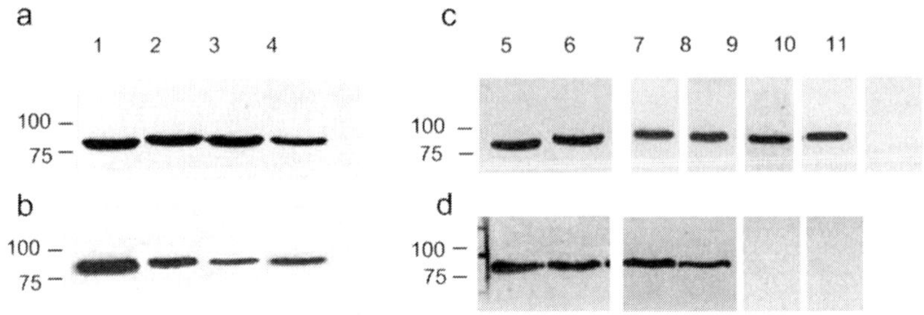

Figure 2. HSP-90 FROM SOME CLADE III NEMATODES BINDS TO GA. Extracts of various nematodes were analysed by immuno-blot (panels a and c) and by GA pull-down assay (panels b and d). Equivalent amounts of protein (determined by Bio-Rad protein assay) were analysed by SDS-PAGE on 10% gels and transferred to NCP. Blots were probed with 1:5000 rabbit anti-*Brugia* Hsp-90 antibody followed by 1:10,000 HRP-labelled anti-rabbit IgG. Bound antibody was detected by chemiluminesence. GA pull-down assays (panels b and d) were carried out with 500 µg of worm lysate using control or GA beads. Control beads were negative in all cases and are not shown. GA pull-down assays were analysed by SDS-PAGE and immuno-blotting as described above. Lane 1, *B. pahangi*, lane 2, *A. viteae*, lane 3 *D. immitis*, lane 4, *O. ochengi*, lane 5, *B. pahangi*, lane 6, *T. cati*, lane 7, *A. suum*, lane 8, *P. equorum*, lane 9, *A. simplex*, lane 10, *P. decipens*, lane 11, *A. crassus*.

Two species belonging to clade I were tested, *T. muris* and *T. spiralis* but neither reacted with the anti-Hsp-90 antibodies. For these species and for *A. crassus*, GA pull-down assays were analysed by Coomassie Blue staining. For *T. spiralis* only, the GA pull-down assay revealed a band of approximately 83 kDa that was not present using control beads. The binding to GA beads of the protein isolated from the *T. spiralis* extract was competed by pre-incubation of the extract with excess free GA. The band was excised from the gel, digested with trypsin and subjected to mass spectrometry at the Sir Henry Wellcome Functional Genomics Unit, University of Glasgow. MASCOT searches revealed multiple peptide hits to Hsp-90 from a variety of species with a high degree of significance ($P < 0.05$). Thus *T. spiralis* Hsp-90 binds to GA, while the GA-binding status of *A. crassus* and *T. muris* Hsp-90 remains uncertain. Table 1 summarises the data from the GA pull-down experiments.

Table 1. Summary of nematode species analysed and GA-binding properties

Species	Clade	Life cycle stage	Life style	GA-binding
C. elegans	V	Mixed stages	Free-living	-
Ce daf-21 mutant	V	Mixed stages	Free-living	-
C. briggsae	V	Mixed stages	Free-living	-
P. pacificus	V	Mixed stages	Free-living	-
O. tipulae	V	Mixed stages	Free-living	-
H. contortus	V	L3	Parasitic, FL larvae	-
T. circumcincta	V	Adult	Parasitic, FL larvae	-
N. brasiliensis	V	Adult	Parasitic, FL larvae	-
H. polygyrus	V	Adult	Parasitic, FL larvae	-
S. ratti	IV	Adult, FL	Parasitic or FL adult, FL larvae	-
P. redivivus	IV	Mixed stages	Free-living	-
G. rostochiensis	IV	J2	Plant parasitic	-
G. pallida	IV	J2	Plant parasitic	-
B. pahangi	III	Adult	Obligate parasite	+
A. viteae	III	Adult	Obligate parasite	+
D. immitis	III	Adult	Obligate parasite	+
O. ochengi	III	Adult, male only	Obligate parasite	+
T. cati	III	Adult	Parasitic, egg in environment	+
P. equorum	III	Adult	Parasitic, egg in environment	+
A. suum	III	Adult	Parasitic, egg in environment	+
A. simplex	III	L3	Marine, larvae in sea	-
P. decipens	III	L3	Marine, larvae in sea	-
A. crassus	III	Adult	Marine, larvae in sea	?
T. spiralis	I	L1	Obligate parasite	+
T. muris	I	Adult	Parasitic, egg in environment	?

FL, free-living; ? uncertain GA-binding status

Bioinformatic analysis of nematode *hsp-90s*

Hsp-90 sequences from a variety of GA-binding and non-binding species of nematode were either retrieved from public databases or were cloned and sequenced as described in Methods. An alignment of the predicted amino acid sequences is given in supplementary data [see Additional file 1] and highlights the degree of conservation of Hsp-90, particularly over the N-terminal ATP-binding domain. Table 2 shows the pairwise similarity and identity between sequences used in the bioinformatic analyses described below. Full genomic and cDNA sequences were available for six species, *C. elegans*, *C. briggsae*, *P. pacificus*, *H. contortus*, *B. pahangi*, and *T. spiralis* allowing the comparison of *hsp-90* gene structure between the different species. Exon-intron boundaries were mapped by comparison of genomic and cDNA sequences and are shown graphically [see Additional file 2]. Three introns are present in *C. elegans* and *C. briggsae* and the position of each is conserved. These are also perfectly conserved with those of *H. contortus* (equivalent to introns 1, 5 and 8 of *Hc-hsp-90*), although here the *Hc-hsp-90* gene contains a total of eight introns. *B. pahangi* contains the largest numbers of introns (eleven), the position of three of which are conserved with *H. contortus*. A single intron (number 8 in *Hc-hsp-90*) is conserved in five of the six species analysed, the exception being *T. spiralis*.

Table 2. Pairwise percent similarities (bold) and identities (italic) between Hsp-90 peptide sequences of fifteen different nematodes calculated under the BLOSUM62 matrix image, generated using MatGAT [45].

	1	2	3	4	5	6	7	8	9	10	11	12	13	14	15
1. B. pahangi		*95.3*	*99.0*	*94.7*	*88.4*	*86.8*	*88.9*	*87.0*	*88.0*	*88.7*	*75.1*	*49.9*	*61.6*	*86.0*	*85.3*
2. T. cati	**98.7**		*95.7*	*97.6*	*88.6*	*87.5*	*88.8*	*86.9*	*87.6*	*88.5*	*74.3*	*49.0*	*60.8*	*85.9*	*85.3*
3. L. sigmondontis	**99.6**	**98.8**		*94.8*	*88.5*	*86.5*	*88.8*	*87.0*	*88.0*	*88.6*	*75.0*	*49.4*	*61.2*	*85.6*	*84.8*
4. A. suum	**97.6**	**98.8**	**97.8**		*88.2*	*86.6*	*88.1*	*86.6*	*87.2*	*88.1*	*73.9*	*48.7*	*60.2*	*85.4*	*84.6*
5. H. contortus	**94.5**	**94.4**	**94.5**	**94.0**		*90.0*	*90.7*	*90.3*	*89.6*	*89.9*	*76.6*	*49.0*	*60.3*	*83.7*	*82.7*
6. P. pacificus	**94.2**	**94.0**	**94.1**	**93.1**	**94.9**		*88.8*	*87.5*	*87.9*	*88.2*	*74.1*	*48.6*	*58.9*	*81.6*	*81.3*
7. C. brenneri	**95.1**	**94.8**	**95.0**	**94.0**	**95.8**	**95.1**		*95.8*	*97.9*	*98.7*	*81.5*	*49.6*	*59.6*	*82.6*	*81.8*
8. C. elegans	**93.9**	**93.7**	**93.8**	**92.9**	**95.6**	**94.2**	**98.5**		*97.0*	*96.0*	*79.3*	*49.8*	*59.9*	*83.8*	*80.8*
9. C. remanei	**94.2**	**94.1**	**94.2**	**93.2**	**95.8**	**94.8**	**99.1**	**99.1**		*97.8*	*80.4*	*49.8*	*59.9*	*82.8*	*80.8*
10. C. briggsae	**95.0**	**94.7**	**94.8**	**93.8**	**96.0**	**95.1**	**100.0**	**98.5**	**99.1**		*81.3*	*49.9*	*59.9*	*83.2*	*81.5*
11. C. japonica	**80.7**	**80.5**	**80.6**	**79.8**	**82.0**	**81.0**	**84.2**	**83.5**	**83.8**	**84.3**		*58.3*	*49.7*	*70.1*	*69.1*
12. G. rostochiensis	**55.1**	**54.7**	**54.9**	**54.0**	**55.0**	**54.3**	**54.9**	**54.7**	**55.2**	**54.9**	**64.7**		*66.3*	*54.3*	*51.9*
13. G. pallida	**68.3**	**68.1**	**68.2**	**67.3**	**68.2**	**67.3**	**68.3**	**68.0**	**68.7**	**68.3**	**65.6**	**70.2**		*67.1*	*64.1*
14. H. glycines	**94.0**	**94.6**	**94.0**	**93.7**	**92.1**	**91.5**	**92.5**	**91.8**	**92.2**	**92.5**	**78.5**	**56.2**	**69.6**		*88.6*
15. M. Hapla	**94.3**	**94.3**	**94.3**	**93.6**	**91.7**	**91.7**	**91.8**	**90.8**	**91.5**	**91.7**	**78.3**	**55.2**	**68.7**	**95.6**	

Identifying sites and lineages subject to adaptive evolution

For the gene-wide analysis, using either set of nested models of codon evolution (M1/M2 or M7/M8) did not identify any evidence of positive selection that was consistent across the phylogenetic tree (see Table 3). However, using the branch-sites methods which allow evolutionary rates to vary across the tree did show some interesting findings. Three branches showed significant evidence of adaptive evolution ($dN/dS > 1$) after Bonferroni correction (see Figure 3, labelled A, B, C). Along each of these branches, sites were detected with posterior probabilities > 0.95 of belonging to a $dN/dS > 1$ class (see Table 4). Along branch A (LRT of $dN/dS > 1$, $p < 0.01$) a single site 659S was identified (amino acid positions correspond to full length B. pahangi Hsp-90 AJ005785). On branch B (LRT of $dN/dS > 1$, $p < 0.05$), three sites with $dN/dS > 1$, 297S, 373S and 464S, were detected. On branch C, nine sites were detected 21S, 29S, 165V, 312S, 458S, 472Q, 485A, 525S and 584S. Importantly ten of these thirteen sites are serine residues which when examined are perfectly or largely conserved across all species in the alignment [see Additional file 1]. The identification of these ten sites as $dN/dS > 1$ results from the serine being encoded by either codons TCn or AGy. As the codon model of evolution assumes that mutations within a codon are independent, the change from TCn-AGy is parsimoniously interpreted as two separate non-synonymous substitutions (Z. Yang personal communication): e.g. TCC (Ser)->ACC (Thr)->AGC (Ser). The biological interpretation of these results suggests that the conservation of serine at these positions is functionally important and that this may be evidence for divergent evolution with rapid convergence back to the functionally stable serine residue. However, this cannot be proven by comparison of the extant species available to us and these sites are not considered further here. The remaining codons 165V, 472Q and 485A are inferred to be evolving rapidly along the lineage separating the plant parasitic nematodes from the animal parasitic and free-living species studied (branch C, Figure 3). Site 472Q is located in the middle region of the Hsp-90 protein and the codon is a conserved cysteine in the four plant parasitic nematodes and glutamine in all other species tested. Site 485A is also located in the middle region of the protein and is glutamine in the plant parasitic species and alanine in all other species. Only a single site, 165V, is predicted to be evolving under positive selection and is also located in the N terminal GA binding region of Hsp-90. Site 165V is a valine residue in the GA-binding species (L. sigmodontis, B. pahangi, A. sum and T. cati) but also in the non-GA binding P. pacificus and H. contortus. The residue is a phenylalanine in C. elegans and tyrosine in other Caenorhabditids analysed. To determine if these initial findings are consistent with the theory that GA-binding is associated with rapid diversifying evolution of the hsp-90 gene, we used a modified branch-site approach [18] to a priori separate the phylogenetic tree into two groups, GA-binders (B. pahangi, L. sigmodontis, A. suum and T. cati) vs non-GA-binders. Repeating the branch-sites analysis failed to provide significant evidence to suggest that an elevated dN/dS ratio associated with adaptive evolution was present in either of the two groups (see Table 5).

Table 3. Statistical analysis of sites subject to adaptive evolution using nested models of codon evolution (M1/M2 or M7/M8)

		LRT statistic		Parameter estimates			Sites under positive selection	
				% Sites with $d_N/d_S > 1$				
Gene	N	M1 vs. M2	M7 vs. M8	M2	M8	d_N/d_S	M2	M8
Hsp90	15	0.00	0.00	n/a	n/a	1.00	None	None

n/a not available

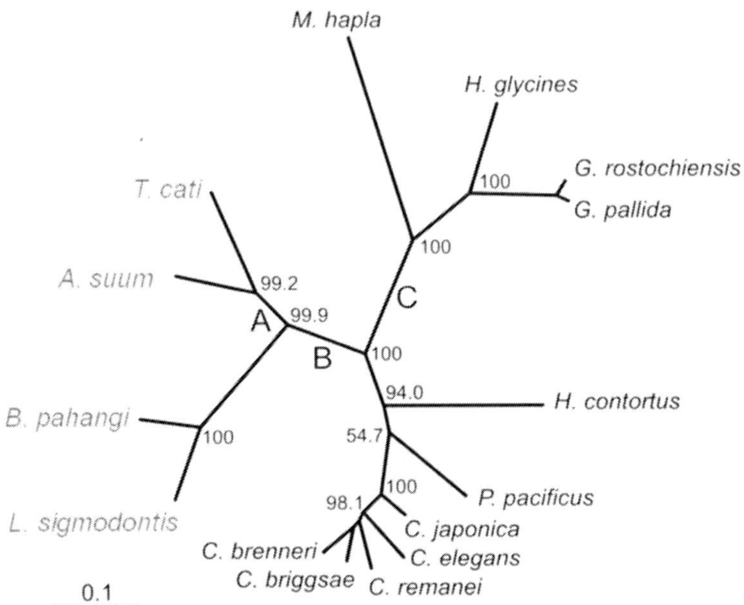

Figure 3. UNROOTED PHYLOGENETIC TREE SHOWING RELATIONSHIP OF ***hsp-90*** SEQUENCES FROM FIFTEEN DIFFERENT NEMATODE SPECIES. The tree was generated using phyML [37] under the HKY substitution model. 1000 bootstrap replicates were performed and node labels represent percent bootstrap support. Branches A, B and C were shown to have significant evidence of positive selection. Scale bar shows substitutions per site.

Table 4. Statistical analysis of branches showing significant evidence of adaptive evolution

			Significance of tests for positive selection		
Gene	Number of sequences	Number of codons	Sites M2	Sites M8	Branch sites All branches
Hsp90	15	687	n/s	n/s	2 branches ($p < 0.05$, B & C Figure 3) 1 branch ($p < 0.01$, A Figure 3

n/s not significant

Table 5. Branch-site analysis of *hsp-90* sequences grouped according to GA-binding status

	Number of branches			Parameter estimates under free model			d_N/d_S		
Clade	All	foreground	LRT statistic	p0[a]	p1[b]	p2[c]	background	foreground	Sites under positive selection
GA binders	27	7	0.00	0.984	0.012	0.004	0.014	1.00	n/a
GA non-binders	27	20	0.00	0.978	0.004	0.18	0.014	1.00	n/a

n/a not available. Number of branches: count of branches partitioned as foreground. LRT statistic: 2 times log likelihood difference of fixed versus free nested models.

[a] Proportion of sites in category p0 where selection is purifying throughout the tree ($0 < \omega_0 < 1$).

[b] Proportion of sites in category p1 where selection is neutral throughout the tree ($\omega_1 = 1$).

[c] Proportion of sites in category p2 where selection is positive in foreground branches; sum of category 2a: purifying selection in background but positive selection in foreground ($0 < \omega_0 < 1$, $\omega_2 > 1$); and category 2b: neutral selection in background but positive selection in foreground ($\omega_1 = 1$, $\omega_2 > 1$

DISCUSSION

The results of our analysis of Hsp-90 from a range of parasitic and free-living nematodes has revealed an interesting heterogeneity in the ability to bind GA. None of the free-living species tested, nor any of the parasitic species which have free-living larval stages in the environment bound to GA. These included animal parasites belonging to clade IV, such as *Strongyloides,* and those such as *Haemonchus* and *Nippostrongylus* belonging to clade V. These parasites are generally gut-dwelling and are transmitted between hosts when eggs, passed out in the faeces of the infected animal, hatch to release free-living larval stages. Following a period of development in the environment, the L3 stage infects a new host. The plant parasites tested, *Globodera* spp., also failed to bind GA. These are commonly known as potato cyst nematodes, and live in the soil where the infective second-stage juvenile worms (J2) attack the roots of susceptible plants. A cyst containing eggs is formed from the body of the dead female worm and under appropriate conditions these will hatch to release the J2 stage worms. These penetrate the roots, start to feed and eventually develop into adults.

The clade III nematodes analysed produced an interesting dichotomy in results, containing both GA-binding and non-binding species. Hsp-90 from the five filarial parasites assayed bound to GA; filarial worms are obligate parasites, transmitted between hosts by the bite of an infected intermediate host, such as mosquitoes or other arthropods. In keeping with this result, Hsp-90 from another obligate parasite *T. spiralis* also bound GA. *T. spiralis* is a clade I species commonly transmitted via ingestion of infected muscle containing the encysted L1 stages. Thus neither filarial worms nor *T. spiralis* have free-living stages in the environment.

The clade III Ascarid parasites also provided an intriguing comparison, as those worms which bound GA (*Toxocara, Ascaris* and *Parascaris*) all have free-living stages outside the host, but are enclosed within a thick impermeable egg-shell while in the environment. Transmission to the new host is by ingestion of the egg containing the infective larva or via a paratenic host (in some species such as *Toxocara*). Interestingly, in the original report on the isolation of GA [2], the in vivo efficacy of GA against the mouse pinworm *Syphacia obvelata* is recorded, suggesting that this parasite may also express a GA-binding Hsp-90. In keeping with this observation, pinworms belong to clade III (Oxyuroids) and are transmitted by ingestion of a resistant egg containing the infective larval stage.

Two of the three clade III species associated with the marine environment (*Anisakis* and *Pseudoterranova*) failed to bind GA. Both parasites have complex life cycles involving one or more intermediate and/or paratenic hosts. The free-living larval stages are ingested by marine crustaceans, which are then eaten by fish, which are subsequently ingested by marine mammals such as seals (*Pseudoterranova*) or cetaceans (*Anisakis*). *A. crassus* belongs to the superfamily Dracunuloidea and parasitises the swim bladder of eels, but also has a free-swimming larval stage. As extracts of *A. crassus* (and *T. muris*) failed to react with either antibody, the GA-binding status of these species remains unresolved. An additional proviso of our data is that in some cases, larval stages were analysed and it is possible that larval and adult worms may display different GA-binding characteristics.

Our data suggest that differences in the ability of nematode Hsp-90 to bind to GA are correlated with particular life histories. Only obligate parasites or those which have free-living stages in the environment enclosed within a protective egg-shell possess a GA-binding Hsp-90, while Hsp-90 from free-living nematodes or from those parasitic species that have free-living larval stages in the environment do not bind GA. The adaptive evolution hypothesis would argue that organisms sharing the same niche as the GA-producing *Streptomyces* should fail to bind GA and the results from the pull-down experiments broadly support this hypothesis. In order to further explore this theory, we analysed *hsp-90* sequences from fifteen different species of nematode in an attempt to identify codons that have undergone adaptive Darwinian evolution within the GA-resistant species. The results of these analyses provided strong evidence for rapid adaptive evolution along three separate lineages and identified a number of residues showing significant evidence of adaptive evolution. However, the model suggests that ten of the thirteen codons identified as sites of adaptive evolution had undergone at least two separate non-synonomous substitutions resulting in the conservation of a serine residue, implying that these sites may be critical for Hsp90 function. In mammalian Hsp-90α and β, both serine and threonine residues are known to be targets of phosphorylation [19, 20], a modification that is essential for some functions of Hsp-90 [21, 22]. Of the remaining three residues (165V, 472Q and 485A) showing significant evidence of adaptive evolution (posterior probability > 0.95 of $d_N/d_S > 1$), only a single site, 165V, was located in the amino terminal section of the Hsp-90 protein, known to be the position of GA binding. However, variation at this site does not differentiate between GA-binding and non-GA binding species. The residue is a valine in the GA binding species, but also in the non-GA binding species *P. pacificus* and *H. contortus*. Comparison of the amino acid sequences of Hsp-90 from GA-binding species and non-binding species identified only four amino acid positions that are conserved within either the GA-binding or the non-binding species, but that differ between the groups. These are: K259 (K GA-binders, T non-binders), F447 (F GA-binders, Y non-binders), F475 (F GA-binders, Y non-binders), V624 (V GA-binders, I non-binders) (all amino acid positions correspond to full length *B. pahangi* Hsp-90 [Genbank:AJ005785]). However, none of these substitutions reside in the N terminal region. Comparison of each of these positions in *Brugia* and *Saccharomyces cerevisiae*, which binds GA [23], shows that the equivalent amino acids in yeast are conserved with the non-GA-binding nematode species. Thus whilst evidence for rapid diversifying selection exists in the evolution of the *hsp-90* gene (as evidenced by the significance of the branch-site tests) and most likely occurred early in nematode divergence, we have been unable to prove with confidence that the selection seen correlates with a change in the GA-binding function of Hsp-90. As Hsp-90 is a multi-functional protein in eukaryotic cells, the observed rapid evolution of the *hsp-90* gene likely correlates with some other key function, rather than the ability to bind GA.

In *S. cerevisiae*, point mutations in *hsp-90* alter the sensitivity to GA and to Radicicol, an alternative inhibitor of Hsp-90 function [23]; e.g. A587T and T101I are hypersensitive to GA, with T22I, G313S and E381K showing a lesser effect. However, additional factors are also known to influence sensitivity to GA in yeast, including the presence of co-chaperones such as Sti1p (mammalian Hop) and Sba1p (mammalian p23). Yeast deficient in Sti1p or Sba1p show increased sensitivity to GA [23, 24]. *C. elegans* expresses homologues of both co-chaperones [25] and a recent study identified STI-1 as an Hsp-90 interacting protein in *C. elegans* [26]. However, whether *sti-1* null animals have an altered response to GA has yet to be explored.

While analysis of *daf-21* mutants and RNAi studies in *C. elegans* clearly show that Hsp-90 is essential in the worm, *C. elegans* is not susceptible to GA and, indeed, in the micro-environment of a Petri dish, can grow perfectly well when fed on a culture of *S. hygroscopicus* [5]. However, it remains unclear whether *C. elegans* would interact with the GA-producing *S. hygroscopicus* in the environment. Although both organisms may be present in the soil, whether they share the same spatial and temporal niches is unknown. Indeed, the ecology of *C. elegans* in the wild is not well understood, despite its extensive use in the laboratory [27]. Although *C. elegans* can be found in soil, it appears that the predominant stage observed in the wild is the dauer larvae, typically found in organic-rich environments such as compost, or in association with invertebrates such as snails and isopods [28]. Dauer stages are resistant and non-feeding and the natural food source of *C. elegans* in the wild is unknown. *C. elegans* possesses a sensitive chemosensory system that, at least in the laboratory setting, allows the worm to differentiate between bacterial food sources and potential pathogens [29]. Thus in the environment, worms could potentially avoid bacteria producing noxious compounds.

Many Hsp90 inhibitors are natural compounds produced by micro-organisms and recent studies have provided an intriguing insight into the mechanisms by which the organisms that produce Hsp90 inhibitors can themselves survive. For example, in the yeast *Humicola fuscoatra*, which produces Radicicol, a single conserved amino acid substitution (L33I) in Hsp-90, is sufficient to alter the binding of Radicicol, without influencing the ability to bind ATP or GA [30]. Here, a leucine to isoleucine substitution alters the size of the hydrophobic drug-binding pocket increasing the binding of H_2O molecules and hydration of the drug. Such findings serve to emphasise that very subtle alterations in amino acid residues can have major implications on drug binding efficacy and highlight the importance of understanding the structural basis of drug binding, which is not always apparent from analysis of DNA sequences.

CONCLUSION

Our results demonstrate that *C. elegans* Hsp-90 is not unique in its low affinity for GA; despite a relatively high level of sequence conservation, nematode Hsp-90s show functional diversity in relation to GA-binding. The ability to bind to GA is associated with the life history of a species; free-living nematodes and those parasitic species which have a free-living larval stage in the environment do not bind GA, while obligate parasites or those parasitic species which have free-living stages enclosed within a resistant egg, do bind GA. As *S. hygroscopicus*, the micro-organism that synthesizes GA, is found in soil, and as Hsp-90 is an essential molecule in all eukaryotes, these observations could be consistent with adaptive evolution of *hsp-90* sequences. Analysis of *hsp-90* sequences from fifteen different nematode species provided evidence of rapid diversifying selection along three separate branches. However, the selection observed did not correlate with the ability to bind GA or not.

METHODS

Worm material

The following species of free-living nematode were used *C. elegans* N2 strain, *C. elegans* strain JT6130, which carries a point mutation in *daf-21* (*hsp-90*), *Caenorhabditis briggsae, Oscheia tipulae, Pristionchus pacificus* (all clade V, obtained from the *C. elegans* Genetics Center) and *Panagrellus redivivus* (clade IV). Parasitic species analysed included *Trichinella spiralis* and *Trichuris muris* (both clade I), and the following clade III species: *Brugia pahangi, Dirofilaria immitis, Onchocerca ochengi, Acanthocheilonema viteae, Litomosoides sigmondontis, Toxocara cati, Ascaris suum, Parascaris equorum, Anisakis simplex, Pseudoterranova decipens* and *Anguillicola crassus.* Clade IV species included *Strongyloides ratti, Globodera rostochiensis* and *G. pallida.* Clade V species included *Nippostrongylus brasilensis, Heligmosomoides polygyrus, Teleodorsagia circumcincta* and *Haemonchus contortus.* Mixed life cycle stages were used for all the free-living nematodes while for parasitic species, usage was determined by the availability of particular life cycle stages; in most cases adult males and females were analysed with the following exceptions: *Globodera* sp. J2 stages; *H. contortus,* L3; *T. spiralis* L1 muscle stage larvae; *A. simplex* and *P. decipens* L3. *S. ratti* adults were free-living rather than parasitic females [31] and for *O. ochengi* only male worms were used.

Maintenance of nematodes

C. elegans N2 strain, *C. briggsae, O. tipulae* and *P. pacificus* were maintained at 20°C on NGM plates with OP50 as a bacterial food source [32]. Worms were washed off plates with M9 buffer (3 g KH_2PO_4, 6 g Na_2HPO_4, 5 g NaCl, 1 ml 1 M $MgSO_4$ in 1 litre ddH_2O) and collected by centrifugation. *P. redivivus* worms were maintained at room temperature on plates containing an oat-based medium. The cooked oats were poured into plates and the worms placed on the surface. As their population blooms, the worms move onto the lid, from where they can be collected by rinsing the lids with M9 buffer. Worms were then washed several times in M9 buffer by centrifugation.

Adult stages of all parasitic species were obtained from their respective hosts and frozen as soon as possible at -80°C. As no signal on a western blot could be obtained using anti- Hsp-90 antibodies to probe extracts of adult *H. contortus*, L3 of this species were obtained by harvesting eggs from infected sheep faeces and allowing these to develop [33]. For *Globodera* species, cysts were rehydrated in tap water for several days and then hatched in tomato root exudate (kindly provided by J. Jones, Scottish Plant Breeding Institute, Invergowrie). Cysts were checked daily for hatched J2 stages. These were removed by pipetting, washed in ddH_20 and purified from debris by flotation on 60% sucrose. For *T. spiralis*, muscle-stage larvae (L1) were obtained by digestion of infected mouse muscle using standard methods [34]. Larval stages (L3) of *A. simplex* and *P. decipens* were obtained from infected fish.

Preparation of worm lysates

Lysates of worms were prepared in ice-cold TNES buffer (50 mM Tris-HCl, pH 7.4, 100 mM NaCl, 2 mM EDTA, 1% Igepal CA-630). Protease inhibitors were added prior to use (Complete Mini Protease Inhibitor Cocktail, Roche Diagnostics). Worms were ground to a fine powder using a pestle and mortar containing liquid Nitrogen. Extracts were resuspended in 1.0 ml TNES, the mortar was washed out with 0.5 ml TNES, and the extract was pelleted at 14000 g for 15 minutes at 4°C. The supernatant was retained and assayed for protein concentration at OD 595 nm using the Bio-Rad Protein assay. Lysates of D. immitis and O. ochengi were prepared using a small glass homogeniser containing 100 μl of ice-cold TNES and the supernatant processed as described above.

Preparation of GA solid support and binding assay

1 mg GA (InvivoGen) was derivatised using 1,6-hexanediamine to yield 17-hexamethylenediamine-17 demethoxygeldanamycin [4]. After aqueous extraction, the mixture was dried under nitrogen then re-dissolved in 400 μl of DMSO. The product was incubated with 400 μl of washed Affi-Beads (Bio-Rad) for 4 h in the dark at room temperature. Beads were washed in TNES and then were rocked overnight at 4°C in 1.0 ml of ethanolamine (1 M, pH 8.3) to block residual binding sites. Prior to use, beads were blocked in 1% BSA in TNES for 1 h at room temperature. Control beads were incubated with DMSO alone.

300-500 μg of total protein in a volume of 300 μl was mixed with 50 μl packed volume of GA or control beads for 2.5 hours at 4°C. The beads were washed extensively in TNES by centrifugation and bound proteins were eluted by boiling in SDS-PAGE sample cocktail (0.125 M Tris, pH 6.8, 0.2 M dithiothreitol, 4% SDS, 20% glycerol, 0.042% bromophenol blue). Aliquots of the input fraction and protein eluted from control and GA beads were run on 10% SDS-PAGE gels and analysed by western blotting. Blots were probed with a 1:5000 dilution of an antibody raised to a C-terminal fragment of recombinant *B. pahangi* Hsp-90, as detailed previously [6], followed by 1:10000 dilution of HRP-labelled anti-rabbit IgG (Sigma). Bound antibody was detected by chemiluminescence using the Super Signal ® West Pico Kit (Thermo Scientific). If no signal or only a weak signal was observed, blots were repeated using the AC88 Hsp-90 monoclonal antibody (a kind gift of D. Toft) as the primary antibody at 1:500 dilution followed by anti-mouse IgG (Sigma) at 1:10000. AC88 recognises an epitope in the C-terminus of Hsp-90 and reacts with Hsp-90 from many different organisms. In all experiments, a lysate of *B. pahangi* was used as a positive control. The specificity of GA binding was determined in subsequent experiments by pre-incubating worm extracts with 20 μM GA prior to incubation with the GA beads.

Sequence and bioinformatic analysis

hsp-90 sequences from *B. pahangi* [GenBank:AJ005785], *H. glycines* [GenBank:AF461150], *M. hapla* [GenBank:AAT35588], *C. elegans* [GenBank :C47E8.5] and *C. briggsae* [GenBank:CBG040560] were obtained from the NCBI. *C. brenneri*, *C. japonica* and *C. remanei hsp-90* sequences were obtained by carrying out a tBLASTn search at http://www. nematode. net using the *Ce*-DAF-21 amino acid sequence to retrieve full length cDNA sequences for each *Caenorhabditis* species.

The full-length sequence of *T. spiralis hsp-90* was identified by a tBLASTn search of the available genome sequence at http://www. genome. wustl.edu with the *Ce*-DAF-21 amino acid sequence. A significant match was obtained to contig 9.2, the sequence of which was obtained from Amy Yu at Washington University School of Medicine. For *P. pacificus*, the full-length *hsp-90* sequence was obtained by tBLASTn search of the sequence available at http://www.pristionchus.org using the *Ce*-DAF-21 amino acid sequence, resulting in the retrieval of contig 4.9. For *Trichinella* and *Pristionchus*, intron and exon sites were identified using Genewise in EXPASY, allowing assembly of the full-length predicted cDNA sequence. For *H. contortus*, the *hsp-90* sequence was assembled by carrying out a tBLASTn search at http://www.sanger.ac.uk using the *Ce*-DAF-21 amino acid sequence to retrieve contig _0076156, as detailed previously [35]. The sequence was submitted to GenBank [GenBank:FJ717747].

For *A. suum* a partial *hsp-90* sequence was obtained by carrying out a tBLASTn search at http://www.nematode.net using the *Ce*-DAF-21 amino acid sequence and retrieving EST sequence kj94a07.y1, which contained

768 bp of 5' *A. suum Hsp-90* sequence. A primer was designed to the 5' end of the gene, 5'-ATGTCTGAGCAACAGCCTGAAGGA-3' and used in combination with a series of degenerate primers, designed over the conserved MEEVD region, to amplify the full length *hsp-90* coding sequence from adult *A. suum* cDNA. One 3' degenerate primer was more successful than others and was used in all subsequent reactions, 5'-ANTCNACNTCNTCCATNCGNGANGC-3'. Full length *A. suum hsp-90* was cloned into pCR2.1 TOPO (Invitrogen) and sequenced by automated DNA sequencing using vector and gene specific primers (MWG Biotech) and assigned an accession number [GenBank:FJ824589].

For *L. sigmodontis* and *T. canis* partial *hsp-90* sequences were obtained by carrying out a tBLASTn search at http://www.ebi.ac.uk using the *Bp*-HSP-90 amino acid sequence. A 721 bp EST was retrieved which contained 676 bp of 3' *L. sigmodontis hsp-90* sequence. As the 5' sequence was missing, an SL1 primer was used (5'-GGTTTAATTACCCAAGTTTGA-3') along with a primer designed to the 3' end of the *L. sigmodontis* sequence, 5'-TTAATCAACTTCTTCCATCCTCG-3'. The full length *L. sigmodontis hsp-90* was amplified from adult *L. sigmondontis* cDNA, cloned into pCR2.1 TOPO (Invitrogen), sequenced as above and assigned an accession number [GenBank:FJ824588].

For *T. canis*, an EST of 647 bp was identified containing the 5' region of the gene. As the only biological material available was from the sister species *T. cati*, PCR was carried out using a primer designed to the 5' end of the *T. canis hsp-90* sequence (5'-ATGTCCGAGTTTCAACAACAACCAG-3') together with the degenerate 3' primer described previously (see above) on cDNA extracted from adult *T. cati* worms. The full-length *T. cati hsp-90* was cloned into pCR2.1 TOPO (Invitrogen), sequenced as above and assigned an accession number [GenBank:FJ824590]. All sequences were compiled from multiple reads in both directions and line-ups and annotations were carried out using Contig express and Vector NTI.

Partial *hsp-90* sequences of *X. index* (1325 bp), *G. pallida* (685 bp) and *G. rostochiensis* (1394 bp) were obtained by carrying out a tBLASTn search at http://www.ebi.ac.uk or http://www.nematode.net using the *Bp*-HSP-90 amino acid sequence. For those species for which both full-length cDNA and genomic *hsp-90* sequences were available, intron/exon boundaries were identified in genomic DNA sequences using Genewise in EXPASY and their positions mapped onto cDNA sequences.

Identifying sites and lineages subject to adaptive evolution

Peptide sequences were generated from translated nucleotide sequences of *hsp-90* from seventeen species of nematode *C. remanei, C. briggsae, C. brenneri, C. elegans, C. japonica, P. pacificus, H. contortus, L. sigmodontis, B.*

pahangi, A. suum, T. cati, M. hapla, Heterodera glycines, G. pallida, G. rostochiensis, T. spiralis and *Xiphinema index*. These were aligned using Prank [36] and the aligned peptide sequence was back translated to generate an aligned nucleotide sequence. A phylogenetic tree, generated with phyML [37] under the HKY substitution model [38], was used as the starting phylogeny for the codeML program of the Phylogenetic Analysis by Maximum Likelihood (PAML) suite [39]. A phylogenetic tree generated under the M0 model inferred that branches leading to two of the species, *T. spiralis* and *X. index*, exhibited much larger divergence (> 3 substitutions per codon). To avoid potential problems of saturation at high divergence these species were removed from further analyses. Sequences from the remaining fifteen species were realigned with Prank and a maximum likelihood phylogenetic tree was again generated using phyML. To determine if evidence for adaptive evolution ($d_N/d_S > 1$) exists in these sequences, three tests of selection were conducted to compare the ratio of non-synonymous to synonymous substitutions (d_N/d_S). (1) Site analysis using a likelihood ratio test (LRT) to compare the codeML nested models M1/M2 and M7/M8 which assume a uniform evolutionary rate across all branches of the tree [40, 41]. (2) Branch-sites methods, which allow for differential rates of evolution across different lineages [7, 8]. Using the branch-sites methods each individual branch was analysed. (3) Species were also grouped into GA-binders or non-binders and compared using the branch-sites method, as such comparisons can improve the power of analysis if the underlying biological assumption to group the branches is sound [18]. In all cases multiple independent runs of the CodeML program were conducted (minimum triplicate) to ensure convergence of maximum likelihood estimates of parameters. Where multiple tests were conducted, the type I error rate was controlled by Bonferroni correction [42] (See Table 3). Hsp-90 amino acid sequences from nematode species used in the bioinformatic analysis were also aligned using ClustalW [43] and coloured using the default residue groupings of CHROMA [44].

NOTES

Nik AIIN Him, Victoria Gillan, Richard D Emes contributed equally to this work.

ACKNOWLEDGEMENTS

This work was supported by grants to ED from the Wellcome Trust and the BBSRC. NH was supported by an award from the Government of Malaysia.

The work would not have been possible without the assistance of many nematode biologists who generously donated material for analysis, including Prof J. Allen, University of Edinburgh, Dr I. Coombs, University of Glasgow, Dr J. Foster, New England BioLabs, Dr Colin Fleming, Agri-Food and Biosciences Institute, Belfast, Prof R. Grencis, University of Manchester, Prof W. Harnett, University of Strathclyde, Dr J. Jones, Scottish Crop Research Institute, Invergowrie, Prof D. Knox, Moredun Research Institute, Dr C. Lawrence, University of Strathclyde, Prof R. Maizels, University of Edinburgh, Dr B. Makepeace, Liverpool School of Tropical Medicine, Dr S. Martin, University of Aberdeen, Prof J. Matthews, Moredun Research Institute, Mr J. McGoldrick, University of Glasgow Veterinary School, Prof M. Viney, University of Bristol. We also thank Professor David Toft for the kind gift of AC88 antibody. We acknowledge Kerry O'Neill and Margaret McFadyen for general laboratory assistance, Gary Saunders for help with scaffolding the *H. contortus* sequence, and our colleagues Dr Collette Britton and Prof. Mark Viney (University of Bristol) for critically reading the manuscript and their helpful discussions. Some nematode strains used in this work were provided by the *Caenorhabditis* Genetics Center, which is funded by the NIH National Center for Research Resources (NCRR).

AUTHORS' CONTRIBUTIONS

NH carried out the pull-down assays; VG identified and sequenced *hsp90* genes; KM helped with the pull-down assays; RDE undertook the d_N/d_S analysis; ED conceived and coordinated the study. NH, VG, RDE and ED contributed to the manuscript. All authors provided comments on the manuscript and read and approved the final draft.

REFERENCES

1. Picard D: Heat-shock protein 90, a chaperone for folding and regulation. Cell Mol Life Sci. 2002, 59 (10): 1640-1648.
2. BeBoer C, Dietz A: The description and antibiotic production of *Streptomyces hygroscopicus* var. Geldanus. J Antibiot (Tokyo). 1976, 29 (11): 1182-1188.
3. Stebbins CE, Russo AA, Schneider C, Rosen N, Hartl FU, Pavletich NP: Crystal structure of an Hsp90-geldanamycin complex: targeting of a protein chaperone by an antitumor agent. Cell. 1997, 89 (2): 239-250.
4. Whitesell L, Mimnaugh EG, De Costa B, Myers CE, Neckers LM: Inhibition of heat shock protein HSP90-pp60v-src heteroprotein complex formation by benzoquinone ansamycins: essential role for stress proteins in oncogenic transformation. Proc Natl Acad Sci USA. 1994, 91 (18): 8324-8328.

5. David CL, Smith HE, Raynes DA, Pulcini EJ, Whitesell L: Expression of a unique drug-resistant Hsp90 ortholog by the nematode *Caenorhabditis elegans*. Cell Stress Chaperones. 2003, 8 (1): 93-104.
6. Devaney E, O'Neill K, Harnett W, Whitesell L, Kinnaird JH: Hsp90 is essential in the filarial nematode *Brugia pahangi*. Int J Parasitol. 2005, 35 (6): 627-636.
7. Yang Z, Nielsen R: Codon-substitution models for detecting molecular adaptation at individual sites along specific lineages. Mol Biol Evol. 2002, 19 (6): 908-917.
8. Zhang J, Nielsen R, Yang Z: Evaluation of an improved branch-site likelihood method for detecting positive selection at the molecular level. Mol Biol Evol. 2005, 22 (12): 2472-2479.
9. Blaxter ML: Nematoda: genes, genomes and the evolution of parasitism. Adv Parasitol. 2003, 54: 101-195. full_text.
10. Dieterich C, Clifton SW, Schuster LN, Chinwalla A, Delehaunty K, Dinkelacker I, Fulton L, Fulton R, Godfrey J, Minx P, et al: The *Pristionchus pacificus* genome provides a unique perspective on nematode lifestyle and parasitism. Nat Genet. 2008, 40 (10): 1193-1198.
11. Mitreva M, Jasmer DP: Biology and genome of *Trichinella spiralis*. (November 23, 2006), WormBook, ed. The C. elegans Research Community, WormBook. 10.1895/wormbook.1.124.1. [http://www.wormbook.org]
12. Ghedin E, Wang S, Spiro D, Caler E, Zhao Q, Crabtree J, Allen JE, Delcher AL, Guiliano DB, Miranda-Saavedra D, et al: Draft genome of the filarial nematode parasite *Brugia malayi*. Science. 2007, 317 (5845): 1756-1760.
13. Opperman CH, Bird DM, Williamson VM, Rokhsar DS, Burke M, Cohn J, Cromer J, Diener S, Gajan J, Graham S, et al: Sequence and genetic map of *Meloidogyne hapla* : A compact nematode genome for plant parasitism. Proc Natl Acad Sci USA. 2008, 105 (39): 14802-14807.
14. Abad P, Gouzy J, Aury JM, Castagnone-Sereno P, Danchin EG, Deleury E, Perfus-Barbeoch L, Anthouard V, Artiguenave F, Blok VC, et al: Genome sequence of the metazoan plant-parasitic nematode *Meloidogyne incognita*. Nat Biotechnol. 2008, 26 (8): 909-915.
15. Parkinson J, Mitreva M, Whitton C, Thomson M, Daub J, Martin J, Schmid R, Hall N, Barrell B, Waterston RH, et al: A transcriptomic analysis of the phylum Nematoda. Nat Genet. 2004, 36 (12): 1259-1267.
16. Blaxter ML, De Ley P, Garey JR, Liu LX, Scheldeman P, Vierstraete A, Vanfleteren JR, Mackey LY, Dorris M, Frisse LM, et al: A molecular evolutionary framework for the phylum Nematoda. Nature. 1998, 392 (6671): 71-75.
17. Birnby DA, Link EM, Vowels JJ, Tian H, Colacurcio PL, Thomas JH: A transmembrane guanylyl cyclase (DAF-11) and Hsp90 (DAF-21) regulate a common set of chemosensory behaviors in *Caenorhabditis elegans*. Genetics. 2000, 155 (1): 85-104.
18. Ramm SA, Oliver PL, Ponting CP, Stockley P, Emes RD: Sexual selection and the adaptive evolution of mammalian ejaculate proteins. Mol Biol Evol. 2008, 25 (1): 207-219.
19. Lees-Miller SP, Anderson CW: The human double-stranded DNA-activated protein kinase phosphorylates the 90-kDa heat-shock protein, hsp90 alpha at two NH2-terminal threonine residues. J Biol Chem. 1989, 264 (29): 17275-17280.
20. Lees-Miller SP, Anderson CW: Two human 90-kDa heat shock proteins are phosphorylated in vivo at conserved serines that are phosphorylated in vitro by casein kinase II. J Biol Chem. 1989, 264 (5): 2431-2437.
21. Kurokawa M, Zhao C, Reya T, Kornbluth S: Inhibition of apoptosome formation by suppression of Hsp90beta phosphorylation in tyrosine kinase-induced leukemias. Mol Cell Biol. 2008, 28 (17): 5494-5506.
22. Zhao YG, Gilmore R, Leone G, Coffey MC, Weber B, Lee PW: Hsp90 phosphorylation is linked to its chaperoning function. Assembly of the reovirus cell attachment protein. J Biol Chem. 2001, 276: 32822-32827.
23. Piper PW, Panaretou B, Millson SH, Trumana A, Mollapour M, Pearl LH, Prodromou C: Yeast is selectively hypersensitised to heat shock protein 90 (Hsp90)-targetting

drugs with heterologous expression of the human Hsp90beta, a property that can be exploited in screens for new Hsp90 chaperone inhibitors. Gene. 2003, 302 (1-2): 165-170.

24. Forafonov F, Toogun OA, Grad I, Suslova E, Freeman BC, Picard D: p23/Sba1p protects against Hsp90 inhibitors independently of its intrinsic chaperone activity. Mol Cell Biol. 2008, 28 (10): 3446-3456.
25. Johnson JL, Brown C: Plasticity of the Hsp90 chaperone machine in divergent eukaryotic organisms. Cell Stress Chaperones. 2009, 14 (1): 83-94.
26. Song HO, Lee W, An K, Lee HS, Cho JH, Park ZY, Ahnn J: *C. elegans* STI-1, the Homolog of Sti1/Hop, Is Involved in Aging and Stress Response. J Mol Biol. 2009, 390 (4): 604-617.
27. Kiontke K, Sudhaus W: Ecology of *Caenorhabditis* species. (January 09, 2006), WormBook, ed. The C. elegans Research Community, WormBook. 10.1895/wormbook.1.37.1. [http://www.wormbook.org]
28. Barriere A, Felix MA: High local genetic diversity and low outcrossing rate in *Caenorhabditis elegans* natural populations. Curr Biol. 2005, 15 (13): 1176-1184.
29. Schulenburg H, Ewbank JJ: The genetics of pathogen avoidance in *Caenorhabditis elegans*. Mol Microbiol. 2007, 66 (3): 563-570.
30. Prodromou C, Nuttall JM, Millson SH, Roe SM, Sim TS, Tan D, Workman P, Pearl LH, Piper PW: Structural basis of the radicicol resistance displayed by a fungal hsp90. ACS Chem Biol. 2009, 4 (4): 289-297.
31. Viney ME: Developmental switching in the parasitic nematode *Strongyloides ratti*. Proc Biol Sci. 1996, 263 (1367): 201-208.
32. Lewis JA, Fleming JT: Basic culture methods. Methods Cell Biol. 1995, 48: 3-29. full_text.
33. Knox DP, Jones DG: Studies on the presence and release of proteolytic enzymes (proteinases) in gastro-intestinal nematodes of ruminants. Int J Parasitol. 1990, 20 (2): 243-249.
34. Knight PA, Wright SH, Lawrence CE, Paterson YY, Miller HR: Delayed expulsion of the nematode *Trichinella spiralis* in mice lacking the mucosal mast cell-specific granule chymase, mouse mast cell protease-1. J Exp Med. 2000, 192 (12): 1849-1856.
35. Gillan V, Maitland K, McCormack G, Nik Him NA, Devaney E: Functional genomics of hsp-90 in parasitic and free-living nematodes. Int J Parasitol. 2009, 39 (10): 1071-1081.
36. Loytynoja A, Goldman N: Phylogeny-aware gap placement prevents errors in sequence alignment and evolutionary analysis. Science. 2008, 320 (5883): 1632-1635.
37. Guindon S, Gascuel O: A simple, fast, and accurate algorithm to estimate large phylogenies by maximum likelihood. Syst Biol. 2003, 52 (5): 696-704.
38. Hasegawa M, Kishino H, Yano T: Dating of the human-ape splitting by a molecular clock of mitochondrial DNA. J Mol Evol. 1985, 22 (2): 160-174.
39. Yang Z: PAML: a program package for phylogenetic analysis by maximum likelihood. Comput Appl Biosci. 1997, 13 (5): 555-556.
40. Nielsen R, Yang Z: Likelihood models for detecting positively selected amino acid sites and applications to the HIV-1 envelope gene. Genetics. 1998, 148 (3): 929-936.
41. Yang Z, Nielsen R, Goldman N, Pedersen AM: Codon-substitution models for heterogeneous selection pressure at amino acid sites. Genetics. 2000, 155 (1): 431-449.
42. Gordi T, Khamis H: Simple solution to a common statistical problem: interpreting multiple tests. Clin Ther. 2004, 26 (5): 780-786.
43. Thompson JD, Higgins DG, Gibson TJ: CLUSTAL W: improving the sensitivity of progressive multiple sequence alignment through sequence weighting, position-specific gap penalties and weight matrix choice. Nucleic Acids Res. 1994, 22 (22): 4673-4680.

44. Goodstadt L, Ponting C: CHROMA: consensus-based colouring of multiple alignments for publication. Bioinformatics. 2001, 17 (9): 845-846.
45. Campanella JJ, Bitincka L, Smalley J: MatGAT: an application that generates similarity/identity matrices using protein or DNA sequences. BMC Bioinformatics. 2003, 4: 29

CHAPTER 6

Engineered Soybean Cyst Nematode Resistance

Vincent P. Klink[1], Prachi D. Matsye[1], Katheryn S. Lawrence[2] and Gary W. Lawrence[3]

[1] Department of Biological Sciences, Mississippi State University, Mississippi State, MS, U.S.A.

[2] Department of Biochemistry, Molecular Biology, Entomology and Plant Pathology, Mississippi State University, Mississippi State, MS,, U.S.A.

[3] Department of Entomology and Plant Pathology, Auburn University, Auburn, AL, U.S.A.

1. INTRODUCTION

A variety of plant parasitic nematodes (PPNs), including the soybean cyst nematode (SCN), elicit the initiation, development and maintenance of a specialized nurse cell from which they derive their nutriment (Figure 1). Remarkably, during parasitism by the PPN, the nurse cell survives the apparently significant resource drain on the root cell that would be expected to detrimentally impact normal physiological processes of the cell. This outcome indicates that the nematode has developed a well tuned apparatus to ensure that the root cell does not collapse and die during parasitism. In contrast, in the soybean-SCN pathosystem, the nurse cell and sometimes the surrounding cells are the sites of the defense response to the parasite (Ross, 1958; Endo, 1965). Therefore, plants have in place a mechanism to overcome the influence of the activities of the nematode. Identifying the factor(s) is of utmost importance in developing resistance to PPNs.

1.2. History

Documented accounts reveal that soybean has been in cultivation for thousands of years (Hymowitz et al. 1970), beginning in Asia perhaps as early as 3,500 B.C. (Liu et al. 1997). While the natural range of soybean is East Asia, after thousands of years of cultivation a true understanding of its native range is complicated at best. However, the extensive range of wild soybean and obvious differences in its growth habit indicates that while environmental cues may be responsible for changes in soybean and plant growth habit in general (Garner Allard 1930; Chapin III et al. 1985; Day et al. 1999), genetic variation that exists in wild populations is of significant benefit to agriculture for production purposes and developing resistance to its many pathogens. This assessment is particularly true for soybean and its most significant pathogen, SCN, as many ecological collections have resulted in the identification of naturally occurring resistance (Ross and Brim, 1957; Ross, 1958; Epps and Hartwig, 1972; Concibido et al. 2004; Ma et al. 2006; Li et al. 2011; Matsye et al. 2012).

While knowledge of soybean's cultivation is long and extensive, scientific information on its dominant pathogen, the SCN, began with its description (Ichinohe, 1952). However, reports going back as early as the 1880's (Noel 1992) and late 1930s (Ichinohe, 1961) reveal knowledge of the nematode and appreciation of its pathogenic capacity. The SCN is a devastating pathogen that causes approximately 7-10% production loss, worldwide, annually and suppresses seed yield more than any other single soybean pathogen (Wrather et al. 1995, 2001a, b, 2003, 2006; Pratt and Wrather, 1998). In contrast, in some fields, as much as a 15% loss in yield has been observed with no visible signs of disease on soybean (Wang et al. 2003). Observations such as these could complicate SCN management since the disease can occur without knowledge of it being present in a particular field.

The near perfect overlap of the agricultural range of soybean with the distribution of SCN, infection creates a scenario where there is a high probability of a widespread and significant effect on yield. Conservative reports have shown that there is approximately 1.1-1.5 billion dollars in agronomic losses, annually, worldwide (Wrather et al. 2001). The overlapping distribution of SCN with that of soybean production was not always the case. Historically, SCN was not found in the U.S., or likely even North America or the New World. That situation changed when the SCN was first identified in the U.S. in North Carolina in 1954 by Winstead et al.

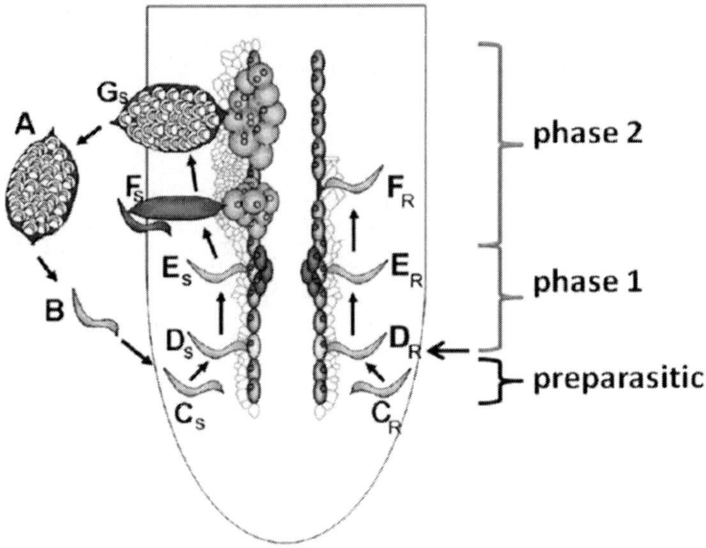

Figure 1. SCN life cycle during susceptible and resistant reactions Figure 1A, cysts with eggs (white). Figure 1B, second stage pre-infective juveniles (pi-J2) (gray) migrate toward the root. SUSCEPTIBLE REACTION: Figure 1C$_S$, the infective-J2 (i-J2) nematodes (light gray) burrow in and migrate toward the stele. Figure 1D$_S$, feeding site selection (yellow). Fig 1E$_S$, p-J2 nematodes molt into J3, then J4. During this time, the original feeding site is incorporating adjacent cells (purple) via cell wall degradation events. Meanwhile, the male discontinues feeding at the end of its J3 stage. Figure 1F$_S$, After maturation, the male and female nematodes copulate. Figure 1G$_S$, the female at ~30 dpi. RESISTANT REACTION: Figure 1C$_R$, Like the susceptible reaction, the infective-J2 (i-J2) nematodes (dark gray) burrow into the root and migrate toward the root stele. Figure 1D$_R$, feeding site selection by the parasitic J2 (p-J2). Fig 1E$_R$, the syncytium begins to develop. Figure 1F$_R$, the syncytium has collapsed resulting in nematode mortality. The right panel shows the initiation phase of infection (black arrow). Phase 1 is the development period. Phase 2 is the maintenance period (adapted from Klink et al. 2011b).

and published a year later (Winstead et al. 1955). Unfortunately for agriculture, SCN is readily transmissible as evidenced by its identification in localities as far away as Mississippi only a few years later by 1957 (Spears, 1957). The SCN now is a registered invasive species in the U.S. Notably, in the U.S., SCN causes more agricultural loss to soybean than the rest of its pathogens combined (Wrather et al. 2001, 2006). Making the problem worse for agriculture is the genetic diversity of the SCN (Golden et al. 1970; Riggs and Schmidt 1988; 1991; Niblack et al. 2002; Bekal et al. 2008). SCN research has determined that the nematode is a species complex originally subdivided into four races (Golden et al. 1970) which later was expanded into 16 races (Riggs and Schmidt 1988, 1991) that have been reorganized, further subdivided and reclassified into distinct populations (Niblack et al. 2002; Niblack and Riggs, 2004). The term population was designated since genetically pure clones are impossible to obtain in the sexually reproducing SCN system (Niblack et al. 2002). The classification scheme of Niblack et al. (2002) is based on the varying ability of SCN populations to infect a panel of 7 soybean genotypes that can resist infection to varying levels. It is noted that some of these designated populations are "strains" that are maintained in the greenhouse and genetically purified through hundreds of generations of single cyst descent (Niblack et al. 1993). Therefore, the genetic background in these "strains" may not resemble the original field-extracted population since allelic forms of the parasitism genes would likely be lost through this purification process.

The genetic diversity found in SCN (Golden et al. 1970; Riggs and Schmidt 1988; 1991; Niblack et al. 2002; Bekal et al. 2008) likely aids in its ability to infect and reproduce on plants other than soybean. Thus, from an ecological standpoint, SCN could pose a threat to plants that grow outside of production areas. This potential problem would be exasperated if those plant species are listed as endangered or threatened species or are a significant component of the ecological community. A number of studies have shown that the SCN reproduces on at least, but certainly is not limited to, 97 legume and 63 non-legume hosts (Epps and Chambers, 1958; Riggs and Hamblen, 1962, 1966a, b) and new SCN hosts are determined on a regular basis (Creech et al. 2006). It has been many years since species range tests have been performed for SCN so it is likely that these lists of hosts are not comprehensive. This virulence capability of SCN poses a problem in terms of its management since SCN populations could be maintained by weedy plants that grow or overwinter in fallow fields or along the boundaries of acreage that is in production (Creech et al. 2006). In addition to these problems, SCN does not even have to reproduce in the plant to still cause damage to the plant. While the genetic diversity of both soybean and SCN may appear to complicate an understanding of the process of infection and the development of resistant cultivars, the natural variation in both the germplasm of soybean (Doyle et al. 1999) and SCN (Bekal et al. 2003; 2008) presents many opportunities to understand the basic machinery of infection of the SCN and the genotype-specific nuances that regulate both susceptibility and defense. These features make the soybean-SCN pathosystem an extremely valuable experimental model (Barker et al. 1993; Opperman and Bird, 1998; Niblack et al. 2006; Klink et al. 2010a).

1.3. Methods To Control Scn Infection

Historically, SCN has been managed through a combination of chemical control, cropping systems, biological control and the identification and use of resistant germplasm. Chemical control for pathogens using methyl bromide first occurred in France (Schneider et al. 2003; Rosskopf et al. 2005) and had subsequently been used for decades for both pre- and post plant nematode control. However, methyl bromide is a chemical that has been phased out of use in 2005 in the U.S. (Schneider et al. 2003; Rosskopf et al. 2005) because it has been classified as a Class 1 (Group VI) stratospheric ozone depletor by the Environmental Protection Agency. Because of the loss of this major control agent for the SCN, even in developing countries by the year 2015 (Rosskopf et al. 2005), it was important to identify other strategies that could be included in the SCN management plan. Biocontrol measures that include bacteria, fungi or even their proteins are feasible (Chen and Dickson, 1996; Kim and Riggs, 1991, 1995; Liu and Chen, 2001; Meyer and Huettel, 1996; Meyer and Meyer, 1996; Timper et al., 1999). Other control methods that had already been used extensively for decades include the time honored crop rotation strategy. This strategy has reduced SCN populations below damaging levels (Francl and Dropkin, 1986; Sasser and Uzzell, 1991; Koenning et al. 1993). Rotating with nonhosts over a 2-3 year period mitigated the undesirable levels of SCN in the field (Ross, 1962; Francl and Dropkin, 1986). Other cropping systems that have had success in SCN control are the use of blending, resistant cultivars and cropping sequence, among others (Niblack and Chen, 2004). While successful, a problem with cropping strategies is that the interval is not long enough to compete with the 9 year cycle that cysts can remain viable, but dormant in production fields (Inagake and Tsutsumi, 1971). With these strategies in place it is possible to develop a tightly managed regime, incorporating some or all of these technologies and principles to mitigate SCN damage. Lastly, genetic engineering has begun to take root with potential as a method to generate resistance (Steeves et al. 2006; McLean et al. 2007; Klink et al. 2009a; Matsye et al. 2012). However, for genetic engineering to be successful, it is first required that candidate genes be identified. The identification of these genes has happened through a series of RNA gene expression studies, employing soybean germplasm that exhibits resistance to SCN.

1.3.1. Available resistant germplasm

Once SCN was identified in the U.S. (Winstead et al. 1955), a very large need existed to determine if soybean germplasm existed that could resist infection. The vast and expansive range of soybean (Morse, 1927) and visually obvious variations in growth form in its various ecological habitats provided the possibility that germplasm that was resistant to SCN existed in its wild populations. Established in 1898, the development of a substantial and publically available seed bank was initiated that is maintained by the USDA-National Plant Germplasm System (USDA-NPGS) (Morse, 1927; Bernard et al. 1987). It now contains approximately 20,000 varieties (accessions) with each accession classified as a plant introduction (PI) through a numbering system. Many of the 7,867 PIs that were already available by 1944, just 10 years before the identification of SCN in the U.S., had been collected in trips to China, Japan, India and Korea in a small window of time between 1924 and 1932 (Bernard et al. 1987). The public availability of the germplasm allowed it to be used in a series of trials to determine if any of the available accessions was resistant to SCN. A number of accessions were determined to be resistant to SCN through two large trials that studied about 5,700 accessions (Ross and Brim, 1957; Ross, 1958; Epps and Hartwig, 1972). Research on these accessions has resulted in the identification of approximately 118 sources of resistance (Concibido et al. 2004). However, only approximately seven sources are used for cultivar development in the U.S. (Shannon et al. 2004). These accessions include the *G. max* PIs known as Peking ($G.\ max_{[\mathrm{Peking}]}$) and $G.\ max_{[\mathrm{PI\ 88788}]}$. Currently, $G.\ max_{[\mathrm{Peking}]}$ and $G.\ max_{[\mathrm{PI\ 88788}]}$ resistance germplasm is present in >97% of all commercial cultivars in the U.S. (Concibido et al. 2004). In addition to these PIs, hundreds of additional accessions that can resist SCN infection have been identified in China (Ma et al. 2006; Li et al. 2011). These banks of germplasm provide an important and substantial genetic resource for understanding the process of parasitism in soybean at the cellular level. This is important to understand because the infection of soybean involves very specific cell types that react in very specific ways to SCN parasitism.

1.4. Cytological Reaction During Resistance

The SCN can remain viable in the soil in eggs ensheathed within the carcass of the dead mother (cyst wall) for up to 9 years (Inagake and Tsutsumi, 1971). However, the devastating interaction of the SCN with soybean begins when it burrows into the root through the epidermal and cortical cells. This has been shown both by cytological studies and gene expression studies of time points collected before the formation of syncytia (Alkharouf et al. 2006; Klink et al. 2009b). The interaction continues through the initiation and subsequent formation a multinucleate nurse cell known as a syncytium from pericycle or neighboring cells (Ross, 1958; Endo, 1964, 1992). The formation of the syncytium is likely to be a very coordinated process, occurring through the injection and subsequent activity of nematode parasitism proteins (Atkinson and Harris, 1989; Smant et al. 1999; Lambert et al. 1999; De Boer et al. 1999, 2002; Wang et al. 1999, 2001, 2003; Gao et al. 2001, 2003; Bekal et al. 2003). These substances likely orchestrate successive waves of interference of the root cell's normal physiological processes and initiate various cell wall dissolving events (Atkinson and Harris, 1989; Smant et al. 1999; Wang et al. 1999, 2001, 2003; Lambert et al. 1999; De Boer et al. 1999, 2002; Gao et al. 2001, 2003; Bekal et al. 2003). The parasitism process merges approximately 200-250 root cells into a common cytoplasm containing as many nuclei, the definition of a syncytium (Jones and Northcote, 1972; Jones, 1981). Additional nematode activities alter the plant cell's physiology (Klink et al. 2005, 2007a; Ithal et al. 2007). The activities benefit the nematode during the sedentary period of its life cycle as they feed and mature (Edens et al. 1995; Hermsmeier et al. 1998; Mahalingam et al. 1999; Vaghchhipawala et al. 2001; Klink et al. 2005, 2007a; Alkharouf et al. 2006; Ithal et al. 2007; Matsye et al. 2011).

Cytological studies of the SCN infection process (Figure 1) have shown that the cellular response of soybean to SCN infection can be divided into an earlier phase (phase 1) and a later phase (phase 2) (Ross, 1958; Endo, 1964, 1965, 1991; Riggs et al. 1973; Kim et al. 1987; Mahalingam and Skorpska, 1996). Phase 1 and 2 span the periods including the initiation, development and maintenance of the syncytium (Figure 1). These observations are not unique to SCN since similar observations have also been made for the cyst nematode *Rotylechulus reniformis* (Robinson et al. 1997), indicating that a basic level of conservation may exist for the process of defense at the site of infection while genotype-specific gene activities also exist (Klink et al. 2011a, Matsye et al. 2011, 2012).

During phase 1, the cellular reactions leading to susceptibility or defense appear the same at the cytological level. The cellular events occurring during the earlier stages of syncytium development include hypertrophy, the dissolution of cell walls, the development of dense cytoplasm, an enlargement of nuclei and an increase in endoplasmic reticulum (ER) and ribosome content (Endo, 1964, 1965; Riggs et al. 1973; Kim et al. 1987; Kim and Riggs, 1992; Mahalingam and Skorpska, 1996). The enlargement of nuclei and increase in ER and ribosome content indicate an increase in gene expression and protein synthesis accompanies the activity of the nematode within the parasitized cells. Therefore, it is likely that the plant cell is being programmed to make specific materials to benefit the nutritional needs of the nematode. It is known that plant parasitic nematodes lack the ability to make materials such as sterols (Chitwood and Lusby, 1990). Therefore, altering the metabolism of the parasitized cell probably would involve the induction of metabolic activity that relates to these processes. Cell fate mapping experiments have demonstrated the metabolism that occurs during these stages of parasitism and some of it relates to enhanced plant sterol production (Klink et al. 2011a; Matsye et al. 2011).

After these earlier events, the cytology of susceptibility and defense become apparent and is referred to as phase 2. Phase 2 of the susceptible reaction is characterized by hypertrophy of nuclei and nucleoli. This process is accompanied by the reduction and dissolution of the vacuole. The reduction and dissolution of the vacuole suggests important events or structural features involved in membrane fusion and/or maintenance are perturbed. This topic will be described in a later section. Other cellular events that have been identified during the susceptible reaction include cell expansion as it incorporates and fuses with adjacent cells (Endo and Veech 1970; Gipson et al 1971; Jones and Northcote, 1972; Riggs et al. 1973; Jones, 1981). Additional activities include the proliferation of cytoplasmic organelles.

In contrast, the cellular aspects of the defense responses occurring during phase 2 depend on the soybean genotype being infected. Information that has been generated through a number of cytological studies have resulted in the development of a system that divides the PIs into cohorts having similar cytological reactions that is based on the cellular characteristics associated with how SCN responds during resistance (Colgrove and Niblack, 2008). Currently, the PIs have been categorized into those genotypes having the *G. max*$_{[\text{Peking}]}$ and *G. max*$_{[\text{PI 88788}]}$-types of defense responses (Colgrove and Niblack, 2008). Much more work in this area of research is required for a comprehensive understanding of the different forms of the defense response. Such knowledge would allow the commonalities of the cytological features to be correlated with the molecular events that are occurring in the parasitized cell types. By doing so, it would allow for the identification of genes that always correlate to resistance, regardless of the cytology or genotype of soybean. It would be likely that these genes are central to all forms of the defense response (Klink et al. 2011a; Matsye et al. 2011, 2012). Among the characteristics that define these cohorts, the *G. max*$_{[\text{Peking}]}$-type of defense includes the development of a necrotic layer that surrounds the head of the nematode (Kim et al. 1987; Endo, 1991). This process is followed by necrosis of the initial cell that the nematode had parasitized. In contrast, in the *G. max*$_{[\text{PI 88788}]}$-type of defense response, the necrotic layer that surrounds the head of the nematode is lacking and the initial cell that the nematode parasitized first experiences necrosis (Kim et al. 1987; Endo, 1991). In addition to these cytological characteristics found in the *G. max*$_{[\text{Peking}]}$ and *G. max*$_{[\text{PI 88788}]}$-types of defense responses are the presence or absence of cell wall appositions (CWAs). CWAs are structures defined as physical and chemical barriers that are designed to prevent cell penetration (Aist et al. 1976, Schmelzer, 2002; An et al. 2006a, b; Hardham et al. 2008). CWAs have been found and studied in other plant-organism pathosystems (Collins et al. 2003; Assaad et al. 2004; Kalde et al. 2007). However, CWAs are not a defining characteristic of all types of defense responses in soybean. CWAs have been found in the *G. max*$_{[Peking]}$-type of resistant reaction and are found in the *G. max*$_{[PI\ 437654]}$ genotype (Mahalingam and Skorpska, 1996). This makes the placement of *G. max*$_{[PI\ 437654]}$ in the *G. max*$_{[Peking]}$ cohort logical (Colgrove and Niblack et al. 2008). In contrast, CWAs are lacking in *G. max*$_{[PI\ 88788]}$. More work is required in the understanding the role(s) that CWAs play, if any, during defense of soybean to SCN. However, the significance and role of CWAs during defense were first demonstrated by Collins et al. (2003), and followed by additional studies performed by Assaad et al. (2004) and Kalde et al. (2007). In those studies, it was shown at the molecular level that CWA formation involves the vesicular transport machinery protein component known as syntaxin. This was a striking discovery since the process of vesicular transport is a conserved cellular process, meaning it has been found in other organisms. The syntaxin gene was first identified in animal systems (Inoue et al. 1992; Bennett et al. 1992) and through a number studies performed in animal and model genetic systems it was shown that syntaxin interacts with other proteins to accomplish specific cellular functions. Unfortunately, while the role of syntaxin in plant defense has been studied (Collins et al. 2003; Assaad et al. 2004; Kalde et al. 2007), the examination of other components of the vesicular transport machinery has received little attention. Until very recently (Matsye et al. 2012), no information existed on how these proteins function or interact with syntaxin during the defense of plants to pathogens. The demonstration that syntaxin plays a role in the defense of plants to pathogens, implicates that other proteins that interact directly or indirectly with syntaxin probably are also involved in the process of defense. A genetic pathway, involving *PEN1*, the β-glycosyl hydrolase *PEN2* and the ABC transporter *PEN3* transports and delivers antimicrobial compounds across the cell membrane to sites where the fungus is attempting to enter (Collins et al. 2003; Lipka et al. 2005; Stein et al. 2006). Other proteins that interact directly with syntaxin have been studied in other experimental systems and include the ATPase known as N-ethylmaleimide-sensitive factor attachment protein (NSF) (Malhotra et al. 1988), the soluble N-ethylmaleimide-sensitive factor attachment receptor protein (SNARE) complex and synaptosomal-associated protein 25 (SNAP25) (Oyler et al. 1989), the soluble N-ethylmaleimide-sensitive factor attachment protein (SNAP) (Weidman et al. 1989; Clary et al, 1990; Collins et al. 2003; Assaad et al. 2004; Kalde et al. 2007), among other proteins. Since these numerous studies have shown very specifically how the protein complex is assembled, it was then possible to determine how specific components of the CWA assembly process that are present during defense of soybean to SCN function (Matsye et al. 2012). However, even though CWAs are lacking in genotypes like *G. max*$_{[PI\ 88788]}$, it does not mean that the proteins are not involved in defense through related activities. Membrane fusion has been shown to play a role in defense through a process known as autophagy. Autophagy is a process known in plants to play crucial roles in defense (Patel and Dinesh-Kumar 2008; Hofius et al. 2009; Lenz et al. 2011; Lai et al. 2011). This knowledge has allowed for a targeted approach in understanding the protein machinery that is involved in defense (Matsye et al. 2012).

1.5. Genomics-Based Studies of Scn

A number of "omics" studies in the soybean-SCN pathosystem have been performed to understand both plant and nematode gene expression at the organismal level. Many of the gene expression studies that relied on the microarray technology were modeled after earlier experiments that were performed in the model plant *Arabidopsis thaliana* that were infected with the beet cyst nematode, *Heterodera schactii* (Puthoff et al. 2003). Studying SCN in *A. thaliana* is complicated by the fact that it is a nonhost to SCN infection so studies investigating the susceptible reaction elicited by SCN in *A. thaliana* cannot be done until suitable mutants or susceptible ecotypes are identified. A number of microarray studies using whole infected soybean roots as a source for the RNA samples have identified genes that are expressed during a susceptible reaction (Alkharouf et al. 2006; Ithal et al. 2007; Klink et al. 2007b).

The parasitism of soybean by SCN begins with and is sustained through the injection of materials that are synthesized in subventral and esophageal glands into the root cell. It is a costly life strategy since typically only about 10% of the nematodes ever make it to maturity in a fully susceptible genotype like *G. max*$_{[Williams\ 82/PI\ 518671]}$. Identifying the genes involved in parasitism would likely occur through collecting the cytoplasm of the cells composing the subventral and esophageal glands. It was hypothesized that these genes would be important for the events of parasitism and would be involved in altering the metabolic processes of the soybean to benefit the nematode. The experiments were performed by microaspirating the cytoplasm of the gland cells, constructing cDNA libraries and sequencing the genes, allowing for downstream bioinformatics analyses to help elucidate what the genes could actually be (Smant et al. 1998). The experiments were then repeated for the SCN, identifying a number of putative parasitism genes (Wang et al. 1999; Gao et al. 2001, 2003). With the development of the Affymetrix® Soybean GeneChip, it was possible to examine the expression of thousands of SCN genes simultaneously. This was made possible because 7,539 *H. glycines* probe sets representing 7,431 transcripts (genes) were printed onto the array. One analysis examined the expression of SCN genes that were expressed specifically during infection of the *G. max*$_{\text{[Williams 82/PI 518671]}}$ genotype that lacks a functional defense response (Ithal et al. 2007). This means that gene expression occurring during a susceptible reaction was monitored. The work examined the expression of previously identified (Wang et al. 1999; Gao et al. 2001, 2003) and analyzed (Bakhetia et al. 2007) putative parasitism genes (Ithal et al. 2007). The remaining genes that were fabricated onto the array were not a focus of the analysis. The experiments confirmed the expression pattern of dozens of putative parasitism genes (Ithal et al. 2007). A gap in the knowledge was that the experiments were not designed to determine what genes were expressed as the nematode experienced a resistant reaction in a soybean genotype that was capable of a defense reaction. This information would be important because it would provide knowledge on the metabolic pathways that may be sensitive to genetically-based control measures. That gap in knowledge was filled in experiments that performed population-specific analyses of gene expression, comparing the susceptible and resistant reactions experienced by SCN as they infected the *G. max*$_{\text{[Peking/PI 548402]}}$ genotype that has a functional defense response to some populations of SCN (Klink et al. 2007a). Thus, from the experiments of Ithal et al. (2007) and Klink et al. (2007a), specific knowledge of gene expression occurring in genotypes both lacking and having functional resistance genes was obtained. It is noted that additional gene expression profiling experiments have also been performed (Elling et al. 2009). In earlier studies, Alkharouf et al. (2007) annotated all of the SCN genes that were available in Genbank and compared them to the genetic model free living nematode, *Caenorhabditis elegans*. The advantage of these comparisons was that the genome of *C. elegans* had been sequenced (C. elegans Sequencing Consortium, 1998), allowing for a substantial annotation process to be executed. In addition, there was a massive amount of functional data obtained through genetic and reverse genetic experiments (Fire et al. 1998; Piano et al., 2000; Kamath et al., 2003; Sonnichsen et al., 2005) that was available for essentially every gene in the genome housed in the *C. elegans* database at http://www.wormbase.org. Since the genomic sequence of *C. elegans* is known, it is possible to find highly conserved and related genes in SCN. The working hypothesis was that if the genes in *C. elegans* and SCN are nearly identical in primary sequence, it would be likely that they have similar function. If the genes have similar function, for example an essential function for survival in *C. elegans*, knocking out that gene in SCN would probably result in lethality for those nematodes if the gene could be knocked out. The annotation of the SCN genes was driven by a homology criterion whereby the SCN genes were pooled into six bins referred to as Groups 1-6 (Alkharouf et al. 2007). The six bins were based on the level of homology the sequence had to *C. elegans* genes. Group 1 had the highest level of homology and Group 6 had the lowest level. For example, Group 1 had *E*-values between 0 and 1*E*-100; Group 2 had *E*-values between 1*E*-100 and 1*E*-80; Group 3 had *E*-values between 1*E*-80 and 1*E*-60; Group 4 had *E*-values between 1*E*-60 and 1*E*-40; Group 5 has *E*-values between 1*E*-40 and 1*E*-20 while Group 6 has *E*-values > 1*E*-20 (Alkharouf et al. 2007). The gene annotation process resulted in taking the nearly 8,334 conserved genes between *H. glycines* and *C. elegans* and identifying 1,508 that have been shown to have lethal phenotypes/phenocopies in *C. elegans* (Alkharouf et al. 2007). The research then was poised to test the function of the 1,508 genes, but it was an unmanageable number of genes. To narrow down the 1,508 genes to a manageable number for functional studies, the genes underwent further annotation procedures (Alkharouf et al. 2007). To do this annotation procedure, firstly, a pool of 150 highly conserved, Group 1, *H. glycines* homologs of genes having lethal mutant phenotypes or phenocopies from the free living nematode *C. elegans* were identified from the pool of 1,508 genes that were fabricated onto the Affymetrix® microarray. Secondly, it was determined that of those 150 genes on the Affymetrix® soybean GeneChip, a subset of 131 genes could have their expression monitored during the parasitic phase of their life cycle. Thirdly, a microarray analyses identified a core set of 32 genes with induced expression occurring during the parasitic stages of infection. The identification of 32 genes that had known expression during the parasitic stages of infection provided a small, but feasible, core set of genes that could be targeted in RNAi-based, reverse genetic screens (Table 1).

1.6. Reverse Genetic Screens to Identify Essential Scn Genes

Unlike *C. elegans*, SCN is not an ideal system for genetic studies because of its obligate endoparasitic life cycle. However, from information learned in *C. elegans*, gene function can be studied by an mRNA nuclease process called RNA interference (RNAi) (Fire et al. 1998). Through this process, a specific mRNA is targeted through a ribonucleoprotein complex for degradation (Hammond et al. 2001; Caudy et al. 2003). The challenge then became demonstrating whether RNAi was functional and reliable in the SCN since the approach does not work in some organisms. However, there are two demonstrated ways that RNAi–based experiments can be performed for SCN, allowing gene function experiments to be performed through the a reverse genetic manner allowed by the RNAi technology.

The first demonstration of RNAi in SCN accomplished the experiment by taking cDNAs for the gene of interest, synthesizing double stranded RNA (dsRNA) *in vitro* and soaking the nematodes in the dsRNA cocktail (Urwin and Atkinson, 2002; Alkharouf et al. 2007). Urwin et al. (2002) examined how the SCN actin gene could be knocked down in its expression. The experiments resembled those performed in *C. elegans* whereby soaking the nematodes in dsRNA resulted in a phenocopy of the normal phenotype generated by the hypomorphic null mutant (Fire et al. 1998; Timmons et al. 2001). Similar experiments that relied on an extensive, but simple, gene annotation pipeline that identified 1,508 candidate genes (Alkharouf et al. 2007) used the cloned genes to synthesize dsRNA from *H. glycines* homologs of small ribosomal protein 3a. Experiments that soaked the SCN with dsRNA resulted in nematode mortality that was demonstrated by vital fluorescent dyes and a phenocopy where the nematodes appeared stiff (Alkharouf et al. 2007). Therefore, RNAi would work in the SCN system. The experiments were then taken a step further in experiments that used the RNAi-soaked nematodes to infect soybean plants to see if the nematodes were impaired in their ability to parasitize soybean. Modeled after the earlier experiments of Urwin et al. (2002), in experiments that used this approach for parasitism genes, it was shown that SCN infection could be suppressed (Bakhetia et al. 2007, 2008). The problem with these experiments, from a nematode biocontrol perspective, is that it would be virtually impossible to synthesize, apply and deliver enough dsRNA to nematodes that are living in the environment to obtain a positive effect even though crude dsRNA extracts can be used (Tenllado et al. 2003). Other problems would be whether the dsRNA remained residually in the soil. Therefore, a second method would be needed that could express the genes as dsRNA in soybean, allowing greater control over the delivery of the dsRNA to SCN.

The second way to perform RNAi experiments for SCN control would be to express the genes in transgenic soybean roots, allowing the nematodes to feed on the genetically engineered roots. The hypothesis is that if the SCN was able to ingest the double stranded RNA manufactured in the plant cells through its stylet in high enough concentrations and if the RNAi metabolic process occurred in SCN, there was a chance that nematode development could be controlled. Prior experiments already demonstrated that the RNAi pathway functioned in SCN (Urwin et al. 2002; Alkharouf et al. 2007). The original experiments that performed host-mediated expression of SCN genes as inverted tandemly duplicated copies for RNAi control in soybean to examine SCN biology were done by Steeves et al. (2006), examining the major sperm protein. Huang et al. (2006) demonstrated the same effect for root knot nematode in the model plant *A. thaliana* so the approach would have broad applicability for PPN control. The experiments were followed by Klink et al. (2009a) that identified many genes from microarray studies that would serve as candidates for RNAi control during parasitism.

Table 1. Annotation of the Affymetrix ® soybean GeneChip in relation to gene pathway analyses.

ProbeSetID	Afx H.g. Accn	Best C.e. Hit	Brief ID
HgAffx.13360.1.S1_at	CB374622	pes-9	Yeast hypothetical 52.9 KD protein
HgAffx.10986.1.S1_at	CA939315	gpd-3	NULL
HgAffx.19591.1.S1_at	CB278666	cpf-1	cleavage stimulation factor like
HgAffx.16755.1.S1_at	CA939427	phb-1	NULL
HgAffx.6532.1.S1_at	CK350603	hsp-6	heat shock 70 protein
HgAffx.19651.1.S1_at	CD748666	phb-2	Prohibitin
HgAffx.17567.1.S1_at	CB825030	ben-1	tubulin
HgAffx.20551.1.S1_at	CB281634	E02H1.1	rRNA methyltransferase
HgAffx.19055.1.S1_at	CB826041	T21B10.2	enolase
HgAffx.24001.2.S1_at	CK351582	ftt-2	14-3-3 protein
HgAffx.22771.2.S1_at	BI749139	uaf-1	NULL
HgAffx.21332.1.S1_at	BI748882	daf-21	heat shock protein (HSP90)
HgAffx.10691.1.S1_at	CD748651	K04D7.1	guanine nucleotide-binding protein
HgAffx.10821.1.S1_at	CB935592	T21B10.7	t-complex protein 1
HgAffx.13633.1.S1_at	CD748017	pyp-1	inorganic pyrophosphatase
HgAffx.20969.1.S1_at	CB379877	rps-1	Ribosomal protein S3a homolog
HgAffx.24120.1.S1_at	CB935135	eft-2	Elongation factor Tu family
HgAffx.22597.1.S1_at	CB826306	kin-19	casein kinase I
HgAffx.11150.1.S1_at	CB378957	D1005.1	ATP citrate lyase
HgAffx.17961.1.S1_at	CB281421	F01G10.1	transketolase
HgAffx.18740.2.S1_at	CA940055	act-4	actin
HgAffx.14431.1.S1_at	CB935363	mdh-1	malate dehydrogenase
HgAffx.19636.2.S1_at	CA939544	rps-4	NULL
HgAffx.18811.1.S1_at	CA940369	F43G9.5	NULL
HgAffx.5490.1.S1_at	CD747934	gpi-1	glucose-6-phosphate isomerase
HgAffx.22868.1.S1_at	BG310682	cpl-1	cathepsin-like protease
HgAffx.13283.1.S1_at	CD748764	K10D6.2	NULL
HgAffx.17866.1.S1_at	CB824474	M03C11.7	NULL
HgAffx.20065.1.S1_at	AF318605	hsp-1	HSP-1 heat shock 70kd protein A
HgAffx.15252.1.S1_at	CK348813	rho-1	p21 ras-related rho (RhoA)
HgAffx.16942.1.S1_at	CK349264	ruvb-2	NULL
HgAffx.23555.2.S1_at	CD748675	Y54E10BR.6	NULL

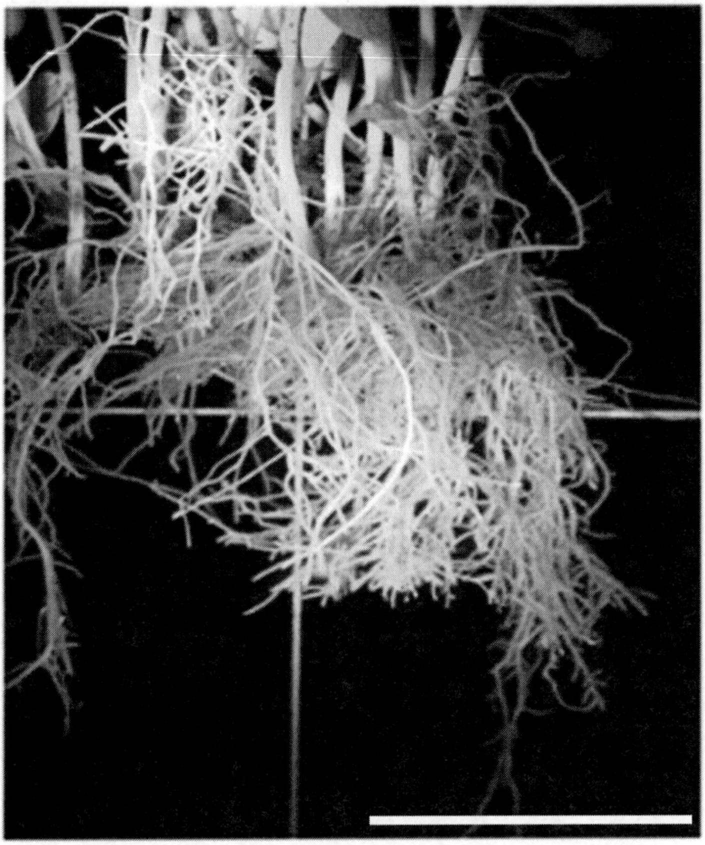

Figure 2. Soybean plants with transgenic roots. The transgenic soybean roots are expressing the enhanced green fluorescent protein (eGFP) (Haseloff et al. 1997) found in the pRAP vectors (Klink et al. 2009c; Matsye et al. 2012). Bar = 10 cm.

The problem with the transgenic approach is that soybean is a difficult to genetically engineer. However, strategies whereby composite plants (Collier et al. 2005) that are chimeras having nontransformed aerial stocks having transgenic root stocks can be readily made in soybean (Klink et al. 2008, 2009a). The simplicity of the approach is evident because the transgenic plants can be made in non-axenic conditions with the use of fluorescent reporter (Collier et al. 2005) (Figure 2). The development of vectors that work in soybean (Klink et al. 2008, 2009a; Ibrahim et al. 2011; Matsye et al. 2012) have made the experiments possible. Further improvements whereby the plant expression vectors are Gateway®-

compatible (Klink et al. 2009a; Ibrahim et al. 2011; Matsye et al. 2012) allows for semi-large reverse genetic screens to be performed. In such experiments, SCN homologs of the small ribosomal protein 3a (Hg-rps-3a) and Hg-rps-4, synaptobrevin (Hg-snb-1) and a spliceosomal SR protein (Hg-spk-1) were tested for functionality in host mediated expression, RNAi-based studies (Klink et al. 2009a). After 8 days of infection, the experiments demonstrated that 81–93% fewer females developed on transgenic roots containing the genes engineered as tandem inverted repeats. Those experiments resulted in lethality for SCN feeding on plants that were expressing the genes as tandemly duplicated inverted repeats (Klink et al. 2009a). The same outcome was shown for root knot nematode in soybean using the same plant expression vector system (Ibrahim et al. 2011). These observations demonstrated that broad spectrum resistance for PPNs in soybean was probable.

1.7. Proteomic Studies of Scn

The prior experiments have discussed gene expression in SCN at the RNA level. These experiments are technologically simplistic to perform, because of major advances in sequencing and detection technologies. However, in these experiments using hybridization to study gene expression, little to no information is obtained as to how much protein is actually synthesized from the RNA or modifications that are known to exist on the protein molecules. Recently, the proteome of SCN was investigated (Chen et al. 2011), resulting in a reference map of protein expression. These experiments add to the already extensive gene expression databases that are available for SCN (Ithal et al. 2007; Klink et al. 2007a, 2009a; Elling et al. 2009). The advantage of the proteomic studies is that it allows for the identification of the relative amounts of the studied proteins to be known. This is in contrast to microarray-based experiments where only different levels of expression can be inferred, but their absolute amounts are not known. Chen et al. (2011) performed experiments using LC-MS/MS on pre-infective J2 SCN. The nematodes were highly pure samples since they had not yet infected soybean roots. The experiments were able to discern 803 spots on 2-D gels (Chen et al. 2011). Of those spots, 426 proteins were identified (Chen et al. 2011). Gene Ontology analyses allowed for the identification of a number of different functional groups, including secreted proteins that may act during parasitism (Chen et al. 2011). While it is likely that the protein list is not comprehensive, the work provides a solid foundation for future work to examine the proteome of SCN and compare with the gene expression studies based on the RNA.

1.8. Soybean Gene Expression

To understand how soybean was reacting to infection, it was going to be imperative to develop ways to monitor gene expression during infection. Unlike the model system, *A. thaliana*, where a number of gene expression microarrays existed (Mussig et al. 2002; Tao et al. 2003) no commercially available microarrays were in place for soybean. The fabrication of soybean microarrays from cDNAs isolated from uninfected and SCN-infected tissues resulted in the identification of genes that are expressed during parasitism by SCN (Alkharouf et al. 2006). Subsequently, after the availability of the Affymetrix soybean GeneChip, a number of gene expression studies have been performed on whole infected soybean roots (Klink et al. 2007b; Ithal et al. 2007). Some of the studies have focused in on expression occurring during the susceptible reaction (Alkharouf et al. 2006; Klink et al. 2007b; Ithal et al. 2007). These studies have resulted in the identification of genes that are highly expressed during the susceptible reaction. Alkharouf et al. (2006) performed experiments that examined the preparasitic stages of infection of the compatible reaction. The experiments identified defense-related genes such as Kunitz trypsin inhibitor (KTI), germin, peroxidase, phospholipase D, 12-oxyphytodienoate reductase (OPR), pathogenesis related-1 (PR1), phospholipase C, lipoxygenase, WRKY6 transcription factor and calmodulin. The experiments demonstrated that multiple defense pathways were induced even early (by 6 hours post infection) during the compatible reaction. This is important to note because the time point at which the sample was collected occurred before the nematode initiated the formation of the syncytium. This meant that soybean was responding in important ways to the presence of the nematode within it s root tissues. Similar lists of genes were identified by Ithal et al. 2007, demonstrating a commonality of expression even though the experiments used different soybean genotypes and populations of SCN. Unfortunately, since only the susceptible reaction was studied, it was unclear whether the expressed genes were specific to the susceptible reaction or would also be differentially expressed in roots if they were undergoing a resistant reaction. This knowledge would be important to identify actively expressed genes that relate specifically to defense.

To distinguish between expression of genes during the susceptible and resistant reactions, an experiment was performed whereby both susceptible and resistant reactions could be obtained in the same soybean genotype (*G. max*$_{\text{[Peking/PI 548402]}}$) (Klink et al. 2007b). The importance in the way the experiment was designed was that it allowed gene expression that pertained specifically to the susceptible or resistant reaction to be identified. Thus, differences in plant genotype could not introduce error into the experiment. The experiments were set up whereby the *G. max*$_{\text{[Peking/PI 548402]}}$genotype was infected with one of two SCN populations that would result in a susceptible or a resistant reaction. Another important feature of the experiment was that the gene expression that occurred as *G. max*$_{[Peking/PI\ 548402]}$, a genotype with functional resistance genes, failed in its effort to defend itself from SCN would be identified. The *G. max*$_{[Peking/PI\ 548402]}$ genotype was infected with *H. glycines*$_{[NL1\text{-}RHg/HG\text{-}type\ 7]}$ (originally called race 3) that obtained a resistant reaction and *H. glycines*$_{[TN8/HG\text{-}type\ 1.3.6.7]}$ (originally called race 14) that obtained a susceptible reaction (Klink et al. 2007b). The experiments revealed induced levels of some genes during different points of the susceptible reaction as compared to the resistant reaction. Some of the genes that were induced in their expression during the susceptible reaction at 12 hours post infection (hpi) were an expansin, peroxidase, plasma membrane intrinsic protein 1C (PIP1C), germin-like protein (GER) 1, beta-Ig-H3 domain-containing protein and chorismate mutase (Klink et al. 2007b). Genes induced during the susceptible reaction at 3 dpi included 4-coumarateCoA ligase family protein, expansin, LTP1, transketolase and a cytochrome P450 (Klink et al. 2007b). Related experiments showing genes that were induced specifically during a susceptible reaction at 8 dpi included 4-coumarate CoA ligase family protein, peroxidase, expansin, matrix metalloproteinase, matrixin family protein and a lipid transfer protein (LTP) (Klink et al. 2007b). All of these proteins were suppressed in their activity during the resistant reaction. However, the problem with the vast amounts of data that was being generated at the time was in obtaining a meaningful annotation that would provide an understanding of the global events occurring in the sample types.

1.9. Improvements in Annotation

The described experiments resulted in the generation of a massive amount of gene expression data and gene lists for the 38,099 genes fabricated on the Affymetrix® soybean GeneChip. Annotated gene lists for soybean genes are very useful because no information is lost from the analysis (Table 1). However, the gene lists do not provide higher order knowledge of how the many genes are functioning during a process under study. It is possible that various metabolic pathways that pertain to a specific process could be identified if the data could be organized into a higher order structure. Since the aforementioned work was done in soybean, often considered a non-model organism, it was difficult to translate the information into gene pathway analyses applications in a manner that would reveal how the gene expression is orchestrated during the process under study. However, an investigation that had been done in *A. thaliana* infected with *Pseudomonas syringae* pv. tomato did show how useful the higher order gene expression knowledge could be in allowing for a visualization of the switch in metabolism from housekeeping to pathogen defense during infection

(Scheideler et al. 2001). The development and presentation of gene pathway information, a procedure that merged the Kyoto Encyclopedia of Genes and Genomes (KEGG) framework (http://www.genome.jp/kegg/catalog/org_list.html) (Goto et al. 1997) with the gene expression data was accomplished through the development of a computer application called Pathway Analysis and Integrated Coloring of Experiments (PAICE) (Paice_v2_90.jar) (http://sourceforge.net/projects/paice/) (Hosseini et al. unpublished; Klink et al. 2011a). This allowed for obtaining higher order cell fate mapping studies to be performed (Klink et al. 2011a; Matsye et al. 2011). Moreover, the sequencing of the soybean genome (Schmutz et al. 2010) made transcriptional mapping experiments that relate to resistance loci possible (Matsye et al. 2011).

1.10. Genomics of The Syncytium

While strides were being made in obtaining a deep analysis of the physiological processes occurring in whole infected roots, the greater challenge would be to identify gene expression that occurred within the syncytium because it would require either drawing the cytoplasm out of the syncytium or a way to physically isolate the cells. The original studies that attempted to determine gene expression in nematode nurse cells was done by Bird et al. (1994) and Wilson et al. (1994). The hypothesis was that by extracting the cytoplasm of the cells that are specifically undergoing the parasitism, it would be possible to determine the gene expression that pertains specifically to parasitism. However, it is noted that gene expression in the cells surrounding the syncytium probably plays some role in the maintenance and development of the susceptible and resistant reactions. This approach to isolate the cytoplasm (Bird et al. 1994; Wilson et al. 1994) would be more challenging for syncytia because it is virtually impossible to determine what cells are infected by SCN. Therefore, instead of collecting the cytoplasm, the collection of the cells would have to occur and it would have to be done through their physical isolation.

The physical isolation of syncytia undergoing a susceptible reaction to the SCN was first described by Klink et al. (2005). The study collected syncytia by a procedure called laser microdissection (Isenberg et al. 1976; Meier Ruge et al. 1976; Emmert-Buck et al. 1996) (Figure 3). The experiments obtained RNA of suitable quality for making cDNA libraries, cloning and sequencing full length genes greater than 1,000 base pairs, making probes for *in situ* hybridization and quantitative PCR (qPCR) and immunocytochemistry which would allow for the visualization of gene expression inside of the infected cells (Klink et al. 2005). These results made it possible to study gene expression occurring within the syncytium at the genome-wide level.

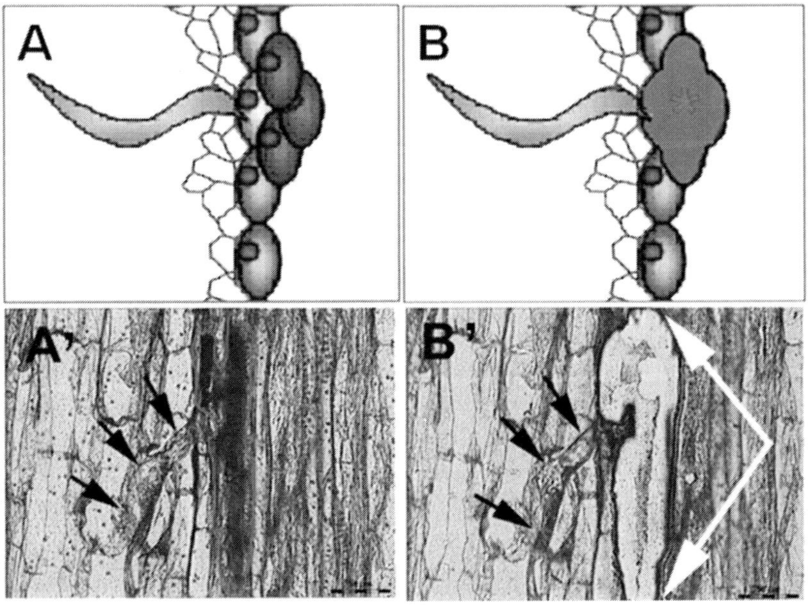

Figure 3. Laser microdissection (LM) of nematode feeding sites. A, cartoon of a nematode (gray) parasitizing a pericycle cell (yellow) that previously was uninfected (green). The parasitism process is resulting in the incorporation of neighboring cells by dissolving their cell walls, forming a syncytium. B, the syncytium (asterisk) was collected after LM. A', an actual root used for LM. The black arrows point to a nematode infecting a root cell (red outline). B', white arrows point to the feeding cell that was collected by LM.

Genomics approaches to syncytium biology resulted in a series of investigations that have focused in on gene expression that occurs during a susceptible reaction in the syncytium (Klink et al. 2005, 2007a, 2009b, 2010b, c, 2011a; Ithal et al. 2007; Matsye et al. 2011, 2012; Kandoth et al. 2011). In these studies, a number of genes were identified. However, to understand gene expression as it specifically pertains to defense, it would be required to study the cells undergoing the defense response. The main obstacle in performing studies with this goal in mind was determining whether the cells undergoing the defense response were already dead at the time of cell collection. The prediction is that cells that were dead would have already halted their physiological processes that pertained to defense and also may not provide RNA of suitable quality for microarray studies. However, it was unlikely that the cells progressing through the earlier stages of defense were dead (Figure 1) since the EM studies revealed very specific progression of cellular architecture during the defense response, suggesting that the cells had to be alive to progress through this developmental process (Endo, 1965; Kim et al. 1987; Endo, 1991). The initial collection of syncytia undergoing the developmental process that leads to their eventual collapse and death were then performed (Klink et al. 2007a). These experiments demonstrated that the cells would be a suitable source for RNA collection and genomics-based analyses. The first set of experiments to use laser microdissected cells undergoing an incompatible reaction for genomics studies determined that the expression of lipoxygenase (LOX), arabinogalactan-protein (AGP18), annexin, a thioesterase family protein heat shock protein (HSP) 70 and superoxidase dismutase (SOD) (Klink et al. 2007a). Many of the genes have very well known roles in plant defense. Subsequent experiments examined more time points occurring during the defense response, spanning phase 1 and phase 2 (Klink et al. 2009b, 2010b, c). The experiments identified a number a genes that were very highly expressed during the resistant reaction, specifically within the syncytium. In contrast, a number of genes were very highly suppressed (>1,000 fold) during the resistant reaction (Klink et al. 2009b, 2010b, c). The experiments were repeated later by Kandoth et al. (2011) in the *G. max*$_{[PI\ 209322]}$ genetic background that either has or lacks the *rhg1* resistance locus. During this time, studies were also performed that examined and compared multiple forms of the resistant reaction that were found in the *G. max*$_{[Peking/PI\ 548402]}$ and *G. max*$_{[PI\ 88788]}$ genotypes (Klink et al. 2009b, 2010b, c, 2011a; Matsye et al. 2011, 2012). These studies were important because the *G. max*$_{[Peking/PI\ 548402]}$ and *G. max*$_{[PI\ 88788]}$ genotypes are well known to undergo different forms of the resistant reaction at the cellular level (Kim et al. 1987; Endo, 1991; Mahalingham and Skorupska, 1996). The *G. max*$_{[Peking/PI\ 548402]}$ and *G. max*$_{[PI\ 88788]}$ PIs are also important genotypes to obtain knowledge from because they are the source of >97% of the resistance germplasm used in commercial breeding programs (Concibido et al. 2004). In some of the earlier studies (Klink et al. 2009b), a number of genes were identified that were induced preferentially in their expression during the resistant reaction. The genes included lipoxygenase, S-adenosylmethionine synthetase, a dnaK domain-containing protein, GRF2 GENERAL REGULATORY FACTOR 2, ACT7 (actin 7), major latex protein-related protein, xyloglucan endotransglucosylase/hydrolase protein 26, cytochrome P450 monooxygenase CYP93D1, pyruvate dehydrogenase E1 beta subunit isoform 2, nitrate transporter (NTP2), endo-1,4-beta-glucancase that were all expressed preferentially between 100 to 383-fold higher in syncytia undergoing the defense response as compared to syncytia undergoing the early stages of a susceptible reaction (Klink et al. 2009b). Additional experiments aided by Illumina® deep sequencing technology which is a sequence by synthesis procedure much like quantitative PCR, but for every gene in the genome simultaneously, identified genes that were expressed only in syncytia undergoing the defense response (Matsye et al. 2011). Some of the genes were expressed at all times during the defense response. Importantly, the Illumina® deep sequencing technology revealed that some of the transcripts that are genes known to be important in defense responses represented between 1 and 17% of the sequenced tags from RNA isolated from the syncytia undergoing the defense response (Table 2) (Matsye et al. 2011). The knowledge gained from these gene expression experiments was then used to select candidate genes whose function during infection could be tested. The cross-comparison of data obtained from the Illumina® sequencing platform with the Affymetrix® microarrays determined the genes within the *rhg1* locus that were expressed specifically during defense (Matsye et al. 2011). Experimentation of these genes in functional tests determined that some of these genes play a role in defense to SCN (Matsye et al. 2012). It was shown that one gene, an α-SNAP allele isolated from the resistant *G. max*$_{\text{[Peking/PI 548402]}}$, provided resistance when genetically engineered into the susceptible *G. max*$_{\text{[Williams 82/PI 518671]}}$ (Matsye et al. 2012). Gene expression and functional studies will be further expanded on in a subsequent section.

Table 2. Group 1 SCN genes expressed during parasitism and used in RNAi studies (Klink et al. 2009)

category	total probe sets	percent
Affymetrix® soybean GeneChip® probe sets (PS)	38,099	
PS with matches to *Arabidopsis thaliana* accessions	23,583	62%
PS with enzyme commission (E.C.) numbers	9,717	29%
PS matching both *A. thaliana* accessions and having E.C. numbers	4,156	11%
PS with chromosomal coordinates	31,188	82%

1.10.1. Soybean resistance clusters

The major SCN resistance trait, *rhg1*, was first identified by Caldwell et al. (1960). In and around the same time, four other major loci, the recessive *rhg2*, *rhg3* (Caldwell et al. 1960) and the dominant *Rhg4* (Matson and Williams, 1965) and *Rhg5* (Rao Arelli 1994) have been identified. In all, there are approximately 61 QTLs that associate with resistance to SCN (Kim et al. 2010). Many of the details of the numerous mapping studies can be found in a review by Concibido et al. (2004). Of all of the loci that associate with resistance to SCN, the best studied is *rhg1*. It is a major resistance locus and has been fine mapped to a region defined in a span of approximately 611,794 nucleotides between the molecular markers ss107914244 and Satt038 on chromosome 18 (Concibido et al. 1994; Mudge et al, 1997; Cregan et al. 1999a; Hyten et al. 2010). It is important to note that the *rhg1* loci found in the different genotypes that exhibit resistance are not the same (Cregan et al. 1999b; Brucker et al. 2005; Matsye et al. 2012). For example, due to the variation in how soybean responds to infection by the SCN, the *rhg1* resistance allele in *G. max*$_{\text{[PI 88788]}}$ is designated *rhg1-b* (Kim et al. 2010). Work by Kim et al. (2010) has resulted in the fine-mapping of the *rhg1-b* locus to within a region of approximately 67 kb. This development was important for the SCN research field because the locus contains approximately 9 genes. However, work in understanding the biological nature of the genes within the locus was not the focus of the Kim et al. (2010) study because the investigation was a genetic mapping analysis. Other resistance loci that are not as well mapped, such as *Rhg4* (Matson and Williams 1965), while providing resistance, account for about a 30% of the resistance of soybean to SCN. In addition to this feature, it functions only against certain populations of SCN.

1.11. Gene Expression Found During Defense at The *Rhg1* Locus

Knowing how and when genes are expressed in syncytia specifically during defense would likely provide knowledge of the genes that regulate or contribute to the process. Matsye et al. (2012) demonstrated in complimentary studies, that an amino acid transporter (AAT) (Glyma18g02580) and an α soluble NSF attachment protein (α-SNAP) (Glyma18g02590) found in the *rhg1* locus, undergo expression specifically in syncytia undergoing defense in both the *G. max*$_{\text{[Peking/PI 548402]}}$ and *G. max*$_{\text{[PI 88788]}}$ genotypes (Matsye et al. 2011). What was notable about the analysis was that AAT and α-SNAP were shown to be expressed throughout the defense response in experiments that sampled time points at 3, 6 and 9 days post infection (dpi), spanning phase 1 and 2 (Matsye et al. 2011). The AAT and α-SNAP genes did not appear to be expressed in syncytia undergoing the susceptible reaction. This difference in expression that was occurring between the resistant and susceptible reaction made it possible that the genes could be involved in the defense response. However, this would only be determined in functional studies that tested how the gene acted during infection (Matsye et al. 2012).

1.12. Genetic Engineering as a Solution for Scn

A number of approaches like conventional breeding programs have been shown for decades to generate resistance to SCN (Brim and Ross, 1966). The resistant cultivars have been shown to result in savings of hundreds of millions of dollars (Bradley and Duffy, 1982). One drawback of conventional breeding programs is that along with the resistance genes that are bred in, a number of genes are also introgressed that could have undesirable characteristics. This is especially a problem when is desirable traits are tightly linked to the undesirable traits. To circumvent this problem, it is possible to genetically engineer in genes of interest. A number of strategies that have been described in this chapter have shown promise in disrupting the soybean-SCN interaction. As noted earlier, RNAi of nematode parasitism genes has been shown in the *Arabidopsis thaliana-Meloidogyne* sp. system to perturb giant cell formation (Huang et al. 2006). This was also shown to work in the soybean-SCN pathosystem (Steeves et al. 2006). Later work that identified highly conserved SCN genes that were expressed during parasitism could be knocked down by RNAi and suppress infection (Klink et al. 2009a; Li et al. 2010). Due to the duplicated nature of the soybean genome (Doyle et al. 1999; Schmutz et al. 2010), RNAi studies of soybean genes may be met with complications and may require methodologies that can knock down entire gene families (Alvarez et al. 2006).

Another procedure to modulate gene expression in soybean to engineer resistance involves the engineering of soybean genes as overexpression constructs (Matsye et al. 2012). To do the studies, genes that are highly expressed during a resistant reaction, identified in accessions of little agronomic value can be expressed to high levels in a soybean genotype that is normally susceptible, but of great economic value. The hypothesis is that if the gene is important in the defense response, the overexpression of that gene in a genotype that is normally susceptible would result in suppressed nematode infection. Such a result was obtained by Matsye et al. (2012) with the overexpression of a naturally occurring truncated allele of an α-SNAP gene. When the α-SNAP gene that was identified in the *G. max*$_{\text{[Peking/PI 548402]}}$ accession was overexpressed in the normally susceptible *G. max*$_{\text{[Williams 82/PI 518671]}}$ genotype, nematode infection was suppressed (Figure 4). The experiments demonstrated the efficacy of the approach, opening up the possibility for large scale reverse genetic screens since the plasmid vectors used to engineer the genes into soybean through the hairy root procedure (Tepfer et al. 1984) was designed with an enhanced green fluorescent reporter (eGFP) (Collins et al. 2005; Klink et al. 2008) was designed using the Gateway® technology for both RNAi and overexpression studies (Klink et al. 2009a; Matsye et al. 2012).

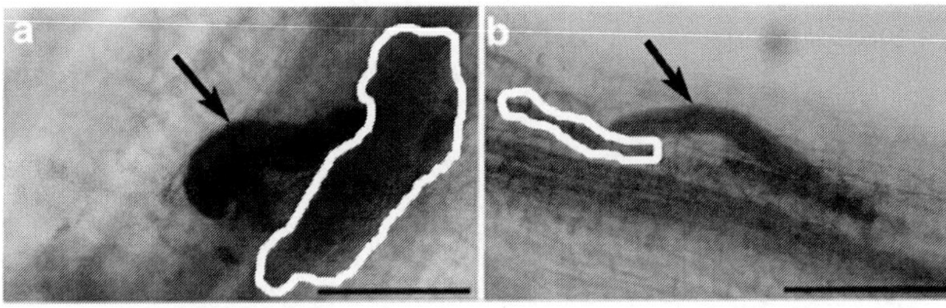

Figure 4. An overexpressed gene affects nematode development. A, a nematode, stained with acid fuchsin for visualization, developing in an experimental control plant. The boundary of the nematode feeding site is encircled in blue. B, a nematode failing to develop in a plant overexpressing a gene identified in the gene expression studies of the syncytium. The boundary of the nematode feeding site is encircled in blue.

2. CONCLUSION

The soybean-SCN pathosystem has been under study for over 60 years. Through a massive amount of basic studies involving agricultural production practices, genomics and genetic engineering, solutions to the chronic and global SCN problem are emerging. The difficulty of studying the system has been met with many improvements in technology that are allowing for basic features of the pathosystem to be exploited so that agricultural practices and economic returns are improved. The basic knowledge gained in this system can now be applied as a model for understanding other recalcitrant pathogens affecting soybean, to obtain a comprehensive understanding of infection and defense.

3. ACKNOWLEDGEMENTS

VPK is thankful for start-up support provided by Mississippi State University and the Department of Biological Sciences; funds in the forms of a competitive Research Improvement Grant; support from the Mississippi Soybean Promotion Board. GWL is thankful to the department of Department of Biochemistry, Molecular Biology, Entomology and Plant Pathology, Mississippi State University, KSL is thankful to the Department of Entomology and Plant Pathology, Auburn University.

REFERENCES

1. Aist JR. 1976. Papillae and related wound plugs of plant cells. Annu Rev Phytopathol 14: 145–163
2. Alkharouf NW, Klink VP, Chouikha IB, Beard HS, MacDonald MH, Meyer S, Knap HT, Khan R, Matthews BF. 2006. Timecourse microarray analyses reveals global changes in gene expression of susceptible *Glycine max* (soybean) roots during infection by *Heterodera glycines* (soybean cyst nematode). Planta 224: 838-852
3. Alkharouf N, Klink VP, Matthews BF. 2007. Identification of *Heterodera glycines* (soybean cyst nematode [SCN]) DNA sequences with high similarity to those of *Caenorhabditis elegans* having lethal mutant or RNAi phenotypes. *Exp Parasitol* 115: 247-258
4. Alvarez JP, Pekker I, Goldshmidt A, Blum E, Amsellem Z, Eshed Y. 2006. Endogenous and synthetic microRNAs stimulate simultaneous, efficient, and localized regulation of multiple targets in diverse species. Plant Cell 18: 1134-1151
5. An Q, Ehlers K, Kogel KH, van Bel AJ, Hückelhoven R. 2006a. Multivesicular compartments proliferate in susceptible and resistant MLA12-barley leaves in response to infection by the biotrophic powdery mildew fungus. New Phytol 172: 563-57
6. An Q, Hückelhoven R, Kogel KH, van Bel AJ. 2006b. Multivesicular bodies participate in a cell wall-associated defence response in barley leaves attacked by the pathogenic powdery mildew fungus. Cell Microbiol 8: 1009-1019
7. Assaad FF, Qiu JL, Youngs H, Ehrhardt D, Zimmerli L, Kalde M, Wanner G, Peck SC, Edwards H, Ramonell K, Somerville CR, Thordal-Christensen H. 2004. The PEN1 syntaxin defines a novel cellular compartment upon fungal attack and is required for the timely assembly of papillae. Mol Biol Cell 15: 5118-5129
8. Atkinson HJ, Harris PD. 1989. Changes in nematode antigens recognized by monoclonal antibodies during early infections of soya bean with cyst nematode Heterodera glycines. Parasitology 98: 479-487
9. Bakhetia M, Urwin PE, Atkinson HJ. 2007. QPCR analysis and RNAi define pharyngeal gland cell-expressed genes of Heterodera glycines required for initial interactions with the host. *Mol Plant Microbe Interact* 20: 306-312
10. Bakhetia M, Urwin PE, Atkinson HJ. 2008. Characterisation by RNAi of pioneer genes expressed in the dorsal pharyngeal gland cell of Heterodera glycines and the effects of combinatorial RNAi. *Int J Parasitol* 38: 1589-1597
11. Barker KR, Koenning SR, Huber SC, Huang JS. 1993. Physiological and structural responses of plants to nematode parasitism with Glycine max-Heterodera glycines as a model system. Pp. 761-771 in DR Buxon R Shibles RA Forsberg BL Blad KH Asay GM Paulsen and RF Wilson, Eds. International Crop Science I: Madison, WI: Crop Science Society of America
12. Bekal S, Niblack TL, Lambert KN. 2003. A chorismate mutase from the soybean cyst nematode Heterodera glycines shows polymorphisms that correlate with virulence. Molecular Plant-Microbe Interactions 16: 439-446
13. Bekal S, Craig JP, Hudson ME, Niblack TL, Domier LL, Lambert KN. 2008. Genomic DNA sequence comparison between two inbred soybean cyst nematode biotypes facilitated by massively parallel 454 micro-bead sequencing. Mol Genet Genomics 279:535-543
14. Bennett MK, Calakos N, Scheller RH. 1992. Syntaxin: a synaptic protein implicated in docking of synaptic vesicles at presynaptic active zones. Science 257: 255-259
15. Bernard RL, Juvik GA, Nelson RL. 1987. USDA Soybean Germplasm Collection Inventory, Vol. 2. INTSOY Series Number 31. IL: International Agriculture Publications, University of Illinois
16. Bird D McK, Wilson MA. 1994. DNA sequence and expression analysis of root-knot nematode-elicited giant cell transcripts. MPMI 7:419-424

17. Bradley EB, Duffy M. 1982. The value of plant resistance to soybean cyst nematode: a case study of Forrest soybeans. National Resource Economics Staff Report No. AGES820929, USDA. Washington DC: U.S. Government Printing Office
18. Brucker E, Carlson S, Wright E, Niblack T, Diers B. 2005. Rhg1 alleles from soybean PI 437654 and PI 88788 respond differently to isolates of Heterodera glycines in the greenhouse. Theor Appl Genet 111: 44-49
19. C. elegans Sequencing Consortium. 1998. Genome sequence of the nematode C. elegans: a platform for investigating biology. Science 282: 2012-2018
20. Caldwell BE, Brim CA, Ross JP. 1960. Inheritance of resistance of soybeans to the soybean cyst nematode, *Heterodera glycines*. Agron J 52: 635-636
21. Caudy AA, Ketting RF, *Hammond* SM, Denli AM, Bathoorn AM, Tops BB, Silva JM, Myers MM, Hannon GJ, Plasterk RH. 2003. A micrococcal nuclease homologue in *RNAi* effector complexes. Nature 425: 411-414
22. Chapin III FS, Shaver GR. 1985. Individualistic growth response of tundra plant species to environmental manipulations in the field. Ecology 66: 564–576
23. Chen W, Chao G, Singh KB. 1996. The promoter of a H2O2-inducible, Arabidopsis glutathione S-transferase gene contains closely linked OBF- and OBP1-binding sites. Plant J 10: 955-966
24. Chen SY, Dickson DW. 1996. Pathogenicity of fungi to eggs of *Heterodera glycines*. Journal of Nematology 28: 148-158
25. Chen X, MacDonald MH. Khan F, Garrett WM, Matthews BF, Natarajan SS. 2011. Two-dimentional proteome reference maps for the soybean cyst nematode Heterodera glycines. Proteomics 11: 4742-4746
26. Chitwood DG, Lusby WR: Metabolism of plant sterols by nematodes. *Lipids* 1990, 26: 619-627
27. Clary DO, Griff IC, Rothman JE. 1990. SNAPs, a family of NSF attachment proteins involved in intracellular membrane fusion in animals and yeast. Cell 61: 709–721
28. Colgrove AL, Niblack TL. 2008. Correlation of female indices from virulence assays on inbred lines and field populations of Heterodera glycines. J Nematol 40: 39–45
29. Collier R, Fuchs B, Walter N, Kevin Lutke W, Taylor CG. 2005. Ex vitro composite plants: an inexpensive, rapid method for root biology. Plant J 43: 449-457
30. Collins NC, Thordal-Christensen H, Lipka V, Bau S, Kombrink E, Qiu JL, Hückelhoven R, Stein M, Freialdenhoven A, Somerville SC, Schulze-Lefert P. 2003. SNARE-protein mediated disease resistance at the plant cell wall. Nature 425: 973–977
31. Concibido VC, Denny RL, Boutin SR, Hautea R, Orf JH, Young ND. 1994. DNA Marker analysis of loci underlying resistance to soybean cyst nematode (*Heterodera glycines* Ichinohe). Crop Sci 34: 240–246
32. Concibido VC, Diers BW, Arelli PR. 2004. A decade of QTL mapping for cyst nematode resistance in soybean. Crop Sci. 44: 1121-1131
33. Creech JE, Johnson WG. 2006. Survey of broadleaf winter weeds in Indiana production fields infested with soybean cyst nematode (*Heteroderaglycines*). Weed Technol. 20: 1066-1075
34. Cregan PB, Mudge J, Fickus EW, Danesh D, Denny R, Young ND. 1999a. Two simple sequence repeat markers to select for soybean cyst nematode resistance conditioned by the rhg1 locus. Theor Appl Genet 99: 811–818
35. Cregan PB, Mudge J, Fickus EW, Marek LF, Danesh D, Denny R, Shoemaker RC, Matthews BF, Jarvik T, Young ND. 1999b. Targeted isolation of simple sequence repeat markers through the use of bacterial artificial chromosomes. Theor Appl Genet 98:919–928
36. Day TA, Ruhland CT, Grobe CW, Xiong F. 1999. Growth and reproduction of Antarctic vascular plants in response to warming and UV radiation reductions in the field. Oecologia 119: 24–35
37. De Boer JM, Yan Y, Wang X, Smant G, ussey RS, Davis EL. 1999. Developmentla expression of secretory beta 1, 4-endonucleases in the subventral esophageal glands of Heterodera glycines. Molecular Plant-Microbe Interactions 12: 663-669

38. De Boer JM, Mc Dermott JP, Davis EL; Husses RS, Popeijus H, Smant G, Baum TJ. 2002. Cloning of a putative pectate lyase gene expressed in the subventral esophageal glands of Heterodera glycines. J. Nematology 34: 9-11
39. Doyle JJ, Doyle JL, Brown AH. 1999. Origins, colonization, and lineage recombination in a widespread perennial soybean polyploid complex. Proc Natl Acad Sci 96: 10741-10745
40. Edens RM, Anand SC, Bolla RI. 1995. Enzymes of the phenylpropanoid pathway in soybean infected with Meloidogyne incognita or Heterodera glycines. J Nematology. 27: 292-303
41. Elling AA, Mitreva M, Gai X, Martin J, Recknor J, Davis EL, Hussey RS, Nettleton D,McCarter JP, Baum TJ. 2009. Sequence mining and transcript profiling to explore cyst nematode parasitism. BMC Genomics 10: 58
42. Emmert-Buck MR, Bonner RF, Smith PD, Chuaqui RF, Zhuang Z, Goldstein SR, Weiss RA, Liotta LA. 1996. Laser capture microdissection. Science 274: 998–1001
43. Endo BY. 1964. Penetration and development of *Heterodera glycines* in soybean roots and related and related anatomical changes. Phytopathology 54: 79–88
44. Endo BY. 1965. Histological responses of resistant and susceptible soybean varieties, and backcross progeny to entry development of *Heterodera glycines*. Phytopathology 55: 375–381
45. Endo BY. 1991. Ultrastructure of initial responses of susceptible and resistant soybean roots to infection by *Heterodera glycines*. Revue Nématol 14: 73-84
46. Endo BY, Veech JA. 1970. Morphology and histochemistry of soybean roots infected with *Heterodera glycines*. Phytopathology 60: 1493–1498
47. Epps JM, Chambers AY. 1958. New host records for Heterodera glycines including one in the Labiate. Plant Disease Reporter 42: 194
48. Epps JM, Hartwig EE. 1972. Reaction of soybean varieties and strains to soybean cyst nematode. J Nematol 4: 222
49. Fire A, Xu S, Montgomery MK, Kostas SA, Driver SE, Mello CC. 1998. Potent and specific genetic interference by doublestranded RNA in Caenohrabditis elegans. Nature 391:806–811
50. Francl LJ, Dropkin VH. 1986. Heterodera glycines population dynamics and relation of initial population density tp soybean yield. Plant Disease: 70: 791-795
51. Gao B, Allen R, Maier T, Davis EL, Baum TJ, Hussey RS. 2001. Identification of putative parasitism genes expressed in the esophageal gland cells of the soybean cyst nematode Heterodera glycines. *Mol Plant Microbe Interact* 2001, 14:1247-1254.
52. Gao B, Allen R, Maier T, Davis EL, Baum TJ, Hussey RS. 2003. The parasitome of the phytonematode Heterodera glycines. Mol Plant Microbe Interact 16: 720-726
53. Garner WW, Allard HA. 1930. Photoperiodic response of soybeans in relation to temperature and other environmental factors. J. Agric. Res. 41:719-735
54. Gipson I, Kim KS, Riggs RD. 1971. An ultrastructural study of syncytium development in soybean roots infected with *Heterodera glycines*. Phytopathology 61: 347-353
55. Golden AM, Epps JM, Riggs RD, Duclos LA, Fox JA, Bernard RL. 1970. Terminology and identity of infraspecific forms of the soybean cyst nematode (Heterodera glycines). Plant Dis Rep 54: 544–546
56. Goto S, Bono H, Ogata H, Fujibuchi W, Nishioka T, Sato K, Kanehisa M. 1997. Organizing and computing metabolic pathway data in terms of binary relations. Pac Symp Biocomput. 1997: 175-186
57. Hammond SM, Boettcher S, Caudy AA, Kobayashi R, Hannon GJ. 2001. Argonaute2, a link between genetic and biochemical analyses of RNAi. Science. 293: 1146-1150
58. Hardham AR, Takemoto D, White RG. 2008. Rapid and dynamic subcellular reorganization following mechanical stimulation of Arabidopsis epidermal cells mimics responses to fungal and oomycete attack. BMC Plant Biol 8: 63
59. Haseloff J, Siemering KR, Prasher DC, Hodge S. 1997. Removal of a cryptic intron and subcellular localization of green fluorescent protein are required to mark transgenic Arabidopsis plants brightly. Proc Natl Acad Sci U S A 94: 2122-2127

60. Hermsmeier D, Mazarei M, Baum TJ. 1998. Differential display analysis of the early compatible interaction between soybean and the soybean cyst nematode. Molecular Plant-Microbe Interactions 11: 1258-1263
61. Hofius D, Schultz-Larsen T, Joensen J, Tsitsigiannis DI, Petersen NH, Mattsson O, Jørgensen LB, Jones JD, Mundy J, Petersen M. 2009. Autophagic components contribute to hypersensitive cell death in Arabidopsis. Cell 137: 773-783
62. Huang G, Allen R, Davis EL, Baum TJ, Hussey RS. 2006. Engineering broad root-knot resistance in transgenic plants by RNAi silencing of a conserved and essential root-knot nematode parasitism gene. Proc Natl Acad Sci USA 103: 14302-14306
63. Hymowitz t. 1970. On the domestication of soybean. Economic Botany 24: 408-421
64. Hyten DL, Choi IY, Song Q, Shoemaker RC, Nelson RI, Costa JM, Specht JE, Cregan PB. 2010. Highly variable patterns of linkage disequilibrium in multiple soybean populations. Genetics 175: 1937-1944
65. Ichinohe M. 1952. On the soybean nematode, Heterodera glycines n. sp., from Japan. Magazine of Applied Zoology 17: 1-4
66. Ichinohe M. 1961. Studies on the soybean cyst nematode, *Heterodera glycines* Hakkaido National Experiment Station Report no. 56
67. Inagaki H, Tsutsumi M. 1971. Survival of the soybean cyst nematode, Heterodera glycines Ichinohe (Tylenchida: Heteroderidae) under certain storage conditions. Appl Entomol Zool (Jpn) 8: 53–63
68. Inoue A, Obata K, Akagawa K. 1992. Cloning and sequence analysis of cDNA for a neuronal cell membrane antigen, HPC-1. J Biol Chem 267: 10613-10619
69. Isenberg G, Bielser W, Meier-Ruge W, Remy E. 1976. Cell surgery by laser micro-dissection: a preparative method. J. Microsc 107: 19–24
70. Ithal N, Recknor J, Nettleston D, Hearne L, Maier T, Baum TJ, Mitchum MG. 2007. Developmental transcript profiling of cyst nematode feeding cells in soybean roots. Molecular Plant-Microbe Interactions 20: 293-305
71. Jones MGK. 1981. The development and function of plant cells modified by endoparasitic nematodes. Pages 255-279 in: Plant Parasitic Nematodes, Vol. III. B. M. Zuckerman and R. A. Rohde, eds. Academic Press, New York, U.S.A.
72. Jones MGK, Northcote DH. 1972. Nematode-induced syncytium-a multinucleate transfer cell. J Cell Sci 10: 789–809
73. Kandoth PK, Ithal N, Recknor J, Maier T, Nettleton D, Baum TJ, Mitchum MG. 2011. The Soybean Rhg1 locus for resistance to the soybean cyst nematode Heterodera glycines regulates the expression of a large number of stress- and defense-related genes in degenerating feeding cells. Plant Physiol 155: 1960-1975
74. Kalde M, Nühse TS, Findlay K, Peck SC. 2007. The syntaxin SYP132 contributes to plant resistance against bacteria and secretion of pathogenesis-related protein 1. Proc Natl Acad Sci U S A 104. 11050-11055
75. Kamath RS, Fraser AG, Dong Y, Poulin G, Durbin R, Gotta M, Kanapin A, Le Bot N, Moreno S, Sohrmann M, Welchman DP, Zipperlen P, Ahringer J. 2003. Systematic functional analysis of the *Caenorhabditis elegans* genome using RNAi. Nature 421: 231–237
76. Kim DG, Riggs RD. 1991. Characteristics and efficacy of a sterile hyphomycete (ARF18), a new biocontrol agent for *Heterodera glycines* and other nematodes. Journal of Nematology 23: 275–282
77. Kim DG, Riggs RD. 1995. Efficacy of the nematophagous fungus ARF18 in alginate-clay pellet formulation against *Heterodera glycines.* Journal of Nematology 23:275-282
78. Kim KS, Riggs RD. 1992. Cytopathological reactions of resistant soybean plants to nematode invasion. Pp. 157–168 *in* J. A. Wrather and R. D. Riggs, eds. Biology and Management of the Soybean Cyst Nematode. St. Paul: APS Press
79. Kim M, Hyten DL, Bent AF, Diers BW. 2010. Fine mapping of the SCN resistance locus *rhg1-b* from PI 88788. The Plant Genome 3: 81-89
80. Kim YH, Riggs RD, Kim KS. 1987. Structural changes associated with resistance of soybean to *Heterodera glycines.* J Nematol 19: 177–187

81. Klink VP, MacDonald M, Alkharouf N, Matthews BF. 2005. Laser capture microdissection (LCM) and expression analyses of *Glycine max* (soybean) syncytium containing root regions formed by the plant pathogen *Heterodera glycines* (soybean cyst nematode). Plant Mol Bio 59: 969-983

82. Klink VP, Overall CC, Alkharouf N, MacDonald MH, Matthews BF. 2007a. Laser capture microdissection (LCM) and comparative microarray expression analysis of syncytial cells isolated from incompatible and compatible soybean roots infected by soybean cyst nematode (*Heterodera glycines*). Planta 226: 1389-1409

83. Klink VP, Overall CC, Alkharouf N, MacDonald MH, Matthews BF. 2007b. A comparative microarray analysis of an incompatible and compatible disease response by soybean (Glycine max) to soybean cyst nematode (Heterodera glycines) infection. Planta 226: 1423-1447

84. Klink VP, MacDonald MH, Martins VE, Park S-C, Kim K-H, Baek S-H, Matthews BF. 2008. MiniMax, a new diminutive Glycine max variety, with a rapid life cycle, embryogenic potential and transformation capabilities. Plant Cell, Tissue and Organ Culture 92: 183-195

85. Klink VP, Kim K-H, Martins VE, MacDonald MH, Beard HS, Alkharouf NW, Lee S-K, Park S-C, Matthews BF. 2009a. A correlation between host-mediated expression of parasite genes as tandem inverted repeats and abrogation of the formation of female Heterodera glycines cysts during infection of Glycine max. Planta 230: 53-71

86. Klink VP, Hosseini P, Matsye P, Alkharouf N, Matthews BF. 2009b. A gene expression analysis of syncytia laser microdissected from the roots of the *Glycine max* (soybean) genotype PI 548402 (Peking) undergoing a resistant reaction after infection by *Heterodera glycines* (soybean cyst nematode) Plant Mol Bio: 71: 525-567

87. Klink VP, Hosseini P, MacDonald MH, Alkharouf N, Matthews BF. 2009c. Population-specific gene expression in the plant pathogenic nematode Heterodera glycines exists prior to infection and during the onset of a resistant or susceptible reaction in the roots of the Glycine max genotype Peking. BMC-Genomics 10: 111

88. Klink VP, Matsye PD, Lawrence GW. 2010a. Developmental Genomics of the Resistant Reaction of Soybean to the Soybean Cyst nematode, Pp. 249-270, In Plant Tissue Culture and Applied Biotechnology. Eds. Kumar A., Roy S. Aavishkar Publishers, Distributors, India.

89. Klink VP, Hosseini P, Matsye P, Alkharouf N, Matthews BF. 2010b. Syncytium gene expression in *Glycine max*[PI 88788] roots undergoing a resistant reaction to the parasitic nematode *Heterodera glycines* Plant Physiology and Biochemistry 48: 176-193

90. Klink VP, Overall CC, Alkharouf N, MacDonald MH, Matthews BF. 2010c. Microarray detection calls as a means to compare transcripts expressed within syncytial cells isolated from incompatible and compatible soybean (Glycine max) roots infected by the soybean cyst nematode (Heterodera glycines). Journal of Biomedicine and Biotechnology 1-30

91. Klink VP, Hosseini P, Matsye PD, Alkharouf N, Matthews BF. 2011a. Differences in gene expression amplitude overlie a conserved transcriptomic program occurring between the rapid and potent localized resistant reaction at the syncytium of the *Glycine max* genotype Peking (PI 548402) as compared to the prolonged and potent resistant reaction of PI 88788. Plant Mol Bio 75: 141-165

92. Klink VP, Matsye PD, Lawrence GW. 2011b. Cell-specific studies of soybean resistance to its major pathogen, the soybean cyst nematode as revealed by laser capture microdissection, gene pathway analyses and functional studies. in *Soybean - Molecular Aspects of Breeding* pp. 397-428. Ed. Aleksandra Sudaric. Intech Publishers

93. Koenning SR, Schmitt DP, Barker KR. 1993. Effects of cropping systems on population density of Heterodera glycines and soybean yield. Plant Disease 77: 780-786

94. Lai Z, Wang F, Zheng Z, Fan B, Chen Z. 2011. A critical role of autophagy in plant resistance to necrotrophic fungal pathogens. Plant J 66: 953-968
95. Lambert KN, Allen KD, Sussex IM. 1999. Cloning and characterization of an esophageal-gland specific chorismate mutase from the phytopathogenic nematode Meloidogyne javanica. Molecular Plant-Microbe Interactions 12: 328-336
96. Lenz HD, Haller E, Melzer E, Kober K, Wurster K, Stahl M, Bassham DC, Vierstra RD, Parker JE, Bautor J, Molina A, Escudero V, Shindo T, van der Hoorn RA, Gust AA, Nürnberger T. 2011. Autophagy differentially controls plant basal immunity to biotrophic and necrotrophic pathogens. Plant J 66: 818-830
97. Li J, Todd TC, Oakley TR, Lee J and Trick HN. 2010. Host derived suppression of nematode reproductive and fitness genes decreases fecundity of *Heterodera glycines*. Planta 232: 775-785
98. Li Y-H, Qi X-T, Chang R and Qiu L-J. 2011. Evaluation and Utilization of Soybean Germplasm for Resistance to Cyst Nematode in China. in *Soybean - Molecular Aspects of Breeding* pp. 373-396. Ed. Aleksandra Sudaric. Intech Publishers
99. Lipka V, Dittgen J, Bednarek P, Bhat R, Wiermer M, Stein M, Landtag J, Brandt W, Rosahl S, Scheel D, Llorente F, Molina A, Parker J, Somerville S, Schulze-Lefert P. 2005. Pre- and postinvasion defenses both contribute to nonhost resistance in Arabidopsis. Science 310: 1180-1183
100. Liu X, Z, Chen SY. 2001. Screening isolates of *Hirsutella* species for biocontrol of *Heterodera glycines.* Biocontrol Science and Technology 11:151-160
101. Liu XZ, Li JQ, Zhang DS. 1997. History and status of soybean cyst nematode in China. International Journal of Nematology 7: 18-25
102. Ma Y, Wang W, Liu X, Ma F, Wang P, Chang R, Qiu L. 2006. Characteristics of soybean genetic diversity and establishment of applied core collection for Chinese soybean cyst nematode resistance. Journal of Intergrative Biology 48: 722-731
103. Mahalingham R, Skorupska HT. 1996. Cytological expression of early response to infection by Heterodera glycines Ichinohe in resistant PI 437654 soybean. Genome 39: 986–998
104. Mahalingam R, Wang G, Knap HT. 1999. Polygalacturonidase and polygalacturonidase inhibitor protein: gene isolation and transcription in Glycine max-Heterodera glycines interactions. Molecular Plant-Microbe Interactions 12: 490-498
105. Malhotra V, Orci L, Glick BS, Block MR, Rothman JE. 1988. Role of an N-ethylmaleimide-sensitive transport component in promoting fusion of transport vesicles with cisternae of the Golgi stack. Cell 54: 221–227
106. Matson AL, Williams LF. 1965. Evidence of a fourth gene for resistance to the soybean cyst nematode. Crop Sci. 5: 477
107. Matsye PD, Kumar R, Hosseini P, Jones CM, Alkharouf N, Matthews BF, Klink VP. 2011. Mapping cell fate decisions that occur during soybean defense responses. Plant Mol Bio 77: 513-528
108. Matsye PD, Lawrence GW, Youssef RM, Kim K-H, Lawrence KS, Matthews BF, Klink VP. 2012. The expression of a naturally occurring mutant of an alpha soluble NSF attachment protein gene in *Glycine max* (soybean) partially suppresses infection by the plant parasitic nematode *Heterodera glycines.* Plant Molecular Biology (in press)
109. McLean MD, Hoover GJ, Bancroft B, Makhmoudova A, Clark SM, Welacky T, Simmonds DH, Shelp BJ. 2007. Identification of the full-length *Hs1pro-1* coding sequence and preliminary evaluation of soybean cyst nematode resistance in soybean transformed with *Hs1pro-1* cDNA. Canadian Journal of Botany 85: 437-441
110. Meier-Ruge W, Bielser W, Remy E, Hillenkamp F, Nitsche R, Unsold R. 1976. The laser in the Lowry technique for microdissection of freeze-dried tissue slices. Histochem J 8: 387-401
111. Meyer SLF, Huettel RN. 1996. Application of a sex pheromone, pheromone analogs, and *Verticillum lecanii* for management of Heterodera glycines. J. Nematology 28: 36-42

112. Meyer SLF, Meyer RJ. 1996. Greenhouse studies comparing strains of the fungus *Verticillium lecanii* for activity against the nematode *Heterodera glycines*. Fundamentals of Applied Nematology 19: 305–308
113. Morse WJ. 1927. Soybeans: culture and varieties. Farmer's bulletin NO. 1520. Washington, D.C.: U.S. Dept. of Agriculture. 38 pp
114. Mudge J, Cregan PB, Kenworthy JP, Kenworthy WJ, Orf JH, Young ND. 1997. Two microsatellite markers that flank the major soybean cyst nematode resistance locus. Crop Sci 37: 1611-1615
115. Müssig C, Fischer S, Altmann T. 2002. Brassinosteroid-regulated gene expression. Plant Physiol 129: 1241-1251
116. Niblack TL, Chen SY. 2004. Cropping systems and crop management practices. Breeding for resistance and tolerance. Pp. 181-206. *in* D. P. Schmitt, J. A. Wrather, and R. D. Riggs, eds. Biology and management of soybean cyst nematode, 2nd ed. Marceline, MO: Schmitt & Associates of Marceline
117. Niblack TL, Heinz RD, Smith GS, Donald PA (1993) Distribution, density, and diversity of Heterodera glycines in Missouri. J Nematol 25:880–886
118. Niblack TL, Arelli PR, Noel GR, Opperman CH, Orf JH, Schmitt DP, Shannon JG, Tylka GL. 2002. A revised classification scheme for genetically diverse populations of *Heterodera glycines*. J Nematol 34: 279-288
119. Niblack TL, Lambert KN, Tylka GL. 2006. A model plant pathogen from the kingdom animalia: Heterodera glycines, the Soybean Cyst Nematode. Annu Rev Phytopathol 44: 283-303
120. Niblack TL, Riggs RD. 2004. Variation in virulence phenotypes. Breeding for resistance and tolerance. Pp. 57-71. *in* D. P. Schmitt, J. A. Wrather, and R. D. Riggs, eds. Biology and management of soybean cyst nematode, 2nd ed. Marceline, MO: Schmitt & Associates of Marceline
121. Noel GR. 1992. History, distribution and economics. Pp 1-13 in RD Riggs and JA Wrather, editors. Biology and Management of the soybean cyst nematode. St. Paul, MN: APS Press
122. Opperman CH, Bird D McK. 1998. The soybean cyst nematode, *Heterodera glycines:* a genetic model system for the study of plant-parasitic nematodes. Current Opinion in Plant Biology 1: 1342-1346
123. Oyler GA, Higgins GA, Hart RA, Battenberg E, Billingsley M, Bloom FE, Wilson MC. 1989. The identification of a novel synaptosomal-associated protein, SNAP-25, differentially expressed by neuronal subpopulations. J Cell Biol 109: 3039-3052
124. Patel S, Dinesh-Kumar SP. 2008. Arabidopsis ATG6 is required to limit the pathogen-associated cell death response. Autophagy 4: 20-27
125. Piano F, Schetter AJ, Mangone M, Stein L, Kemphues KJ. 2000. RNAi analysis of genes expressed in the ovary of *Caenorhabditis elegans*. Current Biology 10: 1619–1622
126. Pratt PW, Wrather JA. 1998. Soybean disease loss estimates for the southern United States, 1994-1996. Plant Disease 82: 114-116
127. Puthoff DP, Nettleton D, Rodermel SR, Baum TJ. 2003. *Arabidopsis* gene expression changes during cyst nematode parasitism revealed by statistical analyses of microarray expression profiles. Plant J 33: 911–921
128. Rao-Arelli AP. 1994. Inheritance of resistance to *Heterodera glycines* race 3 in soybean accessions. Plant Dis. 78: 898-900.
129. Riggs RD, Hamblen ML. 1962. Soybean-cyst nematode host studies in the Leguminosae. Ark Agric Exp Stn Rep Series 110 Fayetteville AR 17p
130. Riggs RD, Hamblen ML. 1966. Additional weed hosts of *Heterodera glycines*. Plant Dis Rep 50: 15-16
131. Riggs RD, Hamblen ML. 1966. Further studies on the host range of the soybean-cyst nematode. Ark Agric Exp Stn Bulletin 718 Fayetteville AR 19p
132. Riggs RD, Schmitt DP. 1988. Complete characterization of the race scheme for Heterodera glycines. J Nematol 20: 392-395
133. Riggs RD, Schmitt DP. 1991. Optimization of the *Heterodera glycines* race test procedure. J Nematol 23: 149-154

134. Riggs RD, Kim KS, Gipson I. 1973. Ultrastructural changes in Peking soybeans infected with *Heterodera glycines*. Phytopathology 63: 76–84
135. Robinson AF, Inserra RN, Caswell-Chen EP, Vovlas N, Troccoli A. 1997. *Rotylenchulus* species: Identification, distribution, host ranges, and crop plant resistance. Nematropica 27: 127-180
136. Ross JP, Brim CA. 1957. Resistance of soybeans to the soybean cyst nematode as determined by a double-row method. Plant Dis Rep 41: 923–924
137. Ross JP. 1958. Host-Parasite relationship of the soybean cyst nematode in resistant soybean roots. Phytopathology 48: 578-579
138. Ross JP. 1962. Crop rotation effects on the soybean cyst nematode population and soybean yields. Phytopathology 52: 815-818
139. Brim CA, Ross JP. 1966. Registration of Pickett soybeans. Crop Science 6: 305
140. Rosskopf EN, Chellemi DO, Kokalis-Burelle N, Church GT. 2005. Alternatives to Methyl Bromide: A Florida Perspective. American Phytopathological Society. APSnet feature, http://www.apsnet.org/publications/apsnetfeatures/Documents/2005/MethylBromideAlternatives.pdf
141. Sasser JN, Uzzell G, Jr. 1991. Control of the soybean cyst nematode by crop rotation in combination with nematicide . J. Nematology 23: 344-347
142. Scheideler M, Schlaich NL, Fellenberg K, Beissbarth T, Hauser NC, Vingron M, Slusarenko AJ, Hoheisel JD. 2001. Monitoring the switch from housekeeping to pathogen defense metabolism in *Arabidopsis thaliana* using cDNA arrays. J Biol Chem 277: 10555–10561
143. Schmutz J, Cannon SB, Schlueter J, Ma J, Mitros T, Nelson W, Hyten DL, Song Q, Thelen JJ, Cheng J, Xu D, Hellsten U, May GD, Yu Y, Sakurai T, Umezawa T, Bhattacharyya MK, Sandhu D, Valliyodan B, Lindquist E, Peto M, Grant D, Shu S, Goodstein D, Barry K, Futrell-Griggs M, Abernathy B, Du J, Tian Z, Zhu L, Gill N, Joshi T, Libault M, Sethuraman A, Zhang XC, Shinozaki K, Nguyen HT, Wing RA, Cregan P, Specht J, Grimwood J, Rokhsar D, Stacey G, Shoemaker RC, Jackson SA. 2010. Genome sequence of the palaeopolyploid soybean. Nature 463: 178-183
144. Schmelzer E. 2002. Cell polarization, a crucial process in fungal defence. Trends Plant Sci 7: 411-415
145. Schneider SM, Rosskopf EN, Leesch JG, Chellemi DO, Bull CT, Mazzola M. 2003. United States Department of Agriculture-Agricultural Research Service research on alternatives to methyl bromide: pre-plant and post-harvest. Pest Manag Sci. 59: 814-826
146. Shannon JG, Arelli PR, Young LD. 2004. Breeding for resistance and tolerance. Pp. 155-180. *in* D. P. Schmitt, J. A. Wrather, and R. D. Riggs, eds. Biology and management of soybean cyst nematode, 2nd ed. Marceline, MO: Schmitt & Associates of Marceline
147. Smant GA, Stokkermans JPWG, Yan Y, De Boer JM, Baum TJ, Wang X, Hussey RS, Gommers FJ, Henrissat B, Davis EL, Helder J, Schots A, Bakker J. 1998. Endogenous cellulases in animals: isolation of 1,4-endoglucanase genes from two species of plant-parasitic nematodes. PNAS USA 95: 4906-4911
148. Sönnichsen B, Koski LB, Walsh A, Marschall P, Neumann B, Brehm M, Alleaume AM, Artelt J, Bettencourt P, Cassin E, Hewitson M, Holz C, Khan M, Lazik S, Martin C, Nitzsche B, Ruer M, Stamford J, Winzi M, Heinkel R, Röder M, Finell J, Häntsch H, Jones SJ, Jones M, Piano F, Gunsalus KC, Oegema K, Gönczy P, Coulson A, Hyman AA, Echeverri CJ. 2005. Full-genome RNAi profiling of early embryogenesis in Caenorhabditis elegans. Nature 434: 462-429
149. Spears JF. 1957. Review of soybean cyst nematode situation for presentation at public hearing on the need for Federal Domestic Plant Quarantine, July 24, 1957
150. Steeves RM, Todd TC, Essig JS, Trick HN. 2006. Transgenic soybeans expressing siRNAs specific to a major sperm protein gene suppress Heterodera glycines reproduction. Funct Plant Biol 33: 991–999

151. Stein M, Dittgen J, Sánchez-Rodríguez C, Hou BH, Molina A, Schulze-Lefert P, Lipka V, Somerville S. 2006. Arabidopsis PEN3/PDR8, an ATP binding cassette transporter, contributes to nonhost resistance to inappropriate pathogens that enter by direct penetration. Plant Cell 18: 731-746
152. Tao Y, Xie Z, Chen W, Glazebrook J, Chang HS, Han B, Zhu T, Zou G, Katagiri F. 2003. Quantitative nature of Arabidopsis responses during compatible and incompatible interactions with the bacterial pathogen Pseudomonas syringae. Plant Cell 15: 317-330
153. Tenllado F, Martı´nez-Garcı´a B, Vargas M, Dı´az-Ruı´z JR. 2003. Crude extracts of bacterially expressed dsRNA can be used to protect plants against virus infections. BMC Biotechnol 3:3
154. Tepfer D. 1984. Transformation of several species of higher plants by Agrobacterium rhizogenes: sexual transmission of the transformed genotype and phenotype. Cell 37: 959-967
155. Timmons L, Donald LC, Andrew F (2001) Ingestion of bacterially expressed dsRNAs can produce specific and potent genetic interference in Caenorhabditis elegans. Gene 263:103–112
156. Timper P, Riggs RD, Crippen DL. 1999. Parasitism of sedentary stages of *Heterodera glycines* by isolates of a sterile nematophagous fungus. Phytopathology 89: 1193-1199
157. Urwin PE, Lilley CJ, Atkinson HJ. 2002. Ingestion of double-stranded RNA by preparasitic juvenile cyst nematodes leads to RNA interference. *Mol Plant Microbe Interact* 15: 747-752
158. Vaghchhipawala Z, Bassuner R, Clayton K, Lewers K, Shoemaker R,m Mackenzie S. 2001. Modulations in gene expression and mapping of genes associated with cyst nematode infection of soybean. Molecular Plant-Microbe Interactions 14: 42-54
159. Wang D, Stravopodis D, Teglund S, Kitazawa J, Ihle JN. 1996. Naturally occurring dominant negative variants of Stat5. Mol Cell Biol 16: 6141-6148
160. Wang X, Allen R, Ding X, Goellner M, Maier T, DeBoer JM, Baum TJ, Hussey RS, Davis EL. 2001. Signal peptide-selection of cDNA cloned directly from the esophageal gland cells of the soybean cyst nematode Heterodera glycines. Molecular Plant-Microbe Interactions 14: 536-544
161. Wang J, Meyers D, Yan Y, Baum T, Smant G, Hussey R. 2003. The soybean cyst nematode reduces soybean yield without causing obvious symptoms. Plant Disease 87: 623-628
162. Wang X, Myers D, Yan Y, Baum T, Smant G, Hussey R, Davis E. 1999. In planta localization of a 1, 4-endoglucanase secreted by Heterodera glycines. Molecular Plant-Microbe Interactions 12: 64-67
163. Weidman PJ, Melançon P, Block MR, Rothman JE. 1989. Binding of an N-ethylmaleimide-sensitive fusion protein to Golgi membranes requires both a soluble protein(s) and an integral membrane receptor. J Cell Biol 108: 1589-1596
164. Wilson MA, Bird D McK van der Knaap E. 1994. A comprehensive subtractive cDNA cloning approach to identify nematode-induced transcripts in tomato. Phytopathology 84: 299-303
165. Winstead NN, Skotland CB Sasser JN. 1955. Soybean cyst nematodes in North Carolina. Plant Disease Reporter 39: 9-11.
166. Wrather JA, Chambers AY, Fox JA, Moore WF, Sciumbato GL. 1995. Soybean disease loss estimates for the southern United States, 1974-1994. Plant Disease 79: 1076-1079
167. Wrather JA, Anderson TR, Arsyad DM, Tan Y, Ploper LD, Porta-Puglia A, Ram HH, Yorinori JT. 2001a. Soybean disease loss estimates for the top ten soybean producing countries in 1998. Canadian Journal of Plant Pathology 23: 115-121
168. Wrather JA, Steinstra WC, Koenning SR. 2001b. Soybean disease loss estimates for the United States from 1996-1998. Canadian Journal of Plant Pathology 23: 122-131
169. Wrather JA, Koenning SR, Anderson TR. 2003. Effect of diseases on soybean yields in the United States and Ontario (1999-2002). Online. Plant Health Progress doi:

10.1094/PHP-2003-0325-01-RV, http://www.plantmanagement network.org/sub/php/review/2003/soybean/>(19 November 2003)

170. Wrather JA, Koenning SR. 2006. Estimates of disease effects on soybean yields in the United States 2003-2005. J Nematology 38: 173-180

CHAPTER 7

Nematode Parasites and Associates of Ants: Past and Present

George Poinar Jr.

Department of Zoology, Oregon State University, Corvallis, OR 97331, USA

ABSTRACT

Ants can serve as developmental, definitive, intermediate, or carrier hosts of a variety of nematodes. Parasitic ant nematodes include members of the families Mermithidae, Tetradonematidae, Allantonematidae, Seuratidae, Physalopteridae, Steinernematidae, and Heterorhabditidae. Those nematodes that are phoretically associated with ants, internally or externally, are represented by the Rhabditidae, Diplogastridae, and Panagrolaimidae. Fossils of mermithids, tetradonematids, allantonematids, and diplogastrids associated with ants show the evolutionary history of these relationships, some of which date back to the Eocene (40 mya).

1. INTRODUCTION

Nematodes are one of the most abundant groups of animals known. Studies on their evolutionary history suggest that they probably arose in the Precambrian, which explains their wide abundance today in the terrestrial and marine environments. While only some 20,000 have been described, their species diversity has been estimated to be as high as 10 million [1].

One would assume that with their strict housekeeping habits, ants would not tolerate nematodes in or around their nests and would quickly dispose of any nest mates that might have become infected. However nematodes have been able to use some astonishingly sophisticated tactics to successfully parasitize these social insects. The present work covers the systematics, life history, pathology, and records of all described extant and fossil nematodes associated with formicids. This includes representatives of the nematode families Mermithidae, Tetradonematidae, Allantonematidae, Seuratidae, Physalopteridae, Steinernematidae, Heterorhabditidae, Rhabditidae, Diplogastridae, and Panagrolaimidae. Fossil records of mermithids, tetradonematids, allantonematids, and diplogastrids associated with ants reveal the evolutionary history of these associations, some of which date back 40 million years.

2. MERMITHIDAE

The family Mermithidae includes parasites of invertebrates, especially insects. Because of their large size, mermithids are easily detected in ants upon dissection (Figure 1) or as they leave their hosts (Figure 2). Most mermithid species, including those that attack ants, parasitize only a specific host species, genus, or family while others can infect representatives of several insect orders. Mermithids that attack aquatic insects, such as midges (Chironomidae, Ceratopogonidae) and mosquitoes (Culicidae), have a direct life cycle. Direct life cycles occur when, after growth and development is completed in the host, the mermithid emerges, molts to the adult stage, mates, and oviposits in the host's environment. The infective stage mermithid emerges from the egg, actively locates and enters a host, and initiates development in the hemocoel.

Some mermithids have an indirect life cycle, which is more complicated but allows hosts to be parasitized in environments hostile to nematodes. In an indirect cycle, the mermithid emerges from the host, molts, mates, and oviposits in the environment. But instead of emerging from the egg to search for a developmental host, the infective stage remains in the egg, waiting to be ingested by an invertebrate that serves as a paratenic host. When mermithid eggs are ingested by a paratenic host, the hatching infective stage penetrates the gut wall and enters the body cavity. But instead of developing, the mermithid encysts and enters a diapause. The encysted nematode can be carried through the different stages of host metamorphosis, but for its cycle to be completed, the paratenic host must be captured and fed to the brood of the developmental host. At the completion of the mermithids growth phase in the development host (like an ant), the latter is attracted to an aquatic or semiaquatic habitat favorable to the nematode. This is when the mermithid exits, leaving the dying host behind. The developmental host is usually not only larger, but usually in a completely different taxonomic category and environment from the paratenic host. While the developmental host can live in a relatively dry habitat, the paratenic host usually inhabits an aquatic, semiaquatic, or damp biome. Also, both hosts can be widely separated taxonomically and may not even belong to the same phylum.

The first written account of a nematode parasite of ants was made by the Reverend William Gould in his 1747 book An account of English Ants (Table 1) [2].

Figure 1: Mermithid exposed in the gaster of Camponotus sp. from the Sierra Nevada, California.

The "white and long kind of worm, which is often met within their bodies" certainly refers to mermithid nematodes. For a number of years, mermithids were listed under "Filaria," "Gordius," or "Mermis," and that is why mermithid systematics can be confusing and why early names for Gould's ant mermithid included Gordius formicarum Diesing [3] and Filaria formicarum von Siebold [4].

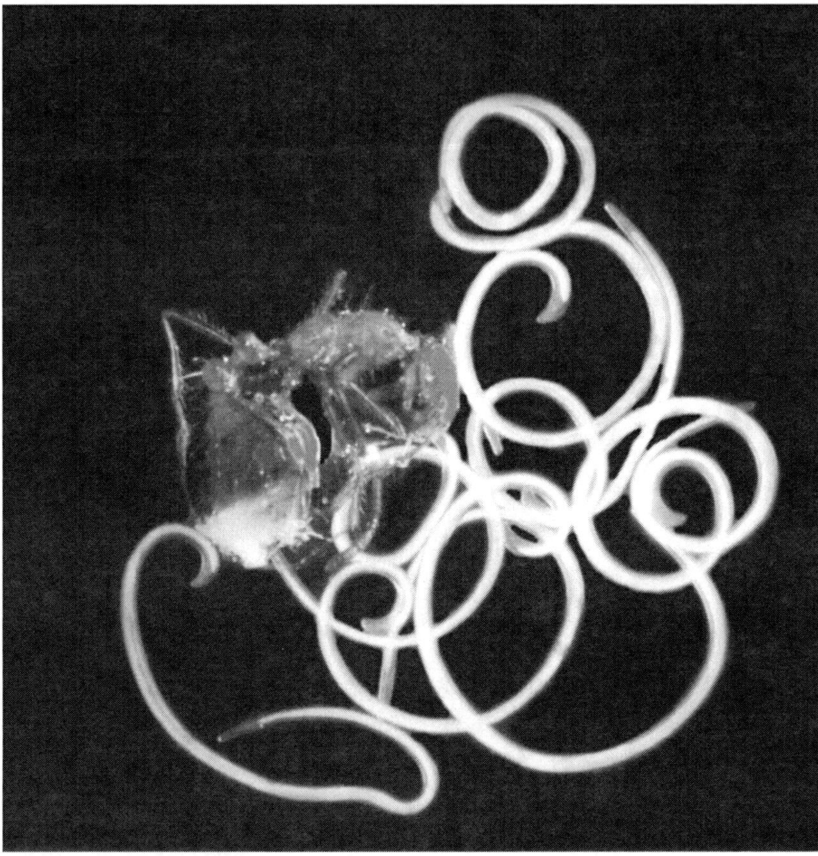

Figure 2: Parasitic juveniles of Allomermis solenopsi removed from the gaster of a fire ant worker. Photo courtesy of S. D. Porter, USDA-ARS.

Table 1: Section from Gould [2] referring to the first reported instance of mermithid parasitism of ants.

Amongſt other Incidents that tend to leſſen and deſtroy Ant-Flies, it is obſervable that abundance of them are demoliſhed by a white and long Kind of Worm, which is often met with in their Bodies. You may frequently take three from the Inſides of the large, but ſeldom more than one from a ſmall Ant-Fly. Theſe Worms lie in a ſpiral Form, and ſome of them may be extended Half an Inch.

The first described ant mermithid was Pheromermis myrmecophila from Lasius spp. [5]. However it was originally described in the genus "Mermis", then assigned to the genus Pheromermis [6], then moved to the genus Allomermis [7] and lastly to the genus Camponotimermis [8]. Its position in the genus Pheromermis was recently confirmed by Kaiser, who showed its similarity with the European ant mermithid, Pheromermis villosa [9]. Over the years, a large number of ant species have been reported parasitized by mermithids. A list of Holarctic parasitized ants was presented by Passera [10] and Neotropical parasitized ants by Poinar et al. [11]. A compilation of all described mermithids from ants is presented in Table 2.

Table 2: Mermithid nematodes described from ants.

Mermithid	Host	Reference
Agamomermis cephaloti	*Cephalotes minutus*	[11]
Agamomermis costaricensis	*Odontomachus hastatus*	[11]
Agamomermis ecitoni	*Eciton burchellii*	[11]
Allomermis solenopsi	*Solenopsis invicta*	[12]
Camponotimermis bifidus	*Camponotus aethiops*	[13]
Comanimermis clujensis	*Formica fusca* *Camponotus aethiops*	[14]
* *Heydenius formicinus*	*Prenolepis henschei*	[15]
* *Heydenius myrmecophila*	*Linepithema* sp.	[11]
Meximermis ectatommi	*Ectatomma ruidum*	[11]
Pheromermis lasiusi	*Lasius niger*	[16]
Pheromermis myrmecophila	*Lasius* spp.	[5]
Pheromermis villosa	*Lasius flavus, L. niger*	[17]

* fossil.

Fossils, such as the postparasitic juvenile of Heydenius formicinus emerging from a male Prenolepis henschei (Figure 3) [15], as well as from a worker ant (Figure 4) in Baltic amber [1] show that ants have been parasitized by mermithids for at least 40 million years and probably much longer. The fossil record of Neotropical mermithid parasites of ants is represented by a parasitic juvenile of Heydenius myrmecophila adjacent to its ant host, Linepithema sp. in 20–30-million-year-old Dominican amber (Figure 5) [11]. It is assumed that the traumatic events of the ant host entering the resin caused the mermithid to emerge prematurely from an opening in the gaster of the ant.

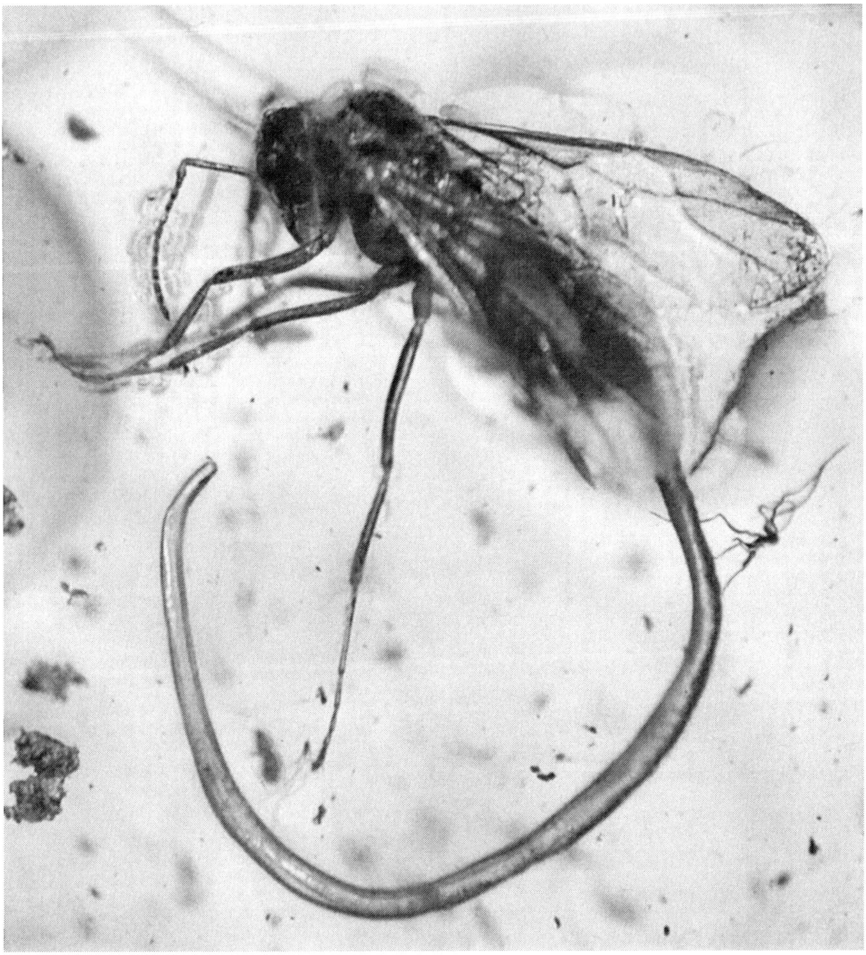

Figure 3: The fossil nematode, Heydenius formicinus, emerging from a male Prenolepis henschei in Baltic amber.

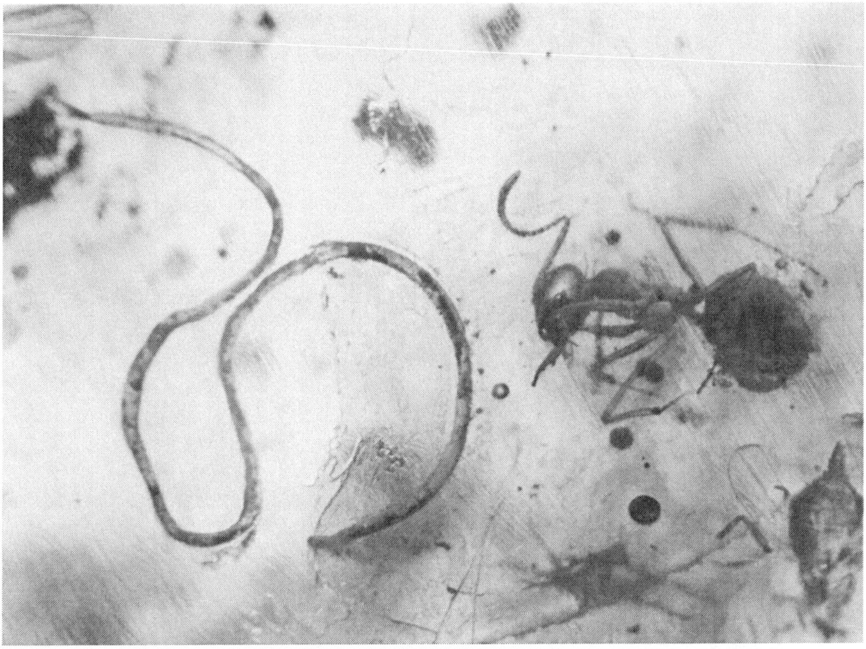

Figure 4: Heydenius formicinus adjacent to its worker ant host in Baltic amber.

Depending on the caste and length of time the mermithid is associated with its host, various degrees of host intercastes and abnormalities appear. Wheeler [18] was the first to provide an explanation for these phenomena by correlating the unusual morphological conditions with mermithid infections (Figure 6). Parasitized queen ants (mermithogynes) are shorter, have a smaller thorax (stenothoracy), reduced wings (brachyptery), enlarged abdomen (physogastry), and smaller head (microcephaly) than their uninfected counterparts. Parasitized worker ants (mermithergates or macroergates) often develop morphological features characteristic of queens and soldiers. Attacked male ants (mermithaners) have shorter wings but enlarged heads, eyes, and gasters. Infected soldiers (mermithostratiotes) have reduced heads, an ocellus, and changes in pilosity (Figure 7) [19–24].

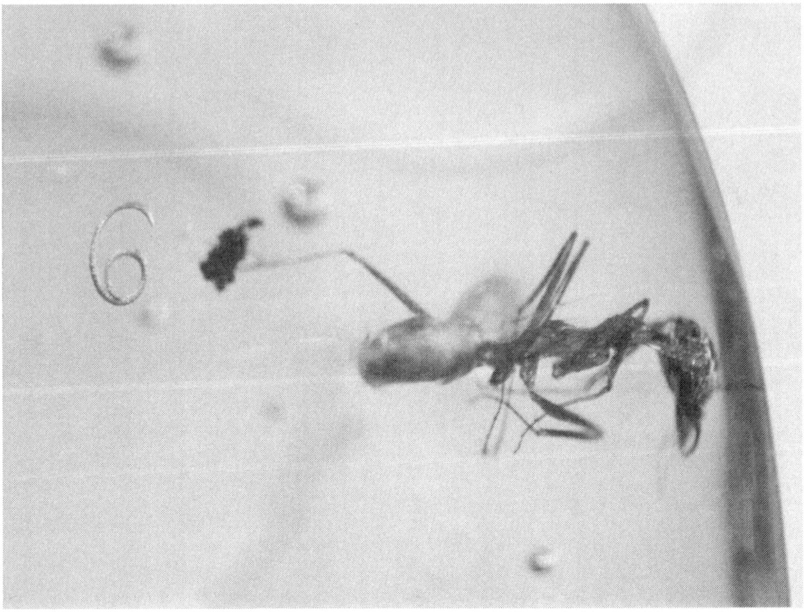

Figure 5: Heydenius myrmecophila adjacent to its Linepithema ant host in Dominican amber.

The life cycle of most ant mermithids remains a mystery. Crawley and Baylis [5] assumed that P. myrmecophila has a direct cycle, where infection is brought about by the eclosing preparasitic mermithid entering the ant host. When development is completed, the postparasitic juvenile emerges, molts to the adult stage in the ant's habitat, mates, oviposits and the cycle continues. However, no one has demonstrated a direct cycle for any mermithid parasite of ants. In 1934, Vandel [25] studied a mermithid parasite of Pheidole pallidula and realized that the infection must be initiated in the ant larva. He assumed that the nematodes were in the soil surrounding the ant colony so the infective stages could penetrate directly into the ant larva; however he was unable to confirm the infection process.

The first life cycle of an ant mermithid was achieved by Kaiser with the European Pheromermis villosa [17, 26] (Figure 8). Kaiser showed that P. villosa had an indirect cycle involving oligochaetes as paratenic hosts. Workers of Lasius spp. collecting protein for the brood capture oligochaetes containing the infective stages of P. villosa and, unknowingly, feed them to the developing larvae. At this point, the nematode becomes active, penetrates into the ant larva's hemocoel, and initiates development. It was a significant discovery and raises the question whether all mermithid infections of ants have indirect life cycles. Other possible paratenic hosts for Pheromermis could be small aquatic insects that ingest mermithid eggs from the bottom debris of seepage areas or the edges of other water sources. The wasp parasite, Pheromermis pachysoma [6], also has an indirect cycle and uses caddis flies as paratenic hosts, which the eusocial wasps (Vespidae) feed to their brood [27].

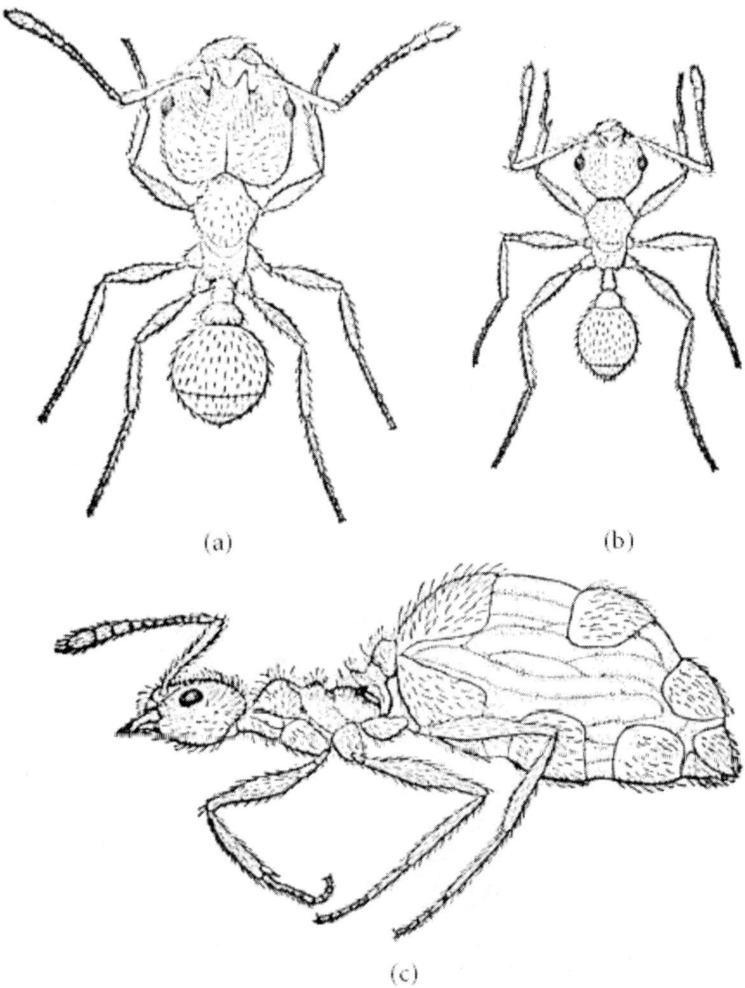

Figure 6: Plate (modified) of Pheidole dentata (referred to as P. commutata) from [18] showing the first evidence that mermithid nematodes could cause intercastes of ants. (a) Normal soldier; (b) normal worker; (c) parasitized worker (mermithergate).

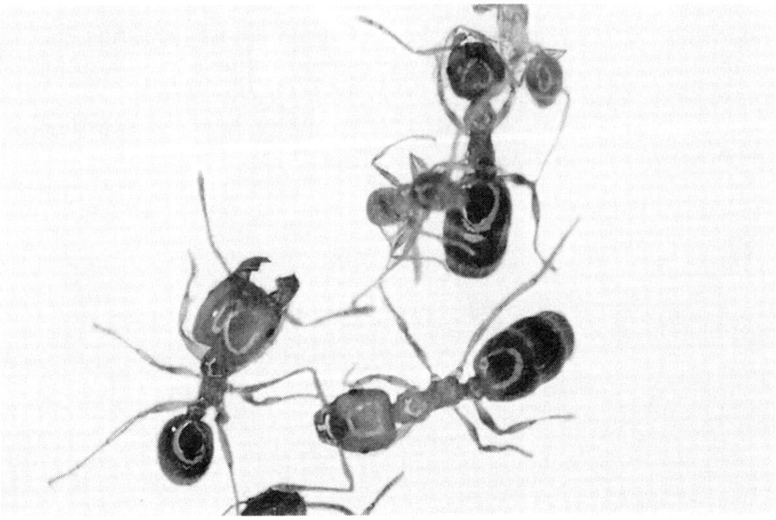

Figure 7: Two mermithid-infected soldiers (mermithostratiotes) (arrows) of Pheidole pallidula adjacent to smaller workers and an uninfected soldier (with large head). Note smaller heads on infected soldiers. Photo courtesy of Luc Passera.

Figure 8: Pheromermis villosa in a carpenter ant from Holland.

Thus far, seven genera of mermithids are known to infect ants, namely, Agamomermis, Allomermis, Camponotimermis, Comanimermis, Heydenius, Meximermis, and Pheromermis (Table 2). All of the ant hosts of these mermithids feed their brood animal protein (in contrast to other genera, such as the leaf cutting ants), and this behavior suggests they have an indirect life cycle involving a paratenic host. The two genera, Agamomermis and Heydenius, are collective group genera for immature extant and fossil mermithids, respectively [1].

There are some morphological and behavioral patterns that characterize mermithids with indirect cycles. They normally have smaller eggs with thicker shells than the eggs of direct development soil or freshwater mermithids. Also their eggs are completely embryonated when laid. Finally, the deposited eggs will not hatch in the environment even though the enclosed parasitic juvenile is fully developed. Hatching only occurs when a potential invertebrate paratenic host ingests the eggs. The eggs of Pheromermis spp. are small, numerous, fully embryonated when laid and do not hatch in the environment. Fully embryonated eggs ensure that the infective stages are ready to enter paratenic hosts as soon as they are ingested [6, 17].

The ant mermithid, Allomermis solenopsi [12], possess an unusual morphological feature on the mature eggs that could play a crucial role in its life cycle. The surface of the eggs is covered with elongate, erect, spiny adhesive processes. How these function in the life cycle is unknown, but the related species, A. trichotopson, possesses similar structures [28]. Since A. solenopsi parasitizes the fire ant, Solenopsis invicta in Brazil (Figure 2), the related A. trichotopson, whose host is unknown, may infect Solenopsis geminata in Jamaica. Could these egg processes somehow be connected with parasitism of Solenopsis spp.?

Can mermithids be manipulated to control ants? Aside from killing the ant host upon emergence, mermithids drain the host of food, reduce the flight muscles and fat body, and cause morphological modifications as mentioned above [9, 17, 24, 26]. Since mermithid-infected Solenopsis has reduced reproductive organs and die shortly after the nematodes emerge [11, 12, 29], it has potential as a biological control agent. However, if the cycle is always indirect as shown for P. villosa, it would be very difficult to artificially infect the ant brood. It would be necessary to first infect the paratenic host and then supply large numbers of these infected invertebrates to worker ants for transport back to the nest. Working with a mermithid that has a direct cycle would be easier; however there is still the problem of raising and disseminating the nematodes.

3. TETRADONEMATIDAE

The tetradonematids are a diverse group of nematodes that have traditionally been aligned with the Mermithidae. However, aside from some distinctive morphological characters, female tetradonematids normally mature, mate, and produce eggs within the host, which does not occur with mermithids. Two tetradonematids have been described from extant ants. Tetradonema solenopsis is a parasite of the red imported fire ant, Solenopsis invicta, in Brazil [30, 31]. Very little is known about this nematode aside from the scant description showing that females contained eggs and worker infection levels reached 12.5%. Parasitized ants that succumbed to the infections could be recognized by their slightly enlarged gaster with scallop-appearing dorsal sclerites.

The second tetradonematid from extant ants is Myrmeconema neotropicum from Cephalotes atratus in Peru and Panama [32]. Myrmeconema is the only nematode that causes its ant host to radically change color (from black to red), which is crucial for completion of its life cycle [33]. This color change was a mystery for early taxonomists and the variety Cephalotes atratus var. rufiventris was erected solely on the basis of its red abdomen, which was later shown to be the result of Myrmeconema infections [32].

Developing females of M. neotropicum occur in ant pupae (Figure 9) but do not produce masses of eggs until they are carried into the adult ant (Figure 10). As the females deteriorate, eggs are released into the ant's hemocoel (Figure 11). At this stage of development, the gasters of the infected worker ants turn from black to red and are held high in the air (Figure 12) [33]. Birds mistake the red gasters for fruits and the nematode eggs are passed through the birds' digestive system and end up in the droppings, which are deposited on leaves and branches. Cephalotes workers collect and feed the infested excreta to their brood, which is how the larvae become infected [33].

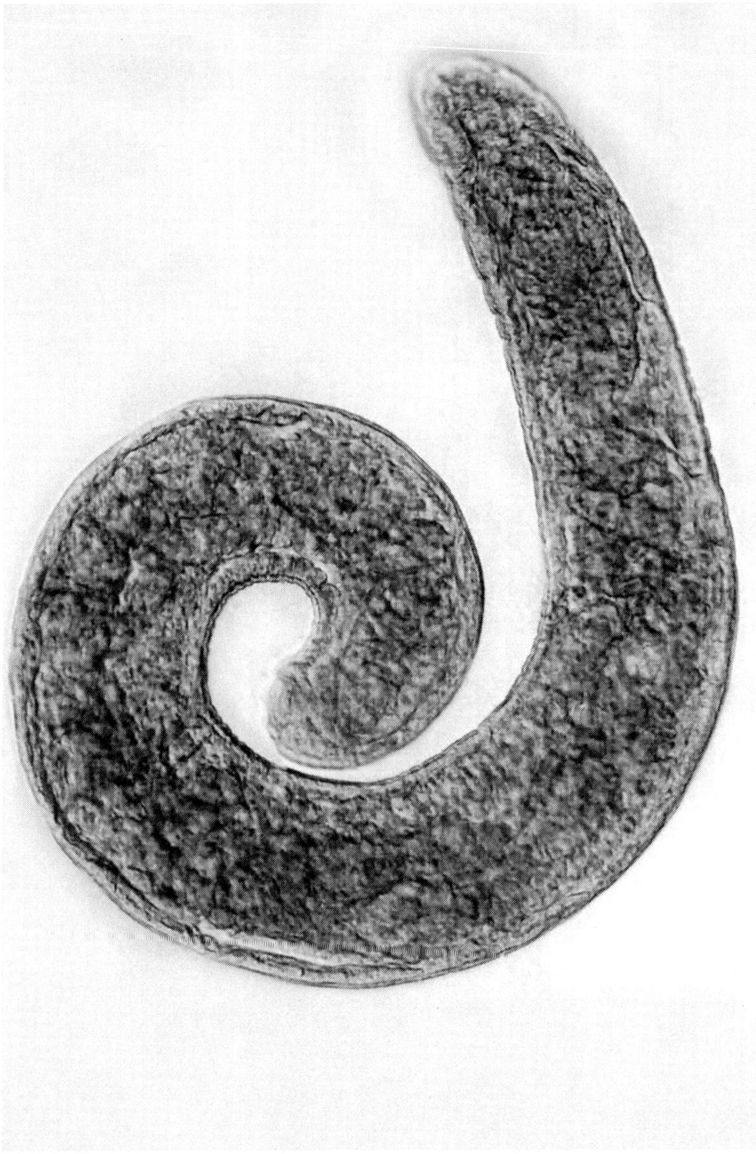

Figure 9: Mature female of Myrmeconema neotropicum in the early stages of egg production removed from a pupa of Cephalotes atratus. Arrow shows position of vulva.

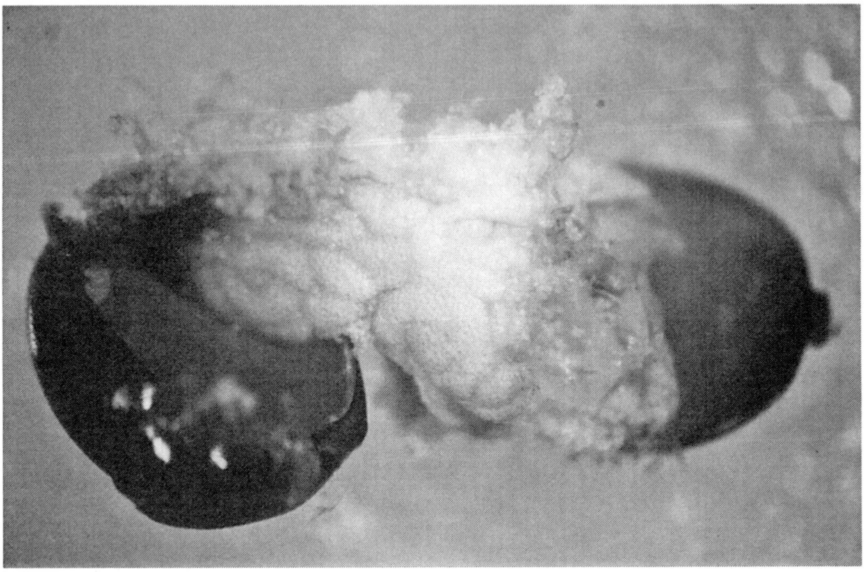

Figure 10: Mature females of Myrmeconema neotropicum packed with eggs in the gaster of a Cephalotes atratus worker in Peru.

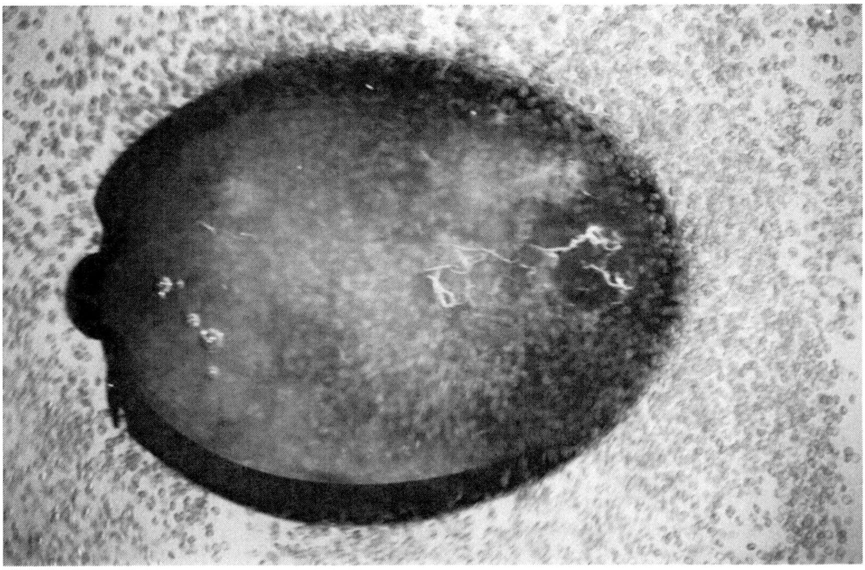

Figure 11: Eggs of Myrmeconema neotropicum released from the gaster of an infected Cephalotes atratus worker in Peru.

Figure 12: Worker of Cephalotes atratus infected with Myrmeconema neotropicum. The raised, red abdomen occurs when the nematode eggs are infective and ready for transport by birds. Photo courtesy of Stephen P. Yanoviak.

Aside from their red gasters, parasitized ants are smaller with reduced head widths. They are sluggish, clumsy, generally less aggressive, and about 40% heavier than nonparasitized workers. They do not bite when handled, and their alarm/defense pheromone supply is significantly reduced or absent.

Myrmeconema is probably widely distributed throughout the Neotropics since this association has been in existence for some 20–30 million years. The fossil worker ant, Cephalotes serratus in Dominican amber, is surrounded by the eggs of Myrmeconema antiqua (Figure 13) [1]. The ant has a hole in its abdomen that quite possibly was made by a bird. Many of the eggs, which closely resemble those of M. neotropicum in size and shape, contain fully developed juveniles (Figure 14). All indications suggest that M. antiqua had a similar life history to the extant M. neotropicum and involved bird carriers.

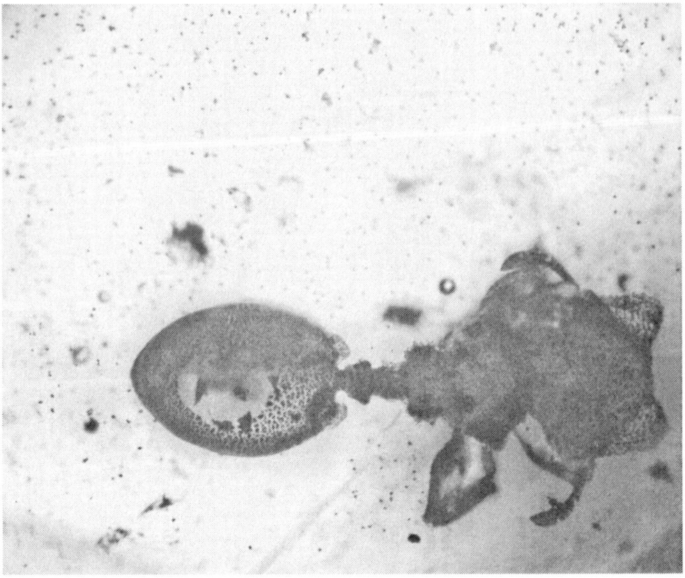

Figure 13: Worker of Cephalotes serratus infected with Myrmeconema antiqua in Dominican amber. Note microscopic eggs widely distributed in the amber that were released from a hole in the ant's gaster.

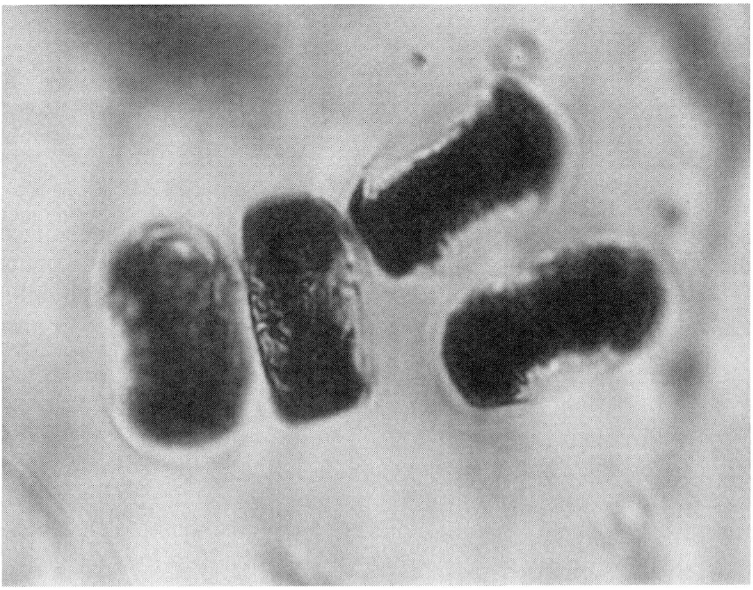

Figure 14: Detail of eggs of Myrmeconema antiqua in various stages of development in Dominican amber.

4. ALLANTONEMATIDAE

It is curious why so few cases of allantonematid infections have been reported in ants. Since ants are probably one of the most investigated insect groups, is the absence of tylenchid parasitism due to a lack of observations or its rarity? The first and only described allantonematid parasite of extant ants is Formicitylenchus oregonensis that was parasitizing a queen Camponotus ant in Western Oregon, USA [34]. The queen had already chewed off her wings and appeared to be searching for a nesting site. There was a single large parasitic female (Figure 15) and 120 third-stage juvenile nematodes in the ant's gaster. The third-stage juveniles exited through the ants reproductive and digestive tracts and molted twice to reach the adult stage. The enlarged pharyngeal glands in the free-living females suggest that they penetrate the cuticle to enter the body cavity of the host, probably ant larvae. Although the complete life cycle is unknown, the nematodes are clearly distributed by infected queen ants. The gonads of the infected ant were greatly reduced, and her eggs were abnormal. Since carpenter ants can be damaging to structures, F. oregonensis can be considered as a potential biological control agent.

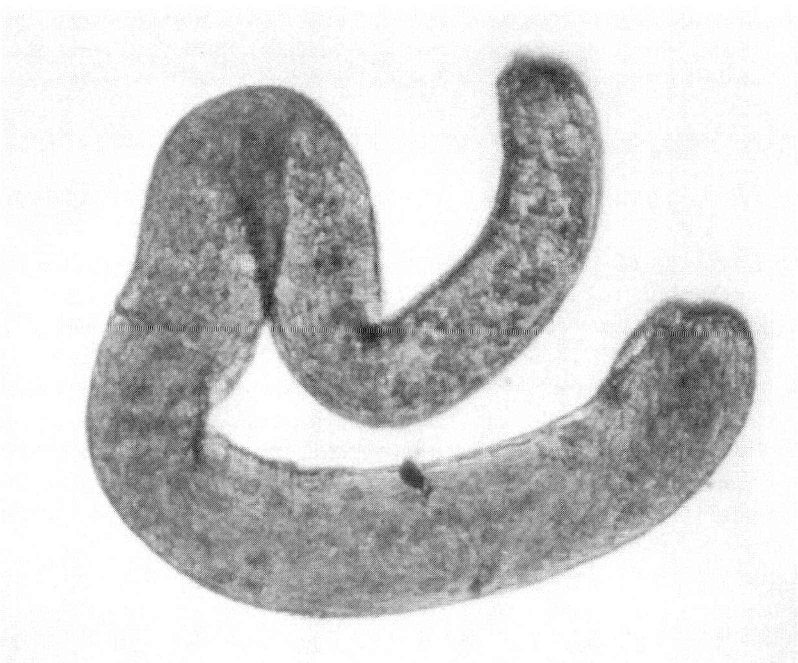

Figure 15: Female of the allantonematid Formicitylenchus oregonensis removed from the body cavity of Camponotus vicinus in Oregon, USA.

Since the original report of this parasite, the present author recovered a worker carpenter ant also infected with F. oregonensis, thus indicating that Formicitylenchus is probably restricted to ant hosts, especially members of the genus Camponotus. Formicitylenchus shows a close relationship with the allantonematid beetle parasite, Metaparasitylenchus [34]. It is possible that their last common ancestor parasitized beetles and the host shift from arboreal beetles to arboreal ants occurred during the anagenesis of Formicitylenchus. The close physical association between wood-boring beetles and Camponotus ants may be significant. Rogers [35] commented that "...the potential parasite would be expected to find its hosts in organisms which occupy the same niche largely independent of their phylogenetic position. In fact the specificity of many parasites is based on the ecological relationship of the hosts, especially in groups which have only recently become parasitic."

Another reason that allantonematid parasitism of ants may be more widespread than presumed is the discovery of juveniles of a fossil allantonematid, Palaeoallantonema cephalotae, in the ant, Cephalotes serratus, in Dominican amber [1] (Figure 16). Just before this fossil was discovered, Steven Yanoviak submitted an extant worker of Cephalotes christopherseni from Peru that was also infected with an allantonematid. The parasitic female (Figure 17) of this still undescribed species and the developing juveniles inside her body (Figure 18) show features typical of the family.

5. SEURATIDAE

The discovery of adults of Rabbium paradoxus [36] inside the gaster of worker Camponotus castaneus in Florida (Figure 19) was a surprise since all known nematodes of the Seuratidae are heteroxenous and develop to the adult stage in the digestive tract of vertebrates [37]. However, in R. paradoxus, the vertebrate host is obviously not required for adult development. The females of R. paradoxus have an anteriorly placed vulva (Figure 20), and the eggs embryonate inside the uterus (Figure 21). Since the other member of the genus, R. caballeroi, occurs in the gut of lizards in the Bahamas [38], it is likely that R. paradoxus originally had (or still has) a lizard definitive host. If the complete life cycle occurs just in ants, then C. castaneus would serve as both intermediate and definitive hosts. C. castaneus is a generalist feeder and will ingest vertebrate feces so it could acquire nematode eggs from lizard droppings. Parasitized worker ants had swollen gasters and showed unusual behavior by foraging during the day instead at night. This would make them easily captured by vertebrate predators.

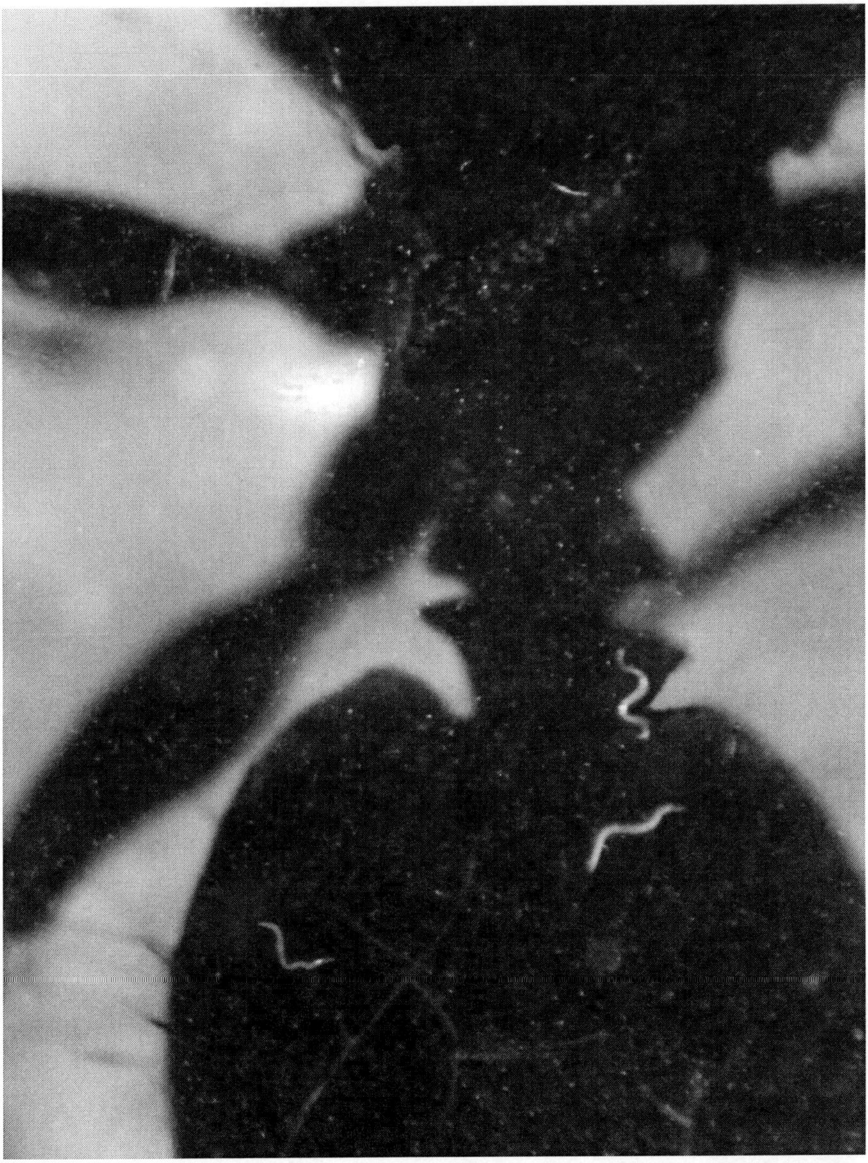

Figure 16: Three juveniles of the allantonematid, Palaeoallantonema cephalotae, that emerged from a worker Cephalotes ant in Dominican amber.

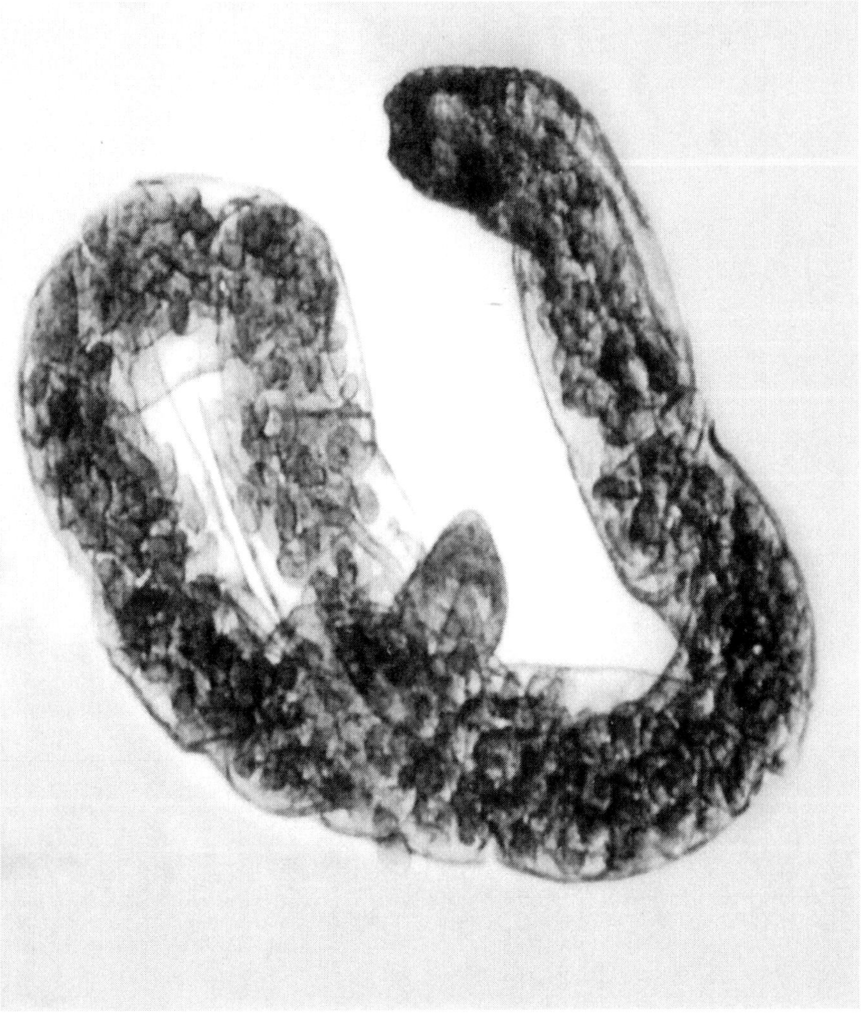

Figure 17: Parasitic female of an undescribed allantonematid from workers of Cephalotes christopherseni in Peru.

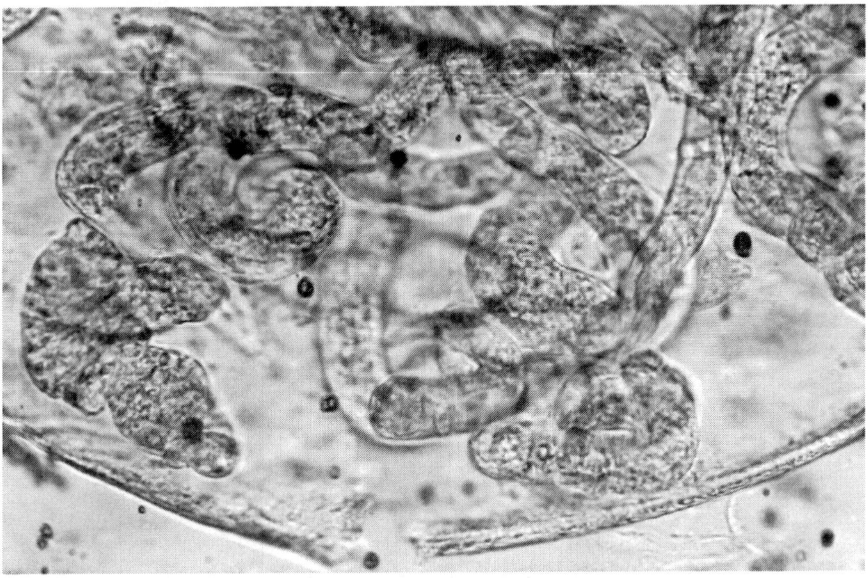

Figure 18: Juveniles developing inside the body of the female allantonematid shown in Figure 17.

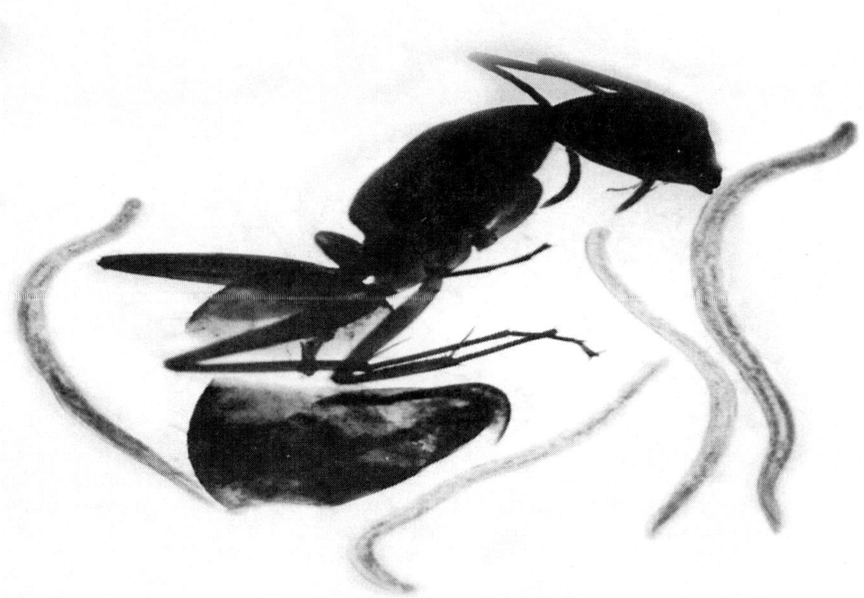

Figure 19: Adults of Rabbium paradoxus adjacent to their ant host, Camponotus castaneus, in Florida.

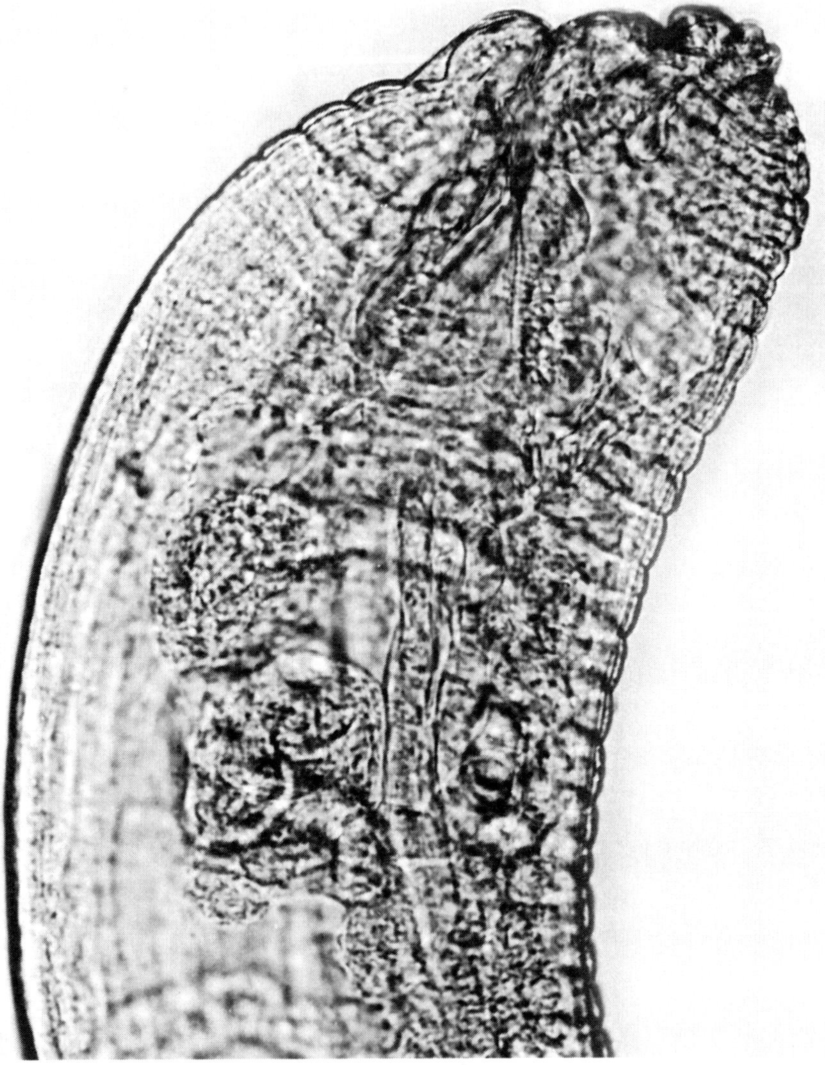

Figure 20: Head of a female of Rabbium paradoxus. Arrow shows anteriorly located vulva.

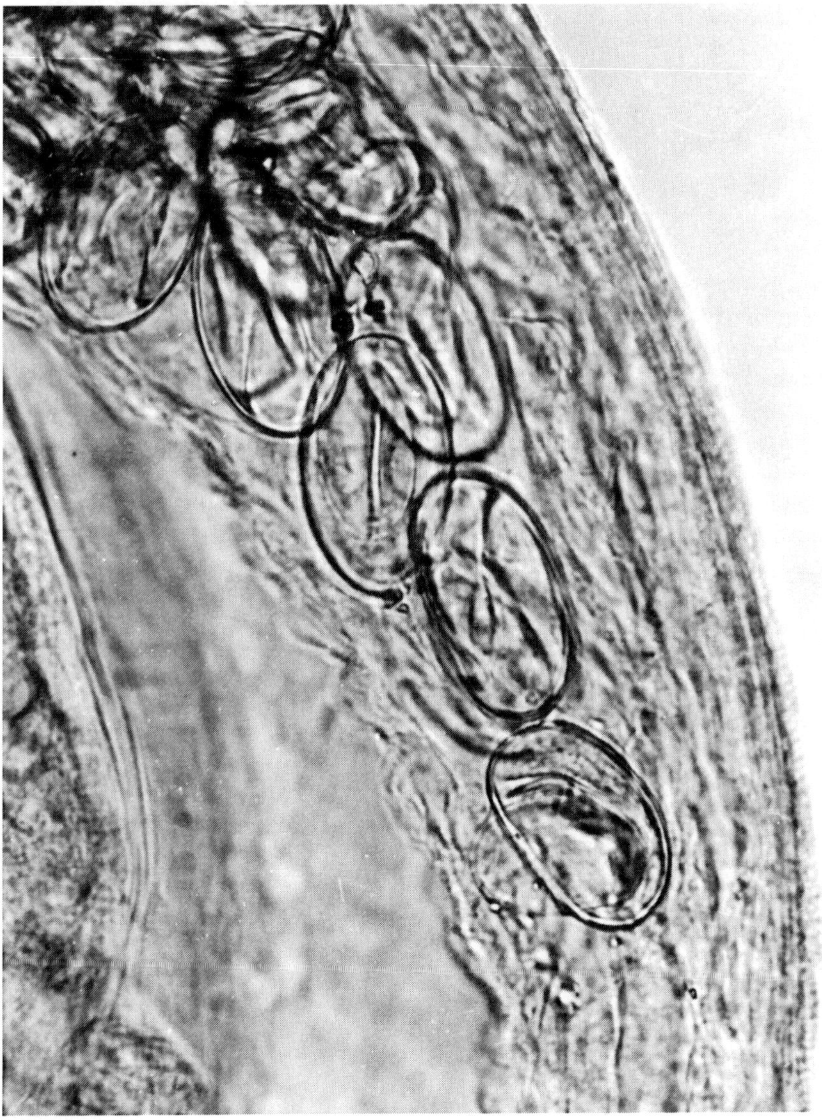

Figure 21: Embryonated eggs in the uterus of Rabbium paradoxus.

The original life cycle of R. paradoxus may have been similar to that of the seuratoid Skrjabinelazia galliardi, a parasite of sphaerodactyline lizards in Brazil [37]. The female nematodes living in the gut of the lizard produce eggs that are passed out and ingested by insects. These eggs hatch in the insect gut and the juveniles enter the body cavity without further development. Growth is resumed when the insect intermediate hosts are eaten by lizards [38]. Unfortunately, the complete life cycle of R. paradoxus remains a mystery, but its precocious development is quite interesting.

6. PHYSALOPTERIDAE

There are few reports of heteroxenous nematodes utilizing ants as intermediate hosts, that is, hosts where the nematodes develop only to the third-stage infective juveniles. Maturity to the adult stages occurs when the intermediate host is eaten by a vertebrate definitive host. One such nematode is the physalopterid, Skrjabinoptera phrynosoma that lives in the stomach of the Texas horned lizard, Phrynosoma cornutum, and uses the harvester ant, Pogonomyrmex barbatus, as an intermediate hosts [39]. However instead of depositing isolated eggs that would pass from the lizard, the gravid nematodes die with the retained eggs enclosed in thick walled capsules. The females with their enclosed eggs pass out of the lizard and are collected by worker ants that feed them to their brood. The nematode eggs hatch in the gut of the ant larvae and the juveniles enter the fat body, where they develop only to the third stage. These juveniles are carried through the pupal and into the adult stage of the ant, where they eventually reside in membranous capsules. The nematodes complete their development to the adult stage when infected ants are eaten by the lizards. Worker ants with more than 10 nematodes were still active but had enlarged, lighter colored gasters. The interesting, pivotal stage in this life cycle is the attractiveness of the dead, egg-laden female nematodes to worker ants.

7. RHABDITIDAE, DIPLOGASTRIDAE, AND PANAGROLAIMIDAE

This category includes juvenile nematodes living in the postpharyngeal glands of ants (internal phoresis) or being carried on the outside surface of ants (external phoresis) (Table 3). While these might not be considered parasites, in some instances where the association has been examined critically [40] damage has been inflicted on the ant's postpharyngeal glands and some of the nematodes increased in size during their stay in this location. Thus at most, they could be considered weak parasites. If they break through the glands and introduce microbes into the body cavity of the ant, they could even be regarded as pathogenic. However the latter scenario has not been documented.

Table 3: Juvenile nematodes of Rhabditida and Tylenchina associated with ants.

Nematode	Family	Host	Reference
Diplogasteroides spengelii	Diplogastridae	*Lasius brunneus*	[41]
Diploscapter lycostoma	Rhabditidae	*Formica* spp., *Lasius* spp.	[42]
"	"	*Myrmica rugulosa*	[42]
"	"	*Linepithema humile*	[40]
* *Formicodiplogaster myrmenema*	Diplogastridae	*Azteca alpha*	[1]
Halicephalobus similigaster	Panagrolaimidae	*Lasius brunneus*	[41]
Koerneria histophora	Diplogastridae	*Lasius* spp.	[42]
"	"	*Lasius brunneus*	[41]
Oscheius dolichurus	Rhabditidae	*Formica rufa,*	[42, 43]
"	"	*Lasius flavus*	[42, 43]
"	"	*Tetramorium caespitum*	[42]
"	"	*Camponotus herculeanus*	[44]
"	"	*Lasius claviger*	[44]
"	"	*Lasius brunneus*	[41]
Pristionchus lheritieri	Diplogastridae	*Formica rufa, Lasius* spp.	[42]
* Unknown	"	*Azteca* spp.	[1]
Unknown	"	*Formica obscuriventris*	Present work (Figure 22)

* Fossil.

Most of the nematodes in the postpharyngeal glands are dauer juveniles of free-living microbotrophs living in the ant's environment. The dauers enter the glands when external conditions become unsuitable (low humidity or diminished food supply). These resistant dauer juveniles can survive for relatively long periods. The nematodes may leave the glands when the environment is more suitable (moist with associated microbes), if the ant dies and the dauer initiates development within the decomposing ant, or when the nematodes are transferred from ant to ant during trophallaxis.

Janet [43] was the first to discover postpharyngeal rhabditids (Oscheius dolichurus) in Lasius flavus and Formica rufa in France. Wahab [42] was the first to systematically study these associations in the ant genera Lasius, Formica, Tetramorium, and Myrmica in Germany (Table 3). More recently Köhler [41] examined nematodes in the heads of ants collected from sap fluxes and rotten wood on trees in Germany. The most common ant that visited these fluxes was Lasius brunneus and, from a total of 262 workers collected, 43.5% carried nematodes, with Koerneria histophora being the most common associate. While most ants carried a single nematode, numbers occasionally reached up to 85 dauers per ant. Köhler [41] also found diplogastrid dauers in 4 males and a queen of L. brunneus. The infection rate of ants associated with L. brunneus workers varied depending on the weather cycle. There were more nematodes in ants during the dry period in August than during the rainy months of April and May. Also important in determining the rate of nematodes being carried by the ants was the location of the nests. Rates of infestation by nematodes in L. brunneus were much higher when the ants were collected from sap fluxes and rotten wood [41], than were collected from under stones and leaf litter [42].

Köhler [41] was able to infest ants by placing them in a Petri dish with rotten wood containing waving dauer stages. Both Wahab [42] and Köhler [41] provided evidence that the dauers can be transmitted from ant to ant via trophallaxis, which was supported in part by the experiments of Naarmann [53] showing that Formica ants mix food with secretions from the postpharyngeal gland before regurgitating it to nest mates. These ant-dauer associations probably occur worldwide since Markin and McCoy [40] reported Diploscapter lycostoma in the postpharyngeal glands of the Argentine Ant, Linepithema humile in California and Nickle & Ayre [44] found Oscheius dolichurus in the head glands of Camponotus herculeanus and Lasius claviger in Ontario, Canada. The author has also found dauer diplogastrids in the postpharyngeal glands of workers of Formica obscuriventris in Oregon (Figure 22).

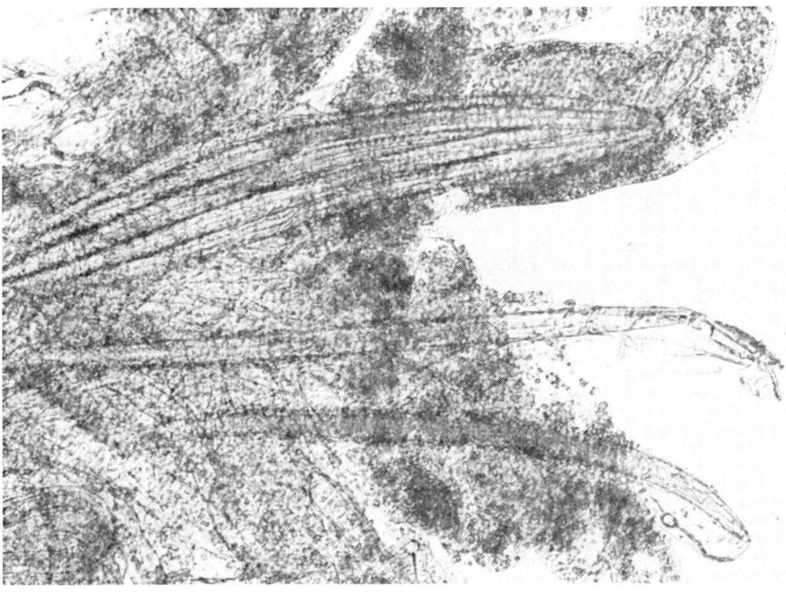

Figure 22: Dauer juveniles of a diplogastrid in the postpharyngeal glands of Formica obscuriventris clivia from Oregon.

Figure 23: Three dauer juveniles of Formicodiplogaster myrmenema adjacent to a worker of Azteca alpha in Dominican amber.

Figure 24: Detail of a dauer juvenile of Formicodiplogaster myrmenema adjacent to a worker of Azteca alpha in Dominican amber.

The association between dauer nematodes and ants is at least 20–30 million years old. Evidence for this is the discovery of dauer juveniles of the fossil diplogastrid, Formicodiplogaster myrmenema, carried by Azteca alpha workers in Dominican amber [1] (Figures 23 and 24). The dauer stages appear to be associated with the abdomen of the ants, suggesting that they were being carried in the segmental membranes of the gaster (external phoresis). None of the fossil stages occurred around the mouthparts of the ants. Also, developing stages of F. myrmenema were associated with nest material adjacent to worker Azteca ants in Dominican amber [1]. This indicates that F. myrmenema was developing in the nests of A. alpha, which is probably the case with extant nematodes in the head glands of ants. Whether the dauers of F. myrmenema were also in the postpharyngeal glands of the fossil ants is unknown.

8. STEINERNEMATIDAE AND HETERORHABDITIDAE

Included in this section are the so-called entomopathogenic nematodes belonging to the genera Steinernema and Heterorhabditis. It is quite likely that entomopathogenic nematodes infect ants under natural conditions, but no reports are known. Infection is initiated by a third-stage infective juvenile that enters the host's body cavity, apparently per os [50]. After reaching the hemocoel, the infective stage initiates development and, in so doing, releases a symbiotic bacterium (Xenorhabdus spp. in Steinernema nematodes and Photorhabdus spp. in Heterorhabditis nematodes) that is carried in the infective stage's gut lumen. The bacterium kills the insect soon after it is released in the body cavity. The nematodes feed on the mixture of bacteria and insect hemolymph and develop to the adult stage in the body cavity. With adequate nourishment, the nematodes undergo a second generation but when nourishment is limited, the juveniles form third-stage infective juveniles. By introducing the bacteria that quickly kill the hosts, these nematodes avoid specific defense responses and have a wide host range, attacking representatives of many insect orders and even other arthropods [54].

Laboratory experiments have shown that these nematodes can infect a number of ant species (Table 4) and they also have been used in the field against pest ants [50, 52, 54–56]. Poole [50] attempted to control field populations of ants (Solenopsis richteri and S. invicta) with Steinernema carpocapsae. Using a dose of 1 million infective stages per mound for S. invicta, the nematodes caused 35% mortality in the fall and 80% mortality in the spring. With S. richteri, the death rate was 80% in the spring and 36% in the fall. Poole [50] noted that workers were infected less than other stages, possibly because of their greater activity and grooming behavior. However, workers regurgitated infective stages to the alates and larvae. Queen ants were more susceptible and up to 3,000 infective stage juveniles could be produced in some infections (Figure 25).

Table 4: Ants infected by entomopathogenic nematodes (Steinernema carpocapsae and Heterorhabditis bacteriophora) under laboratory and/or field conditions.

Ant	Nematode	System	Reference
Acromyrmex octospinosus	*S. carpocapsae*	Aqueous	[45]
Camponotus sp.	*S. carpocapsae*	Sucrose	[46]
Camponotus sp.	*S. carpocapsae*	Aqueous	[47]
Myrmica sp.	*S. carpocapsae*	Aqueous	[47]
Pogonomyrmex sp.	*S. carpocapsae*	Sucrose	[48]
Solenopsis spp.	*S. carpocapsae*	Alginate capsules	[46]
Solenopsis geminata	*S. carpocapsae*	Aqueous	[49]
Solenopsis invicta	*S. carpocapsae*	Aqueous	[50, 51]
Solenopsis invicta	*H. bacteriophora*	Aqueous	[52]
Solenopsis richteri	*H. bacteriophora*	Aqueous	[52]
Solenopsis richteri	*S. carpocapsae*	Aqueous	[50]

Figure 25: Developing stages of Steinernema carpocapsae removed from the body of an infected queen of Solenopsis invicta.

Further field trials of S. carpocapsae and Heterorhabditis bacteriophora against the red imported fire ant, S. invicta, gave control rates of 37.5% with S. carpocapsae but less with H. bacteriophora [52]. In field trials comparing applications of Steinernema carpocapsae and amidinohydrazone against S. invicta, Morris et al. [55] estimated that nematode applications at a rate of 2 million per gallon per mound resulted in 47% mortality.

Controlling fire ants in the field is difficult because of the small mound opening through which the nematodes are introduced. Also, it is desirable to have recycling of the nematodes in the nests, but healthy ants appear to remove infected individuals before the cycle is completed. Since the number of nematodes needed to overwhelm a colony of ants is quite high using inundative methods, consideration was given to the development of baits or other more efficient delivery systems [46, 48, 52, 55, 56]. These other methods are still under investigation.

9. UNKNOWN NEMATODES

Gösswald [57] reported the presence of several encysted nematodes in the flight muscles of a queen Teleutomyrmex schneideri in Germany. The cysts were quite small, being only 25 µm in diameter. Except for their small size, the cysts are similar in appearance to those of the vespid mermithid, Pheromermis pachysoma, formed in the body wall of Trichoptera paratenic hosts [27] and the ant parasite, P. villosa, in the body of oligochete paratenic hosts [26]. However, the Pheromermis cysts are 60–100 µm and 80 µm in diameter, respectively. It is possible that juvenile nematodes of a mermithid parasite were acquired after the queen was fully formed and the nematodes preferred to encyst rather than initiate development. The other likelihood is that the nematodes were the infective stages of a heteroxenous nematode parasite and were waiting for transfer to a vertebrate definitive host. However, the only cysts of heteroxenous nematodes known from ants are those of the physalopterid, S. phrynosoma, the smallest of which measures 633 µm in diameter [39].

In 1907, Janet [58] found nematodes 7-8 mm in length developing in the head cavities and emerging from the labial region of workers of Formica fusca. Just before the nematodes emerged, the infected ants began trembling and eventually died. The head cavities of infected ants were empty upon nematode exit. This behavior of developing in the head of ants is known for some phorid flies but not for nematodes. Whether this was a mermithid with an unusual developmental location or a heteroxenous nematode using the ant as an intermediate host is unknown.

10. IDENTIFICATION KEY TO NEMATODE FAMILIES ASSOCIATED WITH ANTS

See Table 5.

Table 5

(1)	Nematodes represented as dauer or postdauer juveniles in the pharyngeal glands of ants	Rhabditidae, Diplogastridae and Panagrolaimidae
	Nematodes developing in the body cavity of ants	(2)
(2)	Only juvenile nematodes present	(3)
	Adult nematodes with or without juveniles	(4)
(3)	Elongate nematodes normally over 15 mm in length at completion of development; not enclosed in membranous capsules	Mermithidae
	Nematodes under 10 mm in length; enclosed in membranous capsules	Physalopteridae
(4)	Nematodes reproducing in dead ants; infective juveniles produced	(5)
	Nematode adults, eggs and/or juveniles in living ants; infective juveniles absent	(6)
(5)	Males with a bursa; females with a pointed tail	Heterorhabditidae
	Males without a bursa; females with a bluntly rounded tail, often bearing a small point at tip	Steinernematidae
(6)	Eggs and juveniles present	Allantonematidae
	Eggs, but no juveniles present	(7)
(7)	Vulva positioned at middle or lower half of body	Tetradonematidae
	Vulva positioned in upper fourth of body	Seuratidae

11. CONCLUSIONS

Representatives of most invertebrate parasitic nematode families attack ants, with the exception of sphaerulariids, entaphelenchids, and oxyurids. While mermithids are the most commonly encountered nematode parasites of ants, the complete life cycle of only a single species is known. The life cycle of ant mermithids can be quite complicated when it involves paratenic hosts living in completely different habitats. Even less is known about the life cycles of other ant parasitic nematodes, certainly not enough to consider using them as biological control agents. While the inundative application of entomopathogenic nematodes (Steinernema and Heterorhabditis) can control ants in isolated colonies, establishing nematodes for the sustained control of ant populations has not been achieved.

There are probably many additional nematode parasites of vertebrates utilizing ants as intermediate hosts. Reptiles, mammals, and amphibians eat ants, and it follows that nematodes other than Skrjabinoptera phrynosoma would have devised methods of cycling themselves through ants to reach their definitive hosts. In the mysterious case involving Rabbium paradoxus, the presence of adults of a heteroxenous nematode in an ant raises the question of whether formicids can serve as sole hosts or this is just a case of precocity.

Fossils show that mermithids were infecting ants over 40 million years ago and tetradonematids and allantonematids had established parasitic associations with ants some 20–30 million years ago. Such fossils, which can be used to calibrate molecular clocks, provide minimum dates for the occurrence of nematode lineages and show the antiquity of nematode-ant relationships.

ACKNOWLEDGMENTS

The author thanks S. D. Porter, Luc Passera and Stephen P. Yanoviak for supplying photos and Brad Vinson for supplying the queen of Solenopsis invicta infected with Steinernema carpocapsae. Grateful appreciation is extended to Roberta Poinar for commenting on earlier versions of the paper.

REFERENCES

1. G. Poinar Jr., The Evolutionary History of Nematodes, Brill, Leiden, The Netherlands, 2011.
2. W. Gould, An Account of English Ants, A. Millar, London, UK, 1747.
3. C. M. Diesing, Systema Helminthum, vol. 2, Vindobonae, Wien, Austria, 1851.
4. C. T. E. von Siebold, "Über die Fadenwürmer der Insekten," Entomologische Zeitschrift, vol. 4, pp. 78–84, 1843.
5. W. C. Crawley and H. A. Baylis, "Mermis parasitic on ants of the genus Lasius," Journal of the Royal Microscopical Society, vol. 257, pp. 353–372, 1921.
6. G. Poinar Jr., R. S. Lane, and G. M. Thomas, "Biology and redescription of Pheromermis pachysoma (V. Linstow) n. gen., n. comb., a parasite of yellowjackets," Nematologia, vol. 22, pp. 360–370, 1976.
7. I. A. Rubstov, Mermithidae: Classification, Importance and Utilization, Academy of Sciences, Leningrad, Russia, 1978.
8. A. K. Artyukhovsky, Soil Mermithids; Systematics, Biology and Utilization, Voronezh University, Voronezh, Russia, 1990.
9. H. Kaiser, "Terrestrial and semiterrestrial Mermithidae," in Manual of Agricultural Nematology, W. R. Nickle, Ed., pp. 899–965, Marcel Dekker, New York, NY, USA, 1991.
10. L. Passera, "Les fourmis hôtes provisoires ou intermédiaires des helminthes," L'Année Biologique, vol. 14, pp. 227–259, 1975.
11. G. Poinar Jr., J. P. Lachaud, A. Castillo, and F. Infante, "Recent and fossil nematode parasites (Nematoda: Mermithidae) of Neotropical ants," Journal of Invertebrate Pathology, vol. 91, no. 1, pp. 19–26, 2006.
12. G. Poinar Jr., S. D. Porter, S. Tang, and B. C. Hyman, "Allomermis solenopsi n. sp. (Nematoda: Mermithidae) parasitising the fire ant Solenopsis invicta Buren (Hymenoptera: Formicidae) in Argentina," Systematic Parasitology, vol. 68, no. 2, pp. 115–128, 2007.
13. G. V. Ipateva, I. N. Pimenova, and F. V. Mukhamedzyanova, "A new nematode genus and species, Camponotimermis bifidus gen. and sp. n., from ants," in Helminthes of Insects, M. D. Sonin, Ed., pp. 80–87, E. J. Brill, Leiden, The Netherlands, 1990.
14. D. Coman, "Mermithide freatice in fauna Republich Populare Române," Academia Republicii Populare Romîne, Studii şi Cercetări ştiint, Serii, vol. 2, pp. 123–152, 1953.
15. G. Poinar Jr., "First fossil record of nematode parasitism of ants; a 40 million year tale," Parasitology, vol. 125, no. 5, pp. 457–459, 2002.

16. I. A. Rubstov, "A new species of mermithid from ants," Parazitologiya, vol. 4, pp. 338–341, 1970 (Russian).
17. H. Kaiser, "Morphologische Analyse des Ameisen-Parasitioden Pheromermis villosa n. sp. (Nematoda, Mermithidae)," Mitteilungen des Naturwissenschaftlichen Vereins fur Steiermark, vol. 116, pp. 269–294, 1986.
18. W. M. Wheeler, "The parasitic origin of macroergates among ants," The American Naturalist, vol. 35, pp. 877–886, 1901.
19. A. Vandel, "Modifications déterminées par un Nématode du genre Mermis chez les ouvrières et les soldats de la fourmi Pheidole pallidula (Nyl.)," Bulletin biologique de la France et de la Belgique, vol. 61, pp. 38–47, 1927.
20. W. M. Wheeler, "Mermis parasitism and intercastes among ants," Journal of Experimental Zoology, vol. 50, pp. 165–237, 1928.
21. K. Gösswald, "Weitere Beiträge zur Verbreitung der Mermithiden bei Ameisen," Zoologischer Anzeiger, vol. 90, pp. 13–27, 1930.
22. K. Gösswald, "Über bisher unbekannte, durch den Parasitismus der Mermithiden (Nemat.) verursachte Formveränderungen bei Ameisen," Zeitschrift für Parasitenkunde, vol. 10, no. 1, pp. 138–152, 1938.
23. L. Passera, "Présence d'Hexamermis sp. (Nematoda, Mermithidae) dans les reines vierges et les mâles de la fourmi Pheidole pallidula Nyl. (Formicidae, Myrmicinae)," Bulletin de la Société Zoologique de France, vol. 99, pp. 315–324, 1974.
24. L. Passera, "Origine des intercastes dans les sociétés de Pheidole pallidula (Nyl.) (Hymenoptera Formicidae) parasitées par Mermis sp. (Nematoda Mermithidae)," Insectes Sociaux, vol. 23, no. 4, pp. 559–575, 1976.
25. A. Vandel, "Le cycle évolutif d'Hexamermis sp. parasite de la fourmi (Pheidole pallidula)," Annales des Sciences Naturelles, vol. 17, pp. 47–58, 1934.
26. H. Kaiser, "Über Wechselbeziehungen zwischen Nematoden (Mermithidae) und Ameisen," Zoologischer Anzeiger, vol. 217, pp. 156–177, 1986.
27. G. Poinar Jr., "Distribution of Pheromermis pachysoma (Mermithidae) determined by paratenic invertebrate hosts," Journal of Nematology, vol. 13, pp. 421–424, 1981.
28. G. Steiner, "A remarkable new genus and species of mermithid worms from Jamaica," Proceedings of the United States National Museum, vol. 65, pp. 1–4, 1924.
29. D. A. McInnes and W. R. Tschinkel, "Mermithid nematode parasitism of Solenopeis ants (Hymenoptera: Formicidae) of northern Florida," Annals of the Entomological Society of America, vol. 89, no. 2, pp. 231–237, 1996.
30. D. P. Jouvenaz and D. P. Wojcik, "Parasitic nematode observed in the fire ant, Solenopsis richteri, in Argentina," The Florida Entomologist, vol. 73, pp. 674–675, 1990.
31. W. R. Nickle and D. P. Jouvenaz, "Tetradonema solenopsis n. sp. (Nematoda: Tetradonematidae) parasitic on the red imported fire ant Solenopsis invicta Buren from Brazil," Journal of Nematology, vol. 19, pp. 311–313, 1987.
32. G. Poinar Jr. and S. P. Yanoviak, "Myrmeconema neotropicum n. g., n. sp., a new tetradonematid nematode parasitising South American populations of Cephalotes atratus (Hymenoptera: Formicidae), with the discovery of an apparent parasite-induced host morph," Systematic Parasitology, vol. 69, no. 2, pp. 145–153, 2008.
33. S. P. Yanoviak, M. Kaspari, R. Dudley, and G. Poinar Jr., "Parasite-induced fruit mimicry in a tropical canopy ant," The American Naturalist, vol. 171, no. 4, pp. 536–544, 2008.
34. G. Poinar Jr., "Formicitylenchus oregonensis n. g., n. sp. (Allantonematidae: Nematoda), the first tylenchid parasite of ants, with a review of nematodes described from ants," Systematic Parasitology, vol. 56, no. 1, pp. 69–76, 2003.
35. W. P. Rogers, The Nature of Parasitism, Academic Press, New York, NY, USA, 1962.
36. G. Poinar Jr., A. G. Chabaud, and O. Bain, "Rabbium paradoxus sp. n. (Seuratidae: Skrjabinelazienae) maturing in Camponotus castaneus (Hymenoptera: Forimicidae)," Proceedings of the Helminthological Society of Washington, vol. 56, pp. 120–124, 1989.

37. A. G. Chabaud, O. Bain, and G. Poinar Jr., "Skrjabinelazia galliardi (Nematoda, Seuratoidea): compléments morphologiques et cycle biologique," Annales de Parasitologie, vol. 63, no. 4, pp. 278–284, 1988.
38. M. B. Chitwood, "A new spiruroid nematode, Rabbium caballeroi, from the stomach of Leiocephalus carinatus, from the Bahama Islands," in Libro Homenaje al Doctor Eduardo Caballero y Caballero, pp. 471–473, Secretaria de Educacion Publica, Mexico City, Mexico, 1960.
39. S. H. Lee, "The life cycle of Skrjabinoptera phrynosoma (Ortlepp) Schulz, 1927 (Nematoda: Spiruroidea), a gastric nematode of Texas horned toads, Phrynosoma cornutum," The Journal of parasitology, vol. 43, no. 1, pp. 66–75, 1957.
40. G. P. Markin and C. W. McCoy, "The occurrence of a nematode, Diploscapter lycostoma, in the pharyngeal glands of the Argentine Ant, Iridomyrmex humilis," Annals of the Entomological Society of America, vol. 61, pp. 505–509, 1968.
41. A. Köhler, "Nematodes in the heads of ants associated with sap flux and rotten wood," Nematology. In press.
42. A. Wahab, "Untersuchungen über Nematoden in den drüsen des kopfes der Ameisen (Formicidae)," Zeitschrift für Morphologie und Ökologie der Tiere, vol. 52, no. 1, pp. 33–92, 1962. View at Publisher •
43. M.-Ch. Janet, "Sur les nématodes des glandes pharyngiennes des fourmis (Pelodera sp.)," Comptes Rendus de l'Academie des Sciences, vol. 117, pp. 700–703, 1893.
44. W. R. Nickle and G. L. Ayre, "Caenorhabditis dolichura (A. Schneider, 1866) Dougherty (Rhabditidae, Nematoda) in the head glands of the ants Camponotus herculeanus (L.) and Acanthomyops claviger (Roger) in Ontario," Proceedings of the Entomological Society of Ontario, vol. 96, pp. 96–98, 1966.
45. A. Kermarrec, "Étude des relations synécologiques entre les nématodes et la fourmi-manioc: Acromyrmex octospinosus Reich," Annales de Zoologie-Ecologie Animale, vol. 7, pp. 27–44, 1975.
46. R. Georgis and G. Poinar Jr., "Field effectiveness of entomophilic nematodes Neoaplectana and Heterorhabditis," in Integrated Pest Management for Turfgrass and Ornamentals, A. R. Leslie and R. L. Metcalf, Eds., pp. 213–224, United States Environmental Protection Agency Publication, Government Printing Office, Washington, DC, USA, 1989.
47. R. A. Bedding, "Nematode parasites of Hymenoptera," in Plant and Insect Nematodes, W. R. Nickle, Ed., pp. 755–795, Marcel Dekker, New York, NY, USA, 1984.
48. R. Georgis, "Nematodes for biological control of urban insects," Proceedings of the American Chemical Society; Division of Environmental Chemicals, vol. 27, pp. 816–821, 1987.
49. A. Kermarrec, G. Febvay, and M. Decharme, "Protection of leaf-cutting ants from biohazards: is there a future for microbiological control?" in Fire Ants and Leaf-Cutting Ants, C. S. Lofgren and R. K. van der Meer, Eds., pp. 339–356, Westview Press, Boulder, Colo, USA, 1986.
50. M. A. Poole, Survey and control efficacy of endoparasites of Solenopsis richteri Forel and Solenopsis invicta Buren in Mississippi [Ph.D. dissertation], Mississippi State University, Mississippi State, Miss, USA, 1976.
51. C. S. Lofgren, W. A. Banks, and B. M. Glancey, "Biology and control of imported fire ants," Annual Review of Entomology, vol. 20, pp. 1–30, 1975.
52. E. D. Quattlebaum, Evaluation of fungal and nematode pathogens to control the red imported fire ant, Solenopsis invicta Buren [Ph.D. dissertation], Clemson University, Clemson, SC, USA, 1980.
53. H. Naarmann, "Untersuchungen über Bildung und Weitergabe von Drüsensekreten bei Formica (Hymenopt. Formicidae) mit Hilfe der Radioisotopenmethode," Experientia, vol. 19, no. 8, pp. 412–413, 1963.
54. G. Poinar Jr., Nematodes for Biological Control of Insects, CRC Press, Boca Raton, Fla, USA, 1979.

55. J. R. Morris, K. W. Stewart, and R. L. Hassage, "Use of the nematode Steinernema carpocapsae for control of the red imported fire ant (Hymenoptera: Formicidae)," The Florida Entomologist, vol. 73, pp. 675–677, 1990.
56. G. Poinar Jr. and R. Georgis, "Biological control of social insects with nematodes," in Integrated Pest Management for Turfgrass and Ornamentals, A. R. Leslie and R. L. Metcalf, Eds., pp. 255–269, United States Environmental Protection Agency Publication, Government Printing Office, Washington, DC, USA, 1989.
57. K. Gösswald, "Histologische Untersuchungen an der arbeiterlosen Ameise Teleutomyrmex schneideri Kutter (Hym. Formicidae)," Mitteilungen der Schweizerischen Entomologischen Gesellschaft, vol. 26, pp. 81–128, 1953.
58. M.-Ch. Janet, "Sur un Nématode qui se développe dans la tête de la Formica fusca," Mémoires de la Société Académique d'Archéologie, sciences et Arts du Département de l'Oise, vol. 20, pp. 1072–1073, 1907.

CHAPTER 8

Nutritional Requirements of Soybean Cyst Nematodes

Steven C. Goheen[1], James A. Campbell[1] and Patricia Donald[2]

[1] *Pacific Northwest National Laboratory, Richland, Washington, USA*

[2] *U. S. Department of Agriculture/ARS, Jacksonville, Tennessee, USA*

1. INTRODUCTION

Soybeans [*Glycine max*] are the second largest cash crop in US Agriculture, but the soybean yield is compromised by infections from *Heterodera glycines*, also known as Soybean Cyst Nematodes [SCN]. SCN are the most devastating pathogen or plant disease soybean farmers confront. This obligate pathogen requires nutrients from the plant to complete its life cycle. To date, SCN nutritional requirements are not clearly defined. Growth media supporting SCN still contain soy products. Understanding the SCN nutritional requirements and how host plants meet those requirements should lead to the control of SCN infestations. The nutritional requirements of SCN are reviewed in this chapter and those requirements are compared to those of other nematodes. Carbohydrates, vitamins, amino acids, lipids, and other nutritional requirements are discussed.

The survival of parasitic nematodes requires adequate nutrition. These essential nutrients are at least partially supplied by the host. But, availability of nutrients may not alone be sufficient for survival and reproduction. The parasite must also be able to establish a feeding site.

Both the establishment of the feeding site and the presence of adequate nutrients for the soybean cyst nematode [SCN] are discussed below.

1.1. Feeding Site Establishment

Nematodes have differing mouth part structures which are adapted to their food source [1]. In the case of plant-parasitic nematodes, a stylet [analogous to a hypodermic needle], is used to puncture plant cells and a pump mechanism located in the nematode esophagus allows for exchange of fluids between the nematode and plant [1]. Most studies of the economically important root-knot and cyst-forming plant-parasitic nematodes have focused on what fluids are secreted by the nematode and how this facilitates establishment of a feeding site [2-4]. Specific information on the essential nutrients provided by the plant is lacking. In this chapter we focus on what is known about nutrient requirements for soybean cyst nematode, SCN.

The SCN is an obligate parasite requiring a host plant to complete its life cycle (see Figure 1). The cysts are found in the soil and contain eggs and first stage juveniles. The second stage juvenile hatches from the egg and penetrates plant roots. If the roots are a plant that is a host for SCN, the third and fourth stage juveniles molt into an enlarged shape called a sausage once a feeding site is successfully established where the primary goal is removing nutrients from the plant for use by the nematode. After enough nutrients have been obtained by the nematodes, those destined to become males molt into a worm-shape again and migrate out of the roots in search of a female. As the females mature, their size increases breaking root epidermal cells and the nematode is exposed to the soil where she emits pheromones to attract the males already in the soil. Once fertilization of the eggs has occurred, the female dies and her hardened body becomes the cyst which protects the eggs from environmental extremes and organisms which can kill the eggs. Some eggs are extruded into the soil in a gelatinous matrix and these eggs are thought to hatch once conditions favor hatch. The eggs within the cyst go through diapause and can survive within the cyst for more than a dozen years under the right conditions. Juveniles which enter nonhost plant roots may molt into a third stage juvenile but a successful feeding site will not be established and the plant will recognize the nematode as an invader and form necrotic cells surrounding the nematode effectively killing the nematode. Alternatively, some plants are slower to recognize the nematode as an invader and a molt to the third stage may occur but no further development of the nematode will occur. Once the nematode reaches the sausage stage, it lacks the muscles to leave the root and it dies.

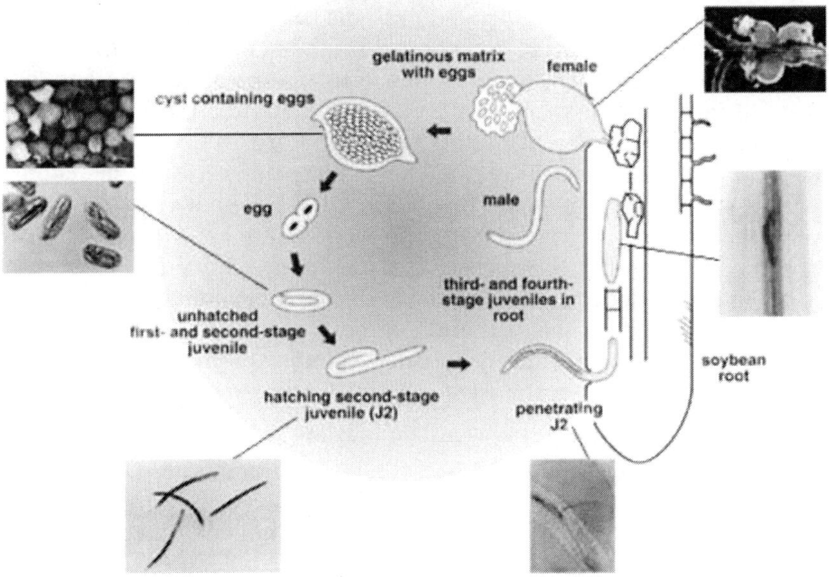

Figure 1. The life cycle of the soybean cyst nematode (SCN) is shown. Soil contains cysts with eggs as well as first stage juveniles. Second stage juveniles hatch from eggs and can then penetrate plant roots. The third and fourth stage juveniles feed off the plant. Males migrate out of the roots in search of a female. Maturing females rupture the root, releasing pheromones to attract males from soil. Females die after egg fertilization and her body becomes the cyst. This figure was obtained with permission from www.extension.umn.edu.

As an important crop in the United States [5], there are over 120 soybean lines which have some level of resistance to SCN [6]. Commercial soybean varieties primarily contain one or more different sources of resistance but 95% of all resistance is found from one source, PI 88788. Peking [PI 548402] and Hartwig [PI 437654] are also found in a few commercial varieties. Genetics of resistance is complex with multiple genes involved and interaction of minor genes or nongenetic sources complicates understanding of the process. In a resistant reaction, cytological changes occur and these have been documented [7-19]. Initial reaction to the nematode during the formation of the syncytium in both susceptible and certain resistant lines is identical for the first 4 days after infection [7. 9. 11]. Resistant reactions can be seen about day 4-5 [7, 9-11].

Cyst nematode juveniles hatch from eggs within the cyst or in the soil and enter plant roots typically in the zone of root elongation. They migrate to the pericycle and establish a feeding site [20]. Cellulases break polysaccharide chains and associated proteins in the plant cell walls. Other enzymes have been shown to be secreted by the nematodes as they move through plant tissue [21]. Rapid response by the plant to the nematode inhibits formation of a successful feeding site. A successful feeding site initiation results when the plant fails to respond or responds slowly to the presence of the nematode. One of the ways plant-parasitic nematodes protect themselves from plant responses to the nematodes is through secretion of peroxiredoxin, glutathione periosidase, and secreted lipid binding proteins within the surface coat of the nematode [22]. Although considerable knowledge is now available on the morphological changes in the plant cells due to the presence of the nematode feeding site and molecular studies have advanced our understanding of the interactions on a molecular level, the details of host specificity are unknown [23].

Information is available on the changes that occur within soybean plants when a compatible interaction between SCN and the plant occur. Information is also present on incompatible reactions when plant resistance inhibits SCN reproduction through either a hypersensitive response or formation of small syncytia which limit SCN reproduction. Infection of plant-parasitic nematodes is thought to alter plant products from the shikimic pathway. Infection by SCN increases the concentration of glucose, K, Ca and Mg in the roots but information is not available on whether these increases are products SCN then extracts from plant cells or whether these are responses by the plant to the presence of the nematode.

1.2. Nutritional Requirements

Heterodera glycines is considered to have a wide host range. Riggs and Hamblen tested 1152 entries from the Leguminosae family and found that 399 of these entries from 23 genera were susceptible. Poor hosts included 270 entries in 12 other genera [24]. Additional host studies have been conducted by Riggs and Hamblen [25-26], Miller and Gray [27-28], Venkatesh et al, [29], and Venkatesh et al [30]. Variability in host status within a plant species potentially makes identification of necessary nutrients required for establishment of the obligate feeding site easier but to date the specifics have eluded scientists.

A summary of the plants invaded by SCN are shown in Table 1. Most hosts of SCN are legumes and are limited to three subfamilies of the Leguminosae; however, approximately 50 genera in 22 families including nonlegumes are also hosts [31-32]. Some plants allow SCN to penetrate plant roots but limit reproduction of SCN [33]. The reason for this could be nutritional, or it could be due to other barriers within the plant. To determine which of those two possibilities are controlling virulence of SCN, nutritional requirements should be investigated more fully.

Table 1. Common names for plants that have been identified as good hosts for soybean cyst nematode [24-31].

Host Common Name	Host Scientific Name	Use
azuki bean	*Vigna angularis*	edible
. bean tree	*Laburnum sp*	ornamental
beans, green, dry	*Phaseolus vulgaris*	edible
beard tongue	*Penstemon digitalis*	ornamental
begger tick	*Desmodium ovalifolium*	weed
bells of ireland	*Mollucella laevis*	ornamental
bitter cress	*Barbarea vulgaris*	spice
bladder senne	*Colutea arborescens*	shrub -ornamental
bush clover	*Lespeza capitata*	prairie plant
California burclover	*Medicago hispida*	weed
common chickweed	*Stellaria media*	weed
common lespedeza	*Lespedeza striata*	weed
coral bells	*Heuchera sangiunea*	ornamental
cranesbill	*Geranium maculatum*	weed
largeflowered beardtongue	*Penstemon gradiflorus*	wildflower
field pea tuberous vetch	*Lathyrus tuberosus*	edible/weed
fennugreek	*Trigonella goenum-gracum*	spice
foxglove	*Digitalis sp.*	weed
. geranium	*Pelargonium sp*	ornamental
gold apple	*Lycopersicon esulentum*	weed
golden chain	*Laburnum anagyroides*	ornamental

Host Common Name	Host Scientific Name	Use
grass pea vine	*Lathyrus sativa*	edible/ornamental
green pea	*Pisum sativum*	edible
hairy vetch	*Vicia villosavillosa*	forage /cover crop
hemp sesbania	*Sesbania exaltata*	weed
henbit	*Lamium amplexicaule*	weed
hog peanut	*Amphicarpa bracteata*	weed
Indian joint vetch	*Aeschynomene virginica*	weed
indigo	*Indigofera parodiana*	shrub/herbaceous/small tree
clover Kenyan clover	*Trifolium*	ornamental
Korean lespedeza	*Lespedeza stiulacea*	forage
lance leaf rattlebox	*Crotalaria lanceolata*	weed
large flowered beard tongue	*Penstemon grandiflorus*	wild flower
large leaf lupine	*Lupinus polyphyllus*	wild flower
licorice milk vetch	*Astragalus glaucophyllus*	forage
little bur clover	*Medicago minima*	weed
milk vetch	*Astragalus canadensis*	forage
milky purslane	*Euphorbia supine*	weed
mouse ear chickweed	*Cerastium vulgatum*	weed
Common mullein	*Verbascum thapsus*	weed
nasturtium	*Tropaelum pergrinum*	ornamental
old field toadflax	*Linaria canadensis*	weed
pigeon pea	*Cajanus cajan*	edible
Americana pokeweed	*Phytolacca*	weed
purple deadnettle	*Lamium purpureum*	weed
purslane	*Portulaca oleracea*	weed
rainbow pink	*Dianthus chinensis*	ornamental
river bank lupine	*Lupinus rivularis*	edible
Rusian sickle milk vetch	*Astragalus falcate*	weed
service lespedeza	*Lespedeza cuneata*	weed
shrub lespedeza	*Lespedeza bicolor*	ornamental

Siberian pea tree	*Caragana arborescens*	ornamental
sicklepod	*Cassia tora*	weed
small flowered buttercress	*Cardamine parviflora*	weed
soybean	*Glycine max*	edible
Spanish broom	*Spartium junceum*	ornamental
speedwell	*Veronica peregrine*	weed
spider flower	*Cleome spinosa*	ornamental
spotted burclover	*Medicago arabica*	forage
stinking clover	*Cleome serrulata*	weed
sweet clover	*Melilotus taurica*	weed
sweet pearl lupine	*Lupinus mutabilis*	edible
tiny vetch	*Vicia hirsute*	ornamental vine
white horsehound	*Marrubium vulgare*	medicinal plant
white lupine	*Lupinus albus*	livestock feed
white pea	*Lathyrus ochrus*	wild flower
Wilcox penstemon	*Penstemon wilcoxi*	wilflower
winged pigweed	*Cycloloma atriplicifolia*	weed
yellow lupine	*Lupinus lateus*	wild flower

In many ways, it is inappropriate to compare humans to nematodes. But, from a nutritional perspective, much more is known about human nutrition than what is known about nutritional requirements of nematodes. For humans, numerous biochemical and mineral components are essential nutrients. But, for nematodes, only a few are known. Yet, nematodes have a comparatively simple digestive system. So, it would be reasonable to predict that nutritional requirements for these organisms are more extensive than what is currently known.

It is also inappropriate to generalize nutritional needs from studies on one nematode to all the nematodes within the various trophic categories. Certainly there should be similarities, but it is clear from the literature that animal parasitic nematodes have different needs from the plant parasites. And, it may also be that those plant parasites infecting specific organisms, such as SCN might have nutritional needs that synergize with the contents of the host soybean plant.

Survival is best understood when chemically defined culture media can be shown to not only sustain life, but also to promote reproduction. Chemically defined media have been identified for the survival of some nematodes and this work has recently been reviewed [34]. The successful media originally included all the amino acids in *Escherichia coli*, and in the amino acid ratios found in *E. coli*. Nematode growth media has been since modified to include a greater number of constituents including glucose, minerals, growth factors, nucleic acid precursors, vitamins, a sterol and heme source. However, SCN has not yet been shown to survive or reproduce on these media. Currently, the only growth media known to sustain SCN includes soy products [35].

Articles published on the nutritional requirements of a wide range of nematodes, generally do not specify SCN [1. 36-37]. While a few nutritional requirements for individual nematode species have been studied, these requirements are limited and their applicability to SCN is unknown. It is assumed that plant- and animal-parasitic nematodes may have different nutritional requirements from entomopathogenic, and microbivorous nematodes.

2. LIPIDS

Lipids consist of many non-water soluble components including free fatty acids, phospholipids, triglycerides, sterols, and other species. Many of these classes have been studied at least in one host-nematode relationship and are the most studied with the exception of nucleic acids due to their great structural variety and importance as food reserves. For example, Krusberg [38] reported the total lipids and fatty acids from 5 species of plant parasitic nematodes, and their common hosts. They found that the nematodes had the same fatty acids as the hosts, with the exception of the polyunsaturated fatty acids. These appeared to be synthesized by the nematodes. There was also some speculation that nematode fatty acid synthesis resembled that of bacterial pathways rather than that of higher animals. It was not clear from the study whether intestinal flora of the nematode could have been at least partially responsible for this difference, or whether the nematode itself synthesized the fatty acids. Some nematodes are clearly capable of synthesizing longer chain fatty acids from shorter chain precursors. They are also capable of desaturating the fatty acids [39].

Entomopathogenic nematodes infecting locusts consume host fat and protein [40]. A decrease in lipid reserves has been seen in starved nematodes which can be related to decreased infectivity [41]. Lipid content is also known to decrease when nematodes come out of anhydrobiosis [42]. Lipids associated with the nematode surface [cuticle] are triacylglycerols, sterols, specific phospholipids, and other glycolipids [43-45].

The most widely known class of essential nutrients for nematodes is sterol [36,46]. This nutritional requirement was first discovered by Dutky et al.

[47] and thought to be potentially a means for control of plant parasitic nematodes. A recent review further confirms this nutritional sterol requirement for the nematode *C. elegans* [48]. Nematode parasites of animals also require sterol for larval development [49]. The biochemical mechanism which converts sitosterol to cholesterol appears to be lacking in nematodes [50]. Nematodes are capable of modifying sterols obtained from their diet [46] but degradation of sterols to CO_2 by nematodes is not clear [51]. More than 63 sterols have been identified from free-living and plant-parasitic nematodes. Characteristics of sterols which can be used by nematodes include those which have a hydroxyl group at C-3, a trans-A/B ring system and an intact nonhydroxylated side chain but lack methyl groups at C-4 [52]. Plant sterols are different than animal sterols with plants being unique in methyl, ethyl or related alkyl groups at the C-24 position of the sterol side chain [52]. There are also differences between plant sterols and plant-parasitic nematode sterols. These findings suggest that nematodes ingest plant sterols and remove the C-24 side chain. In addition, the nematode saturates the double bonds in the four-membered ring system to produce stannols [52]. Steroid hormones are important in development processes and in transition to different life stages [53]. Most likely genetic and biochemical methods will be needed to determine the function of hormones found in nematodes [54]. Novel genes involved in the production of 17β-hydroxysteroid dehydrogenase in the soybean cyst nematode have been reported [55].

Sterols were first reported in soy oil by Kraybill et al. [56]. Formononetin is an o-methyl-isoflavone mainly produced in legumes, including soybean plants [57]. It helps stimulate the production of steroids in mammals, and possibly also in nematodes. Research in this area by the USDA was reviewed by Chitwood [58].

3. AMINO ACIDS AND PROTEINS

There are no clearly defined requirements for proteins, amino acids, or peptides for SCN. However, it is unlikely that nematodes synthesize all the amino acids. For humans, there are 9 essential amino acids [phenylalanine, valine, threonine, tryptophan, isoleucine, methionine, leucine, lysine, and histidine]. Some others are required under special circumstances [arginine, cysteine, glutamine, proline, serine, tyrosine, and asparagiene]. Cysteine, tyrosine, and arginine are required during rapid growth, such as in infancy. And, arginine, cysteine, glycine, glutamine, histidine, proline, serine and tyrosine are required by some individuals because these amino acids are not adequately synthesized by these individuals. These are essential components for the synthesis of many essential enzymes and structural proteins ; it is anticipated there are similar needs in the nematode diet.

Protein consumed by parasitic nematodes can severely damage the host. Juveniles have high protein requirements and consuming the host protein can severely weaken the plant [46].

There have been efforts to identify the essential amino acids of nematodes [59-61], but so far common requirements have not been identified. However, protein synthesis in cotton roots is modified when the root-knot nematode [RKN] infects susceptible plants. These plant-parasitic nematodes influence the distribution of amino acids in cotton root galls [61]. Also, there is one genetic modification of the cotton plant which makes them less susceptible to infection by the RKN. This modification is responsible for the synthesis of a 14 kDa protein [60].

For the snail parasitic nematode, *Rhabditis maupasi*, five essential amino acids have been identified. These include lysine, methionine, phenylalanine, tryptophane, and valine [62]. In the entomophilic locust parasite, *M. migrescens*, essential nutrients include protein nitrogen [63]. Essential amino acids have also been identified for the nematode *C. briggsae* [64].

4. VITAMINS

There are 13 essential vitamins required by humans. These include Vitamin A [Retinol] Vitamin B_1[Thiamine] Vitamin C [Ascorbic acid] Vitamin D [Calciferol] Vitamin B_2 [Riboflavin] Vitamin E [Tocopherol] Vitamin B_{12} [Cobalamins] Vitamin K_1 [Phylloquinone] Vitamin B_5 [Pantothenic acid] Vitamin B_7 [Biotin] Vitamin B_6 [Pyridoxine] Vitamin B_3 [Niacin] Vitamin B_9 [Folic acid]. Of these, vitamin E is known to be a nutritional requirement for the gastrointestinal parasite, *Heligmosomoides bakeri* [65], and several of the B vitamins are known to be essential nutrients of *C. elegans* [66-68].

For SCN, DNA sequences responsible for the biosynthesis of enzymes that can produce some of the B vitamins *de novo* have been discovered [69]. Therefore, SCN may not need the same B vitamins as *H. bakeri*, for example. And, it is likely that there are other differences in vitamin and supplement requirements across all nematodes.

5. MINERALS

Considerable research on mineral requirements for nematodes has been reported in mammalian parasites. For example, the gastrointestinal nematode, *H. bakeri*, requires boron [70], zinc [71], and selenium [65] for survival. And, other nematodes have similar mineral requirements [72-74]. For example, magnesium, sodium, potassium, manganese, calcium and copper are required nutrients of *C. elegans* [5]. However, SCN mineral requirements remain unclear.

Whether minerals, influence nematode survival may not help in their control if necessary minerals are readily available in soil, and essential to the host organisms. But, elements not essential to survival of the host could be controlled in soils to help control SCN survival.

6. CARBOHYDRATES

Nematodes require carbohydrates for energy, usually in the form of glycogen. One study showed that several different carbohydrates were sufficient to provide a carbon, or energy source for *C. elegans*, and that glucose was more effective than fructose or sucrose [76]. For *C. elegans*, glucose along with cytochrome c and β-sitosterol were sufficient to sustain a healthy population.

One of the most striking features of soybean chemistry is the abundance of pinitol [77-79]. Pinitol is a carbohydrate with unusual nutritional properties [77]. Figure 2 shows a total ion chromatogram of a derivatized extract of soybean roots. It is unusual for a plant to have so much pinitol. The levels shown in this study indicate pinitol is present at a concentration of 26 mg/g (dry weight) compared to peanuts with only 4.7 mg/g or clover with 14 mg/g [79]. However, there is no evidence that pinitol, or any of the related inositols are needed for SCN survival [79].

7. OTHER NUTRIENTS OR FEEDING REQUIREMENTS

The nematode *Rhabditis maupasi* requires hemin or another iron porphyrin for survival [62]. Similarly, *C elegans* also requires a heme source for survival [34]. It is likely that many other nematodes require heme, or a closely related hemin. There is also good evidence that SCN requires a heme source [80].

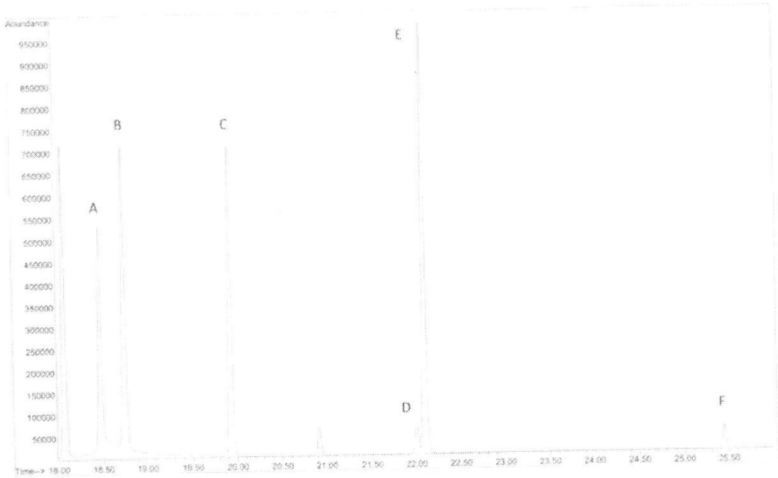

FIGURE 2. A total ion chromatogram of derivatized soybean root extract is shown. A = D-(-)-Fructose, B = D-Pinitol, C = D-(+)-Glucose, D = D-*chiro*-inositol, E = β-D-(+)-Glucose, F = *Myo*-inositol. Reproduced with permission from [79].

8. DISCUSSION

In comparison to our knowledge of human nutrition, our understanding of nutritional requirements of SCN is in its infancy. Limited information is available for members of the Nematoda Phyllum, but such a small amount of information is available that extrapolation across trophic groups and even within genera may be misleading. Finding a successful artificial diet would be a reasonable first step in defining the nutritional needs of SCN. But, this data needs to be coupled with a good understanding of feeding site establishment and plant responses to SCN infections.

Studying biochemical pathways would be a valuable approach, and could also help identify pathways that could be blocked to help minimize SCN survival. Our laboratory began by examining the chemistry of the plant to identify unique nutrients necessary for SCN survival, but that approach was not immediately successful. Another approach is to continue to use DNA mapping to better understand potential plant and parasite pathways. While this approach is less direct, it is currently a very active area of investigation, and can reveal more information than simply nutritional requirements.

Details of the SCN host-parasite responses during infection and feeding site establishment have been more extensively investigated than nutritional requirements. Relationships between the available nutrients from host plants compared to non-hosts could provide valuable clues on these requirements. And, once an adequate media for SCN survival has been well defined, methods to control this pest should follow.

NAN. ACKNOWLEDGEMENTS

The authors acknowledge support from the USDA and Battelle.

REFERENCES

1. Munn, E.A. Munn, P. D. Feeding and digestion in Lee, D. L. (ed). The Biology of Nematodes. Taylor and Frances. New York; 2002. p211-232.
2. Davis, E. L., Hussey, R. S., Baum, T. J. Getting to the roots of parasitism by nematodes. *Trends in Parasitology*. 2004; 20:134-141.
3. Hussey, R. S., Mims, C. W. Ultrastructure of esophageal glands and their secretory granules in the root-knot nematode, *Meloidogyne incognita. Protoplasma*. 1990; 156: 9-18.
4. Wyss, U. Feeding behavior of plant-parasitic nematodes.. In: Lee, D. L. (ed.). The Biology of Nematodes. Taylor and Francis: New York; 2002. p233-259.
5. Wrather, A., Koenning, S. Effects of Diseases on Soybean Yields in the United States 1996 to 2007. *Plant Management Network*. 2009; (http://www.plantmanagementnetwork.org/pub/php/research/2009/yields/).
6. Shannon, J.G., Arelli, P. R., Young, *L. D. Breeding for resistance and tolerance. In Schmitt, D. P., Wrather, J. A., Riggs, R. D. eds, Biology and Management of Soybean Cyst Nematode, Ed 2. Schmitt & Associates of Marceline, Marceline, MO,* 2004: p155–180.
7. Endo, B.Y. Histological responses of resistant and susceptible soybean varieties, and backcross progeny to entry development of *Heterodera glycines. Phytopathology*. 1965; 55:3.75-381.
8. Endo, B.Y. Ultrastructure of initial responses of resistant and susceptible soybean roots to infection by *Heterodera glycines. Revue of Nematology*. 1991; 14:73-94.
9. Riggs, R.D., Kim, K.S., Gipson, I. Ultrastructural changes in Peking soybeans infected with *Heterodera glycines. Phytopathology*. 1973; 63:76-84.
10. Acido, J. R., Dropkin, V. H., Luedders, V. D.. Nematode population attrition and histopathology of *Heterodera glycines*-soybean associations. *Journal of Nematology*. 1984; 16:48-57.
11. Kim, Y. H., Riggs, R. D., Kim, K. S. Structural changes associated with resistance of soybean to *Heterodera glycines. Journal of Nematology*. 1987; 19:177-187.
12. Halbrendt, J.M., Lewis, S.A., Shipe, E. R.. A technique for evaluating *Heterodera glycines* development in susceptible and resistant soybean. *Journal of Nematology*. 1992; 24:84-91.
13. Kim, Y. H., Riggs, R. D. Cyclopathological reactions of resistant soybean plants to nematode invasion. in Wrather, J.A., Riggs, R. D. (eds). Biology and management of the soybean cyst nematode. APS Press. St. Paul. 1992; p157-168.

14. Mahalingam, R., Skorupska, H.T.. Cytological expression of early response to infection by *Heterodera glycines* Ichinohe in resistant PI 437654 soybean. *Genome*. 1996; 39:986-998.

15. Klink, V.P. Overall, C. C., Alkharouf, N., MacDonald, M. H., Matthews, B. F. Laser capture microdissection (LCM) and comparative microarray expression analysis of synctial cells isolated from incompatible and compatible soybean roots infected by soybean cyst nematode (*Heterodera glycines*). *Planta*. 2007; 226:1389-1409.

16. Klink, V.P., Overall, C. C., Alkharouf, N., MacDonald, M. H., Matthews, B. F. A comparative microarray analysis of an incompatible and compatible disease response by soybean (*Glycine max*) to soybean cyst nematode (*Heterodera glycines*) infection. *Planta*. 2007; 226:1423-1447.

17. Klink, V.P., Hosseini, P., Matsye, P., Alkharouf, N., Matthews, B. F. A gene expression analysis of syncytia laser microdissected from the roots of the *Glycine max* (soybean) genotype PI 548402 (Peking) undergoing a resistant reaction after infection by *Heterodera glycines* (soybean cyst nematode). *Plant Molecular Biology*. 2009; 71:525-567.

18. Klink, V.P., Hosseini, P., Matsye, P., Alkharouf, N. W. Matthews, B. F. Syncytium gene expression in *Glycine max* (PI 88788) roots undergoing a resistant reaction to the parasitic nematode *Heterodera glycines*. *Plant Physiological and Biochemistry*. 2010; 48:176-193.

19. Klink, V.P., Overall, C. C., Alkharouf, N., MacDonald, M. H., Matthews, B. F. Microarray detection calls as a means to compare transcripts expressed within synctial cells isolated from incompatible and compatible soybean (*Glycine max*) roots infected by the soybean cyst nematode (*Heterodera glycines*). *Journal of Biomedicine and Biotechnology*. 2010:491217 (1-30).

20. Wyss, U., Stender, C., Lehmann, H.. Ultrastructure of feeding sites of the cyst nematode *Heterodera schachtii* Schmidt in roots of susceptible and resistant *Raphanus sativus* L. var. *oleiformis*. Pers. cultivars. *Physiological Plant Pathology*. 1984; 25: 21-37.

21. Davis, E. L., Hussey, R. S. Baum T. J., Bakker, J., Schots, A. Nematode parasitism genes. *Annual Review of Phytopathology*. 2000; 38:365-396.

22. Gheysen, G., Jones, J. T. Molecular aspects of plant-nematode interactions. In Perry, R. N., Moens, M. (eds). *Plant Nematology*. CABI. 2006; p234-254.

23. Vanholme, B., De Meutter, J., Tytgat, T., Van Montagu, M., Coomans, A. Gheysen, G. Secretions of plant-parasitic nematodes: a molecular update. *Gene*. 2004; 332: 13-27.

24. Riggs, R.D., Hamblen, M.L. Soybean cyst nematode host studies in the family Leguminosae. Arkansas Agricultural Experiment Station Report Series 110. Fayetteville, AR 1962; 17p.

25. Riggs, R.D., Hamblen, M.L. Additional weed hosts of *Heterodera glycines*. *Plant Disease Reporter*. 1966; 50:15-16.

26. Riggs, R.D., Hamblen, M.L. Further studies on the host range of the soybean cyst nematode. Arkansas Agricultural Experiment Station Bulletin 718. Fayetteville, AR 1966; 19p.

27. Miller, L.I., Gray, B. J. Reaction of lambsquarter, Swiss chard, and spinach to eleven isolates of the soybean cyst nematode. *Virginia Journal of Science*. 1966; 17: 246.

28. Miller, L.I., Gray, B. J. Development of eleven isolates of *Heterodera glycines* on six legumes. *Phytopathology*. 1967; 57:647.

29. Venkatesh, R., Harrison, S. K., Riedel, R. M. Weed hosts of soybean cyst nematode (*Heterodera glycines*) in Ohio. *Weed Technology*. 2000; 14:156-160.

30. Venkatesh, R., Harrison, S. K., Regnier, E. E., Riedel, R. M. Purple deadnettle effects on soybean cyst nematode populations in no-till soybean. North Central Weed Science. Society. 2004; 59:56.

31. Riggs, R.D., Host Range. In Riggs, R. D., Wrather, J. A. (eds). Biology and management of the soybean cyst nematode. APS Press, St Paul, MN. 1992; p107-114.

32. Mock, V.A., Creech, J.E., Davis, V.M., Johnson. W. G. Plant growth and soybean cyst nematode response to purple deadnettle (*Lamium pupureum*), annual ryegrass, and soybean combinations. *Weed Science*. 2009; 57:489-493.
33. Riggs, R.D., Nonhost root penetrations by soybean cyst nematode. *Journal of Nematology*. 1987; 19:251-254.
34. Braeckman, B. P., Houthoofd, K., Vanfleteren, J. R. Intermediary metabolism, WormBook, (ed. The C. elegans Research Community), WormBook, doi/10.1895/wormbook. 2009; 1.146.1, http://www.wormbook.org.
35. Krusberg, L.R. Studies on the culturing and parasitism of plant-parasitic nematodes, in particular *Ditylenchus dipsaci* and *Aphelenchoidesrizemabosi* on alfalfa tissues. *Nematologica*. 1961; 6:181-200.
36. Chitwood, D. J. Biochemistry and function of nematode steroids. *Critical Reviews in Biochemistry and Molecular Biology*. 1999; 34(4) 273-284.
37. DeLey, P., Mundo-Ocampo, M. Cultivation of nematodes.. In Nematology: Advances and Perspectives Vol 1. Nematode morphology, physiology and ecology. CABI Publishing, Cambridge, Mass. 2004; p541-608.
38. Krusberg, L. R. Analyses of total lipids and fatty acids of plant-parasitic nematodes and host tissues. *Comparative Biochemistry and Physiology*. 1967; 21: 83-90.
39. Ruess, L., Haggblom, M. H., Zapta, E. J. G., Dighton, J. Fatty acids of fungi and nematodes – possible biomarkers in the soil food chain?" *Soil Biology & Biochemistry*. 2002; 34: 745-756.
40. Gordon, R, Webster, J. M., Hislop, T. G. Mermithid parasitism, protein turnover and vetillogenesis in the desert locust, *Schistocerca gregaria* Forskal. *Comparative Biochemistry and Physiology*. 1973; 46B: 575-593.
41. Barrett, J., Wright, D. J. Intermediary metabolism. In Perry, R. N., Wright, D. J. (eds). The physiology and biochemistry of free-living and plant-parasitic nematodes. CABI. London. 1998; p331-353.
42. Crowe, J. H., Madin, K. A. C., Loomis, S. H. Anhydrobiosis in nematodes: metabolism during resumption of activity. *Journal of Experimental Zoology*. 1977; 201:57-63.
43. Blaxter, M. L. Cuticle surface proteins of wild type and mutant *Caenorhabditis elegans*. *Journal of Biological Chemistry*. 1993; 2638: 6600-6609.
44. Bird, A.F., Bird, J. The Structure of Nematodes. 2nd edition. Academic Press, San Diego. 1991.
45. Spiegel, Y. and M.A. McClure.. The surface coat of plant-parasitic nematodes: chemical composition, origin, and biological role- a review. *Journal of Nematology*. 1995; 27:127-134.
46. Chitwood, D. J., Lusby, W. R. Metabolism of plant sterols by nematodes. *Lipids*. 1991; 26 (8): 619-627.
47. Dutky, S. R., Robbins, W. E., Thompson, J. V. The demonstration of sterols as requirements for the growth, development, and reproduction of the DD-136 nematode. *Nematologica*. 1967; 13: 140.
48. Entchev, E., Kurzchalia, T. V. Requirement of sterols in the life cycle of the nematode *Caenorhabditis elegans*" *Seminars in Cell & Developmental Biology*. 2005; 16: 175-182.
49. Bolla, R.I, Weinstein, P.P., Cain, G.D. Fine structure of the coelomocyte of adult *Ascaris suum*. *Journal of Parasitology*. 1972; 58: 1025-1036.
50. Cheong, C. M., Na, K., Kim, H., Jeong, S., Joo, H., Chitwood, D. J., Paik, Y. A potential biochemical mechanism underlying the influence of sterol deprivation stress on *Caenorhabditis elegans* longevity. *Journal of Biological Chemistry*. 2011; 286: 7248-7256.
51. Rothstein, M. Nematode biochemistry. IX. Lack of sterol biosynthesis in free-living nematodes. *Comparative Biochemistry and Physiology*. 1968; 27: 309-317.
52. Chitwood, D. J. Biosynthesis in Perry, R. N., Wright, D. J. (eds). The physiology and biochemistry of free-living and plant-parasitic nematodes. CABI. London. 1998; p303-330.

53. Riddle, D.L. and P.S. Albert. Genetic and environmental regulation of dauer larva development. In *C. elegans* II. Riddle, D., Meyer, B., Priess, J., and Blumenthal, T. (ed). Cold Spring Harbor: Cold Spring Harbor Press. 1997; p739-768.
54. Chervitz, S.A., Aravind, L., Herloc, G. S., Ball, C. A., Koonin, E. V., Dwight, S. S.. Comparison of the complete protein sets of worm and yeast: orthology and divergence. *Science*. 1998; 282: 2022-2028.
55. Skantar, A. M., Guimond, N. A., Chitwood, D. J. Molecular characterization of two novel 17β-hydroxysteroid dehydrogenase genes from the soybean cyst nematode *Heterodera glycines*. *Nematology*. 2006; 8(3) 321-333.
56. Kraybill, H. R., Thornton, M. H., Eldridge, K. E. Sterols from crude soybean oil. *Industrial and Engineering Chemistry*. August, 1940:1138-1139.
57. Medjakovic, S. and Jungbauer, A. "Red Clover Isoflavones Biochanin A and Formononetin are Potent Ligands of the Human Aryl Hydrocarbon Receptor". *The Journal of Steroid Biochemistry and Molecular Biology*. 2008; 108 (1–2) 171–177.
58. Chitwood, D. J. Research on plant-parasitic nematode biology conducted by the United States Department of Agriculture-Agricultural Research Service." *Pest Management Science*. 2003; 59: 748-753.
59. Young, J. R., Riggs, R. D. Identification of the free amino acids of nematode resistant and susceptible soybeans. *Arkansas Academy of Science Proceedings*. 1964; 18: 46-49.
60. Hedin, P. A., Creech, R. G. Altered amino acid metabolism in root-knot nematode inoculated cotton plants. *Journal of Agriculture and Food Chemistry*. 1998; 46: 4413-4415.
61. Hedin, P. A., Creech, R. G. Effects of root-knot nematodes on distribution of amino acids in cotton root galls. Mississippi State University Experiment Station Bulletin 1103 Office of Agricultural Communications, Mississippi. 2001; p1-6.
62. Brockelman, C. R., Jackson, G. J. Amino acid, heme, and sterol requirements of the nematode, *Rhabditis maupasi Journal of Parasitology*. 1978; 64(5): 803-809.
63. Baylis, H. A. Observations on the nematode *Mermis migrescens* and related species. *Parasitology*. 1944; 36: 122-132.
64. Vanfleteren, J. R. Amino acid requirements of the free-living nematode *Caenorhabditis briggsae*. *Nematologica*. 1973; 19: 93-99.
65. Smith, A., Madden, K. B., Yeung, K. J. A., Zhao A., Elfrey, Finkelman, , F. J., Lavender, O., Shea-Donohue, T., Uban, J. F. Deficiencies in selenium and/or vitamin E lower the resistance of mice to *Heligmosomides polygyrus* infections. *Journal of Nutrition*. 2005; 135: 830-836.
66. Liu , C., Lu, N. Biotin requirement and its biosynthesis blockage in the free-living nematode, *Caenorhabditis elegans*. *Federation of American Societies for Experimental Biology*. 2008; 22: 1102-1104.
67. Nicholas, W. L. Hansen, E., Dougherty, E. C. The B-vitamins required by *Caenorhabditis briggsae (Rhabditidae)*. *Nematologicia*. 1962; 8: 129-135.
68. Szewczyk, N. J. , Kozak, E., Conley, C. A. Chemically defined medium and *Caenorhabditis elegans*. *BMC Biotechnology*. 2003; 3: 19.
69. Craig, J. P., Bekal, S. , Niblack, T. , Domier, L., Lambert, K. N. Evidence for horizontally transferred genes involved in the biosynthesis of vitamin b1, b5, and b7 in *Heterodera glycines*. *Journal of Nematology*. 2009; 41(4) 281–290.
70. Bourgeois, A. C., Scott, M. E., Sabally, K., Koski, K. G. Low dietary boron reduces parasite (nematoda) survival and alters cytokine profiles but the infection modifies liver minerals in mice. *Journal of Nutrition*. 2007; 137(9): 2080-2086.
71. Shi, H. N., Scott, M. E., Koski, K. G., Boulay, M., Stevenson, M. Energy restriction and severe zinc deficiency influence growth, survival and reproduction of *Heligmosomoides polygyrus* (Nematoda) during primary and challenge infections in mice. *Parasitology*. 1995; 110: 599-609.
72. Coop, R. L., Kyriazakis, I. Influence of host nutrition on the development and consequences of nematode parasitism in ruminants. *Trends in Parasitology*. 2001; 17(7) 325-330.

73. Koski, K. G., Scott, M. E. Gastrointestinal nematodes, trace elements, and immunity. *Journal of Trace Elements in Experimental Medicine.* 2003; 16: 237-51.
74. McClure, S. J., McClure, T. J., Emery, D. I. Effects of molybdenum intake on primary infection and subsequent challenge by the nematode parasite *Trichostromgylus colubriformis* in weaned Merino lambs. *Research in Veterinary Science.* 1999; 67: 17-22.
75. Lu, N. C., Cheng A. C., Briggs, G. M. A Study of Mineral Requirements in *Caenorhabditis elegans. Nematologica.* 1983; 29: 425-434.
76. Lu, N. C., Goetsch, K. M. Carbohydrate Requirement of *Caenorhabditis elegans* and the Final Development of a Chemically Defined Medium. *Nematologica.* 1993; 39: 303-311.
77. Campbell, J.A., Goheen, S.C., Donald, P. Extraction and Analysis of Inositols and Other Carbohydrates from Soybean Plant Tissues. in Recent Trends for Enhancing the Diversity and Quality of Soybean Products, Krezhova, D. ed. InTech publisher http://www.intechopen.com/articles/show/title/extraction-and-analysis-of-inositols-and-other-carbohydrates-from-soybean-plant-tissues. 2011; p421 – 446.
78. Garland, S., Goheen, S., Donald, P.A., Campbell, J. Application of derivatization gas chromatography/mass spectrometry for the identification and quantitation of pinitol in plant roots. *Analytical Letters.* 2009; 42: 2096-2105.
79. McDonald IV, L. W., Goheen, S. C., Donald, P. A., Campbell J. A. Identification and quantitation of various inositols and o-methylinositols present in plant roots related to soybean cyst nematode host status. *Nematropica.* 2012; 42: 1-8.
80. Ko, M. P., Huang, P., Huang, J., Barker, K. R. The occurrence of phytoferritin and its relationship to effectiveness of soybean nodules. *Plant Physiology.* 1987; 83: 299-305.

CHAPTER 9

Evolution of Embryonic Development in Nematodes

Jens Schulze and Einhard Schierenberg*

University of Cologne, Biocenter, Zuelpicher Str. 47b 50967 Köln, Germany

ABSTRACT

Background

Nematodes can be subdivided into basal Enoplea (clades 1 and 2) and more derived Chromadorea (clades 3 to 12). Embryogenesis of *Caenorhabditis elegans* (clade 9) has been analyzed in most detail. Their establishment of polarity and asymmetric cleavage requires the differential localization of PAR proteins. Earlier studies on selected other nematodes revealed that embryonic development of nematodes is more diverse than the essentially invariant development of *C. elegans* and the classic study object *Ascaris* had suggested. To obtain a more detailed picture of variations and evolutionary trends we compared embryonic cell lineages and pattern formation in embryos of all 12 nematode clades.

Methods

The study was conducted using 4-D microscopy and 3-D modeling of developing embryos.

Results

We found dramatic differences compared to *C. elegans* in Enoplea but also considerable variations among Chromadorea. We discovered 'Polarity Organizing Centers' (POCs) that orient cleavage spindles along the anterior-posterior axis in distinct cells over consecutive cell generations. The resulting lineally arranged blastomeres represent a starting point for the establishment of bilateral symmetry within individual lineages. We can discern six different early cleavage types and suggest that these variations are due to modifications in the activity of the POCs in conjunction with changes in the distribution of PAR proteins. In addition, our studies indicate that lineage complexity advanced considerably during evolution, that is we observe trends towards an increase of somatic founder cells, from monoclonal to polyclonal lineages and from a variable (position-dependent) to an invariable (lineage-dependent) way of cell fate specification. In contrast to the early phase of embryogenesis, the second half ('morphogenesis') appears similar in all studied nematodes. Comparison of early cleavage between the basal nematode *Tobrilus stefanskii* and the tardigrade *Hypsibius dujardini* revealed surprising similarities indicating that the presence of POCs is not restricted to nematode embryos.

Conclusions

The pattern of cleavage, spatial arrangement and differentiation of cells diverged dramatically during the history of the phylum Nematoda without corresponding changes in the phenotype. While in all studied representatives the same distinctive developmental steps need to be taken, cell behavior leading to these is not conserved.

Keywords: nematode embryogenesis cell lineage polarity symmetry formation cell specification evolution *Tobrilus Prionchulus C. elegans*

BACKGROUND

Over many decades various suggestions have been made concerning phylogenetic relationships among nematodes. With the availability of an increasing number of gene sequences the phylogeny of this phylum was put on a more objective basis and resulted in major revisions of previous classifications [1, 2, 3, 4]. In this work we refer to the phylogeny of Holterman *et al.*[5], dividing the phylum Nematoda into 12 different clades (Figure 1a). We followed the proposal of De Ley and Blaxter [2] based on

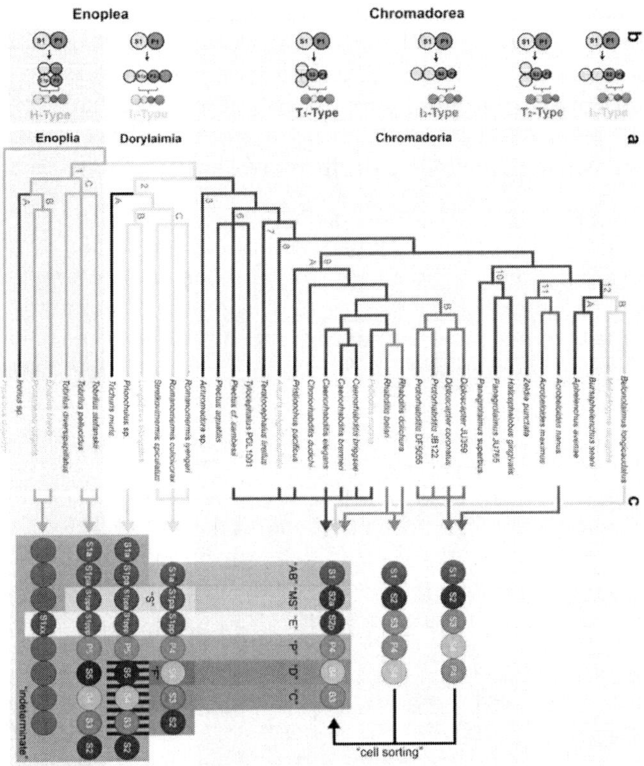

Figure 1. PHYLOGENY AND DEVELOPMENT. **A.** Phylogenetic tree of nematodes (after [5]; clade 9 after [4]) with species whose embryogenesis is compared here. Different colors of lineage branches indicate distinct early cleavage patterns as indicated in Figure 1b. Individual clades (1 to 12) are marked with red numbers. Clades 4 and 5 are not included due to fragmentary data (see results). Based on distinct developmental characters described in the text some clades are further subdivided (A, B, C). Data of species written in grey have been extracted from the literature. References are given in the text. **B.** Six early cleavage types. Color of lettering corresponds to lineage branches where this pattern is found. Division of 2 to 4 cells (top) and origin and relative position of alimentary tract (Enoplea) or gut (Chromadorea) precursor (yellow) and germline cell P3 (red) generated with the next division are shown. **C.** Alignment of early blastomeres (colored circles) along the a-p axis. Colored columns indicate fate assignments ('AB' = AB-like and so on, for definition, see materials and methods). From clade 2C onward all early blastomeres can be assigned one of the six basic fates, however the position in the sequence of cells varies between species. Striated area indicates that in *Prionchulus* S3-S5 form largely bilaterally symmetric clones (for further description, see text). In some Chromadorea the initial order of founder cells is different to that in *C. elegans* due to the absence of PR in the germline. After cellular rearrangements ('cell sorting') they all merge into a single, standard pattern prior to the onset of gastrulation. PR, polarity reversal in the germline.

molecular and morphological criteria and subdivide nematodes into the two classes Enoplea (clades 1 and 2) and Chromadorea (clades 3 to 12). The former consists of two subclasses Enoplia (clade 1) and Dorylaimia (clade 2). Molecular and morphological data indicate that clade 1 comprises representatives closest to the common ancestor of nematodes while Chromadorea include phylogenetically more derived species [2, 5, 6].

Our current picture of embryonic development in nematodes is essentially shaped by the striking similarity between the classic model system *Ascaris megalocephala* (Clade 8; Figure 1a; [7, 8]) and *Caenorhabditis elegans* (clade 9; [9]; http://www.wormbook.org). Work on *C. elegans* and its closer relatives has provided an initial insight into the recent evolution of embryonic and postembryonic development in clades 8 to 10 and discloses wide homologies in features, phenotypes and cell lineages [4, 10, 11, 12, 13, 14, 15, 16, 17, 18, 19]. Nevertheless, our understanding of the evolution of embryogenesis in the nematode phylum is still fragmentary. Species studied so far were usually chosen because of easy accessibility and amenable breeding conditions, and therefore represent a biased minority of the taxon. But generalizations from developmental characters of model organisms have to be taken with caution, because these organisms are often highly derived [20].

Embryonic studies revealed distinct developmental characters of nematode species and higher taxa that can be related to their phylogenetic position [14, 21, 22, 23, 24, 25, 26]. But not only inter-species but also intra-species variations have been uncovered, for example plasticity in pattern formation in *Acrobeloides nanus*[21] and *Diploscapter coronatus*[27] or postembryonic mouth dimorphism in *Pristionchus pacificus* (clade 9; [28, 29]). A high regulative potential was demonstrated by the hierarchy of somatic cell fate transformations after cell ablation in the early embryo of *A. nanus* (clade 11; [30]). Even more dramatic peculiarities are found in Enoplea. Asymmetric cleavages and distinct cell lineages are initially missing in clade 1B (*Enoplus brevis, Pontonema vulgaris*) and only a gut lineage is present [22, 31]. Development of *Tobrilus diversipapillatus* (clade 1C) is characterized by a prominent coeloblastula [32], a developmental character thought to be absent in nematodes. Compared to *C. elegans, Romanomermis culicivorax* (clade 2C) displays major differences in the establishment of embryonic polarity, pattern formation, programming of somatic founders and cell lineage complexity [33, 34]. Hence, embryonic development of nematodes is much more diverse than the essentially invariant development of *C. elegans* and its closer relatives indicates.

Although in the Enoplea, development is usually much slower than in *C. elegans*, and their embryos are less transparent, we performed an extended analysis of development in selected species of this poorly studied basal group. Additionally we studied development in more detail in those clades of Chromadorea where only limited embryonic data have been available so far (Figure 1a). With these results, we now can address to what extent (i) nematodes follow a common general developmental program to generate the body plan typical for this phylum and (ii) developmental differences can be related to phylogenetic position (that is whether specific points can be defined where during evolution certain characters first appeared). The question which type of early cleavage was followed by the last common ancestor of nematodes has been controversially discussed in the past [35, 36]. Therefore, we explored the notion that a nematode species with invariant polyclonal lineages generated by a fixed set of founder cells, such as *C. elegans* may have evolved from an ancestor such as *Enoplus* with just a single monoclonal lineage and a predominantly variable early embryogenesis.

METHODS

Strains and culture

The strains *Acrobeloides maximus* (DF5048), *Acrobeloides nanus* (ES501), *Caenorhabditis brenneri* (SB280), *Caenorhabditis briggsae* (AF16), *Caenorhabditis elegans* (N2), *Choriorhabditis dudichi* (SB122), *Diploscapter* sp. (JU359), *Diploscapter coronatus* (PDL0010), *Halicephalobus gingivalis* (JB128), *Panagrolaimus* sp. (JU765), *Panagrolaimus superbus* (DF5050), *Plectus aquatilis* (PDL0018), *Plectus cf. sambesii* (ES601), *Pristionchulus pacificus* (PS312), *Protorhabditis* sp. (DF5055), *Protorhabditis* sp. (JB122), *Rhabditis belari* (ES103), *Rhabditis dolichura* (ES101), *Teratocephalus lirellus* (JB049), *Tylocephalus* sp. (PDL1001), *Zeldia punctata* (PDL0003) are cultured at 23°C on minimal agar plates essentially as described in [26]. Gravid animals of *Achromadora sp.*, *Aphelenchus avenae, Bursaphelenchus seani, Ironus sp.*, *Prionchulus sp.*, *Tobrilus diversipapillatus, Tobrilus pellucidus* and *Tobrilus stefanskii* are isolated from soil samples of various origins essentially as described in [21]. *Romanomermis culicivorax, Romanomermis iyengari* and *Strelkovimermis spiculatus* were kindly provided by Dr. Edward Platzer, University of California, Riverside, CA, USA. *Trichuris muris* was kindly provided by Dr. Heinz Mehlhorn, Heinrich-Heine University, Düsseldorf, Germany. *Belonolaimus longicaudatus* was analyzed in the laboratory of Dr. Ole Becker, University of California, Riverside, CA, USA. Some gravid *Enoplus brevis* were kindly provided by Dr. V. Malakhov, Moscow State University, Russia, others were isolated from salt marsh soil supplied by Dr. W. Armonies, AWI List, Germany. Additional strains were obtained from Paul de Ley and Jim Baldwin, University of California,

Riverside, CA, USA; Marie-Anne Felix, University Jacques Monod, Paris, France; Wouter Houthoofd and Wim Bert, University of Ghent, Belgium; Walter Sudhaus, Freie Universität Berlin, Germany; and Ralf Sommer, Max-Planck-Institute for Developmental Biology, Tübingen, Germany.

Cell nomenclature and cell fate assignments

Projection of the *C. elegans* standard cell nomenclature [9, 37] onto other nematodes implies the presence of similar cell differentiation patterns, which is not necessarily true. Therefore, we apply neutral lineage names (S1-S4/S5, somatic founder cells; P1-P4/P5, germline). We use the standard nomenclature (AB, MS, C, D) to indicate cell fates that correspond to those in *C. elegans* as described in [34]. In this species but not necessarily in other nematodes S1 = AB, S2 = EMS, S3 = C, S4 = D.

Although we traced cell divisions only up to a few hundred cells, we could assert the degree of variation from the pattern found in *C. elegans* (fixed lineages with unambiguous fate assignments). As we did not observe prominent rearrangements of cell clones after gastrulation we were able to assign fate categories ('AB'-like, 'S'-like, and so on; Figure 1c) to each of the founder cells based on position, structure and behavior of their descendants [34]. These cell fate categories imply similarities to *C. elegans* in terms of cell types derived from this lineage (for example 'AB', generates at least the majority of neurons; 'S' contributes to the pharynx) but may differ in detail. 'E' stands exclusively for gut fate in all studied species.

Microscopy

With the exception of Enoplus (see below) 1-cell and 2-cell stage embryos collected from culture plates or cut out of gravid adults were mounted on slides carrying a thin 3% agarose layer as a mechanical cushion. The cover slip was sealed with melted petroleum jelly. Embryonic development was studied with DIC optics using a 100 × objective. Stacks of optical sections were digitally recorded at 30 to 60 second intervals and at 23°C with a 4-D microscope (Zeiss, Axioscope 2 mot). 3-D tracing of cell behavior and generation of cell lineages were software-supported (Simi Biocell; Unterschleißheim, Germany) essentially as described in [34]. Due to a sticky surface *Enoplus* 1-cell stages adhered reliably to an untreated microscope slide. This was immersed into a petri dish filled with brackish water (ratio sea water to tap water, 2:1) and then covered with aluminum foil. Development was recorded on time lapse video tape with a 20 × water immersion objective. Evaporated water was replaced daily. If not stated otherwise the development of at least three embryos per species has been analyzed.

RESULTS

Embryogenesis of *Enoplus brevis*

We confirmed, with photo documentation, the observations of Russian researchers (see introduction). First cleavages are equal (Figure 2a1-a2) and lead to variable spatial arrangements. In the 8-cell stage one blastomere divides after a delay (Figure 2a3, black asterisk). This cell is the founder of the gut lineage. During gastrulation its two daughter cells are translocated into the interior of the embryo (Figure 2a4-a5) as in *C. elegans*.

Embryogenesis of *Tobrilus stefanskii*

While in the early *E. brevis* embryo, only a single fixed cell lineage exists [31], in *Romanomermis culicivorax* already six fixed lineages are present as in *C. elegans* even though with different fate assignments [34]. Searching for species occupying intermediate evolutionary steps between these two extremes, we analyzed early embryogenesis of the basal nematode *Tobrilus stefanskii* (n = 11; clade 1C). With its early symmetric cleavages, the absence of visible cell lineages and its prominent coeloblastula, first described for another member of the genus [32], embryogenesis differs strikingly from *C. elegans* (clade 9A; [9]) and *R. culicivorax* (clade 2C; [33, 34]).

T. stefanskii is a gonochoristic species with oviposition at the 1-cell stage preceding pronuclear fusion (Figure 2b1). The first division generates cells of equal size (Figure 2b2). Both spindles in the 2-cell stage orient at right angles to the first division, following the 'centriolic principle' (that is during consecutive cleavages spindles form at right angles to each other due to changing positions of centrioles; [38, 39]). We named this early cleavage pattern 'H-Type' (Figure 1b). The first two cell division rounds produce descendants of equal size (Figure 2b2-b3). In the 4-cell stage various spatial arrangements are found [22, 32]. Then the first asymmetric cleavages are observed. Based on these and subsequent asymmetries, and on specific cell behavior and lineage analysis described below, cells could be assigned

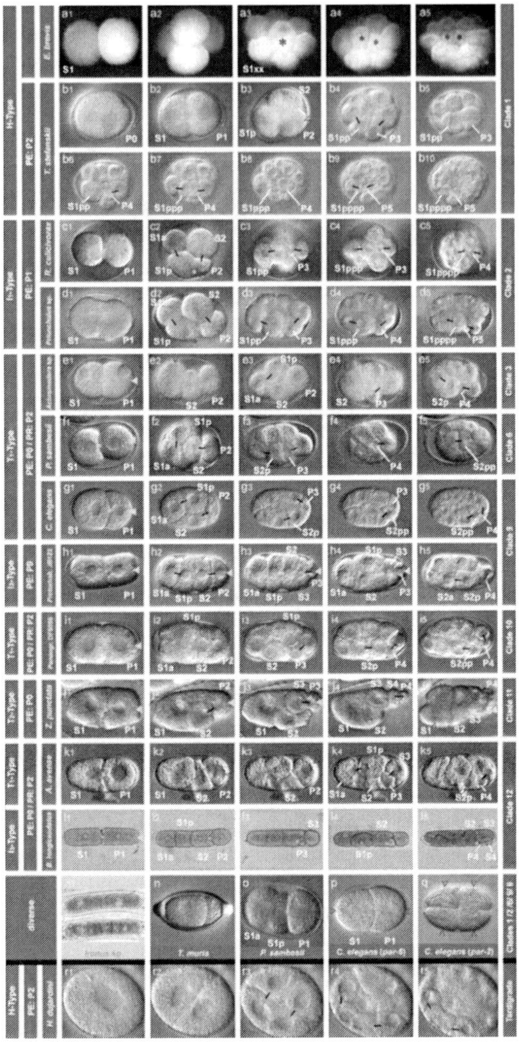

Figure 2. CLEAVAGE AND POLARITY. Early cleavage patterns in 12 nematode species (**A**-**L**) and the tardigrade *Hybsibius dujardini* (**R**). In addition, single images of selected nematode embryos are shown (**M**-**Q**). Representatives are ordered according to clades (right margin). On the left margin the type of early cleavage, stage when visible polarity is established (PE) and polarity reversal in the germline (PR) are indicated where applicable. Position and action of POCs deduced from cell behavior are marked. Green dot, primary POC acting on two adjacent cells; green arrowhead, primary POC acting on P1 only; purple dot, secondary POC acting on two adjacent cells; yellow arrowhead, tertiary POC, acting on germline cells; open red arrowhead, orientation of cleavage spindle; black bars connect sister cells. Figure 2r was taken from [48] with permission from Elsevier; Figure 2q was taken from [67].

lineage names in retrospect (for cell nomenclature, see materials and methods). One blastomere of the 4-cell stage shows the stem cell behavior and additional features typical for a germline cell in other nematodes (for example permanent contact with gut precursors; see below). Hence, we named it P2 and its sister S2. The remaining two cells are then daughters of S1. For better comparison with other species we named them S1a and S1p, although position relative to each other is variable. The nuclei in the cell cousins S1p and P2 leave their central position and become located side by side at the cell membrane (not shown). This is as found in *R. culicivorax* where, however, cell asymmetry starts to form already in the 2-cell stage (Figure 2c1; [33]). With the onset of division, the forming mitotic spindles in S1p and P2 become oriented and asymmetrically displaced toward a common point (green square, Figure 2b3). The orientation of these two spindles corresponds to the anterior-posterior (a-p) axis of the embryo. The resulting P3 is much smaller than its sibling S3, while size differences between S1pa and S1pp are less prominent (Figure 2b4).

The events described above are repeated when the cousins S1pp and P3 (Figure 2b4 to b6) and S1ppp and P4 (Figure 2b7 to b9) cleave asymmetrically (marked in Figure 3m). Spindles in S1pppp and P5 are oriented along the a-p axis as well but the divisions of these cells are not obviously asymmetric. The soma/germline cell pairs named above remain firmly attached to each other and their mitotic spindle microtubules appear to find distinct anchorage sites in the area of cell attachment (Figure 2b3 to b10) leading to a linear orientation of spindles. Consequently an array of cells forms along the future midline as a prerequisite for the establishment of bilateral symmetry (see paragraph below). This strongly resembles cell behavior in the early embryo of *R. culicivorax* (Figure 2c1 to c5; [33]) induced by the 'Region of First Midbody' (RFM). As we do not know whether the underlying mechanism is the same in both species we named it more generally 'Polarity Organizing Center' (POC). To distinguish it from other centers found in the higher-numbered clades (see below) we call it 'primary POC' (marked in green in Figures 2 to 5).

After having identified a germline, we traced the fate of the remaining cells. In the 8-cell stage S1pp and P3 are neighbors with nuclei asymmetrically positioned adjacent to the primary POC (Figure 3a), and both divide with delay (Figure 3m). S1pp turns out to give rise to the alimentary tract (that is gut+pharynx; Figure 3a to 3f). This is different from *R. culicivorax*[34] and *Enoplus*[31, 40], where the founder cell for the alimentary tract is already present one cell generation earlier (Figure 1c). In Chromadorea the differences are more dramatic as there the gut is generated by S2p (Figure 1c and below; [7, 9, 21, 22]).

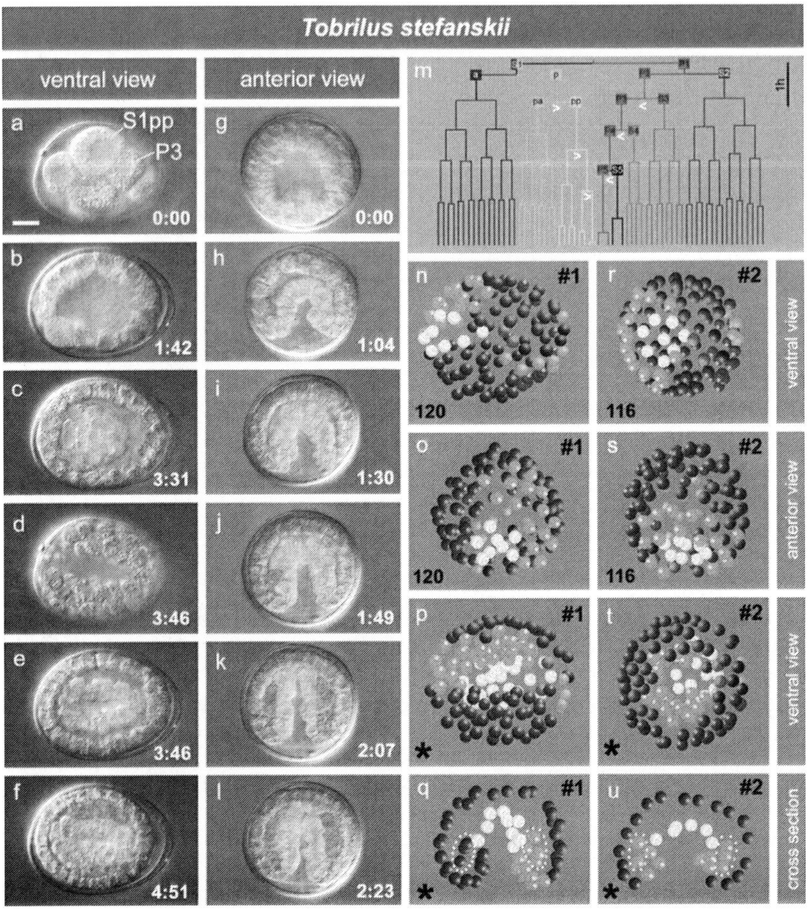

Figure 3. EMBRYOGENESIS OF ***Tobrilus stefanskii.*** A-F; G-L, cleavage, blastocoel formation, gastrulation and tissue formation of two embryos. Yellow, alimentary tract precursor(s); green, blastocoel; blue, slit-like blastopore; red, body muscle cells, green dot, primary POC; time in hours and minutes; **M**, cell lineage with seven somatic lineage branches and the germline (P0-P5); white arrowheads, asymmetric divisions; **N-Q** and **R-U**, 3-D reconstructions of two embryos with similar stages indicating different spatial arrangements of cells. Each sphere represents a nucleus, color code as shown in m; p, t, partial reconstruction (*), not all cells could be traced, total cell number > 200; q, u, same stages as p and t, respectively, optical cross section; dotted areas, cells contributing to mesoderm. Scale bar, 10 µm.

As in other nematodes [21, 41] the germline in *T. stefanskii* remains in permanent contact with the gut founder and its descendants. In contrast to *C. elegans*, in *T. stefanskii* P4 performs another asymmetric division resulting in S5 and P5 (Figure 2b10). The latter behaves like P4 in *C. elegans*, in that it divides symmetrically. Its daughters follow the gut precursors into the center of the embryo (Figure 3p, t).

Typical for all three of the *Tobrilus* species we looked at, is the formation of a prominent coeloblastula (Figure 3b). They undergo a 'canonical' gastrulation, with the invagination of a layer of alimentary tract precursor cells (Figure 3c, g to j, n to 3u; [32]). The blastocoel starts to form as early as the 4-cell stage and grows to its full size before the 100-cell stage. Gastrulation starts with the invagination of the eight S1pp descendants (Figure 3g, o, s). Other cell groups join them but their lineage origin varies among the analyzed embryos. As divisions continue, the immigrated cells fill the blastocoel (Figure 3c, g to 3i), leaving an oval-shaped blastopore furrow on the ventral side (Figure 3d) that reaches deep into the alimentary tract primordium (Figure 3k). Eventually it closes from the center in both directions along the a-p axis like a zipper, resulting in future mouth and anus at its ends (not shown) similar to *R.culicivorax* and other basal nematodes [22, 34]. After immigration of future alimentary tract cells (leaving only small remnants of the blastocoel), body muscle precursors of various lineage origin move between these and the outer layer of cells (Figure 3f, j to 3l).

Lineage analysis and 3-D reconstructions (n = 9) revealed that in contrast to members of clade 1B in the early *T. stefanskii* embryo, three distinct cell lineages can be defined (Figure 1c), for gut ('E'), pharynx ('S'; additional cells contribute variably to this organ) and germline ('P'). For the remainder no fixed correlation between lineage and fate exists, that is the spatial arrangement of these blastomeres and their contributions to tissues is variable. This is exemplified in the two embryos shown in Figure 3. In embryo #1 (Figure 3p) descendants of S1pa (orange; left side) and S2 (brown; right side) form separate clones. Cells of both origins migrate in (Figure 3q) to contribute to the mesoderm. In embryo #2 (Figure 3t) S1pa descendants are located on both sides and at this stage cells between the inner and outer cell layer are all members of this lineage branch (Figure 3u). Among the other specimens of *T. stefanskii* we found various combinations of S1a, S1p and S2 lineages contributing to body muscle cells.

Embryogenesis of *Prionchulus* sp

To narrow this gap between *Tobrilus* with three and *Romanomermis* with six lineages we studied a second representative of clade 2B, *Prionchulus* sp. (Mononchida, n = 4). In earlier studies contradictory conclusions had been drawn concerning the existence of early asymmetric cleavages and distinct cell lineages in this genus [42, 43].

We found several basic similarities to *R. culicivorax* (Figure 2c1 to c5; [33, 34]) in that the first division in *Prionchulus* is symmetric, nuclei of the 2-cell stage occupy adjacent positions (Figures 2d1, 4b), and both spindles become oriented along the a-p axis (Figure 4c to 4d). The latter indicates the action of a primary POC already in the 2-cell stage. We call this the 'I-Type' of cleavage. As it differs from similar I-Type cleavage patterns seen in higher-numbered clades (see below) we named it 'I_1-Type (Figure 1b). In *Prionchulus*, S1p is definitely larger than S1a while size differences between S2 and P2 are less prominent (Figures 2d2, 4e). A similar pattern can be identified in (*Longidorus elongatus* Figure 1a; [44])

Restricted by the limited space in the eggshell, 4-cell stage blastomeres arrange themselves in two alternative rhomboid variants, whereby either S2 or P2 occupy a position adjacent to S1a (Figure 2d2, n = 2; Figure 4f, n = 2). In both of these S1p and P2 touch each other and form lineally oriented spindles (Figure 4f) resulting in daughter blastomeres arranged in tandem (Figure 2d3). The same process is repeated with the division of the cell pairs S1pp-P3 (Figures 2d3 to d4, g to 4h) and S1ppp-P4 (Figures 2d4 to d5, 4h to 4i). The fact that members of each pair do not move relative to each other underlines the continuous presence of a centrally located POC.

Gastrulation starts with the immigration of S1pppa+S1pppp in the absence of a coeloblastula (Figures 2d5, 4i). Descendants of S1ppa follow somewhat later and contribute to the pharynx. Thus, they show some resemblance to the behavior of MS (S2a) in *C. elegans*. However, as we could not find any contribution to body muscles we marked their fate with 'S' in Figure 1c. The same applies to *Tobrilus* (see above) and *Romanomermis*[34]. Contacts remain between P5, still positioned on the surface, and S1pppp and its posterior daughter S1ppppp (Figure 4i to 4k). *Prionchulus* shows high similarities to *Tobrilus* with respect to blastopore formation (Figure 5l) and the behavior of S1a, S1pa and S2 descendants. A further similarity, revealed by 3-D reconstructions of two embryos (Figure 4n to u) is the considerable positional variability of blastomeres.

As in *Tobrilus* we could only identify lineages for the alimentary tract ('S' + 'E') and the germline ('P'). However, here additional cells (S3-S5) occupy essentially invariant positions. They participate in the formation of bilaterally symmetric structures (Figures 1c, 4 p, q, t, u; see below) and appear to contribute to body muscles and hypodermis roughly corresponding to the C and D lineages in *C. elegans*. Due to limited transparency we could not ascertain whether S3-S5 make fixed contributions to the developing embryo. In any case these differ from the pattern found in *Romanomermis* and in *C. elegans*.

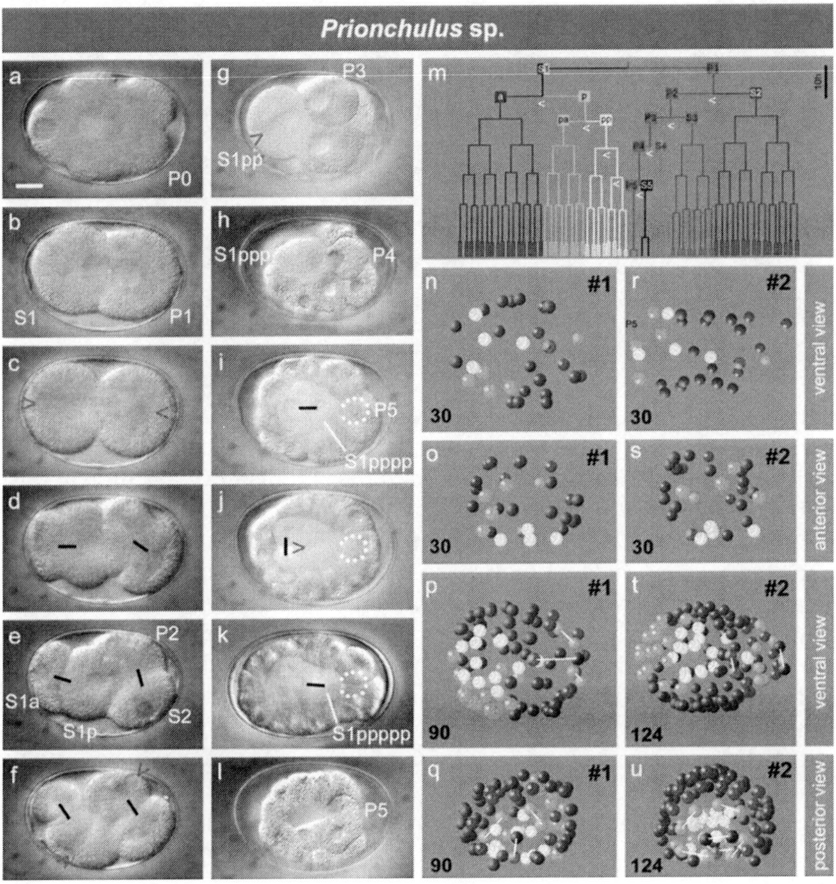

Figure 4. EMBRYOGENESIS OF ***Prionchulus*** SP. A-L, cleavage, gastrulation and blastopore formation. Yellow, alimentary tract precursor(s); blue, slit-like blastopore; dotted circle, position of P5 (out of focus); green dot, primary POC; red arrowhead, orientation of cleavage spindle; black bars connect sister cells; **M**, cell lineage with seven somatic lineage branches and a germline (P0-P5); white arrowheads, asymmetric divisions; **N-Q** and **R-U**, 3-D reconstructions of two embryos; p-q, t-u, differences in cell numbers (lower left corner) intentionally to visualize +/- bilateral symmetry in descendants of S4-S5 and to a lesser degree also of S3 in contrast to variable spatial arrangements of S1 and S2 descendants. Sister cells marked with white bars. Each sphere represents a nucleus; color code as shown in m. Scale bar, 10 μm.

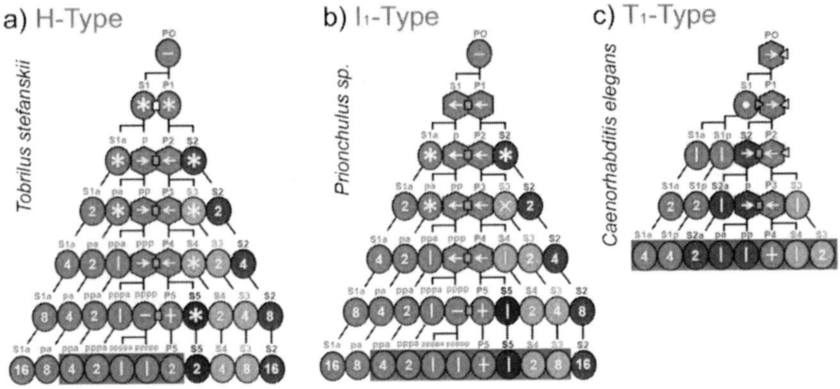

Figure 5. EARLY CLEAVAGE AND MODEL OF POC ACTION. A-C, Simplified early lineage trees of three nematode species with different cleavage types. A series of unequal cleavages attributed to the action of POCs results in cells arranged in tandem along the a-p axis. For symbols and color codes of POCs, see Figure 2. Note that cells performing l-r divisions belong to different generations. These cells or their descendants are shown in grey boxes to indicate lineage branches involved in the establishment of bilateral symmetry. Circles, blastomeres performing symmetric cleavages; hexagons, blastomeres performing asymmetric cleavages. Orientation of cleavage spindles: asterisk, variable; arrow, longitudinal, pointing toward smaller daughter cell; horizontal bar, longitudinal with daughters of equal size, vertical bar, left-right; circle, dorsal-ventral; 'X' somewhat variable generating imperfect symmetric clones; '+', germline cell giving rise to two mirror image gonadal arms in the anterior and posterior half of the adult.

Embryogenesis in *Prionchulus* is much slower than that in *Tobrilus*. Initially, in both species cell cycles of all blastomeres are close to synchronous. Later, descendants of P3 and S1pp divide more slowly than other blastomeres. This is more obvious in *Tobrilus* than in *Prionchulus* (Figures 3m and 4m).

Our finding that cell lineages exist in the early *Prionchulus* embryo accords with the report of Drozdovskiy [42], however, we did not find the *Ascaris*-like invariant cleavage pattern with strict early bilateral symmetry and invariant cell positioning as shown in his sketches.

Early embryogenesis in Chromadorea: 4 different cleavage types

T_1-type, Clades 3, 4, 5, 6, 7, 8, 9A, 10, 12A (Figure 1b, blue lineage branches)

In the standard *C. elegans* (clade 9A; Figure 2g1 to g5), the first division is unequal, generating a larger S1 and a smaller P1 cell. While AB divides with transverse spindle orientation (following the centriolic principle; see above) P1 reorients its spindle to divide into a larger anterior S2 and a smaller posterior P2 (Figure 2g1 to g2). We named this early cleavage pattern 'T-typ'. In P2 a reversal of cleavage polarity (PR) takes place [41]. To distinguish this cleavage pattern we call it 'T_1-type'.

In *Achromadora* (clade 3; Figure 2e1 to e5), the first division is more or less equal in size (Figure 2e1) otherwise early cell behavior is similar to *C. elegans*. Our limited observations in representatives of clades 4 and 5 (Desmodorida and Monhysterida) correspond well with descriptions by Malakhov [22] and make clear that, like *C. elegans*, early cleavages follow the T_1-type pattern (data not shown). The same is true for species of clade 6 (*Plectus*; Figure 2f1 to f5 and *Tylocephalus*; [26]), clade 7 (*Teratocephalus*; [26]), clade 8 (*Ascaris*; [7]), clade 10 (*Panagrolaimus*; Figure 2i1 to i5) and clade 12A (*Aphelenchus*; Figure 2k1-k5). A slight variation of the T_1-type cleavage pattern (green lineage branches) was detected in two representatives of the genus *Rhabditis*, where PR was found in either P2 or P3 [45].

I_2-type, Clade 9B (Figure 1b, red lineage branches)

More prominent is a modification found in clade 9B with the genera *Protorhabditis* (Figure 2h1 to h5) and *Diploscapter*[27]. There, both first blastomeres divide with a-p oriented spindles resulting in a tandem arrangement of blastomeres [14] resembling the I_1-type cleavage pattern found in *Romanomermis* and *Prionchulus* (Figures 2c1 to c5; d1 to d5). The elongate eggshell allows a strictly linear order of blastomeres after the divisions of S1, P1 and P2 (Figure 2h2, h3). The germline cells, due to the absence of PR, end up occupying positions posterior to their somatic sisters. To establish the d-v axis and to reach the typical neighborhood between the primordial germ cell P4 and the gut precursor S2p, prominent rearrangements ('cell sorting'; Figure 1c) are required (Figure 2h3 to h5) as described for *D. coronatus*[27].

T_2-type, Clade 11 (Figure 1, purple lineage branches)

Acrobeloides nanus[21] and *Zeldia punctata* (Figure 2 j1-j5) follow the T-type of cleavage; however, in contrast to the T_1-type PR is absent and therefore germline cells occupy the most posterior positions. Consequently, cell rearrangements among P2 descendants are required to reach a *C. elegans*-like pattern prior to the onset of gastrulation as described above for the 'I_2-type'.

I_3-type, Clade 12 (Figure 1, pink lineage branches)

Belonolaimus longicaudatus (Figure 2 l1 to l5) and *Meloidogyne incognita*[14] follow the I-Type of cleavage. However, this differs from the two 'I-types' introduced above (PR in P1 or absent) in that PR takes place in P2.

In summary, in nematodes we found six different early cleavage patterns, two among Enoplea and four among Chromadorea. In the phylogenetic tree (Figure 1a) I_2-, T_2- and I_3-types are restricted to distinct branches indicating they are apomorphic modifications of the prevalent T_1-type.

The POC, a general developmental principle in nematodes?

Having detected the action of a POC in three representatives of Enoplea we wanted to determine whether such a mechanism is also characteristic for early embryogenesis of Chromadorea and thus may constitute a general developmental principle in nematodes.

In contrast to *Tobrilus, Prionchulus* and *Romanomermis*, three POCs can be defined in *C. elegans* which are involved in longitudinal spindle orientation and serial arrangement of founder cells. Laser ablation experiments in 2-cell embryos revealed a microtubule-pulling force at the anterior pole of P1 [46]. Located in the RFM, it resembles the 'primary POC' described above, although normally it only acts in P1 (Figure 2g1; green arrowhead). However, in embryos with defects in the expression of *par* genes, spindle orientations in P1 and S1 are altered (Figure 2p, q), suggesting an interaction between PAR proteins and the primary POC (see discussion). Another POC ('secondary POC'; marked in purple in Figure 2) positioned in the 'region of the second midbody' (RSM), initially orients spindles in the two sister cells S2 and P2 and subsequently in the cell cousins S2p+P3 and S2pp+P4 (Figure 2g2 to g5). This differs from the situation in Enoplea described above where S1, not S2, descendants are involved. For *C. elegans* it has been shown that the polarizing function of the secondary POC depends on the presence of MES-1/SRC-1 proteins (see discussion). A third POC ('tertiary POC'; marked in yellow in Figure 2) is established at the posterior pole of the 1-cell stage as a consequence of sperm entry [47]. It polarizes the fertilized egg, is crucial for asymmetric divisions of germline cells and depends on the polar distribution of PAR proteins (Figure 5c; see discussion).

As the majority of Chromadorea studied follow the T_1-type of cleavage (Figure 1a; blue branches; Figure 2e to 2g, i, k) and early cell behavior is very similar to *C. elegans*, it indicates the activity of all three POC types. The other cleavage types (Figure 1b) can be explained with an altered expression pattern of PAR proteins affecting the function of the primary POC and the presence or absence of a PR in the germline (Figures 2, 5 and discussion).

POC-based establishment of polarity beyond nematodes?

Our studies on early embryogenesis in nematodes revealed a characteristic cleavage pattern where, through several cell generations, two blastomeres orient their spindles towards a POC resulting in a series of asymmetric cleavages (Figure 2). We wondered whether this pattern is unique for nematodes and therefore compared it to early development of the tardigrade *Hybsibius dujardini* studied by Gabriel *et al.*[48]. According to the Ecdysozoa hypothesis [49] tardigrades are a sister group of nematodes. We found that early development of *Hybsibius* shows unexpected similarities to *Tobrilus* (Figure 2b). The first two divisions are equal with subsequent spindles perpendicular to each other (Figure 2r1 to r3), typical for the 'H-Type' of cleavage (Figure 1b). Then, tandem orientation of spindles in two neighboring cell cousins and subsequent asymmetric divisions indicate the presence of a POC (Figure 2r3, r4). As in *Tobrilus*, two adjacent cell cousins perform asymmetric, longitudinally oriented cleavages (Figure 2r5). Thus, a distinct POC-controlled division pattern is obviously shared between nematodes of clade 1C and at least one representative of tardigrades. In a species belonging to a different branch of tardigrades the H-type of early cleavage was observed, too, but not the reproducible unequal divisions [50].

Establishment of bilateral symmetry

We found that despite considerable differences with respect to early development (for example H-, I-, T-cleavage types; PR in P1, P2, P3 or absent; Figures 1b, 2, 3m, 4m, 5), in all studied embryos with the exception of *Enoplus* (clade 1B; Figure 2a), a linear sequence of cells with different fates is generated along the a-p body axis due to the action of one or more POCs. Some or all of these divide into left and right descendants with equivalent fates on both sides of the midline. This way bilateral symmetry is established within individual cell lineages. To determine to what extent nematodes with different cleavage patterns vary in the way they develop bilateral symmetry, we compared early pattern formation in all 12 clades (Figure 1). Three representative species for the H-, I-, and T-types are shown in Figure 5.

In *T. stefanskii* (Figure 5a) a visible polarity is established in the 4cell stage (see above) and descendants of S1p and P2-P5 are aligned along the future midline. However, only three somatic cells (S1pppa, S1ppppa, S1ppppp) perform left-right oriented cleavages resulting in bilateral symmetric clones. Adhering to the centriolic principle (see above) the remaining blastomeres divide with variable spindle orientations giving rise to clones with indeterminate spatial positions.

In *Prionchulus* sp. (Figure 5b) in addition to three descendants of S1p also S4 and S5 execute divisions with transverse spindle orientation and subsequently generate bilaterally symmetric clones. S3 divides with oblique spindle orientation forming an imperfect early l-r symmetry (Figure 4p to 4u). The other cells show variable arrangements.

Thus, in *Tobrilus* and *Prionchulus* part of bilateral body symmetry must be established later, probably in a position-dependent manner as a result of cell-cell interactions.

In contrast, in *C. elegans* (Figure 5c) and all other studied members of clades 2C-12 all somatic founder cells or their early descendants form bilaterally symmetric clones (for modification of this principle in S1a cells, see [9]), independent of whether they follow the T_1-(Figure 5c), T_2-, I_2-(Figure 5d) or I_3-type of cleavage.

Comparing the three examples shown in Figure 5 a tendency can be observed towards an earlier and complete fixation of bilateral symmetry. The start of asymmetric divisions shifts from the 4-cell stage (*Tobrilus*) to the 1-cell stage (*C. elegans*) while the number of cell generations needed to perform all the l-r divisions described above decreases from seven (*Tobrilus*) to five (*C. elegans*) and the number of somatic lineages involved in early symmetry formation (Figure 1c) increases from two (*Tobrilus*) to five (*C. elegans*).

Differences and similarities during ongoing embryogenesis

Our study makes clear that early embryogenesis differs dramatically among species, particularly within Enoplea. In addition, tissue formation varies during later stages. In a previous publication we showed that hypodermis is generated in *Romanomermis* very differently from that in *C. elegans*[34]. Preliminary data indicate that hypodermis formation in *Prionchulus* does not follow the pattern found in *Romanomermis*, in spite of both being members of the same clade. The peculiarities of gastrulation in *Tobrilus* (Figure 3) including the way cells assemble to form the gut are not only different compared to *C. elegans* but also to *Enoplus* (see above). We also found indications that founder cells contribute differently to the pharynx [9, 34] and how this organ is formed (data not shown). However, eventually all variants seem to merge into a common pattern. The process of transforming a ball of cells consisting of three germ layers into an elongated worm during the "morphogenesis phase" starts with a ventral indentation separating head and tail regions in all studied nematodes. The progressive elongation of the embryo looks similar to *C. elegans*[51], although the degree of elongation varies considerably (that is juveniles may be longer and thinner). From this we conclude that the developmental constraints during the second half of embryogenesis are higher than during the early phase.

DISCUSSION

In this paper representatives of all 12 nematode clades (according to the phylogeny by [5]) have been compared with respect to their early embryogenesis. Our data document that the very similar and reproducible development found in the reference systems *Ascaris* and *C. elegans* (see introduction) exemplifies only one of many ways to generate a nematode worm from a 1-cell embryo. They indicate that the cleavage and differentiation program of blastomeres diverged dramatically during evolution. This is particularly obvious in Enoplia (clade 1) and Dorylaimia (clade 2). In each of these taxa developmental peculiarities and variations appear to be higher than in all Chromadorea (clades 3 to 12) combined. Because of distinct developmental features (Figures 1b, 2 to 5) it appears reasonable to break down clades 1, 2, 9 and 12 into subgroups (Figure 1a).

We believe that major developmental characters found in Enoplea but absent in Chromadorea (clades 3 to 12; Figures 1, 2) are plesiomorphic. These include the 'canonical' gastrulation in the genus *Tobrilus* with its large blastocoel (Figure 3; [32]) found in many other animal phyla including Nematomorpha (our unpublished results), the nearest phylogenetic neighbors of nematodes, and the similarity of early cleavage patterns between *Tobrilus* (Figure 2b) and the tardigrade *Hybsibius* (Figure 2r) which according to the Ecdysozoa hypothesis [49] belongs to a neighboring phylum. Thus, our findings support the positioning of Enoplea close to the base of the phylogenetic tree of nematodes postulated on grounds of molecular sequence data [2, 5, 6]. This implies that the route of lineage evolution went from an S1 to an S2 origin of the gut (Figure 1c; [22, 40]), which required considerable modifications in cleavage pattern and fate assignment [34]. Furthermore, comparison of embryogenesis between nematodes, nematomorphs and tardigrades suggests that certain features (for example the absence of initial asymmetric cleavages) were shared by their last common ancestor.

Variations, evolutionary trends and developmental system drift

Our data including those on *Romanomermis* (clade 2C) indicate a boost in lineage complexity, that is a stepwise increase from a single to five somatic lineages (Figures 1c, 5) and a change from a monoclonal to a polyclonal fate assignment [34]. This change, which appears to have coincided with the transition from Enoplea to Chromadorea (Figure 1), is correlated with an increase in early fate decisions, which in turn reduces the amount of cell migration necessary for proper tissue formation [12, 34, 52].

Comparing embryogenesis of representatives along the nematode phylogenetic tree (Figure 1a) we find that intra-species variation of early cell patterns decreases due to an increase of founder cells that generate descendants occupying fixed positions. Hence, our data are in favor of the conception that invariant development following distinct lineage programs is a derived and not an original feature (see introduction).

From the degree of individual variation within a species we can deduce how strongly fixed the developmental program must be and thus how a change in cleavage type (Figure 1b) may have been established during evolution. In *Plectus* (clade 6) we observed rare cases (2/< 100 embryos) where S1 cleaved with longitudinal (Figure 2o) rather than transverse spindle orientation (Figure 2f2), meaning a switch from a T_1- to an I_2-cleavage type. As this deviation was found to be compatible with normal embryogenesis it demonstrates that the developmental program must possess a sufficient degree of plasticity to allow the disregard of the 'centriolic principle' (see results section). Such an embryo must potentially cope with altered segregation of cytoplasmic components and changes in relative cell positions which in turn may affect inductive signaling. In addition, our observation suggests that the change from one to another cleavage type started with a modification that was initially rare in the population.

The divergence of developmental pathways without corresponding changes in the emerging phenotype ('developmental system drift'; [53]) seems to be a widely spread phenomenon in the animal kingdom. The present study gives additional examples for this. Moreover, it indicates that embryogenesis in nematodes includes general distinctive steps that need to be taken. However, cellular events leading to these are not conserved among species suggesting that evolutionary constraints have been low on cell behavior but high not only on structure and function of the juvenile but also on intermediate embryonic stopovers. Two examples may suffice to put our idea across. (i) The contact beween gut and germline (a critical feature in many systems; [54]) can be achieved via PR in either P1, P2 or P3 or alternatively via 'cell sorting' in species where PR is absent. (ii) Cell fate assignment can be reached in the absence of a fixed early cleavage program or alternatively via monoclonal or polyclonal cell lineages. Recent findings by Lin *et al.*[19] show that a similar relationship can be found between cells and molecules. In two closely related nematode species behavior of early blastomeres with alternative fates is identical while the underlying signaling network differs.

The enormous differences in genomes among even closely related species [55] in contrast to the conserved morphology indicate a particularly relaxed relationship between genotype and phenotype in nematodes. This discrepancy can be attributed to the special construction of nematodes including a single chamber hydroskeleton which allows adaptation to very diverse habitats but leaves little room for modifications of the body plan.

Polarity organizing centers (POCs) and embryonic pattern formation

One of our central findings is the general presence of one or more POCs in early nematode embryos. While in the Enoplea *Romanomermis*[33], *Tobrilus* and *Prionchulus* (Figures 2b to 2d; 5a to 5b) we found evidence that early embryogenesis involves just a single (primary) POC, in Chromadorea up to three (primary, secondary and tertiary) POCs appear to be active in organizing orientation and asymmetry of divisions. Only for *C. elegans* do we have information about the molecular basis of the secondary and tertiary POC.

Laser ablation experiments in *C. elegans*[46] revealed a 'cortical pulling site' at the anterior pole of P1. Observations by Keating and White [56] support the view that the midbody between AB and P1 specifies a region of the cortex that directs rotational alignment of the centrosome-nucleus complex. This region of the first midbody (RFM; [33]) corresponds to our primary POC.

Most prominent is the activity of the tertiary POC at the posterior pole. Sperm entry initiates asymmetric distribution of PAR and LET-99 proteins [47, 57, 58, 59, 60] and subsequently anterior-posterior (a-p)-oriented spindles and asymmetric divisions in the germline [61, 62]. However, in a member of clade 11 (Figure 1A) axis polarity was found to be independent of the sperm entry point [63].

Mutants and RNAi phenotypes of *pkc-3, par-2, par-3* and *par-6* in *C. elegans* demonstrate that a knockout of any of these genes changes the orientation of the cleavage spindle in the 2-cell stage and this way leads from the T-type to H- or I-types of cleavage (Figure 2p, q), even though followed by abnormal development [64, 65, 66, 67, 68].

To explain the different cleavage types in Chromadorea we propose that in contrast to Enoplea (where the primary POC always induces a-p spindle orientation in two adjacent cells; Figures 2b to d, 5a, b), the polarizing function in Chromadorea is controlled by PAR proteins [27]. This view is supported by mutations and RNAi experiments in *C. elegans* leading to abnormal cleavage patterns. The *par-2/par-3* double mutant follows the I-type of cleavage, demonstrating that neither of these genes is required for longitudinal orientation of cleavage spindles [64]. According to our model the prevalent T-type (Figure 1b) can be explained with the anterior PAR complex suppressing the activity of the primary POC in S1. At the same time it gives a simple explanation why uniform distribution of anterior PAR proteins (after knockout of *par-2*) results in transverse spindle orientation in S1 and P1 (H-type; Figure 2q) and uniform distribution of posterior PAR proteins (after knockout of anterior *par* genes) leads to longitudinal spindle orientation in both cells (I-type; Figure 2p).

In *C. elegans* PAR proteins switch function after the 4-cell stage has been reached and are involved in the establishment of apical-basal polarity [69]. Instead the MES-1/SRC-1 system is required to continue a-p polarization of blastomeres [70, 71, 72] and to induce the reversal of cleavage polarity (PR) in P2 [41] as is the case in the T_1- and I_3-cleavage types (Figures 1b, 2). The expression domain of MES-1/SRC-1 corresponds to our secondary POC. Berkowitz and Strome [72] suggested that species without PR may have lost the MES-1/SRC-1 system (accordingly a secondary POC is not indicated in Figure 2h, j). However, the question then remains how a-p divisions in the germline can be maintained in the I_2- and T_2-types. A straightforward explanation could be a prolonged activity of posterior PAR proteins. Indeed, in *Protorhabditis* (clade 9B, Figure 2h) an extended posterior expression of PAR-1 in the germline was observed by Brauchle *et al.*[18].

The peculiarities of early development in those Enoplea that we followed in some detail (Figure 2a to d) gave no indications that *par* genes are involved in cell polarization along the a-p axis as found in *C. elegans*. Although it would appear that this function may have been newly acquired in Chromadorea, our fragmentary data from two other Enoplida indicate that the situation is less clear. *Trichuris muris* (clade 2A) performs an extremely asymmetric first division (Figure 2n; see also [22]), and in *Ironus* (clade 1A) three consecutive longitudinally oriented cleavages result in a tandem of eight cells (Figure 2m). Both patterns cannot be readily explained with the activity of a primary POC alone. As our preliminary analysis of the *R.*

culicivorax (clade 2C) genome indicates that *par* genes are present it remains to be determined whether at least in some Enoplea they may play a role in establishing the primary body axis.

Consecutive asymmetric divisions following a POC-like principle are not restricted to invertebrates like nematodes and tardigrades but have been described for posterior blastomeres of the ascidian embryo as well [73]. In equivalent cells in the left and right half of the bilaterally symmetric embryo asymmetric positioning of cleavage spindles is induced by centrosome attracting bodies (CAB; [74, 75]). The CAB contains proteins that are homologous to the anterior PAR proteins in *C. elegans*[76]. It remains to be determined whether this asymmetry generating mechanism can be traced back to the last common ancestor or has been acquired independently in all three taxa.

Midline formation and bilateral symmetry

A bilaterally symmetric body is typical for all higher animals. For *C. elegans* it has been shown that this symmetry is generated early within individual lineages [9]. The construction of a bilateral body plan requires the presence of a midline, separating left from right. Our studies revealed that in all analyzed nematodes, except clade 1B, such a midline can be defined during early embryogenesis.

In contrast to the models proposed by Meinhardt [77] for planarians, insects and vertebrates, in nematodes bilateral symmetry is generated in a surprisingly simple way. Due to the actions of one or more POCs (described above) which cause longitudinal spindle orientations and cleavage asymmetries over consecutive rounds of division, an array of cells is formed along the a-p axis. Depending on the species, part or all of these divide into left and right daughters (Figure 5) from which bilaterally symmetric clones arise. It appears likely that such a strategy requires specific conditions, for example a low number of blastomeres and their early specification on a single cell basis, features typical for nematodes.

Cell lineage and cell specification

The case of *C. elegans* demonstrates that even in a system with essentially invariant development and complex polyclonal lineages, inductive interactions are an integral part of the developmental program which requires specific cell contacts during narrow time windows [78, 79, 80, 81]. The different early cleavage patterns described above (Figures 1b, 2) result in different cell neighborhoods. Either changes in cell contacts did not pose a problem during nematode evolution because interactions like those in *C.*

elegans did not exist yet, or the network of interactions was adjusted simultaneously. Our data are in accordance with a stepwise establishment of such interactions going along with increasing invariance of cell positions. Experimental interference like recombining blastomeres [70, 82] in basal representatives could give us a better idea to what extent the pattern and relevance of embryonic inductions changed during evolution. Analysis of cell-specification patterns and their molecular underpinnings, particularly in Enoplea, should help to discern between two alternative visions. (i) Inductive interactions in early nematode embryos are historic remnants. They reach back to times when fixed lineages where mainly absent (for example *Enoplus*; [31]) or only simple, monoclonal lineages existed (for example *Romanomermis*; [34]) and were required for diversification of cell fate. (ii) Alternatively, they are more recent acquisitions made possible after fixed lineages assured invariant cell neighborhoods.

CONCLUSIONS

Nematodes are suitable objects to study evolution of development because species from all branches of the phylogenetic tree can be analyzed, embryos develop outside the mothers and most of them are transparent enough to perform cellular analysis *in vivo*. Our findings that early embryogenesis varies considerably among species indicates that constraints are high on the preservation of crucial developmental steps but not on cellular behavior leading to these. We argue that the direction of evolution went from indeterminate early cleavage without initial polarity to invariant development with establishment of polarity before division of the zygote. The observed action of a primary POC gives a clue how polarity in certain nematodes and other related taxa like tardigrades can be established in a way that differs from *C. elegans*, that is independent of the sperm entry point.

ACKNOWLEDGEMENTS

We are indebted to various colleagues for sharing strains (listed in materials and methods) and to Vladimir Malakhov for introduction into enoplid development and Oleksandr Holovachov for identification of *P. cf. sambesii*. We thank Bob Goldstein for Figure 2r, July Ahringer for Figure 2q and Jay Burr for thoughtful comments on the manuscript. JS was supported by a grant of the Deutsche Forschungsgemeinschaft (Schi 214/16-1).

COMPETING INTERESTS

The authors declare that they have no competing interests.

AUTHORS' CONTRIBUTIONS

Both authors contributed to the conception and design of the study, were involved in acquisition of data, its analysis and interpretation. Both authors drafted the manuscript and read and approved the final version.

REFERENCES

1. Blaxter ML, De Ley P, Garey JR, Liu LX, Scheldeman P, Vierstraete A, Vanfleteren JR, Mackey LY, Dorris M, Frisse LM, Vida JT, Thomas WK: A molecular evolutionary framework for the phylum Nematoda. Nature. 1998, 392: 71-75.
2. De Ley P, Blaxter ML: Systematic Position and Phylogeny. The Biology of Nematodes. Edited by: Lee DL. 2002, Taylor and Francis, London, 1-30
3. Meldal BH, Debenham NJ, De Ley P, De Ley IT, Vanfleteren JR, Vierstraete AR, Bert W, Borgonie G, Moens T, Tyler PA, Austen MC, Blaxter ML, Rogers AD, Lambshead PJ: An improved molecular phylogeny of the Nematoda with special emphasis on marine taxa. Mol Phylogenet Evol. 2007, 42: 622-636.
4. Kiontke K, Barriere A, Kolotuev I, Podbilewicz B, Sommer R, Fitch DH, Felix MA: Trends, stasis, and drift in the evolution of nematode vulva development. Curr Biol. 2007, 17: 1925-1937.
5. Holterman M, van der Wurff A, van den Elsen S, van Megen H, Bongers T, Holovachov O, Bakker J, Helder J: Phylum-wide analysis of SSU rDNA reveals deep phylogenetic relationships among nematodes and accelerated evolution toward crown clades. Mol Biol Evol. 2006, 23: 1792-1800.
6. Aleshin VV, Kedrova OS, Milyutina IA, Vladychenskaya NS, Petrov NB: Relationships among nematodes based on the analysis of 18S rRNA gene sequences: molecular evidence for monophyly of Chromadorian and Secernentean nematodes. Russ J Nematol. 1998, 6: 175-184.
7. Boveri T: Die Entwicklung von *Ascaris megalocephala* mit besonderer Rücksicht auf die Kernverhältnisse [in German]. Festschrift für Carl von Kupffer. 1899, Gustav Fischer Verlag, Jena, 383-430.
8. Müller H: Beitrag zur Embryonalentwicklung von *Ascaris megalocephala* [in German]. Zoologica. 1903, 17: 1-30.
9. Sulston JE, Schierenberg E, White JG, Thomson JN: The embryonic cell lineage of the nematode *Caenorhabditis elegans*. Dev Biol. 1983, 100: 64-119.
10. Schnabel R, Hutter H, Moerman D, Schnabel H: Assessing normal embryogenesis in *Caenorhabditis elegans* using a 4D microscope: variability of development and regional specification. Dev Biol. 1997, 184: 234-265.
11. Houthoofd W, Borgonie G: The embryonic cell lineage of the nematode *Halicephalobus gingivalis* (Nematoda: Cephalobina: Panagrolaimoidea). Nematology. 2007, 9: 573-584.

12. Houthoofd W, Jacobsen K, Mertens C, Vangestel S, Coomans A, Borgonie G: Embryonic cell lineage of the marine nematode *Pellioditis marina*. Dev Biol. 2003, 258: 57-69.
13. Houthoofd W, Willems M, Jacobsen K, Coomans A, Borgonie G: The embryonic cell lineage of the nematode *Rhabditophanes sp*. Int J Dev Biol. 2008, 52: 963-967.
14. Dolinski C, Baldwin JG, Thomas WK: Comparative survey of early embryogenesis of Secernentea (Nematoda), with phylogenetic implications. Can J Zool. 2001, 79: 82-94.
15. Vangestel S, Houthoofd W, Bert W, Borgonie G: The early embryonic development of the satellite organism *Pristionchus pacificus*: differences and similarities with *Caenorhabditis elegans*. Nematology. 2008, 10: 301-312.
16. Zhao Z, Boyle TJ, Bao Z, Murray JI, Mericle B, Waterston RH: Comparative analysis of embryonic cell lineage between *Caenorhabditis briggsae* and *Caenorhabditis elegans*. Dev Biol. 2008, 314: 93-99.
17. Sommer RJ: Evolution of development in nematodes related to *C. elegans*. WormBook. 2005, 'The C. elegans Resarch Community', [http://www.wormbook.org]
18. Brauchle M, Kiontke K, MacMenamin P, Fitch DH, Piano F: Evolution of early embryogenesis in rhabditid nematodes. Dev Biol. 2009, 335: 253-262.
19. Lin KT, Broitman-Maduro G, Hung WW, Cervantes S, Maduro MF: Knockdown of SKN-1 and the Wnt effector TCF/POP-1 reveals differences in endomesoderm specification in *C. briggsae* as compared with *C. elegans*. Dev Biol. 2009, 325: 296-306.
20. Bolker JA: Model systems in developmental biology. Bioessays. 1995, 17: 451-455.
21. Skiba F, Schierenberg E: Cell lineages, developmental timing, and spatial pattern formation in embryos of free-living soil nematodes. Dev Biol. 1992, 151: 597-610.
22. Malakhov VV: Nematodes Structure, Development, Classification and Phylogeny. Washington Smithsonian Institution Press;. 1994
23. Voronov DA: The embryonic development of *Pontonema vulgare* (Enoplida: Oncholaimidae) with a discussion of nematode phylogeny. Russ J Nematol. 1999, 7: 105-114.
24. Goldstein B: On the evolution of early development in the Nematoda. Philos Trans R Soc Lond B Biol Sci. 2001, 356: 1521-1531.
25. Schierenberg E: Three sons of fortune: early embryogenesis, evolution and ecology of nematodes. Bioessays. 2001, 23: 841-847.
26. Lahl V, Halama C, Schierenberg E: Comparative and experimental embryogenesis of Plectidae (Nematoda). Dev Genes Evol. 2003, 213: 18-27.
27. Lahl V, Schulze J, Schierenberg E: Differences in embryonic pattern formation between *Caenorhabditis elegans* and its close parthenogenetic relative Diploscapter coronatus. Int J Dev Biol. 2009, 53: 507-515.
28. Bento G, Ogawa A, Sommer RJ: Co-option of the hormone-signalling module dafachronic acid-DAF-12 in nematode evolution. Nature. 2010, 466: 494-497.
29. Kiontke K, Fitch DH: Phenotypic plasticity: different teeth for different feasts. Curr Biol. 2010, 20: 710-712.
30. Wiegner O, Schierenberg E: Regulative development in a nematode embryo: a hierarchy of cell fate transformations. Dev Biol. 1999, 215: 1-12.
31. Voronov DA, Panchin YV: Cell lineage in marine nematode *Enoplus brevis. Development*. 1998, 125: 143-150.
32. Schierenberg E: Unusual cleavage and gastrulation in a freshwater nematode: developmental and phylogenetic implications. Dev Genes Evol. 2005, 215: 103-108.
33. Schulze J, Schierenberg E: Cellular pattern formation, establishment of polarity and segregation of colored cytoplasm in embryos of the nematode *Romanomermis culicivorax*. Dev Biol. 2008, 315: 426-436.
34. Schulze J, Schierenberg E: Embryogenesis of *Romanomermis culicivorax*: an alternative way to construct a nematode. Dev Biol. 2009, 334: 10-21.

35. Voronov DA: Comparative embryology of nematodes and the law of embryo similarity [in Russian]. Zh Obshch Biol. 2001, 62: 34-48.
36. Aleshin VV: Whether variable cleavage of Enoplida (Nematoda) is primitive? Notes to D.A. Voronov article "Comparative embryology of Nematoda and the law of embryologic similarity" [in Russian]. Zh Obshch Biol. 2004, 65: 74-80.
37. Deppe U, Schierenberg E, Cole T, Krieg C, Schmitt D, Yoder B, von Ehrenstein G: Cell lineages of the embryo of the nematode *Caenorhabditis elegans*. Proc Natl Acad Sci USA. 1978, 75: 376-380.
38. Costello DP: On the orientation of centrioles in dividing cells, and its significance: a new contribution to spindle mechanics. Biol Bull. 1961, 120: 285-312.
39. Thery M, Bornens M: Cell shape and cell division. Curr Opin Cell Biol. 2006, 18: 648-657.
40. Voronov DA, Panchin YV, Spiridonov SE: Nematode phylogeny and embryology. Nature. 1998, 395: 28-
41. Schierenberg E: Reversal of cellular polarity and early cell-cell interaction in the embryos of *Caenorhabditis elegans*. Dev Biol. 1987, 122: 452-463.
42. Drozdovskiy EM: Contribution to the Analysis of the Embryogenesis of certain Adenophorea (Nematoda) [in Russian]. *Reports of the Academy of Sciences of the USSR*. 1969, 186: 720-723
43. Borgonie G, Jacobsen K, Coomans A: Embryonic lineage evolution in nematodes. Nematology. 2000, 2: 65-69.
44. Wyss U:*Longidorus elongatus (Nematoda) Embryonalentwicklung.*1973, Institut für den Wissenschaftlichen Film, Göttingen, Germany, Film E2046,
45. Laugsch M, Schierenberg E: Differences in maternal supply and early development of closely related nematode species. Int J Dev Biol. 2004, 48: 655-662. 1
46. Hyman AA: Centrosome movement in the early divisions of *Caenorhabditis elegans*: a cortical site determining centrosome position. J Cell Biol. 1989, 109: 1185-1193.
47. Gönczy P, Rose LS: Asymmetric cell division and axis formation in the embryo. WormBook. 2005, 'The C. elegans Resarch Community', [http://www.wormbook.org]
48. Gabriel WN, McNuff R, Patel SK, Gregory TR, Jeck WR, Jones CD, Goldstein B: The tardigrade *Hypsibius dujardini*, a new model for studying the evolution of development. Dev Biol. 2007, 312: 545-559.
49. Aguinaldo AM, Turbeville JM, Linford LS, Rivera MC, Garey JR, Raff RA, Lake JA: Evidence for a clade of nematodes, arthropods and other moulting animals. Nature. 1997, 387: 489-493.
50. Hejnol A, Schnabel R: The eutardigrade *Thulinia stephaniae* has an indeterminate development and the potential to regulate early blastomere ablations. Development. 2005, 132: 1349-1361.
51. Priess JR, Hirsh DI: *Caenorhabditis elegans* morphogenesis: the role of the cytoskeleton in elongation of the embryo. Dev Biol. 1986, 117: 156-173.
52. Azevedo RB, Lohaus R, Braun V, Gumbel M, Umamaheshwar M, Agapow PM, Houthoofd W, Platzer U, Borgonie G, Meinzer HP, Leroi AM: The simplicity of metazoan cell lineages. Nature. 2005, 433: 152-156.
53. True JR, Haag ES: Developmental system drift and flexibility in evolutionary trajectories. Evol Dev. 2001, 3: 109-119.
54. Wylie C: Germ cells. Cell. 1999, 96: 165-174.
55. Kiontke K, Gavin NP, Raynes Y, Roehrig C, Piano F, Fitch DH: *Caenorhabditis* phylogeny predicts convergence of hermaphroditism and extensive intron loss. Proc Natl Acad Sci USA. 2004, 101: 9003-9008.
56. Keating HH, White JG: Centrosome dynamics in early embryos of *Caenorhabditis elegans*. J Cell Sci. 1998, 111 (Pt 20): 3027-3033.
57. Kemphues K, Strome S: Fertilization and establishment of polarity in the embryo. C elegans II. Edited by: Riddle DL, Blumenthal T, Meyer BJ, Priess J. 1997, New York: Cold Spring Harbor Laboratory Press

58. Cowan CR, Hyman AA: Asymmetric cell division in *C. elegans*: cortical polarity and spindle positioning. Annu Rev Cell Dev Biol. 2004, 20: 427-453.
59. Krueger LE, Wu JC, Tsou MF, Rose LS: LET-99 inhibits lateral posterior pulling forces during asymmetric spindle elongation in *C. elegans* embryos. J Cell Biol. 2010, 189: 481-495.
60. Zonies S, Motegi F, Hao Y, Seydoux G: Symmetry breaking and polarization of the *C. elegans* zygote by the polarity protein PAR-2. Development. 2010, 137: 1669-1677.
61. Grill SW, Hyman AA: Spindle positioning by cortical pulling forces. Dev Cell. 2005, 8: 461-465.
62. Nguyen-Ngoc T, Afshar K, Gonczy P: Coupling of cortical dynein and G alpha proteins mediates spindle positioning in *Caenorhabditis elegans*. Nat Cell Biol. 2007, 9: 1294-1302. 10.1038/ncb1649.
63. Goldstein B, Frisse LM, Thomas WK: Embryonic axis specification in nematodes: evolution of the first step in development. Curr Biol. 1998, 8: 157-160.
64. Cheng NN, Kirby CM, Kemphues KJ: Control of cleavage spindle orientation in *Caenorhabditis elegans*: the role of the genes par-2 and par-3. Genetics. 1995, 139: 549-559.
65. Tabuse Y, Izumi Y, Piano F, Kemphues KJ, Miwa J, Ohno S: Atypical protein kinase C cooperates with PAR-3 to establish embryonic polarity in *Caenorhabditis elegans*. Development. 1998, 125: 3607-3614.
66. Hung TJ, Kemphues KJ: PAR-6 is a conserved PDZ domain-containing protein that colocalizes with PAR-3 in *Caenorhabditis elegans* embryos. Development. 1999, 126: 127-135.
67. Kamath RS, Martinez-Campos M, Zipperlen P, Fraser AG, Ahringer J: Effectiveness of specific RNA-mediated interference through ingested double-stranded RNA in *Caenorhabditis elegans*. Genome Biol. 2001, 2: 1-10.
68. Hao Y, Boyd L, Seydoux G: Stabilization of cell polarity by the *C. elegans* RING protein PAR-2. Dev Cell. 2006, 10: 199-208.
69. Goldstein B, Macara IG: The PAR proteins: fundamental players in animal cell polarization. Dev Cell. 2007, 13: 609-622. 10.1016/j.devcel.2007.
70. Arata Y, Lee JY, Goldstein B, Sawa H: Extracellular control of PAR protein localization during asymmetric cell division in the *C. elegans* embryo. Development. 2010, 137: 3337-3345.
71. Bei Y, Hogan J, Berkowitz LA, Soto M, Rocheleau CE, Pang KM, Collins J, Mello CC: SRC-1 and Wnt signaling act together to specify endoderm and to control cleavage orientation in early *C. elegans* embryos. Dev Cell. 2002, 3: 113-125.
72. Berkowitz LA, Strome S: MES-1, a protein required for unequal divisions of the germline in early *C. elegans* embryos, resembles receptor tyrosine kinases and is localized to the boundary between the germline and gut cells. Development. 2000, 127: 4419-4431.
73. Conklin EG: The organization and cell lineage of the ascidian egg. J Acad Nat Sci. 1905, Philadelphia, 13: 1-119.
74. Hibino T, Nishikata T, Nishida H: Centrosome-attracting body: a novel structure closely related to unequal cleavages in the ascidian embryo. Dev Growth Differ. 1998, 40: 85-95.
75. Nishikata T, Hibino T, Nishida H: The centrosome-attracting body, microtubule system, and posterior egg cytoplasm are involved in positioning of cleavage planes in the ascidian embryo. Dev Biol. 1999, 209: 72-85.
76. Patalano S, Pruliere G, Prodon F, Paix A, Dru P, Sardet C, Chenevert J: The aPKC-PAR-6-PAR-3 cell polarity complex localizes to the centrosome attracting body, a macroscopic cortical structure responsible for asymmetric divisions in the early ascidian embryo. J Cell Sci. 2006, 119: 1592-1603.
77. Meinhardt H: Different strategies for midline formation in bilaterians. Nat Rev Neurosci. 2004, 5: 502-510.

78. Schnabel R, Priess JR: Specification of Cell Fates in the Early Embryo. C elegans II. Edited by: Riddle DL, Blumenthal T, Meyer BJ, Priess J. 1997, New York: Cold Spring Harbor Laboratory Press
79. Goldstein B: An analysis of the response to gut induction in the *C. elegans* embryo. Development. 1995, 121: 1227-1236.
80. Eisenmann DM: Wnt signaling. Wormbook. 2005, 'The C. elegans Resarch Community', [http://www.wormbook.org]
81. Priess JR: Notch signaling in the *C. elegans* embryo. WormBook. 2005, 'The C. elegans Resarch Community', [http://www.wormbook.org]
82. Goldstein B: Establishment of gut fate in the E lineage of *C. elegans*: the roles of lineage-dependent mechanisms and cell interactions. Development. 1993, 118: 1267-1277.

CHAPTER 10

Characterization of Root-Knot Nematodes (Meloidogyne Spp.) Associated with Abelmoschus Esculentus, Celosia Argentea and Corchorus Olitorius

F.Y. Daramola, J.O. Popoola, A.O. Eni and O. Sulaiman

Department of Biological Sciences, School of Natural and Applied Sciences, College of Science and Technology, Covenant University, P.M.B. 1023, Canaanland Ota, Ogun State, Nigeria

ABSTRACT

Vegetable crops are highly susceptible to a wide range of pests and diseases among which are the root-knot nematodes (Meloidogyne species). Traditionally, identification of Meloidogyne species had been based on use of morphological characters such as the perennial pattern. In recent times, accuracy of nematode identification using only morphological tools has been challenged due to similarities in morphological characters of some nematode species. The aim of this study was to identify the root-knot nematodes associated with some vegetable crops cultivated on Covenant University farm, Ota, South-west, Nigeria using molecular tools and to determine the population densities of Meloidogyne spp. on the selected vegetable crops. Plant-parasitic nematodes were extracted from soil and roots of Abelmoschus esculentus, Celosia argentea and Corchorus olitorius cultivated on Covenant University farm. The nematode species were identified and counted under a compound microscope. The molecular characterization of the Meloidogyne species was done using single adult female nematodes and eggs which were picked out for DNA extraction and amplified with specie-specific primers through Polymerase Chain Reaction (PCR) and separated on 0.5% agarose gel. High population of plant-parasitic nematodes was recorded on the vegetable crops cultivated on covenant university farm. Also significantly higher population ($p<0.005$) of Meloidogyne spp. were found in association with C. argentea and Corchorus olitorius than those recorded on Abelmoschus esculentus from the vegetable farm. The molecular characterization of the Meloidogyne species from the farm indicates Meloidogyne incognita as the nematode species associated with the vegetable crops.

INTRODUCTION

In Africa diets, vegetables are important sources of food and medicine (Probst *et al.*, 2010; Padulosi *et al.*, 2011). Farmers have cultivated vegetable crops for generations as additional nutrient sources that supply the body with minerals, vitamins and certain hormone precursors in addition to protein and energy (Mathieu and Meissa, 2007; Wujisguleng and Khasbagen, 2010). Over the years, there has been significant reduction in the production and yield of vegetable crops due to various pest and diseases among which is the root-knot disease caused by *Meloidogyne* species. Globally, plant-parasitic nematodes have been described as important pests of vegetable and horticultural crops (Johnson *et al.*, 1996; Rius *et al.*, 2007). Damage caused by nematodes accounts for about 40% reductions in quantity and quality of yield of harvested vegetables (Anwar and Mckenry, 2010). These nematodes are predominantly damaging on vegetable crops in tropical and subtropical countries (Sikora and Fernandez, 2005).

Vegetable crops are particularly susceptible to root-knot nematodes which cause the characteristic swelling and disruption of vascular tissues of the plant. They reduce plant vigor, cause root lesions, rotting and deformation. Damage due to the feeding activity of the nematodes on the vegetables could result in the reduction in quality and quantity of the harvested vegetable crops (Schippers, 2000). Affected plants may show slow or stunted growth, yellowing of the leaves, wilting of the plant and seedlings produce few roots. Swelling or galls develop on the root of infected plants as a result of nematode expansion of root cells and damage the vascular tissues of root, thereby interfering with the normal movement of water and nutrients through the plant which eventually lead to reduction in yield or total crop failure.

About ninety species of the genus *Meloidogyne* have been reported worldwide (Hunt *et al.*, 2005), the four most common and frequently occurring species, *M. incognita, M. javanica, M. arenaria, M. hapla* and their known races that readily attack vegetable crops in outdoor as well as in indoor cultivations (Zijlstra *et al.*, 2000) have also been variously reported in association with several cultivated horticultural and vegetable crops in Nigeria (Enopka *et al.*, 1996; CABI and EPPO., 2002; Agwu and Ezigbo, 2005; Bello *et al.*, 2014).

Distinct morphological characters that have been used for identification of *Meloidogyne* spp. includes, the second stage juveniles, morphology of the female adult and males, body length, perineal pattern, head and tail, the excretory pore, dorsal esophageal gland opening, plasmid and spicule (Eisenback and Triantaphyllou, 1991). In recent times, various molecular approaches have been designed for accurate identification of various *Meloidogyne* species. This is primarily because DNA-based methods are rapid and reliable compared to the morphological methods (Powers *et al.*, 2005).

Various methods based on DNA and Polymerase Chain Reaction (PCR) have been developed and successfully used for identifying a large number of *Meloidogyne* species (Singh, 2009). The most popularly used DNA-based methods include: mitochondrial DNA, Restriction Fragment Length Polymorphism (RFLP), Amplified Fragment Length Polymorphism (AFLP), Random Amplified Polymorphism DNA (RAPD), Sequence Characterized Region Markers (SCAR-PCR) and Ribosomal DNA (Blok and Powers, 2009).

One of the major challenges facing root-knot nematode taxonomy is the need to establish an accurate, efficient and reliable method of identifying various closely related *Meloidogyne* species. Thus, DNA-based diagnostic methods have gained much acceptance due to their accuracy, reliability and robust nature (Powers, 2004). Besides, they have greater advantage over morphological method of identifying since they can be applied on various stages of development, discriminate individual species from mixed populations and also utilize DNA voucher specimens that have been stored for years (De Ley *et al.*, 2002). This present study seeks to identify the *Meloidogyne* species associated with *Corchorus olitorius, Abelmoschus esculentus* and *Celosia argentea* and to determine the population densities of the associated plant-parasitic nematodes of the selected vegetable crops cultivated on Covenant University farm in Nigeria.

MATERIALS AND METHODS

Morphological characterization of *Meloidogyne* spp. of selected vegetable crops

Study site: Soil and root samples were collected from vegetable farm in Covenant University, Ota, Ogun State, Nigeria, located on a latitude of 6°40' 14.37"N and longitude 3°9' 32.31"E.

Sample collection: Soil and root samples were randomly collected from the rhizosphere of the selected vegetables; *Abelmoschus esculentus*, *Celosia argentea* and *Corchorus olitorius* using a hand trowel to a depth of about 15 cm and within a 25 cm radius from the base of the plants. Three replicates per sample were collected in bags and bulked together to form a composite sample which gave a representation of the nematode types and population on the field. Each composite sample was properly labeled and assigned name, collection date, location and time.

Extraction of nematodes from the samples: The Extraction Tray method of Whitehead and Hemming (1965) was employed for the extraction of vermiform nematodes from the samples. Two hundred grams (200 g) each, of the composite samples were placed in the upper sieve of a modified Baermann tray set-up which was made up of a double-ply facial tissue, sandwiched between a pair of plastic sieves and placed in a bowl of water with about 500 mL of water in it. The setup was allowed to remain undisturbed for 24 h, after which the sieves were gently lifted off. The resulting nematode suspension in the bowl were poured into a 500 mL Nalgene wash bottle and left undisturbed for 5 h after which the supernatant was decanted.

Individual adult females and egg mass of RKN were picked from teased galled-root sample under a stereo-microscope and placed PCR tubes containing 8 μL sterile ddH_2O water.

Identification of nematodes: The identification of *Meloidogyne* juveniles in the aqueous suspension was determined using a compound microscope. Two milliliters of the nematode suspension was pipetted after bubbling air through the suspension for homogeneity and dispensed into a counting dish for nematode identification. Nematode identification was done with the aid of pictorial key of Mai and Lyon (1975).

Molecular characterization of *Meloidogyne* spp.
DNA extraction protocols: Two extraction protocols were employed to quantify and ascertain the protocol that yielded higher DNA quality. The protocols are: Qiagen Extraction Kit protocol (Qiagen ®) and a rapid

method for nematode identification described by Subbotin *et al.* (2001) was modified for the extraction of DNA from adult females nematodes and eggs.

DNA extraction using qiagen extraction kit: DNA was extracted according to the manufacturer instruction (Qiagen®). DNA Extraction Protocol of Subbotin *et al.* (2001).

Single adult females, juveniles and egg mass were picked alone and were transferred into Eppendorf tubes containing 8 µL distilled water and 10 µL nematode lysis buffer (500 mM KCl, 100 mM Tris-HCl pH 8.0, 15 mM $MgCl_2$; 1:0 mM DTT, 4.5% Tween 20) and crushed with an microhomogeniser Vibro Mixer (Zürich, Switzerland) for 2.5-3 min. Two microlitres proteinase K were added and the tubes were incubated at 65°C (1 h) and 95°C (10 min) consecutively and finally centrifuged (1 min; 16 000 g). The DNA suspension was stored at -20°C and used for further studies.

Table 1: Root-Knot Nematodes (Meloidogyne sp.) primer specific and sequences for PCR reactions

Primer code	Sequence
C2F3/1180	5'-GGTACAATGTTCAGAAATTTGTGG-3'
Mitochondrial	5'-TACCTTTGACCAATCAGGCT-3'
Far/Rar (*M. arenaria*) SCAR	5'-TCGGCGATAGAGGTAAATGAC-3'
	5'-TCGGCGATAGACACTACAAACT-3'
Fjav/Rjav (*M. javanica*) SCAR	5'-GGTGCGCGATTGAACTGAGC-3'
	5'-CAGGCCCTTCAGTGGAACTATAC-3'
MI-F/MI-R (*M. incognita*) SCAR	5'-GTGAGGATTCAGCTCCCCAG-3'
	5'-ACGAGGAACATACTTCTCCGTCC-3'

DNA quantification: The concentration and purity of the extracted DNA was monitored using NanoDrop ND-1000 Spectrophotometer at a wavelength of 260 and 280 nm. The DNA was then stored at -40°C until used for DNA-PCR.

Primers for nematode identification: Four *Meloidogyne* species specific primers; C2F3/1108 (mitochondrial), Far/Rar (*M. arenaria*) SCAR, Fjav/Rjav (*M. javanica*) SCAR and MI-F/MI-R (*M. incognita*) SCAR were used in this study. Table 1 shows the primer codes and their sequences.

Polymerase Chain Reactions (PCR): The PCR reactions were carried out in a total volume of 25 µL containing; PCR master mix (12.5 µL), sterilized water (10.5 µL), primer forward (0.5 µL), primer reverse (0.5 µL) and template DNA (1.0 µL). The mixture was placed in a thermal cycler already heated at 94°C. PCR amplification conditions are as follows: denaturation at 94°C for 1 min, annealing at 48°C for 1 min and extension at 70°C for 2 min, repeated for 45 cycles. A 5 min incubation period at 72°C followed the last cycle in order to complete any partially synthesized second strand.

Agarose gel electrophoresis: The PCR reaction products were separated on 0.5% agarose gel with 0.01% ethidium bromide incorporated into the gel. The DNA amplification products were visualized using the automated trans-illuminator (UVP) with digital camera compatible with Window/Vista.

Statistical analysis: The data on nematode population was subjected to analysis of variance and the means partitioned by Duncan's multiple range tests at 5% level of probability.

RESULT

Plant-parasitic nematodes associated with the selected vegetable crops: Root-knot nematodes (*Meloidogyne* species) and some other plant-parasitic nematodes were observed in soil and root samples collected from the selected vegetable crops. Other Plant-Parasitic Nematodes (PPN) found in association with the selected vegetable crops include: *Pratylenchus* sp., *Rotylenchulus* sp., *Scutellonema* sp. and *Helicotylenchus* sp.

Population density of nematodes associated with the selected vegetable crops: The population densities of the *Meloidogyne* species and other parasitic nematodes extracted from the rhizosphere of the selected vegetable crops is presented in Table 2. Significantly higher populations of the *Meloidogyne* species were recorded on the vegetable crops when compared with the population of the other plant parasitic nematodes.

Table 2: Mean population densities of nematodes associated with vegetable crops on covenant university, ota, farm

Vegetable crop	*Meloidogyne* spp./ 200 mL soil	Other plant-parasitic nematodes/200 mL soil	Non-parasitic nematodes/200 mL soil	Total nematodes/ 200 mL soil
Abelmoschus esculentus	482.67	133.67	33.67	650.01
Corchorus olitorius	626.67	283.00	62.33	972.00
Celosia argentea	642.33	249.00	72.67	964.00

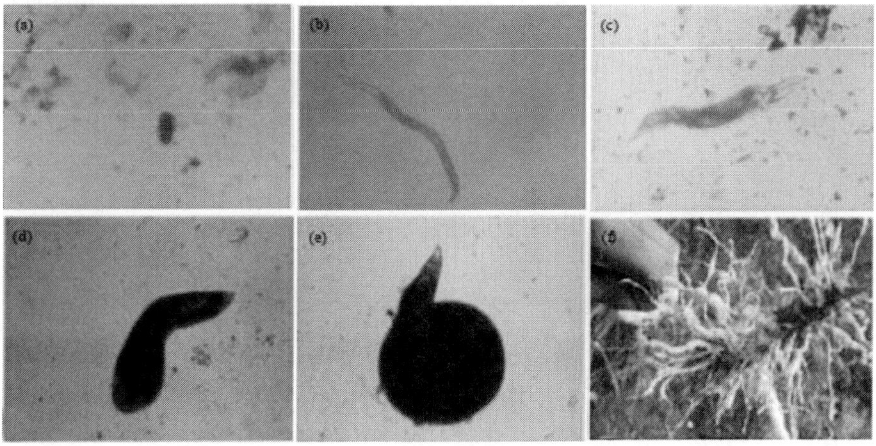

Figure 1. (a-f): Micrographs of the developmental stages and the characteristic root-galling symptom associated with plants infected with of Meloidogyne spp., (a) Single egg, (b) 2nd stage juvenile, (c) 3rd stage juvenile, (d) 4th stage juvenile, (e) Adult female and (f) Root galling symptom on Celosia argentea root

Table 3: Mean squares of nematode species associated with vegetable crops in covenant university

Source of variation	SS	df	MS	F	p-value	F crit
Rows	22481.25	2	11240.63	7.397355	0.045295*	6.944272
Columns	436921.6	2	218460.8	143.7671	0.000188**	6.944272
Error	6078.187	4	1519.547			
Total	465481	8				

**Significant at $p<0.001$

The population of root-knot nematodes recorded on *Celosia argentea*, *Corchorus olitorius* and *Abelmoschus esculentus* are 642.33, 626.67 and 482.67, respectively (Table 2 and 3). The result indicates lower populations of RKN on *Abelmoschus esculentus*.

Significantly higher population of plant parasitic nematodes (PPN) were also were also recorded on the selected vegetable crops than non-parasitic nematodes (Table 2).

Morphological characterization of *Meloidogyne* spp.: Micrographs of the developmental stages *Meloidogyne* spp. obtained from roots of the vegetable crops is described in Figure 1a-e. The observed developmental stages are single egg, second stage juvenile (J2), third stage juvenile (J3), fourth stage juvenile (J4) and adult female. The characteristic root galling symptom associated with vegetable plants infected with *Meloidogyne* spp. was also shown in Figure 1f.

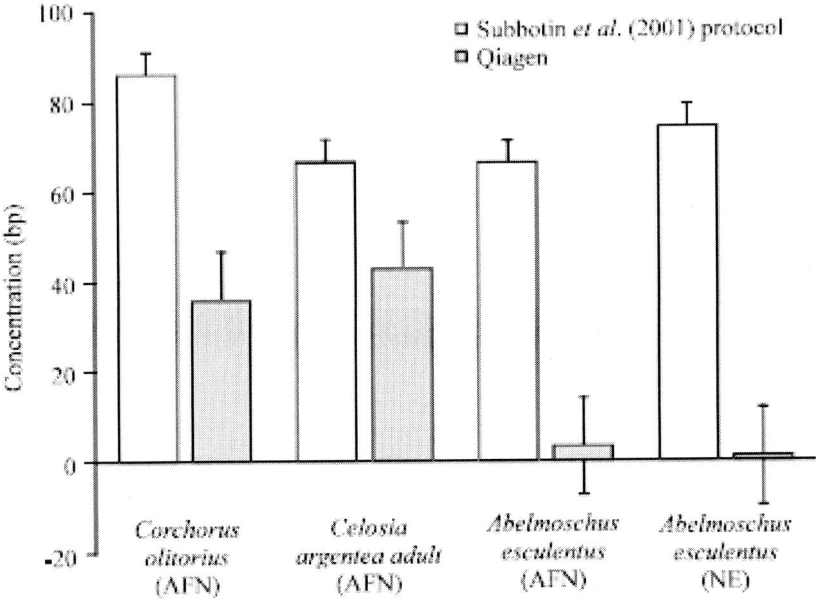

Figure 2: Comparative DNA concentrations of Subbotin et al. (2001) and Qiagen extraction kit protocols

DNA quantification and PCR amplification reactions: Modified Subbotin *et al.* (2001) protocol yielded higher concentration of DNA than the Qiagen kit Protocol (Qiagen ®). Figure 2 shows the comparative DNA concentrations by both protocols. Primer C2F3/1108 (mitochondrial) generated amplified bands of approximately (400 bp), primers MIF-MIR (*M. incognita*) generated bands of (100-200 bp), no amplified band was observed for both Far/Rar (*M. arenaria*) and Fjav/Rjav (*M. javanica*). Gel electrophoresis of amplified bands is shown in Figure 3.

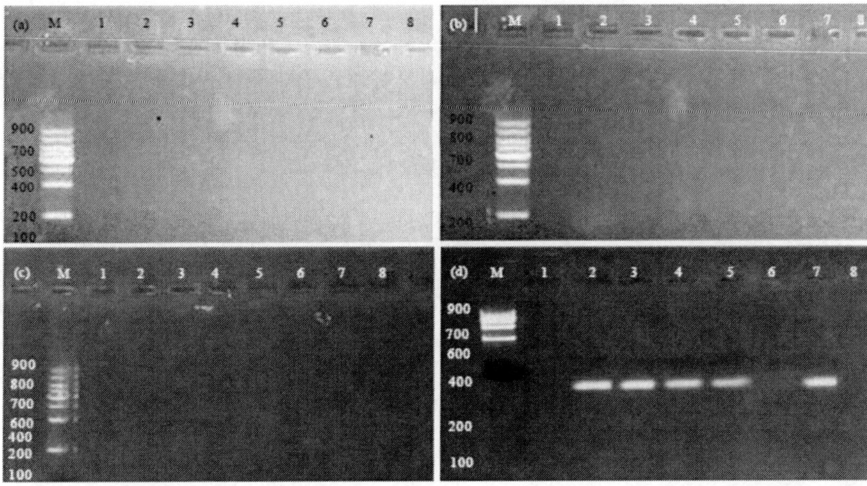

Figure 3(a-d): Gel electrophoresis of Root- Knot Nematodes (Meloidogyne sp.) Species specific primer (a) Far/Rar for M. arenaria with no visible band, (b) Fjav/Rjav for M. javanica with no visible bands, (c) MIF-MIR for M. incognita, bands were visible and (d) Mitochondrial specific primers C2F3/1108 for Meloidogyne sp.

DISCUSSION

In this study, Root-Knot Nematodes (*Meloidogyne* species) were found in association with soil and root samples of *Celosia argentea*, *Corchorus olitorius* and *Abelmoschus esculentus* cultivated on Covenant University (CU) Farm, Ota, Nigeria. Similarly, *Pratylenchus* sp., *Rotylenchulus* sp., Scutellonema sp and Helicotylenchus sp were observed in association with the selected vegetable crops. The higher population of parasitic nematodes recorded on the crops indicate that the selected vegetable crops in Covenant University are highly susceptible and good hosts of root-knot nematodes. Susceptibility of vegetable crops to *Meloidogyne* species has been variously reported worldwide (CABI and EPPO., 2002; Sikora and Fernandez, 2005; Noling, 2012). They have been described as good hosts and highly susceptible to pathogenic nematodes, particularly *Meloidogyne* species (Picard *et al.*, 2004; Presswell *et al.*, 2014; Anwar and Van Gundy, 1989).

In tropical countries, vegetables are usually susceptible to nematode attack and their production depends on the correct management of these pathogens (Sikora and Fernandez, 2005). According to Noling (2012), plant symptoms and yield reductions are often directly related to preplant infestation levels in soil and as infestation levels increases, the amount of damage and yield loss will also increase. The mere presence of these nematodes on the farm is a potential risk to the cultivated vegetables, therefore routine and timely nematode management is essential in order to ensure good plant health and maximize productivity.

The relatively high populations of root-knot nematode recovered in both *Celosia argentea* and *Corchorus olitorius* were indicative of their higher susceptibility to RKN compare to *Abelmoschus esculentus* on CU farm. The presence of various motile stages of developmental stages particularly from infective J2 (second juvenile stage) to adult is an indication that *Meloidogyne* sp. are actively feeding on the roots of the vegetables. These observations are in tandem with the observed stunted growth, chlorotic features at harvest, identification of root galling and brown lesions on the vegetables, particularly, *Celosia argentea* during field observation.

According to Thomas *et al.* (2012) and Kaur and Attri (2013), precision and accuracy in nematode identification is of great importance in order to proffer effective management strategies for nematode control. Identification based on morphological characters is time consuming, labour intensive labor, requires experience and expertise, even then, has not been sufficient for identification and has led to the acceptance of molecular studies.

Rapid and reliable identification of nematode species is required for a number of reasons such as utilizing appropriate crop rotations, managing resistance effectively, developing bio-control strategies, studying virulence, plant-nematode interactions and for quarantine purposes (Zijlstra *et al.*, 2000, 2004).

It was observed in the course of this study that the sensitivity of a DNA extraction assay on nematodes from root and soil extracts requires fresh and live nematode samples. Primers optimized successfully resulted in amplification of DNA obtained from all the nematode sources including egg masses and females. The modified extraction protocol of Subbotin *et al.* (2001) for the extraction of genomic DNA of the *Meloidogyne* species also proved to be more effective than the extraction kit which is more expensive and not readily available to researchers especially in the **DEVELOPING COUNTRIES**.

In this study, the nematode samples tested positive via amplification to the primer C2F3/1108 (mitochondrial), there was no amplification for both Far/Rar (*M. arenaria*) and Fjav/Rjav (*M. javanica*) which implies that none of the nematode species belong *M. arenaria* or *M. javanica* but they show amplification for MI-F/MI-R (*M. incognita*) and this suggests that the species of the root-knot nematodes found in association with the selected vegetables crops are *Meloidogyne incognita*.

CONCLUSION

Farmers are often not aware of nematode problems and the damage that nematodes can cause to crops. Covenant university farm was acquired from local settlers in Ota that were predominantly engaged in subsistence agriculture as a means of livelihood. The occurrence of *Meloidogyne* species and other plant-parasitic nematodes in this farmland implies that these nematodes are likely to be wide spread and abundant in the neighboring farmlands. It therefore becomes imperative that resistant plant varieties are cultivated and that effective management strategies are embraced in order to maximize crop yield and also to reduce the spread of nematodes in the local environment.

REFERENCES

1. Agwu, J.E. and F.C. Ezigbo, 2005. Effect of Meloidogyne incognita (root-knot nematode) on the development of Abelmoschus esculentus (okra). Anim. Res. Int., 2: 358-362.
2. Anwar, S.A. and M.V. McKenry, 2010. Incidence and reproduction of Meloidogyne incognita on vegetable crop genotypes. Pak. J. Zool., 42: 135-141.
3. Anwar, S.A. and S.D. van Gundy, 1989. Influence of four nematodes on root and shoot growth parameters in grape. J. Nematol., 21: 276-283.
4. Bello, T.T., B. Fawole and A. Claudius-Cole, 2014. Management of root-knot nematodes (Meloidogyne Spp) on tomato using antagonistic plants. J. Biol. Agric. Healthcare, 4: 97-100.
5. Blok, V.C and T.O. Powers, 2009. Biochemical and Molecular Identification. In: Root-Knot Nematodes, Perry, R.N., M. Moens and J.L. Starr (Eds.). CAB, USA., ISBN: 9781845934934, pp: 98-112.
6. CABI and EPPO., 2002. Meloidogyne incognita. Distribution Maps of Plant Diseases No. 854, CAB International, Wallingford, UK.
7. De Ley, I.T., P. de Ley, A. Vierstraete, G. Karssen, M. Moens and J. Vanfleteren, 2002. Phylogenetic analyses of Meloidogyne small subunit rDNA. J. Nematol. Dis., 34: 319-327.
8. Eisenback, D.E. and H.H. Triantaphyllou, 1991. Root-Knot Nematodes: Meloidogyne Species and Races. In: Manual of Agricultural Nematology, Nickle, W.R. (Eds.). Marcel Dekker, Inc., New York, ISBN: 9780824783976, pp: 191- 274.
9. Enopka, E.N., I.A. Okwujiako and B.E. Madunagu, 1996. Control of root-knot nematodes in tomato with furadan. Global J. Pure Applied Sci., 2: 131-136.
10. Hunt, G.I., M. Luc and R.H. Manzanilla-Lopez, 2005. Identification, Morphologys and Biology of Plant-Parasitic Nematodes. In: Plant Parasitic Nematodes in Subtropical and Tropical Agriculture, Luc, M., R.A. Sikora and J. Bridge (Eds.). CAB International, USA., ISBN: 9781845931445, pp: 11-52.

11. Johnson, A.W., C.C. Dowler, N.C. Glaze and Z.A. Handoo, 1996. Role of nematodes, nematicides and crop rotation on the productivity and quality of potato, sweet potato, peanut and grain sorghum. J. Nematol., 28: 389-399.
12. Kaur, H. and R. Attri, 2013. Morphological and morphometrical characterization of Meloidogyne incognita from different host plants in four districts of Punjab, India. J. Nematol., 45: 122-127.
13. Mai, W.F. and H.H. Lyon, 1975. Pictorial Key to Genera of Plant-Parasitic Nematodes. 4th Edn., Comstock Publishing Associates a Division of Cornell University Press, Ithaca and London.
14. Mathieu, G. and D. Meissa, 2007. Traditional leafy vegetables in senegal: Diversity and medicinal uses. Afr. J. Tradit. Complement. Altern. Med., 4: 469-475.
15. Noling, J.W., 2012. Nematode management in tomatoes, peppers and eggplant. ENY-032/NG032, Entomology and Nematology Department, Florida Cooperative Extension Service, Institute of Food and Agricultural Sciences, University of Florida, pp: 1-15.
16. Padulosi, S., V. Heywood, D. Hunter, A. Jarvis and S.S. Yadav et al., 2011. Underutilized Species and Climate Change: Current Status and Outlook. In: Crop Adaptation to Climate Change, Yadav, S.S., R. Redden, J.L. Hatfield, H. Lotze-Campen and A.J.W. Hall (Eds.). John Wiley and Sons, New York, ISBN: 9780470960905, pp: 507-521.
17. Picard, D., O. Plantard, M. Scurrah and D. Mugniery, 2004. Inbreeding and population structure of the potato cyst nematode (Globodera pallida) in its native area (Peru). Mol. Ecol., 13: 2899-2908.
18. Powers, T., 2004. Nematode molecular diagnostics: From bands to barcodes. Annu. Rev. Phytopathol., 42: 367-383.
19. Powers, T.O., P.G. Mullin, T.S. Harris, L.A. Sutton and R.S. Higgins, 2005. Incorporating molecular identification of Meloidogyne spp. into a large-scale regional nematode survey. J. Nematol., 37: 226-235.
20. Presswell, B., S. Evans, R. Poulin and F. Jorge, 2014. Morphological and molecular characterization of Mermis nigrescens Dujardin, (Nematoda: Mermithidae) parasitizing the introduced European earwig (Dermaptera: Forficulidae) in New Zealand. J. Helminthol.
21. Probst, L., L. Aigelsperger and M. Hauser, 2010. Consumer attitudes towards vegetable attributes: Potential buyers of pesticide-free vegetables in Accra and Kumasi, Ghana. Ecol. Food Nutr., 49: 228-245.
22. Rius, J.E.P., N. Vovlas, A. Troccoli, G. Liebanas, B.B. Landa and P. Castillo, 2007. A new root-knot nematode parasitizing sea rocket from Spanish Mediterranean coastal dunes: Meloidogyne dunensis n. sp. (Nematoda: Meloidogynidae). J. Nematol., 39: 190-202.
23. Schippers, R.R., 2000. African Indigenous Vegetables: An Overview of the Cultivated Species. Natural Resources Institute (Great Britain), Chatham, UK., ISBN-13: 9780859545150, Pages: 214.
24. Sikora, R.A. and E. Fernandez, 2005. Nematode Parasites of Vegetables. In: Plant Parasitic Nematodes in Subtropical and Tropical Agriculture, Luc, M., R.A. Sikora and J. Bridge (Eds.). 2nd Edn., CAB International, USA., ISBN: 9781845931445, pp: 319-392.
25. Singh, S.K., 2009. Morphological and molecular characterization of root-knot nematode (Meloidogyne Spp.) diversity in Fiji. Master's Thesis, University of the South Pacific, Fiji.
26. Subbotin, S.A., D. Peng and M. Moens, 2001. A rapid method for the identification of the soybean cyst nematode Heterodera glycines using duplex PCR. Nematology, 3: 365-371.
27. Thomas, V.P., S.L. Fudali, J.E. Schaff, Q. Liu and E.H. Scholl et al., 2012. A sequence-anchored linkage map of the plant-parasitic nematode Meloidogyne hapla reveals exceptionally high genome-wide recombination. Genes Genomes Genet. (G3), 2: 815-824.

28. Whitehead, A.G. and J.R. Hemming, 1965. A comparison of some quantitative methods of extracting small vermiform nematodes from soil. Ann. Applied Biol., 55: 25-38.
29. Wujisguleng, W. and K. Khasbagen, 2010. An integrated assessment of wild vegetable resources in inner mongolian autonomous region, China. J. Ethnobiol. Ethnomedicine, Vol. 6.
30. Zijlstra, C., D.H.T.M Donkers-Venne and M. Fargette, 2000. Identification of Meloidogyne incognita, M. javanica and M. arenaria using Sequence Characterised Amplified Region (SCAR) based PCR assays. Nematology, 2: 847-853.
31. Zijlstra, C., R. Van Hoof and D. Donkers-Venne, 2004. A PCR test to detect the cereal root-knot nematode Meloidogyne naasi. Eur. J. Plant Pathol., 110: 855-860.

CHAPTER 11

Phylogenetic Analysis of Entomoparasitic Nematodes, Potential Control Agents of Flea Populations in Natural Foci of Plague

E. I. Koshel,[1] V. V. Aleshin,[2,3,4] G. A. Eroshenko,[1] and V. V. Kutyrev[1]

[1]Russian Research Anti-Plague Institute "Microbe", Saratov 410005, Russia
[2]Belozersky Institute of Physical-Chemical Biology, Lomonosov Moscow State University, Moscow 119991, Russia
[3]Institute for Information Transmission Problems, Russian Academy of Sciences, Moscow 127994, Russia
[4]National Research Institute of Physiology, Biochemistry, and Nutrition of Farm Animals, Russian Academy of Agricultural Sciences, Kaluga Region, Borovsk 249013, Russia

ABSTRACT

Entomoparasitic nematodes are natural control agents for many insect pests, including fleas that transmit Yersinia pestis, a causative agent of plague, in the natural foci of this extremely dangerous zoonosis. We examined the flea samples from the Volga-Ural natural focus of plague for their infestation with nematodes. Among the six flea species feeding on different rodent hosts (Citellus pygmaeus, Microtus socialis, and Allactaga major), the rate of infestation varied from 0 to 21%. The propagation rate of parasitic nematodes in the haemocoel of infected fleas was very high; in some cases, we observed up to 1,000 juveniles per flea specimen. Our study of morphology, life cycle, and rDNA sequences of these parasites revealed that they belong to three distinct species differing in the host specificity. On SSU and LSU rRNA phylogenies, these species representing three genera (Rubzovinema, Psyllotylenchus, and Spilotylenchus), constitute a monophyletic group close to Allantonema and Parasitylenchus, the type genera of the families Allantonematidae and Parasitylenchidae (Nematoda: Tylenchida). We discuss the SSU-ITS1-5.8S-LSU rDNA phylogeny of the Tylenchida with a special emphasis on the suborder Hexatylina.

1. INTRODUCTION

More than 150 species of fleas feeding on different mammalian hosts, primarily rodents, are vectors of the bacterium Yersinia pestis, a causative agent of plague [1, 2]. In natural foci of plague, the dynamics of flea populations are among the main factors controlling the incidence of epizootics that pose a threat to humans inhabiting the areas [3–5]. Entomoparasitic nematodes of the order Tylenchida are known to control populations of various insect hosts [6–9]. The rate of tylenchid infestation in fleas reaches 50–60% in some cases [10, 11], when the nematodes cause castration and early death of the flea hosts [9, 12, 13].

Despite high importance of the Tylenchida as a nematode order harboring entomoparasites and notorious crop pests, their reliable phylogeny is still a challenge. Tylenchid nematodes differ widely in life cycle, parasitic strategies, and the host range that spans plants, fungi, and invertebrates. Phylogenies obtained from SSU and partial LSU rDNA data often disagree with classifications based on morphology and life cycle [14–21]. Phylogenetic resolution inside the order is far from being clear, which in many respects results from the insufficiency of data available to adequately describe its diversity. As for tylenchid parasites of fleas, only 31 species are described to date [9, 22–31], with no molecular vouchering. Here we present a study of parasitic nematodes isolated from fleas sampled from different rodent hosts in a natural focus of plague.

2. MATERIALS AND METHODS

2.1. Collection of Samples

Samples were collected in 2012 (spring and autumn) and 2013 (spring) in the Volga-Ural natural focus of plague (Figure 1). The sampled rodents included sousliks (Citellus pygmaeus), mouse-like rodents (Microtus socialis and Apodemus uralensis), and jerboas (Allactaga major). Three flea species (Citellophilus tesquorum, Neopsylla setosa, and Frontopsylla semura) were sampled on sousliks; two species (Amphipsylla rossica and Ctenophthalmus secundus) were on M. socialis voles; and one species (Mesopsylla hebes) was on jerboas. Fleas were examined for nematode infestation (Table 1). Examination and dissection of fleas were carried out using the dissecting microscope MBS-2 (LOMO, Russia). A half of parasitic nematodes sampled from each flea was preserved for subsequent DNA extraction, and another half was used for morphological analysis. Live fleas infected with nematodes were placed in glass flasks with river sand to obtain free-living forms. Insects were kept in a KBF 720 (E5.2) climate chamber (Binder, Germany) at 26°C and 80% humidity.

2.2. Morphological Analysis

Fixation and clarification of nematode preparations were performed using standard techniques described by De Grisse [32]. Material was mounted on slides in a drop of glycerin, bound by a paraffin circlet (http://pest.cabweb.org). Color staining of preparations was not performed. Morphometric analysis was conducted using the light microscope "Leica DM 1000" (Leica, Germany) with an eyepiece micrometer. Pictures of nematodes were taken with the microscope "DFC 425" (Leica, Germany). Published data on morphometrics [23, 25, 26] were used for comparison.

Table 1: Number of fleas studied and the percentage of fleas infected with nematodes.

Time of sampling	Host rodent species	Flea species	Number of collected fleas	Number of infected fleas	Percentage of infected fleas
April 2012	*Citellus pygmaeus*	*Citellophilus tesquorum*	41	7	17.1%
		Neopsylla setosa	73	5	6.8%
		Frontopsylla semura	54	7	13%
October 2012	*Microtus socialis*	*Amphipsylla rossica*	135	9	6.7%
		Ctenophthabnus secundus	88	1	1.1%
April 2013	*Citellus pygmaeus*	*Citellophilus tesquorum*	34	0	0
		Neopsylla setosa	271	22	8.1%
		Frontopsylla semura	19	4	21%
	Microtus socialis and *Apodemus uralensis*	*Amphipsylla rossica*	6	0	0
		Ctenophthabnus secundus	52	0	0
	Allactaga major	*Mesopsylla hebes*	34	2	5.9%

Figure 1: The sampling region on the map of Europe.

2.3. DNA Extraction, PCR, and Sequencing

DNA samples were extracted with a Diatom DNA Prep (IsoGen Lab, Russia). rDNA fragments were amplified using an Encyclo PCR kit (Evrogen, Russia) and primers given in Table 2. The amplified rDNA fragments were sequenced using an Applied Biosystems 3500xL DNA analyzer. Sequence reads were assembled with the CAP contig assembly program [33] and proofread with the BioEdit software [34]. For three isolates, almost complete sequences of 18S and 28S rRNA and complete sequences of 5.8 rRNA, internal transcribed spacers ITS1 and ITS2 were assembled. The sequences were submitted to GenBank under accession nos. KF155281–KF155283. For the rest of isolates, partial (750–800 bp) sequences of 18S and 28S rRNA genes were submitted to GenBank under accession nos. KF373731–KF373740.

Table 2: Nucleotide sequences of primers used in this study.

Primer	Sequence	Orientation	References
Nik22	tmycygrttgatyctgyc	F	This study
A	gtatctggttgatcctgccagt	F	[35]
Q5nemCh	gccgcgaayggctcattayaac	F	This study
G18SU	gcttgtctcaaagattaagcc	F	[36]
Ves18-d9	gtcgtaacaaggtatccgtaggtgaac	F	This study
R18Tyl1	ggtccaagaatttcacctctc	R	[36]
B	gtaggtgaacctgcagaaggatca	R	[35]
Q39nem	gaaaccttgttacgacttttrcbygg	R	This study
58d1	rcatcgatgaagaacgywg	F	[37]
58r nem	gcwgcgttcttcatcgacyc	R	This study
28d3	gtcttgaaacacggaccaagg	F	[37]
28d6	ggtyagtcgrtcctrag	F	[37]
D2A	acaagtaccgtgagggaaagttg	F	[38]
28r4	gctatcctgagggaaacttcgg	R	[37]
28r2nem	cggtacttgttcgctatcg	R	This study
28r7	agccaatccttwtcccgaagttac	R	[37]
28r12	ttctgacttagaggcgttcag	R	[37]
D3B	tcggaaggaaccagctacta	R	[38]

2.4. Phylogenetic Analysis

The newly obtained rDNA sequences of tylenchid parasites of fleas were aligned with a selected set of other tylenchid sequences obtained from the GenBank. The main selection criterion was to sample representatives of all clades that occur in published SSU and LSU rDNA phylogenies of the Tylenchida [16–21, 39]. Apart from the D2-D3 LSU rDNA expansion segment commonly used in previous studies, we included all LSU rDNA sequence data available for the Tylenchida, with the exception of Basiria sp. SAN-2005 (accession nos. DQ145619, DQ145667) that in our preliminary analyses (data not shown) demonstrated a disputable affinity to the Tylenchida. For the species Anguina tritici, Globodera pallida, Heterodera glycines, Pratylenchus vulnus, and Radopholus similes the nearly complete rDNA sequences were assembled with appropriate cDNA fragments identified with BLAST [40]. Partial LSU rDNA sequence of Ditylenchus dipsaci was combined with the soil environmental clone NTS_28S_061A_2_b4 (accession no. KC558346), as the clone sequence appeared to represent a close tylenchid relative of D. dipsaci. Chimeric sequences were also created in some cases when closely related partial rDNA sequences were found in the database. All sequences and their accession numbers are listed in Table 3. Cephalobidae and Chambersiellidae were chosen as the outgroup. Alignments were constructed with the MUSCLE program [41] and refined manually using the MEGA 5.0 software package [42]. Three alignments were generated: (1) SSU rDNA, (2) D3 region of LSU rDNA, and (3) concatenated rDNA data including SSU, LSU, 5.8S rDNA, and highly conserved regions of ITS1. After discarding ambiguously aligned positions, the alignments length was 1,723,

592, and 4,930 positions, respectively. Bayesian reconstruction of phylogeny was done with the PhyloBayes software, version 3.2 [43] under the GTR + CAT + DP model [44]. Eight independent runs were performed with 4,000,000 cycles each; the first 3,000,000 cycles were discarded. A consensus tree with Bayesian posterior probabilities was constructed for the remained tree sample. Bayesian reconstruction was also performed using the MrBayes software [45] under the GTR + G8 + I model [46] in two independent runs, each with four Markov chains. The chains were run for 5,000,000 generations, with trees sampling every 1,000th generation. The consensus posterior probabilities were calculated after discarding the first 3,000,000 generations. Partitioning "by genes" was used for the concatenated alignment with all parameters unlinked, except for the topology and branch lengths. In addition, node support was estimated with maximum likelihood bootstrap as implemented in the RAxML software, version 7.2.6 [47], under the GTR + G + I model with 1,000 bootstrap replicates. Alternative topologies were tested using the approximately unbiased (AU) [48] and Kishino and Hasegawa [49] tests implemented in the CONSEL software [50] and the expected likelihood weight test [51] implemented in the TREE-PUZZLE software [52]. TREEVIEW [53] was used as the tree viewer and editor, and site-wise log-likelihoods were computed with TREE-PUZZLE under the GTR + G8 + I model with substitution matrix parameters estimated by MrBayes.

Table 3: List of OTUs and accession numbers of sequences.

Name	18S rRNA	ITS1-5.8S rRNA	28S rRNA	%, SSU-ITS1-5.8S-LSU/D3	Reference	Family by [8]
			Chambersiellidae+			
Fescia grossa	KC242218	—	DQ145636 DQ145684	87.1/—	[54] [55]	Chambersiellidae
Geraldius sp. SAN-2010a	—	—	GU062821	17.8/—	[56]	Chambersiellidae
			Cephalobidae			
Acrobeloides maximus	EU196016	JX026706	EU195987	94.8/—	[57] [58] [57]	
Cephalobus cubaensis	AF202161	AF202161	EU253570	89.8/—	[59] [57]	Cephalobidae
Panagrolobus sp. SN-2010	—	—	HM439771	51.9/—	[60]	
Cephalobidae Gen. sp. MHMH-2008	FJ040406	—	—		Holterman et al., 2008, unpublished.	
Zeldia punctata	—	DQ146426	EU195988	96.6/—	[61] [57]	
Zeldia sp.	AY284675	—	—			
			Aphelenchidae			
Aphelenchus avenae	JQ348399	AF119048	—	96.9/—	[62] [63]	
Aphelenchus sp.	—	—	DQ145664 DQ145714		[55]	Aphelenchidae
Paraphelenchus acontioides	—	—	HQ218322	45.5/—	[64]	
Paraphelenchus sp.	AY284642	—	—		[18]	
			Hexatylina + "Anguinata (part)": Iotonchioidae			
Allantonema mirable	—	—	JX291132	10.6/85.8	[39]	
Bradynema listronoti	DQ915805		DQ915804	45.6/96.8	[65]	Allantonematidae
Bradynema rigidum			DQ328730	10.4/86.3	[20]	
Contortylenchus sp.	—	—	DQ328731	—/85.4	[20]	
Deladenus durus	JQ957898	—	—	34.0/—	[66]	
Deladenus proximus	JF304744	JF304744	—	35.2/—	[67]	
Deladenus siricidicola isolate 354	AY633447		AY633444	45.8/98.1	[68]	
Deladenus siricidicola isolate 466	FJ004890	FJ004890	—	41.7/—	[69]	
Deladenus siricidicola isolate 1093	FJ004889	FJ004889	—	42.0/—	[69]	Neotylenchidae
Fergusobia camaldulensae	AY589294	—	AY589346	45.7/98.0	[68]	
Fergusobia sp. 444	EF011667	—	EF011675	45.7/97.3	[68]	
Fergusobia sp. SBG	FJ393270	—	FJ386996	45.7/98.3	[70]	
cf. *Gymnotylenchus* sp. TSH-2005	AY912040	—	—	12.9/—	Powers et al., unpublished.	

Howardula aoronymphium	AY589304	AY589304	AY589395	49.7/96.1	[68]	
Howardula dominicki	AF519234	AF519234	—	37.4/—	[71]	
Howardula neocosmis	AF519226	AF519226	—	38.2/—	[71]	
Howardula phyllotretae	JX291137	—	DQ328728	41.9/86.1	[39] [20]	Allantonematidae
Howardula sp. CD353	—	—	JX291131	—/93.9	[39]	
Howardula sp. SP-A	AF519232	AF519232	—	37.7/—	[71]	
Howardula sp. SP-F	AF519222	AF519222	—	38.2/—	[71]	
Howardula sp. SP-MA	AF519233	AF519233	—	38.1/—	[71]	
Howardula sp. SP-PS	AF519231	AF519231	—	38.1/—	[71]	
Parasitylenchus bifurcatus	KC875397	—		44.0/85.3	[72]	
Parasitylenchus sp.	—	—	DQ328729		[20]	
Psyllotylenchus sp. ex *Frontopsylla semura*	KF373734	—	KF373739	27.1/93.7	This study	Parasitylenchidae
Psyllotylenchus sp. ex *Neopsylla setosa*	KF373733	—	KF373738	27.1/93.7	This study	
Rubzovinema sp. ex *Amphipsylla rossica*	KF155281	KF155281	KF155281	90.0/100.0	This study	
Rubzovinema sp. ex *Ctenophthalmus cecandus*	KF155282	KF155282	KF155282	89.8/100.0	This study	Neotylenchidae
Rubzovinema sp. ex *Citellophilus tesquorum*	KF155283	KF155283	KF155283	93.2/100.0	This study	
Rubzovinema sp. ex *Frontopsylla semura*	KF373732	—	KF373737	27.1/93.7	This study	
Rubzovinema sp. ex *Neopsylla setosa*	KF373731	—	KF373736	27.1/93.7	This study	
Skarbilovinema laumondi	—	—	JX291136	10.9/91.0	[39]	Iotonchioidea
Skarbilovinema lyoni	JX291138	—	DQ328733	41.8/86.3	[39] [20]	
Spilotylenchus sp. ex *Mesopsylla hebes*	KF373735	—	KF373740	27.1/93.4	This study	Parasitylenchidae
cf. *Sychnotylenchus* sp. CSP1-09	DQ080531	—	—	12.9/—	Powers et al., unpublished.	Sychnotylenchidae
Wachekitylenchus bovieni	—	—	DQ328732	—/85.9	[20]	Parasitylenchidae
Unidentified Allantonematidae HaMW	JQ941710	—	—	18.5/—	Rhule, unpublished.	
Unidentified Allantonematidae NK2011_2	AB663183	—	—	12.0/—	[73]	Allantonematidae
Unidentified Allantonematidae NK2011_3	AB663184	—	—	12.0/—	[73]	
Unidentified nematode 804U-025	EU880149	—	—	12.0/—	[74]	
Unidentified nematode CD289	—	—	JX291133	—/84.1	[39]	
Unidentified nematode RGD591T12	AB455970	—	—	12.0/—	[73]	
Unidentified nematode WY2009_BAR-1	—	—	FJ661075	—/96.3	[75]	
Unidentified parasite ex *Chrysobothris affinis*	—	—	DQ202658	—/51.0	Hunt et al., unpublished.	
Hexatylina + "Anguinata (part)": Sphaerularioidea						
Deladenus sp. PDL-2005	AJ966481	—	—	35.0/—	[16]	Neotylenchidae
cf. *Helionema* sp. MHMH-2008	EU669913	—	—	34.0/—	[19]	Parasitylenchidae (genera dubia in Hexatylina)
cf. *Hexatylus* sp. Westplace	AY912050	—	—	12.9/—	Powers et al., unpublished.	Neotylenchidae
Nothotylenchus acris	AY593914	—	—	34.0/—	[76]	Anguinidae
Sphaerularia bombi	AB250212	—	DQ328726	56.7/100.0	Takahashi, unpublished. [20]	Sphaerulariidae
Sphaerularia vespae	AB300595	AB300595	AB300596	54.7/100.0	[77]	
Unidentified nematode 801L-022	EU880129	—	—	12.1/—	[74]	
Anguinata						
Anguina tritici	AY593913	JF826515	HO058555 DQ328723	57.6/92.9	Holterman et al., unpublished. Rao and Rao, unpublished. Rao et al., unpublished. [20]	
Ditylenchus adasi	EU669909	—	—	34.6/—	[19]	
Ditylenchus angustus	AJ966483	—	—	34.6/—	[16]	
Ditylenchus destructor		JX162205		50.0/99.5	[78]	Anguinidae
Ditylenchus dipsaci	AY593911	AY593911	JF327759	60.9/100.0	[76] Zhao 2011, unpublished.	
clone NTS_28S_061A_2_b4			KC558346		[79]	
Ditylenchus drepanocercus	JQ429768	JQ429774	JQ429772	48.7/89.3	[80]	
Ditylenchus halictus	AY589297			52.8/97.3	[68]	
Ficotylus congestae	EU018048			45.6/97.5	[81]	
Halenchus fucicola	EU669912	—	—	34.6/—	[19]	
Pseudhalenchus minutus	AY284638			34.6/—	[19]	
Unidentified entomoparasitic nematode SAS-2006 "*Neotylenchus*" sp.	—	—	DQ328725	—/85.6	[20]	
"Tylenchina": Tylenchidae						
Aglenchus agricola	FJ969113	—	—	46.0/—	van Megen et al., unpublished.	Tylenchidae
Aglenchus sp.	—	—	JQ004996		[82]	

Coslenchus costatus	AY284581	—	—	45.5/—	[18]	
Coslenchus sp.	—	—	JQ005007		[82]	
Filenchus annulatus	JQ814880	—	JQ005017	46.4/—	[82]	
Tylenchus davainei	AY284588	—	—	33.9/—	[18]	
"Tylenchina": Tylodoridae						
Eutylenchus excretorius	EU915487	EU915500	EU915490	35.8/—	[83]	Atylenchidae
Cephalenchus hexalineatus	AY284594	—	—	44.1/—	[18]	Tylodoridae
"Tylenchina": Boleodoridae						
Basiria gracilis	EU130839	—	DQ328717	44.6/—	[84] [20]	
Basiria sp. 3 TJP-2012	—	—	JQ004998	12.0/—	[82]	
Boleodorus thylactus	AY993976	—	—	46.7/—	[16]	Tylenchidae
Boleodorus sp.	—	—	JQ005001		[18]	
Neopsilenchus magnidens	AY284585	—	—	45.6/—	[18]	
Neopsilenchus sp. 3 TJP-2012	—	—	JQ005020		[82]	
Neopsilenchus sp. 1 TJP-2012	—	—	JQ005018	11.9/—	[82]	
"Hoplolaimina": Merliniidae						
Nagelus leptus	—	—	DQ328715	45.2/—	[20]	Telotylenchidae
Nagelus obscurus	EU306350	—	—		[17]	
Pratylenchoides ritteri	AJ966497	—	JX261964	48.7/—	[16] [85]	Pratylenchidae
Psilenchus cf. *hilarulus*	AY284593	—	EU915489	44.1/—	[18] [83]	Psilenchidae
Scutylenchus quadrifer	AY284599	—	—	41.5/—	[18]	Telotylenchidae
Scutylenchus sp.	—	JQ069956	—		[86]	
"Tylenchina": Ecphyadophoridae						
Ecphyadophora sp. JH-2004	AY593917	—	—	33.7/—	[76]	Ecphyadophoridae
"Ditylenchus" brevicauda	AY284635	—	—	33.9/—	[18]	Anguinidae
Malenchus andrassyi	AY284587	—	—	32.3/—	[18]	Tylenchidae
Ottolenchus discrepans	AY284590	—	—	33.7/—	[18]	
Criconematina						
Hemicriconemoides gaddi	—	KC520471	KC520470	55.6/—	[87]	Criconematidae
Hemicriconemoides pseudobrachyurus	AY284622	—	—		[18]	
Hemicycliophora lutosa	—	GQ406237	GQ406240	53.2/—	[88]	Hemicycliophoridae
Hemicycliophora thienemanni	AY284628	—	—		[18]	
Meloidoderita kirjanovae	—	DQ768427	DQ768428	50.8/—	[89]	
Sphaeronema alni	FJ969127	—	—		van Megen, unpublished. [90]	Sphaeronematidae
Meloidoderita sp.	GU253916	GU253917	JQ771954	50.8/—	Cudejkova and Cermak, unpublished.	
Tylenchulus semipenetrans	AJ966511	FJ588909	FJ969710	57.5/—	[16] [91] [92]	Tylenchulidae
"Hoplolaimina": Belonolaimidae						
Belonolaimus longicaudatus	AY633449	DQ672366	GQ896548	55.8/—	[68] [93] [94]	Belonolaimidae
Ibipora lobi	JQ771535	—	—	30.9/—	[95]	
"Hoplolaimina": Hoplolaimidae						
Carphodorus sp.	JQ771538	—	JQ771550	41.3/—	[95]	
Globodera pallida	EU855119	EU85511	BM415342 BM415248 CV577211 CV577977 CV579301E U85511	93.6/—	Nowaczyk et al., unpublished. Opperman, unpublished [96].	Heteroderidae
Heterodera glycines	AF216579 BI704127 BI748392 CA940548 CB379240 CB379263 CB379850 CB380242 CB825296 CB825409 CB825970 CB935610 CK348871 CK348904 CK349175 CK352112	AF216579	AF133304 AF216579 BI704144 BI704144 BI749520 CA940190 CA940212 CA940243 CA940406 CA940424 CA940429 CA940589 CB238697 CB279977 CB299455 CB373844 CB373981 CB379125 CB379140 CB379219 CB379312 CB379439 CB379505 CB379696 CB379707 CB379996 CB380091 CB380241 CB824788 CB824878 CB825995 CB934877 CB934931 CB934950 CB934954 CK348525 CO036619 HM560850 JN684906	98.3/—	[97] [96]. [98] Yan and Davis, unpublished. [99] Ye et al., unpublished. Wei et al., unpublished.	
Morulaimus sp.	JQ771540	—	—	31.5/—	[95]	Belonolaimidae

Radopholus similis	AJ966502 AY912509 EF384224 EY190988 EY191076 EY191697 EY191883 EY192786 EY192788 EY193123 EY193253 EY194340 EY194464 EY194646 EY195472 FJ040398	AY912509 EF384224	EU555409 EY189839 EY190550 EY190620 EY190961 EY191066 EY191073 EY191135 EY191160 EY191173 EY191237 EY192021 EY192028 EY192080 EY192091 EY192247 EY192381 EY192472 EY192501 EY192526 EY192892 EY192907 EY193005 EY193037 EY193249 EY193314 EY193798 EY193897 EY193971 EY194395 EY194454 EY194530 EY195146 EY195204 EY195406	97.5/—	[16] [100] Long et al., unpublished. [101] Holterman et al., unpublished. [102] [100] Zhao unpublished. [86]	Pratylenchidae
			EY195408 EY195580 EY195889 EY195943 GQ281471 JN091962 JQ782249			
Rotylenchulus reniformis	JX406356	FJ374686	HM131884 FJ906072	59.4/—	[103] Rahman et al., unpublished. [104]	Rotylenchulidae
"Hoplolaimina": Pratylenchidae						
Dolichodorus sp. WY-2006	DQ912918	—	—	33.9/—	[105]	Dolichodoridae
Hirschmanniella loofi	EU306353	EU620472	EU620469	51.6/—	[17] [106]	Pratylenchidae
Macrotrophurus arbusticola	AY284595	—		33.9/—	[18]	Telotylenchidae
Meloidogyne arenaria	U42342	U42342	U42342 AF023855 AF023856	99.2/—	Georgi and Abbott. unpublished.	Meloidogynidae
Meloidogyne artiellia	AF248477	AF248477	AF248477	99.2/—	[107]	
Nacobbus aberrans	AJ966494	DQ017473	U47557	49.0/—	[16] [108] [109]	Pratylenchidae
Pratylenchus vulnus	EU669955	JQ966892	BQ580554 CV198923 CV198995 CV199233 CV199349 CV199490 CV200136 CV200423 CV200464 CV200467 CV200471 CV200530 CV200687 CV200896 CV201004 CV201135 EL887566 EL887705 EL888035 EL888060 EL888174 EL888269 EL888739 EL888778 EL889241 EL889472 EL889797 EL889934 EL889934 EL889977 EL889994 EL890380 EL890701 JQ003993 JQ003994 JX047008	100.0/—	[19] [110] [96] [96] [111] Zhao, unpublished. [112]	
Tylenchorhynchus dubius	EU306352	—	DQ328707	53.2/—	[17] [20]	Telotylenchidae
Tylenchorhynchus zeae	—	EF519711	—		[113]	

*Clades of the tree, marked by boldface.

3. RESULTS

3.1. Infestation of Fleas with Nematodes

The infestation rate is shown in Table 1 (in total, 807 flea specimens were studied). Among the six flea species studied, the population size and the percentage of infected fleas varied depending on the season. Three flea species sampled on sousliks (Citellophilus tesquorum, Neopsylla setosa, and Frontopsylla semura) exhibited a stable population density. In the two species, N. setosa and F. semura, the infestation rate was moderate to high in the spring seasons of 2012 and 2013. In C. tesquorum, no infected fleas were detected in spring 2013, whereas in spring 2012 the fleas were highly infested (17.1%). The vole flea Amphipsylla rossica was abundant and moderately infested in autumn, whereas being less abundant in spring, which may explain the absence of infected fleas in the spring sample. Another vole flea, Ctenophthalmus secundus, exhibited a consistently high population density and low infestation rate in both spring and autumn samples.

Adult parasitic females and their progeny were found in the haemocoel of infected fleas. In the infected fleas C. tesquorum, A. rossica, C. secundus, and Mesopsylla hebes, only one generation of parasitic females was observed. Their amount in a flea specimen is determined by the number of free-living infective females that penetrate into the flea larva. We observed 1 to 2 or 1 to 4 adult parasitic females per flea specimen in spring and autumn, respectively. An additional parthenogenetic generation of parasitic females was found in some fleas of N. setosa and F. semura, where up to 16 specimens per flea were observed. As in other entomoparasitic nematodes, the propagation rate depends on the host age. Thus, in young fleas up to 10 juveniles was found per flea specimen, whereas up to 1,000 juveniles of different stages were contained in some old fleas (Figure 2). After the 2nd molt the number of juveniles is maximal, and 3rd stage juveniles massively migrate to the rectal section of the flea intestine for exit to the environment. In some cases, the observed infestation level was so high that nematodes penetrated distal segments of the flea legs, from where they have no way to the environment.

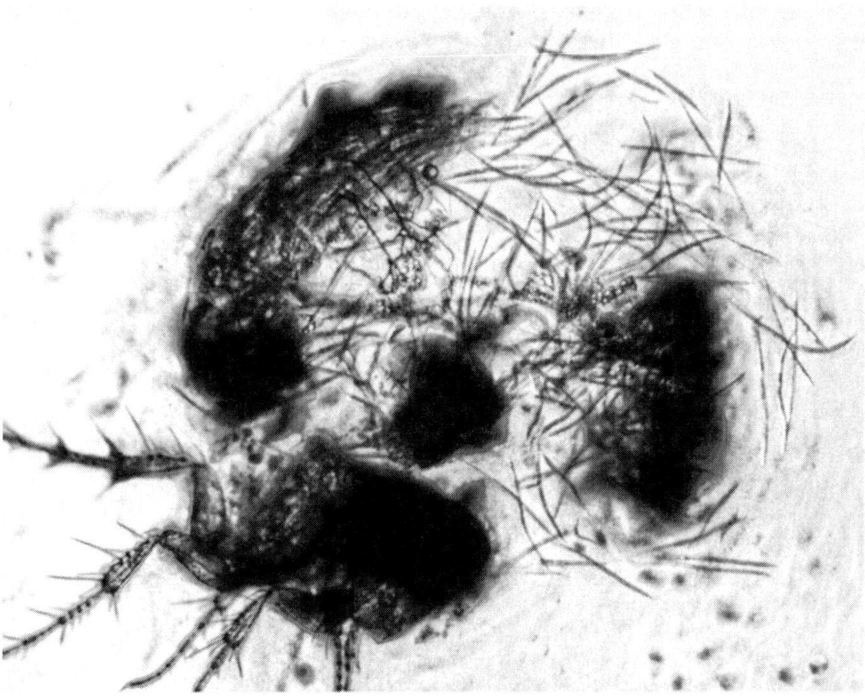

Figure 2: Numerous juveniles of Rubzovinema sp. extracted from the dissected body of a Citellophilus tesquorum flea.

3.2. Morphological Analysis of Entomoparasitic Stages in Nematode Isolates and Their Taxonomic Identification

Analysis of morphology of entomoparasitic stages suggests that the studied nematode isolates from three distinct groups. A single generation of parasitic females was observed in the first two groups and an additional parthenogenetic generation—in the third group. According to morphometric data on adult parasitic females (Tables 4–6), the first two groups belong to the genera Rubzovinema or Spilotylenchus and the third group to the genus Psyllotylenchus. Photographs of parasitic females of Rubzovinema sp., Spilotylenchus sp., and Psyllotylenchus sp. are depicted in Figure 3. Figure 4 shows their distribution among flea samples studied.

Table 4: Comparison of morphometrics in parasitic females of Rubzovinema sp. and Rubzovinema ceratophylla.

Character	*Rubzovinema* sp. (this study)	*Rubzovinema ceratophylla* [26]
N	29	27
L	1278,6 (840–1570)	1265,1 (810–1840)
D	120,8 (85–145)	137,3 (62–200)
A	11,19 (7,9–16,1)	9,51 (6,4–16,8)
C	65,4 (31,4–100)	44,10 (10–86,4)
V%	96,4 (93,1–97,9)	95,44 (92–98,9)
Total length of stylet (St)	18,5 (14–22)	19,5 (18–21)
Length of distal edge of stylet	7,2 (5–8,7)	—
Distance between anterior end and excretory pore (Ex)	20,7 (10–31)	—
Distance between anterior end and nerve ring	61,2 (50–74,5)	
Total length of tail (Cd)	21,9 (10–42)	26,35 (14–47,5)
Distance between vulva and tail end	46,1 (23–75)	—
Distance between vulva and anus (V–A)	26,9 (13–40)	—

All measurements are in μm and in the form mean (range).

Table 5: Comparison of morphometrics of parasitic females in Spilotylenchus sp. and Spilotylenchus maisonabei.

Characters	*Spilotylenchus* sp. (this study)	*Spilotylenchus maisonabei* [23]
N	2	6
L	1,600–1,840	1,244 (1,200–1,320)
D	155–160	125 (107–160)
A	10.3–11.5	10.3 (7.5–12)
C	167.3–177.8	84.4 (64.5–121)
V%	97.4–97.7	96.2 (95.8–96.5)
Total length of stylet (St)	9.5–9.8	9-10
Distance between anterior end and excretory pore	1.5–15.5	23.3 (20–28)
Distance between anterior end and nerve ring	—	52–54
Total length of tail (Cd)	9–11	15.4 (10–19)
Distance between vulva and tail end	41.5–43	47 (42–52)
Distance between vulva and anus (V–A)	32-33	—

All measurements are in μm and in the form mean (range).

Table 6: Comparison of morphometrics of parasitic females in Psyllotylenchus sp. and Psyllotylenchus viviparous.

Character	*Psyllotylenchus* sp. (this study)		*Psyllotylenchus viviparous* [25]	
	Gamogenetic	Parthenogenetic	Gamogenetic	Parthenogenetic
N	3	7	8	10
L	1,016.7 (900–1,100)	446 (420–500)	1,000 (840–1,480)	500 (360–840)
D	81.3 (79–84)	70 (60–80)	77 (62–115)	60 (54–100)
A	12.5 (11.1–13.3)	6.25 (5.6–7)	—	—
C	64.3 (60–68.2)	40.15 (37.1–43.5)	—	—
V%	95.1 (95–95.4)	93.3 (90–95.3)	—	—
Total length of stylet (St)	17.5 (17–18,5)	5.25 (4–6)	17 (15–20)	7 (5–8)
Length of the distal edge of stylet	8.6 (8-9)	—	—	—
Distance between anterior end and excretory pore	26.5 (25–31.5)	17.5 (15–19.5)	23 (13–33)	22 (14–46)
Distance between anterior end and nerve ring	—	51.7 (50–55)	—	—
Total length of tail (Cd)	15.8 (15–17)	11.1 (10.5–11.5)	25 (17–35)	9 (1–17)
Distance between vulva and tail end	48 (45–51)	30.5 (19.7–55)	56 (37–71)	52 (40–104)
Distance between vulva and anus (V–A)	30.8 (29–31.5)	13.5 (11.7–21.6)	—	—

All measurements are in μm and in the form mean (range).

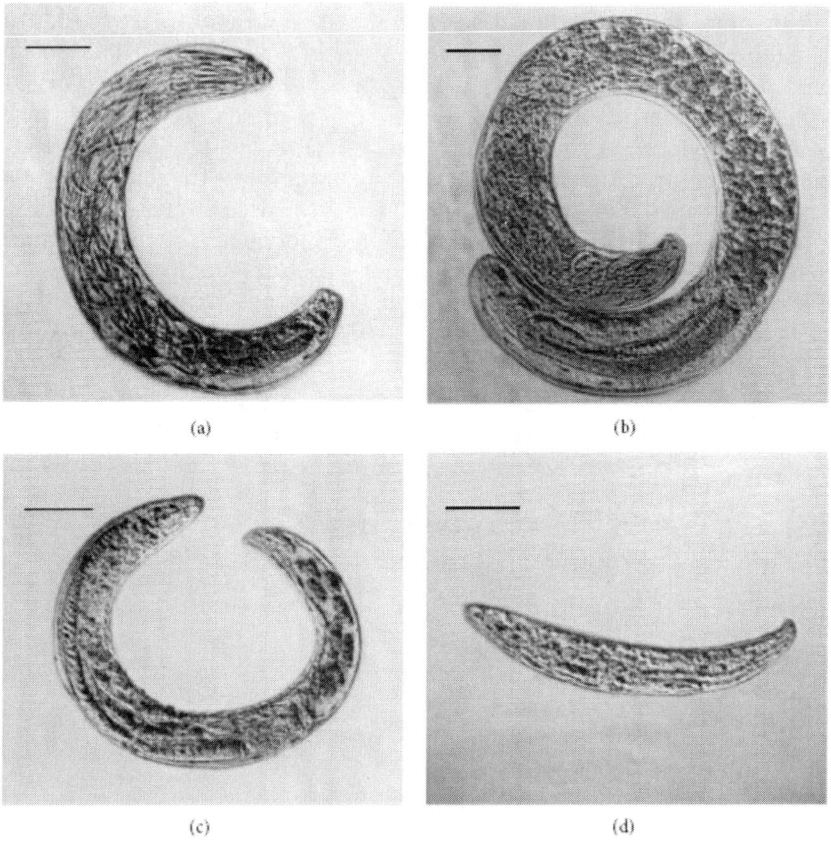

Figure 3: Parasitic females of the studied nematode species. (a) Rubzovinema sp., heterogeneous female; (b) Spilotylenchus sp., heterogeneous female; (c) Psillotylenchus sp., heterogeneous female of the first generation; (d) (c): Psillotylenchus sp., parthenogenetic female of the second generation. Scale bar—200 μm.

According to morphometric evidence, parasitic females and juveniles of the genera Rubzovinema and Spilotylenchus are very similar. However, in the first two groups of isolates we found characters bearing discriminative and identificational value. In particular, the oesophageal glands in juveniles III of the first group are poorly developed. This is a distinctive feature of the genus Rubzovinema, where males and females have shortened oesophageal glands located close to the nerve ring. In the second group of isolates, oesophageal glands are well developed and elongated, which is characteristic of the genus Spilotylenchus. In the first group, the stylet possesses a heavily sclerotized distal spear with a length of approximately half the total stylet length and has a stem with a weaker sclerotization and widening to the base. This stylet structure is characteristic of the genus Rubzovinema, and stylet length (18.5 (14–22) µm) is in accordance with morphometrics given in the description of this genus [26]. In the genus Spilotylenchus, the stylet varies in shape but always possesses a shortened conical distal spear. In the second group of isolates, the stylet structure was similar to that of Spilotylenchus. Also, the vulval lips of the first group are more protruded than in Spilotylenchus. Other features, including the morphometrics, vary widely in both genera, which hampers taxonomic identification. Nevertheless, based on distinctive traits, we identified the first and second group of isolates as Rubzovinema sp. and Spilotylenchus sp., respectively.

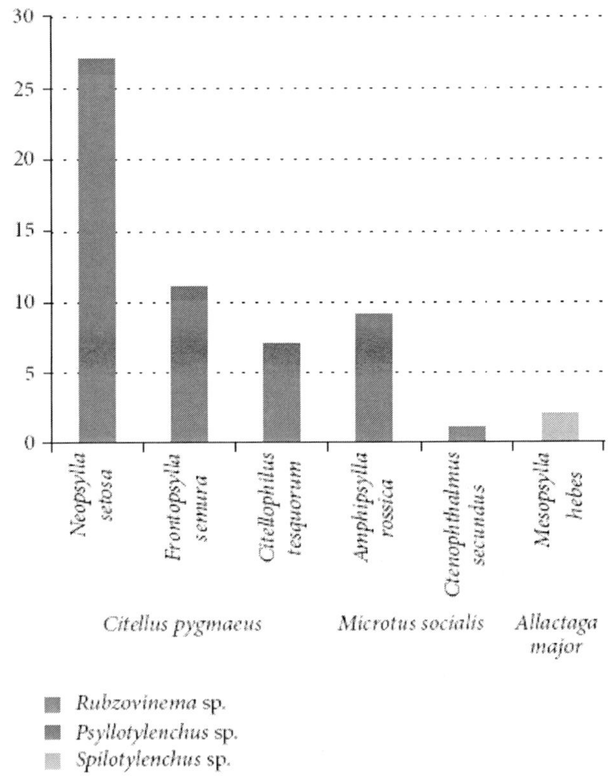

Figure 4: Distribution of the studied nematode species among the flea species studied, whose rodent hosts are given below. The vertical axis shows the numbers of infected fleas.

In the genus Rubzovinema, the single species described to date is Rubzovinema ceratophylla [26]. This species is known to parasitize exclusively the flea Citellophilus tesquorum that feeds on sousliks. The specimens of Rubzovinema studied in this work were isolated from five flea species, C. tesquorum, Neopsylla setosa, Frontopsylla semura, Amphipsylla rossica, and Ctenophthalmus secundus, of which the latter two were sampled on mouse-like rodents. Also, the parasitic females of Rubzovinema sp. differed from R. ceratophylla by morphology; they have a shorter tail and more protruded vulval lips. A morphometric comparison of Rubzovinema sp. and R. ceratophylla is given in Table 4.

The parasitic females of Spilotylenchus sp. were isolated from the flea Mesopsylla hebes associated with jerboas. The females were not identified to the species level because of a small number of available specimens and the lack of a free-living stage. A morphometric comparison of Spilotylenchus sp. and the morphologically closest species Spilotylenchus maisonabei [23] is given in Table 5.

In the genus Psyllotylenchus, descriptions of most species are fragmentary and incomplete, which precluded the species identification of the Psyllotylenchus isolates from the fleas N. setosa and F. semura feeding on sousliks. A morphometric comparison of Psyllotylenchus sp. and the type species of this genus, Psyllotylenchus viviparous [25], is given in Table 6.

The 18S and 28S rDNA sequences of Rubzovinema sp. specimens from A. rossica and C. secundus were 100% identical, which indicates that the isolates belong to the same species. The sequences of Rubzovinema sp. ex C. tesquorum, Rubzovinema sp. ex N. setosa, and Rubzovinema sp. ex F. semura diverged from one another and from the gene sequences of Rubzovinema sp. ex A. rossica and Rubzovinema sp. ex C. secundus by 0.4–0.7%, which corresponds to the levels of intraspecific variation [14, 114–119]. The 18S and 28S rDNA sequences of Psyllotylenchus sp. ex N. setosa and Psyllotylenchus sp. ex F. semura were 100% identical, indicating that they belong to the same species. The 18S and 28S rDNA sequences of Rubzovinema sp. and Psyllotylenchus sp. diverge by 1.2% and 1.9%, respectively. Those of Spilotylenchus sp. ex M. hebes were found to be more divergent. The degree of divergence of the 18S rDNA sequence of Spilotylenchus sp. ex M. hebes from those of either Rubzovinema sp. or Psyllotylenchus sp. was 2.4%; the D3 expansion segment of 28S rDNA diverged by 13.1% and 12.0%, respectively. The observed divergence rate of rDNA sequences agrees well with published evidence on entomoparasitic nematodes [14, 114–118]. Thus, intraspecific divergence of 18S rDNA in Deladenus siricidicola is 1% [120], of D2 and D3 expansion segments in the phytoparasite Bursaphelenchus xylophilus is from 0% to 0.6%, and the interspecific variation between the phytoparasites B.

xylophilus and Bursaphelenchus mucronatus is from 1.7% to 3.7%. The spacers ITS1 and ITS2 are generally more diverged; the intra- and interspecific variation for these species is from 0 to 3.1% and 11.2 to 13.4%, respectively [121–123].

Molecular vouchering is proved to efficiently complement morphological species identification in nematodes [73, 122, 124–128]. Combining the rDNA and morphological data confirms the species identity within each of the three studied groups of isolates.

3.3. Phylogenetic Analysis

In phylogenetic analyses of rDNA we used a dataset with extensive species and gene sampling (SSU-ITS1-5.8S-LSU) compared to earlier published tylenchid phylogenies, most of which were based on SSU rDNA or D2-D3 expansion segments [17, 19–21, 39, 129]. The SSU-ITS1-5.8S-LSU rDNA tree topology (Figure 5) is highly similar to other published phylogenies of tylenchids. In this tree, tylenchomorphs are represented by the sister groups Aphelenchidae and Tylenchida. Most of the tylenchid clades occur in published trees but often contradict classifications based on morphology, as it was also noted by other authors [17, 19–21, 39, 129]. The three robust major branches in the SSU-ITS1-5.8S-LSU rDNA tree (Bayesian posterior probabilities of 0.99–1.0) are (1) the clade includes representatives of the suborders Hoplolaimina, Criconematina, and Tylenchina (excluding Anguinoidea); (2) the majority of classic Anguinata; (3) the suborder Hexatylina. The studied parasites of fleas form a monophyletic group (bootstrap support of 100%) within the Hexatylina.

The nonredundant rDNA data on the Hexatylina in GenBank mostly represents the D2-D3 expansion segments of LSU rDNA. To maximize species sampling of the Hexatylina, we chose the D3 expansion segment as the molecular marker. The phylogenetic tree with the Anguinoidea as an outgroup is shown in Figure 6. In this tree, the suborder Hexatylina consists of two well-supported clades, in accordance with previously published D2-D3 rDNA phylogenies [19, 20, 39]. The clade of the studied flea parasites is placed within the largest branch of the Hexatylina, similarly to the result of the concatenated rDNA analysis.

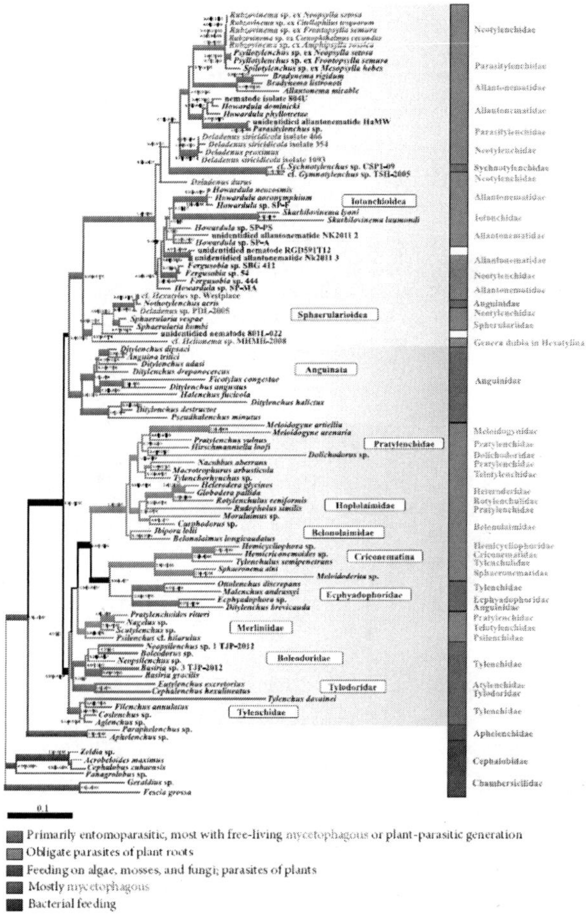

Figure 5: Phylogenetic tree of Tylenchida, inferred from SSU-ITS1-5.8S-LSU rDNA sequences. Topology was inferred using the PhyloBayes software (maxdiff = 0.36). Node support values are shown as follows: the first two values are Bayesian posterior probability assessed using the PhyloBayes and MrBayes software, respectively, and the third is bootstrap support assessed by the ML method. Thick lines lead to the nodes, in which at least one support value of posterior probability is 0.95 and higher. Names of clades (framed) are mainly given by type genera included in them (with the exception of Iotonchioidea). Formal taxonomic position (family by [8]) is shown on the right to the color bar. Colors indicate the ecologies (see the legend). Names of the species of Hexatylina that have a mycetophagous stage in their life cycle are shown in blue. The three robust major branches of Tylenchida are marked by gradient.

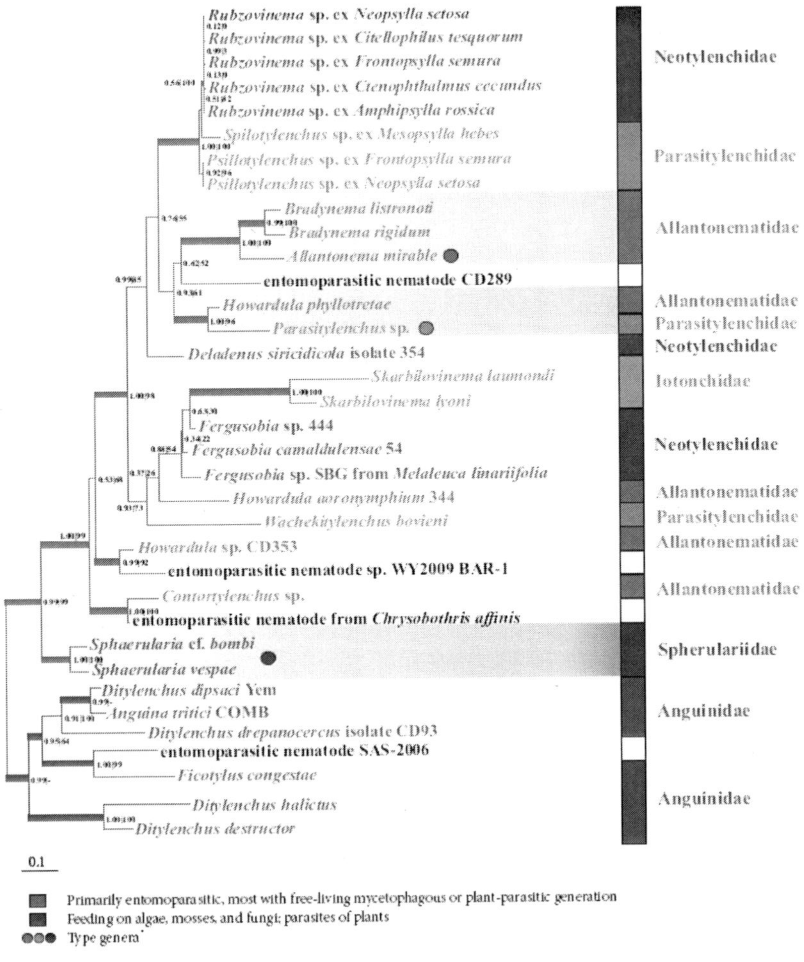

Figure 6: Phylogenetic tree of Hexatylina, inferred from D3 expansion segment of LSU rDNA. Topology was inferred using the PhyloBayes software. Node support values are shown as follows: Bayesian posterior probability/bootstrap support assessed by the ML method. Thick lines indicate the nodes supported at the level of 0.95 and higher. Color of lines indicates the ecologies (see the legend). Names of species were shown in different colors indicating their taxonomic position. Three families that include their type genera (shown as circles) are marked by gradient.

The three alternative relationships between the three major branches of Tylenchida (Figure 5) are not discriminated by the AU and Kishino and Hasegawa tests, and only the basal position of the Hexatylina is rejected by the expected-likelihood weights test (Table 7). All three tests do not discriminate between the alternative placement of the flea parasites as closest to the Allantonema, Parasitylenchus, or Deladenus branches; however, its positioning outside this grouping is not rejected only by a less conservative Shimodaira-Hasegawa test [50].

Table 7: Results of tree topology tests for alternative hypotheses on () the initial divergence of Tylenchida (Figure 4) and on () the relationships within the monophyletic branch that includes the studied group of nematodes parasitizing fleas (designated by asterisk).

Topology	Rank	obs	au	np	bp	pp	kh	sh	c-ELW
				1					
(((H,An),T),o)	1	−1.8	0.787	0.415	0.402	0.804	0.663	0.969	0.4197
((An,(H,T)),o)	2	4.1	0.326	0.198	0.205	0.013	0.254	0.623	0.1848
((H,(An,T)),o)	3	6.9	0.061	0.013	0.014	0.001	0.101	0.492	0.0186
				2					
((((*,Al),P),Ds),o)	1	−1.8	0.787	0.415	0.402	0.804	0.663	0.969	0.4197
((((*,P),Al),Ds),o)	2	1.8	0.495	0.242	0.247	0.130	0.337	0.813	0.2249
(((*,(Al,P)),Ds),o)	3	2.7	0.371	0.110	0.105	0.052	0.243	0.824	0.1209
((*,((Al,P),Ds)),o)	6	15.7	0.063	0.024	0.025	1e − 007	0.053	0.153	0.0272
(((*,Ds),(Al,P)),o)	7	18.3	0.013	0.002	0.002	9e − 009	0.020	0.096	0.0028

Al: Allantonematidae, An: Anguinata, Ds: *Deladenus siricidicola—D. proximus* group, H: Hexatylina, P: Parasitylenchidae, T: Tylenchina, o: outgroup.

4. DISCUSSION

4.1. Ribosomal DNA Phylogeny of the Tylenchida and Relationships within the Suborder Hexatylina

Phylogenetic analyses of SSU [16, 17, 19, 39] and D2-D3 [20, 39] rDNA data using various methods and species sampling generally agree on the monophyly of most tylenchid clades and contradict classic morphology based classifications. In the SSU-ITS1-5.8S-LSU tree (Figure 5), the monophyletic Tylenchida consists of three major robust clades. The first clade diverges into six groups: (1) the "Tylenchidae (part 2)" (by [17]), (2) the Tylodoridae (represented by the two genera, Cephalenchus and Eutylenchus [83]), (3) Boleodorinae + "Tylenchidae (part 1)" (by [Bert]), (4) the Merliniidae [130], (5) Criconematina + Sphaeronematidae + selected Tylenchina, and (6) Belonolaimidae + "Hoplolaimina." The Merliniidae group corresponds to

Clade C in [19] and includes partially the polyphyletic "Telotylenchinae" [131], "Pratylenchidae", and "Hoplolaimina" (Psilenchus cf. hilarulus). Group (5) corresponds to Clade 12A in [129], where Sphaeronematidae (Sphaeronema and Meloidoderita) were earlier shown to be closely related to Criconematina [20, 89], and selected Ecphyadophoridae + Ottolenchus + Malenchus were found to represent a monophyletic clade within the paraphyletic Tylenchina likely to be related to the Criconematina [18, 82]. Group (6) corresponds to Clade VII in [20], Clade 12B in [129], and Clade A + Clade B in [19]. Belonolaimidae (the genera Belonolaimus and Ibipora) tend to occupy the basal position. Clade A in [19] contains a "long branch" of the burrowing nematode Radopholus similes ("Pratylenchidae") in sister position to the Hoplolaimidae [17, 19]. This nematode occupies a similar position relative to the Hoplolaimidae in the SSU-ITS1-5.8S-LSU tree, and we consider this unlikely to be an LBA artefact. Similarly to [95], Carphodorus and Morulaimus that belong to the classic Belonolaimidae comprise the basal branch of Clade A sensu [19]. The clade corresponding to Clade B in [19] contains Meloidogynidae, Dolichodoridae, paraphyletic Pratylenchidae, and a part of Telotylenchidae.

The second major clade of the Tylenchida includes representatives of the classic infraorder Anguinata, with a well-supported monophyletic origin, except for a few species. They belong outside the second clade and may initially have been wrongly identified.

The third major clade includes representatives of the classic suborder Hexatylina and consists of two groups. The smaller one unites the three species of Sphaerularia, Helionema sp., cf. Hexatylus sp., Deladenus sp. PDL-2005, and Nothotylenchus acris (Anguinata: Nothotylenchidae). It is further referred to as the Sphaerularioidea according to the type genus. The larger group contains the clade of studied flea parasites and members of the superfamilies Iotonchioidea (Skarbilovinema spp., Parasitylenchus spp., and Wachekitylenchus bovieni) and Sphaerularioidea (Allantonema mirable, Bradynema spp., Howardula spp., and Contortylenchus sp. (fam. Allantonematidae); Deladenus durus, Deladenus proximus, Deladenus siricidicola, Fergusobia spp., and Gymnotylenchus sp. (fam. Neotylenchidae)). One species of the Anguinata, Sychnotylenchus sp., also joins the larger group. Our study renders the genera Howardula and Deladenus paraphyletic, as was earlier shown in [19, 39, 71, 119].

The genus Howardula is paraphyletic in published rDNA and mitochondrial COI phylogenies [71]. Such characters of Howardula as the degeneration of oesophagus, tail shape, and the absence of stylet in males seem to have evolved independently by convergence. The paraphyletic genus Deladenus is more closely related to either ancestral forms of the Hexatylina or forms typical to the Anguinata. The infraorder Anguinata includes soil-dwelling nematodes, mostly mycetophagous or parasitizing various parts of plants. However, an unidentified entomoparasitic nematode was also grouped within the Anguinoidea [39]. The life cycle of Deladenus spp. is an irregular alternation of free-living and entomoparasitic forms. The nematode D. siricidicola is able of producing an unlimited number of free-living generations in the absence of the host larvae of siricid pine-killing wood wasps [132]. Like in Anguinata, the free-living forms of Deladenus spp. are fungal feeding. Such characters of Deladenus asthe mycetophagy, enlargement of subventral glands in entomoparasitic females versus their reduction in free-living forms, the hypertrophy of dorsal glands, and stylet reduction in free-living forms seem to be symplesiomorphic. Resemblance with the Anguinata is also typical of other mycetophagous free-living forms: Hexatylus (Neotylenchidae), Rubzovinema (Neotylenchidae), Prothal lonema (Sphaerularioidae) Helionema (Hexatylina dubia), and Paurodontidae. For the latter, the entomoparasitic stage is expected but has never been observed. The relationship between the Hexatylina and Anguinata was earlier hypothesized based on morphology [7, 8, 130, 133, 134]. On rDNA phylogenies of tylenchids, the monophyly of the Hexatylina + Anguinata is either supported [19] or not rejected [20]. In the SSU-ITS1-5.8S-LSUrDNA tree obtained in this study, the monophyly of the Hexatylina + Anguinata has the Bayesian posterior probability of 0.91, but the maximum-likelihood bootstrap support is low; the AU and Kishino and Hasegawa tests did not discriminate between alternative hypotheses.

According to our SSU-ITS1-5.8S-LSU rDNA phylogeny (Figure 5), the major robust branches of the Tylenchida are incongruent with morphology-based classifications suggesting three rather than four suborders (the rank is adopted from morphological systems of tylenchids). Among them, the Hexatylina and Anguinata (both are monophyletic) are likely to be sister groups. The third emerged suborder includes representatives of three classic suborders: Tylenchina, Hoplolaimina, and Criconematina, among which only the latter does not contradict morphology-based classifications.

Considering ecological traits coded in Figure 5, the mycetophagy and/or facultative ectophytoparasitism are likely to be ancestral in the Tylenchida. Sedentary phytoparasites (root-knot species of Meloidogyne, the false root-knot genus Nacobbus, and cyst-forming Heterodera and Globodera) and other obligate endoparasites of plants evolved several times from free-living or facultative sedentary forms, as it was previously hypothesized in accordance with the concept of evolutionary trend to endoparasitism in phytonematodes [135]. Similarly, obligate endoparasites of insects from the Hexatylina are likely to have evolved from mycetophagous forms, with some species retaining the ancestral mycetophageous stage in the life cycle (e.g., species of the paraphyletic genus Deladenus and flea nematodes of the genus Rubzovinema). An interesting specific case in the Hexatylina is the genus Fergusobia that includes plant parasites associated with insects [68, 70], which may have transited to plant parasitism via entomoparasitism [39].

4.2. Ribosomal DNA Phylogeny of the Flea Nematodes and Their Classification

The nematodes of fleas do not group with the families known as their relatives in morphology-based systems, as these families do not form monophyletic groups in the tree. However, they do group with both type genera of the families Parasitylenchidae and Allantonematidae (Parasitylenchus and Allantonema, resp.). This grouping is preceded by a successive divergence of Deladenus durus and Deladenus siricidicola (Figure 5). As mentioned above, the pronounced free-living form in Deladenus seems to be ancestral to this group.

Only 31 tylenchid species that parasitize in fleas have been described to date. They differ by morphology, life cycle, and the host specificity, and belong to the five genera: Spilotylenchus (8 species), Psyllotylenchus (20 species), Incurvinema (1 species) Kurochkinitylenchus (1 species), and Rubzovinema (1 species). According to the classification of Siddiqi [8], the genera Spilotylenchus and Psyllotylenchus belong to the family Parasitylenchidae, whereas the genus Rubzovinema is a member of the Neotylenchidae. The two families represent two superfamilies, Iotonchioidea and Sphaerularioidea, respectively. All rDNA phylogenies published to date suggest that these superfamilies are paraphyletic [19, 20, 39], which is also inferred in our study with an extensive gene and taxon sampling.

A high degree of rDNA similarity in the three studied species suggests a closer relationship of these species than that assumed by the accepted system of classification. Earlier, Slobodyanyuk proposed to unite all known flea parasites into one family, the Spilotylenchidae. Its four subfamilies, Spilotylenchinae, Rubzovinematinae, Psyllotylenchinae, and Kurochkinitylenchinae, are discriminated based on the life cycle features [28]. In Spilotylenchinae and Rubzovinematinae, the entomoparasitic stage is represented by parasitic females of one heterosexual generation. In Psyllotylenchinae, in addition to the heterosexual generation, a parthenogenetic generation occurs in the flea haemocoel. In Kurochkinitylenchinae, two heterosexual generations exist in the haemocoel: the first generation produces parasitic females and the second generation produces both females and males [28]. Siddiqi also considered the unification of all flea tylenchids into one family but observed the need for further evidence in support [8].

Our results strongly suggest the inclusion of the three genera, Rubzovinema, Psyllotylenchus, and Spilotylenchus, in one family, the Spilotylenchidae [28]. The ribosomal DNA genetic distance within the family Spilotylenchidae is much smaller than that of certain tylenchid genera, for example, Meloidogyne (Figure 4) or Pratylenchus [19, 84].

4.3. Host Specificity of Flea Nematodes

The majority of tylenchid nematodes are monoxenous or oligoxenous; in particular, flea parasites were thought to be strictly host specific. Earlier papers suggested the lack of strict host specificity in Psyllotylenchus pawlowskyi and Psyllotylenchus viviparous [13, 25]. However, later these species were found to be heterogeneous and sustained revision [9, 27–29]. Spilotylenchus pawlowskyi and Spilotylenchus caspius were referred to as single-host parasites of the flea Coptopsylla lamellifer [27, 136]. Kurochkinitylenchus laevicepsi and Spilotylenchus ivashkini also share the same flea host, Nosopsyllus laeviceps [28, 29]. Before our study, the genus Rubzovinema was known to contain a single species, Rubzovinema ceratophylla, which parasitizes exclusively the flea Citellophilus tesquorum.

We found that at least two out of the three studied species are not single-host parasites. Psyllotylenchus sp. was shown to parasitize two flea species feeding on sousliks, Frontopsylla semura and Neopsylla setosa. Rubzovinema sp. was found on five flea species feeding on different rodent hosts: C. tesquorum, F. semura, N. setosa (all sampled from sousliks), Ctenophtalamus secundus, and Amphipsylla rossica (all sampled from voles). A. rossica, F. semura, and C. tesquorum belong to different families of the superfamily Ceratophylloidea (Leptopsyllidae and Ceratophyllidae), whereas C. secundus and N. setosa belong to the superfamily Hystrichopsylloidea. Unlike the host-specific R. ceratophylla, the studied Rubzovinema sp. parasitizes taxonomically distant fleas feeding on different rodents. Thus, the common opinion that flea nematodes are strictly host specific should be revisited.

As the two species of Rubzovinema demonstrate, even closely related parasites may exhibit different host range size. Among other known examples are the entomoparasitic nematodes of the genus Howardula parasitizing various beetles and flies [71, 137, 138], many phytonematodes [8], sibling species of parasitoid flies [128], and herbivorous insects [139]. The host range of parasites is an indicator of their evolutionary strategy in the ecosystem. Multihost parasites can be considered ecological generalists, in contrast to specialists that coevolve with a particular host. Generalists and specialists play different roles in the ecosystem [140], where they keep in balance, taking advantages and disadvantages of the two strategies. The advantages of generalization are yet to be explained by evolutionary biologists, whereas advantages of specialization are obvious, and it is generally accepted that evolution favors specialism [141, 142]. In the flea parasites, this trend is demonstrated by a greater species diversity of ecological specialists, the genera Spilotylenchus and Psyllotylenchus.

Nevertheless, the generalist Rubzovinema sp. was most abundant in the studied samples, which indicates that extending the host range may be evolutionarily successful. Besides the need to combat the immune response of several hosts, which is a requirement to widen the hosts range [143], the free-living stage of Rubzovinema sp. is to adapt to diverse microbioclimatic conditions of complex environments of rodent habitats. Multihost parasites pay a cost of adapting to alternative conditions [141, 144] compensated by stable survival of the species. Considering the spatial and temporal dynamics of flea populations feeding on a particular rodent host (one or two flea species usually dominate over a sampling season), multihost nematode parasites gain an advantage of their relative independence of population waves of either flea hosts or their rodent hosts. A higher infestation rate observed for Rubzovinema sp., compared to the two other studied species, may be an indicator of a greater ecological plasticity of this multihost parasite.

4.4. Entomoparasitic Nematodes in Natural Foci of Plague

In natural foci of plague, the epizootic dynamics are influenced by numerous climatic and biotic factors. The spatial and temporal population dynamics of the plague agent, Y. pestis, affect the population dynamics of the flea vectors and their mammalian hosts. Members of the transmission route of the plague agent also closely interact with other living organisms. For example, parasites of fleas that in turn feed on rodents are hyperparasites that play the role of high-level control agents on the ecosystem level, the role that entomoparasitic nematodes share with the bacterial plague agent. High-level control agents render the epidemiological state of a natural focus of disease less predictable. On the one hand, a lower density of the flea vector population reduces the plaque transmission rate; on the other, its growth causes an exponential decay of the host rodent population [145] below its epidemiological threshold, above which there is a threat of spillover of plague infection into human population [145]. Hypothetically, nematode-induced decrease of flea population is able to increase the number of rodents above the threshold and thus trigger an epidemic. The dual effect of high-level control agents is well exemplified by cases, when during plague episodes the extermination of rodents by humans causes the return of infection through stimulating the migration of fleas, the plaque vectors [5].

The studied entomoparasitic nematodes possess high potential as control agents of the flea vectors of plague owing to their high propagation rate within the flea host (Figure 2) and high infestation level (up to 21% observed in this study and from 50 to 60%, as estimated by other authors [10, 11]). One of the studied nematode species, Rubzovinema sp., is a multihost parasite. Host-specific parasites reach the optimal level of pathogenicity by maintaining the trade-off between pathogenicity and transmissibility. Adding of a new host to a multihost system makes the model more complicated [141]. The multihost parasite Rubzovinema sp. is expected to exhibit different levels of pathogenicity with respect to different flea hosts which, in turn, play different roles in the transmission of plague. Epizootics cause sporadic mortality in local populations of all members involved in the interaction with the plague agent, and their survival is contingent on migrations within a metapopulation. It is the case when the Cope's law [139, 146] governs the extinction of specialists on a shorter time scale rather than a geological period, and evolution may favor the ecological generalists, such as Rubzovinema sp.

Some authors surmised the involvement of entomoparasitic nematodes in the transmission of the plague agent [4], as it was observed that biofilms of Yersinia pestis adhere to cuticle receptors of Caenorhabditis elegans [147–149]. In this perspective, nematodes parasitizing fleas in natural foci of plague take on greater importance, as they may provide for the transmission route that does not include a mammal [4]. Further studies will clarify the role of flea nematodes in the transmission of plague infection.

CONFLICT OF INTERESTS

The authors declare that there is no conflict of interests regarding the publication of this paper.

ACKNOWLEDGMENTS

The authors thank G. S. Mirzaeva for help with PCR amplification and rDNA fragment analysis, N. V. Popov and the staff of the Laboratory of Epizootics Monitoring for advice and assistance in collecting and processing rodent samples, and, particularly, A. N. Porshakov for help in identification of flea specimens. They also thank O. V. Slobodyanyuk for helpful discussions of results, S. E. Spiridonov for advice on cultivation of entomoparasitic nematodes, S. A. Subbotin for valuable comments on the earlier version of the paper, and E. Yu. Talanova and L. Yu. Rusin for discussions of its final version, and E. A. Musatkina for assistance with the manuscript preparation. They are grateful to the Supercomputer Center of Moscow State University (http://parallel.ru/cluster) and the Bioportal of the University of Oslo (http://www.bioportal.uio.no) for providing computing resources.

REFERENCES

1. A. W. Bacot and C. J. Martin, "Observations on the mechanism of the transmission of plague by fleas," The Journal of Hygiene, vol. 13, supplement 3, pp. 423–439, 1914.
2. V. V. Kutyrev, G. A. Eroshenko, N. V. Popov, N. A. Vidyaeva, and N. P. Konnov, "Molecular mechanisms of interaction of the plague agent with invertebrates," Molecular Genetics, Microbiology and Virology, vol. 24, no. 4, pp. 7–12, 2009.
3. E. N. Pavlovsky, Natural Focality of Transmissive Diseases in Connection with Landscape Epidemiology of Zooanthroponoses, Moscow, Russia, 1964.
4. N. V. Popov, E. I. Koshel, G. A. Eroshenko, and V. V. Kutyrev, "Formation of modern concepts on the mechanism of plague enzooty," Problems of Particularly Dangerous Infections, vol. 3, no. 109, pp. 5–8, 2011.
5. M. J. Keeling and C. A. Gilligan, "Bubonic plague: a metapopulation model of a zoonosis," Proceedings of the Royal Society B, vol. 267, no. 1458, pp. 2219–2230, 2000.
6. R. A. Bedding, R. J. Akhurst, and H. K. Kaya, Nematodes and the Biological Control of Insect Pests, CSIRO Press, Melbourne, Australia, 1993.
7. M. R. Siddiqi, Tylenchida Parasites of Plants and Insects, vol. 645 of Commonwealth Agricultural Bureaux, Farnham Royal, Slough, UK, 1986.
8. M. R. Siddiqi, Tylenchida: Parasites of Plants and Insects, CABI, Wallingford, UK, 2nd edition, 2000.
9. I. A. Rubtsov, Parasites and Enemies of Fleas, Nauka, Leningrad, Russia, 1981.
10. Y. A. Morozov, "About infestation with fleas great gerbils different ages," in Proceedings of the Conference Anti-Plague Facilities in Central Asia and Kazakhstan, pp. 337–338, Alma-Ata, 1974.
11. Y. A. Morozov, "Effect of infestation of nematode on reproduction of fleas gerbils in Muyunkum," in Proceedings of the Conference Anti-Plague Facilities in Central Asia and Kazakhstan, pp. 338–340, Alma-Ata, 1974.
12. J. Deunff, Parasites de Siphonaptères. Étude de la systématique, de la biologie et du pouvoir pathogène des Tylenchides (Nematodea) dans une perspective de lutte biologique [Ph.D. thesis], Etat Sc. Pham, Rennes, France, 1984.
13. Y. V. Kurochhkin, "The nematode Heterotylenchus pawlowskyi sp. n., castrating flea-vectors of plague," Doklady Akademyi Nauk SSSR, vol. 135, pp. 1281–1284, 1960.
14. M. L. Blaxter, P. De Ley, J. R. Garey et al., "A molecular evolutionary framework for the phylum Nematoda," Nature, vol. 392, no. 6671, pp. 71–75, 1998.
15. P. De Ley and M. L. Blaxter, "Systematic Position and Phylogeny," in The Biology of Nematodes, D. L. Lee, Ed., Taylor & Francis, London, UK, 2002.
16. B. H. M. Meldal, N. J. Debenham, P. De Ley et al., "An improved molecular phylogeny of the Nematoda with special emphasis on marine taxa," Molecular Phylogenetics and Evolution, vol. 42, no. 3, pp. 622–636, 2007.
17. W. Bert, F. Leliaert, A. R. Vierstraete, J. R. Vanfleteren, and G. Borgonie, "Molecular phylogeny of the Tylenchina and evolution of the female gonoduct (Nematoda:

Rhabditida)," Molecular Phylogenetics and Evolution, vol. 48, no. 2, pp. 728–744, 2008.

18. M. Holterman, A. Van Der Wurff, S. Van Den Elsen et al., "Phylum-wide analysis of SSU rDNA reveals deep phylogenetic relationships among nematodes and accelerated evolution toward crown clades," Molecular Biology and Evolution, vol. 23, no. 9, pp. 1792–1800, 2006.
19. M. Holterman, G. Karssen, S. Van Den Elsen, H. Van Megen, J. Bakker, and J. Helder, "Small subunit rDNA-based phylogeny of the tylenchida sheds light on relationships among some high-impact plant-parasitic nematodes and the evolution of plant feeding," Phytopathology, vol. 99, no. 3, pp. 227–235, 2009.
20. S. A. Subbotin, D. Sturhan, V. N. Chizhov, N. Vovlas, and J. G. Baldwin, "Phylogenetic analysis of Tylenchida Thorne, 1949 as inferred from D2 and D3 expansion fragments of the 28S rRNA gene sequences," Nematology, vol. 8, no. 3, pp. 455–474, 2006.
21. S. A. Subbotin, D. Sturhan, B. J. Adams et al., "Molecular phylogeny of the order Tylenchida: analysis of nuclear ribosomal RNA genes," Journal of Nematology, vol. 38, no. 2, pp. 296–296, 2006.
22. H. Launay, J. Deunff, and O. Bain, "Spilotylenchus arthuri, gen. n., sp. n. (Nematodea, Tylenchida: Allantonematidae), parasite of Spilopsyllus cuniculi (Dale, 1878) (Siphonaptera: Pulicidae)," Annales de Parasitologie Humaine et Comparee, vol. 58, no. 2, pp. 141–150, 1983.
23. H. Launay and J. Deunff, "Spilotylenchus maisonabei n. sp. (Nematoda: Allantonematidae) parasite of the European rabbit flea Spilopsyllus cuniculi (Dale, 1878) (Siphonaptera: Pulicidae)," Annales de Parasitologie Humaine et Comparee, vol. 13, pp. 293–296, 1990.
24. C. Laumond and J. C. Beaucournu, "Neoparasitylenchus megabothridis n. sp. (Tylenchida: Allantonematidae) parasite of Megabothris tubidus (Siphonaptera: Ceratophyllidae); observations on fleas tylenchides in the S.W. of Europa (author's transl)," Annales de Parasitologie Humaine et Comparee, vol. 53, no. 3, pp. 291–302, 1978.
25. G. O. Poinar Jr. and B. C. Nelson, "Psyllotylenchus viviparus, n. gen., n. sp. (Nematodea: Tylenchida: Allantonematidae) parasitizing fleas (Siphonaptera) in California," Journal of Medical Entomology, vol. 10, no. 4, pp. 349–354, 1973.
26. V. O. Slobodyanyuk, "Validation of genus Rubzovinema gen.n. (Sphaerularioidea) and redescription of species Rubzovinema ceratophylla comb.n. parasite of fleas Citellophillus tesquorum," Zoological Journal, vol. 70, no. 9, pp. 33–43, 1991.
27. O. V. Slobodyanyuk, "Revision of the species Psyllotylenchus pawlowskyi (Kurochkin, 1960) Poinar & Nelson, 1973. I. Redescription of Spilotylenchus pawlowskyi (sensu stricto) comb. n. (Tylenchida: Allantonematidae)," Russian Journal of Nematology, vol. 5, no. 2, pp. 103–112, 1997.
28. O. V. Slobodyanyuk, "Revision of the species Psyllotylenchus pawlowskyi (Kurochkin, 1960) Poinar & Nelson, 1973. II. Description of Kurochkinitylenchus laevicepsi gen. n., sp. n. and Spilotylenchidae fam. n," Russian Journal of Nematology, vol. 7, no. 1, pp. 1–18, 1999.
29. O. V. Slobodyanyuk, "Revision of the species Psyllotylenchus pawlowskyi (Kurochkin, 1960) Poinar & Nelson, 1973. III. Description of Spilotylenchus ivashkini sp. n," Russian Journal of Nematology, vol. 8, no. 1, pp. 45–55, 2000.
30. J. Deunff and H. Launay, "Psyllotylenchus chabaudi, n. sp. (Nematodea, Tylenchida: Allantonematidae), a parasite of Nosopsyllus fasciatus (Bosc) (Siphonaptera: Ceratophyllidae)," Annales de Parasitologie Humaine et Comparee, vol. 59, no. 3, pp. 263–270, 1984.
31. J. Deunff, H. Launay, and J. C. Beaucournu, "Incurvinema helicoides n. gen., n. sp. Nematodea, Tylenchida: allantonematidae parasite de Rhadinopsylla pentacantha (Rothschild, 1897) (Siphonaptera : Hystrichopsyllidae)," Annales de Parasitologie Humaine Et Comparée, vol. 60, pp. 739–746, 1985.

32. A. T. De Grisse, "Redescriptions ou modifications de quelques techniques utilisées dans l'étude des nématodes phytoparasites," Mededelingen Rijksfakulteit Landbouwwetenschappen Gent, vol. 34, pp. 351–369, 1969.
33. X. Huang, "A contig assembly program based on sensitive detection of fragment overlaps," Genomics, vol. 14, no. 1, pp. 18–25, 1992.
34. T. A. Hall, "BioEdit: a user-friendly biological sequence alignment editor and analysis program for Windows 95/98/NT," Nucleic Acids Symposium Series, vol. 41, pp. 95–98, 1999.
35. L. Medlin, H. J. Elwood, S. Stickel, and M. L. Sogin, "The characterization of enzymatically amplified eukaryotic 16S-like rRNA-coding regions," Gene, vol. 71, no. 2, pp. 491–499, 1988.
36. V. N. Chizhov, O. A. Chumakova, S. A. Subbotin, and J. G. Baldwin, "Morphological and molecular characterization of foliar nematodes of the genus Aphelenchoides: A. fragariae and A. ritzemabosi (Nematoda: Aphelenchoididae) from the Main Botanical Garden of the Russian Academy of Sciences, Moscow," Russian Journal of Nematology, vol. 14, no. 2, pp. 179–184, 2006.
37. G. Van der Auwera, S. Chapelle, and R. De Wächter, "Structure of the large ribosomal subunit RNA of Phytophthora megasperma, and phylogeny of the oomycetes," FEBS Letters, vol. 338, no. 2, pp. 133–136, 1994.
38. S. A. Subbotin, N. Vovlas, R. Crozzoli et al., "Phylogeny of Criconematina Siddiqi, 1980 (Nematoda: Tylenchida) based on morphology and D2-D3 expansion segments of the 28S rDNA gene sequences with application of a secondary structure model," Nematology, vol. 7, pp. 927–944, 2005.
39. V. N. Chizhov, N. N. Butorina, and S. A. Subbotin, "Entomoparasitic nematodes of the genus Skarbilovinema: S. laumondi and S. lyoni (Nematoda: Tylenchida), parasites of the flies of the family Syrphidae (Diptera), with phylogeny of the suborder Hexatylina," Russian Journal of Nematology, vol. 20, no. 2, pp. 141–155, 2012.
40. S. F. Altschul, T. L. Madden, A. A. Schäffer et al., "Gapped BLAST and PSI-BLAST: a new generation of protein database search programs," Nucleic Acids Research, vol. 25, no. 17, pp. 3389–3402, 1997.
41. R. C. Edgar, "MUSCLE: multiple sequence alignment with high accuracy and high throughput," Nucleic Acids Research, vol. 32, no. 5, pp. 1792–1797, 2004.
42. M. Nei and S. Kumar, Molecular Evolution and Phylogenetics, Oxford University Press, Oxford, UK, 2000.
43. N. Lartillot, T. Lepage, and S. Blanquart, "PhyloBayes 3: a Bayesian software package for phylogenetic reconstruction and molecular dating," Bioinformatics, vol. 25, no. 17, pp. 2286–2288, 2009.
44. N. Lartillot and H. Philippe, "A Bayesian mixture model for across-site heterogeneities in the amino-acid replacement process," Molecular Biology and Evolution, vol. 21, no. 6, pp. 1095–1109, 2004.
45. J. P. Huelsenbeck and F. Ronquist, "MRBAYES: bayesian inference of phylogenetic trees," Bioinformatics, vol. 17, no. 8, pp. 754–755, 2001.
46. C. Lanave, G. Preparata, C. Saccone, and G. Serio, "A new method for calculating evolutionary substitution rates," Journal of Molecular Evolution, vol. 20, no. 1, pp. 86–93, 1984. • View at Scopus
47. A. Stamatakis, "Maximum likelihood-based phylogenetic analyses with thousands of taxa and mixed models," Bioinformatics, vol. 22, pp. 2688–2690, 2006.
48. H. Shimodaira, "An approximately unbiased test of phylogenetic tree selection," Systematic Biology, vol. 51, no. 3, pp. 492–508, 2002.
49. H. Kishino and M. Hasegawa, "Evaluation of the maximum likelihood estimate of the evolutionary tree topologies from DNA sequence data, and the branching order in Hominoide," Journal of Molecular Evolution, vol. 29, no. 2, pp. 170–179, 1989.
50. H. Shimodaira and M. Hasegawa, "CONSEL: for assessing the confidence of phylogenetic tree selection," Bioinformatics, vol. 17, no. 12, pp. 1246–1247, 2002.

51. K. Strimmer and A. Rambaut, "Inferring confidence sets of possibly misspecified gene trees," Proceedings of the Royal Society B, vol. 269, no. 1487, pp. 137–142, 2002.
52. H. A. Schmidt, K. Strimmer, M. Vingron, and A. Von Haeseler, "TREE-PUZZLE: maximum likelihood phylogenetic analysis using quartets and parallel computing," Bioinformatics, vol. 18, no. 3, pp. 502–504, 2002.
53. R. D. M. Page, "TreeView: an application to display phylogenetic trees on personal computers," Computer Applications in the Biosciences, vol. 12, no. 4, pp. 357–358, 1996.
54. O. Holovachov, S. Bostrom, I. T. De Ley et al., "Morphology, molecular characterisation and systematic position of the genus Cynura Cobb, 1920 (Nematoda: Plectida)," Nematology, vol. 15, pp. 611–627, 2013.
55. S. A. Nadler, P. De Ley, M. Mundo-Ocampo et al., "Phylogeny of Cephalobina (Nematoda): molecular evidence for recurrent evolution of probolae and incongruence with traditional classifications," Molecular Phylogenetics and Evolution, vol. 40, no. 3, pp. 696–711, 2006.• View at Scopus
56. O. Holovachov, S. Bostrom, S. A. Nadler, and P. De Ley, "Systematics and phylogenetic position of the genus Tricirronema Siddiqi, 1993 (Cephalobomorpha)," Journal of Nematode Morphology and Systematics, vol. 12, no. 2, 2009.
57. K. Kiontke, A. Barrière, I. Kolotuev et al., "Trends, stasis, and drift in the evolution of nematode vulva development," Current Biology, vol. 17, no. 22, pp. 1925–1937, 2007.
58. R. Campos-Herrera, F. E. El-Borai, and L. W. Duncan, "Wide interguild relationships among entomopathogenic and free-living nematodes in soil as measured by real time qPCR," Journal of Invertebrate Pathology, vol. 111, no. 2, pp. 126–135, 2012.
59. M.-A. Félix, P. De Ley, R. J. Sommer et al., "Evolution of vulva development in the Cephalobina (Nematoda)," Developmental Biology, vol. 221, no. 1, pp. 68–86, 2000.
60. S. Boström, O. Holovachov, and S. A. Nadler, "Description of Scottnema lindsayae Timm, 1971 (Rhabditida: Cephalobidae) from Taylor Valley, Antarctica and its phylogenetic relationship," Polar Biology, vol. 34, no. 1, pp. 1–12, 2011.
61. J. W. Waceke, D. J. Bumbarger, M. Mundo-Ocampo, S. A. Subbotin, and J. G. Baldwin, "Zeldia spannata n. sp. (Nematoda: Cephalobidae) from the Mojave Desert, California," Journal of Nematode Morphology and Systematics, vol. 8, no. 1, pp. 57–67, 2005.
62. S. Kumari, "Aphelenchus avenae (Nematoda: Aphelenchidae) under the rhizosphere of Brassica napus," Helminthologia, vol. 49, no. 1, pp. 57–59, 2012.
63. H. Iwahori, K. Tsuda, N. Kanzaki, K. Izui, and K. Futai, "PCR-RFLP and sequencing analysis of ribosomal DNA of Bursaphelenchus nematodes related to pine wilt disease," Fundamental and Applied Nematology, vol. 21, no. 6, pp. 655–666, 1998.
64. L. K. Carta, A. M. Skantar, Z. A. Handoo, and M. A. Baynes, "Supplemental description of Paraphelenchus acontioides (Tylenchida: Aphelenchidae, Paraphelenchinae), with ribosomal DNA trees and a morphometric compendium of female Paraphelenchus," Nematology, vol. 13, no. 8, pp. 887–899, 2011.
65. Y. Zeng, R. M. Giblin-Davis, Y. Weimin, G. Bélair, G. Boivin, and K. W. Thomas, "Bradynema listronoti n. sp. (Nematoda: Allantonematidae), a parasite of the carrot weevil Listronotus oregonensis (Coleoptera: Curculionidae) in Quebec, Canada," Nematology, vol. 9, no. 5, pp. 609–623, 2007.
66. K. Rybarczyk-Mydłowska, P. Mooyman, H. van Megen, and Setal, "Small subunit ribosomal DNA-based phylogenetic analysis of foliar nematodes (Aphelenchoides spp.) and their quantitative detection in complex DNA backgrounds," Phytopathology, vol. 102, no. 12, pp. 1153–1160, 2012.
67. Q. Yu, P. de Groot, I. Leal, C. Davis, W. Ye, and B. Foord, "First report and characterization of Deladenus proximus (Nematoda: Neotylenchidae) associated with Sirex nigricornis (Hymenoptera: Siricidae) in Canada," International Journal of Nematology, vol. 21, no. 2, pp. 139–146, 2012.

68. W. Ye, R. M. Giblin-Davis, K. A. Davies et al., "Molecular phylogenetics and the evolution of host plant associations in the nematode genus Fergusobia (Tylenchida: Fergusobiinae)," Molecular Phylogenetics and Evolution, vol. 45, no. 1, pp. 123–141, 2007.
69. I. Leal, B. Foord, C. Davis, P. de Groot, X. O. Mlonyeni, and B. Slippers, "Distinguishing isolates of Deladenus siricidicola ,a biological control agent of Sirex noctilio, from North America and the Southern Hemisphere using PCR- RFLP," Canadian Journal of Forest Research, vol. 42, pp. 1173–1177, 2012.
70. K. A. Davies, W. Ye, R. M. Giblin-Davis, G. S. Taylor, S. Scheffer, and W. K. Thomas, "The nematode genus Fergusobia (Nematoda: Neotylenchidae): molecular hylogeny, descriptions of clades and associated galls, host plants and Fergusonina fly larvae," Zootaxa, no. 2633, pp. 1–66, 2010.
71. S. J. Perlman, G. S. Spicer, D. Dewayne Shoemaker, and J. Jaenike, "Associations between mycophagous Drosophila and their Howardula nematode parasites: a worldwide phylogenetic shuffle," Molecular Ecology, vol. 12, no. 1, pp. 237–249, 2003.
72. C. L. Raak-van den Berg, P. S. van Wielink, P. W. de Jong et al., "Invasive alien species under attack: natural enemies of Harmonia axyridis in the Netherlands," BioControl. In press.
73. N. Kanzaki, R. M. Giblin-Davis, R. H. Scheffrahn et al., "Reverse taxonomy for elucidating diversity of insect-associated nematodes: a case study with termites," PLoS ONE, vol. 7, no. 8, Article ID e43865, 2012.
74. T. O. Powers, D. A. Neher, P. Mullin et al., "Tropical nematode diversity: vertical stratification of nematode communities in a Costa Rican humid lowland rainforest," Molecular Ecology, vol. 18, no. 5, pp. 985–996, 2009.
75. C. Hazir, R. M. Giblin-Davis, N. Keskin et al., "Diversity and distribution of nematodes associated with wild bees in Turkey," Nematology, vol. 12, no. 1, pp. 65–80, 2010.
76. M. Holterman, S. van den Elsen, H. van Megen et al., "Evolutionary relationships between members of the phylum Nematoda based on small subunit ribosomal DNA sequences," 2004.
77. N. Kanzaki, H. Kosaka, K. Sayama, J.-I. Takahashi, and S. Makino, "Sphaerularia vespae sp. nov. (Nematoda, Tylenchomorpha, Sphaerularioidea), an endoparasite of a common Japanese hornet, Vespa simillima Smith (Insecta, Hymenoptera, Vespidae)," Zoological Science, vol. 24, no. 11, pp. 1134–1142, 2007.
78. Q. Yu, "Ditylenchus destructor Thorne, 1945 (Tylenchida: Anguinidae) in Canada," Journal of Nematology, vol. 44, no. 4, pp. 500–500, 2012.
79. B. Steven, L. V. Gallegos-Graves, C. Yeager, J. Belnap, and C. R. Kuske, "Common and distinguishing features of the bacterial and fungal communities in biological soil crusts and shrub root zone soils," Soil Biology and Biochemistry, vol. 69, pp. 302–312, 2014.
80. R. D. L. Oliveira, A. M. Santin, D. J. Seni et al., "Ditylenchus gallaeformans sp. n. (Tylenchida: Anguinidae)—a neotropical gall forming nematode with biocontrol potential against weedy Malastomataceae," Nematology, vol. 15, no. 2, pp. 179–196, 2013.
81. K. A. Davies, W. Ye, R. M. Giblin-Davis, and K. W. Thomas, "Ficotylus congestae gen. n., sp. n. (Anguinata), from ficus congesta (moraceae) sycones in Australia," Nematology, vol. 11, no. 1, pp. 63–75, 2009.
82. M. R. Atighi, E. Pourajam, T. J. Pereira et al., "Redescription of Filenchus annulatus (Siddiqui & Khan, 1983) Siddiqi, 1986 based on specimens from Iran with contributions to the molecular phylogeny of the Tylenchida," Nematology, vol. 15, no. 2, pp. 129–141, 2013.
83. J. E. Palomares-Rius, S. A. Subbotin, G. Liébanas, B. B. Landa, and P. Castillo, "Eutylenchus excretorius Ebsary & Eveleigh, 1981 Nematoda: Tylodorinae) from Spain with approaches to olecular phylogeny of related genera," Nematology, vol. 11, no. 3, pp. 343–354, 2009.• View at Scopus

84. S. A. Subbotin, E. J. Ragsdale, T. Mullens, P. A. Roberts, M. Mundo-Ocampo, and J. G. Baldwin, "A phylogenetic framework for root lesion nematodes of the genus Pratylenchus (Nematoda): evidence from 18S and D2-D3 expansion segments of 28S ribosomal RNA genes and morphological characters," Molecular Phylogenetics and Evolution, vol. 48, no. 2, pp. 491–505, 2008.
85. Z. Majd Taheri, Z. Tanha Maafi, S. A. Subbotin, E. Pourjam, and A. Eskandari, "Molecular and phylogenetic studies on Pratylenchidae from Iran with additional data on Pratylenchus delattrei, Pratylenchoides alkani and two unknown species of Hirschmanniella and Pratylenchus," Nematology, vol. 15, pp. 633–651, 2013.
86. C. Xu, H. Xie, C. Zhao, S. Zhang, and X. Su, "Review of the genus Scutylenchus Jairajpuri, 1971 (Nematoda: Tylenchida), with description of Scutylenchus dongtingensis n. sp. from rhizosphere soil of grass in China," Zootaxa, vol. 3437, pp. 32–42.
87. Y. Yang and S. Zhang, "Identification of Hemicriconemoides gaddi from the Rhizosphere of Longan," Chinese Journal of Tropical Crops, vol. 5, pp. 935–941, 2013.
88. E. Van den Berg, S. A. Subbotin, and L. R. Tiedt, "Morphological and molecular characterisation of Hemicycliophora lutosa Loof & Heyns, 1969 and H. typica de Man, 1921 from South Africa (Nematoda: Hemicycliophoridae)," Nematology, vol. 12, no. 2, pp. 303–308, 2010.
89. N. Vovlas, B. B. Landa, G. Liébanas, Z. A. Handoo, S. A. Subbotin, and P. Castillo, "Characterization of the cystoid nematode Meloidoderita kirjanovae (Nemata: Sphaeronematidae) from Southern Italy," Journal of Nematology, vol. 38, no. 3, pp. 376–382, 2006.
90. J. E. Palomares-Rius, N. Vovlas, S. A. Subbotin et al., "Molecular and morphological characterisation of Sphaeronema alni Turkina & Chizhov, 1986 (Nematoda: Sphaeronematidae) from Spain compared with a topotype population from Russia," Nematology, vol. 12, no. 4, pp. 649–659, 2010.
91. G. Liu, J. Chen, S. Xiao, D. Pan, and S. Zhang, "Intraspecific variability of Tylenchulus semipenetrans populations on citrus and Chinese fir," Scientia Agricultura Sinica, vol. 9, 2011.
92. B. Y. Park, S. N. Park, J. K. Lee, and C. H. Bae, "Morphometric and genetic variability among Tylenchulus semipenetrans populations from citrus growing area in Korea," Plant Pathology Journal, vol. 25, no. 3, pp. 236–240, 2009.
93. U. Gozel, B. J. Adams, K. B. Nguyen, R. N. Inserra, R. M. Giblin-Davis, and L. W. Duncan, "A phylogeny of Belonolaimus populations in Florida inferred from DNA sequences," Nematropica, vol. 36, no. 2, pp. 155–171, 2006.
94. Z. A. Handoo, A. M. Skantar, and P. Mulrooney, "First report of the sting nematode Belonolaimus longicaudatus on soybean in Delaware," Plant Disease, vol. 94, no. 1, p. 133, 2010.
95. G. R. Stirling, A. M. Stirling, R. M. Giblin-Davis et al., "Distribution of southern sting nematode, Ibipora lolii (Nematoda: Belonolaimidae), on turfgrass in Australia and its taxonomic relationship to other belonolaimids," Nematology, vol. 15, pp. 401–415, 2013.
96. J. Parkinson, M. Mitreva, N. Hall, M. Blaxter, and J. P. McCarter, "400000 nematode ESTs on the Net," Trends in Parasitology, vol. 19, no. 7, pp. 283–286, 2003.
97. D. D. Sui, N. Atibalentja, G. R. Noel, and L. L. Domier, "Genetic diversity of the rDNA locus among 27 populations and 8 races of Heterodera glycines from China, Japan, and the United States, as revealed by PCR-RFLP," Journal of Nematology. In press.
98. B. Gao, R. Allen, T. Maier, E. L. Davis, T. J. Baum, and R. S. Hussey, "Identification of putative parasitism genes expressed in the esophageal gland cells of the soybean cyst nematode Heterodera glycines," Molecular Plant-Microbe Interactions, vol. 14, no. 10, pp. 1247–1254, 2001.
99. D. P. Puthoff, M. L. Ehrenfried, B. T. Vinyard, and M. L. Tucker, "GeneChip profiling of transcriptional responses to soybean cyst nematode, Heterodera glycines,

colonization of soybean roots," Journal of Experimental Botany, vol. 58, no. 12, pp. 3407–3418, 2007.

100. G. E. Múnera Uribe, W. Bert, A. R. Vierstraete, E. de la Peña, M. Moens, and W. Decraemer, "Burrowing nematodes from Colombia and their relationship with Radopholus similis populations, R. arabocoffeae and R. duriophilus," Nematology, vol. 12, pp. 619–629, 2010.

101. J. Jacob, M. Mitreva, B. Vanholme, and G. Gheysen, "Exploring the transcriptome of the burrowing nematode Radopholus similis," Molecular Genetics and Genomics, vol. 280, no. 1, pp. 1–17, 2008.

102. J. Li, D. Peng, and W. Huang, "Phylogentic analysis of Radopholus similis from D2 and D3 fragments of the 28S rRNA gene sequences," Journal of Huazhong Agricultural University, vol. 5, 2008.

103. S. T. Nyaku, V. R. Sripathi, R. V. Kantety, Y. Q. Gu, K. Lawrence, and G. C. Sharma, "Characterization of the two intra-individual sequence variants in the 18S rRNA gene in the plant parasitic nematode, Rotylenchulus reniformis," PLoS ONE, vol. 8, no. 4, Article ID E60891, 2013.

104. Y. Zhan, A. Matafeo, H. Shi, and J. Zheng, "Morphological and molecular characterization and host range of Rotylenchulus reniformis population occurring in Hangzhou, Zhejiang, China," Acta Phytopathologica Sinica, vol. 41, no. 1, pp. 37–43, 2011.

105. Y. Zeng, R. M. Giblin-Davis, and W. Ye, "Two new species of Schistonchus (Nematoda: Aphelenchoididae) associated with Ficus hispida in China," Nematology, vol. 9, no. 2, pp. 169–187, 2007.

106. E. Van Den Berg, S. A. Subbotin, Z. A. Handoo, and L. R. Tiedt, "Hirschmanniella kwazuna sp. n. from South Africa with notes on a new record of H. spinicaudata (Schuurmans Stekhoven, 1944) Luc & goodey, 1964 (nematoda: Pratylenchidae) and on the molecular phylogeny of Hirschmanniella Luc & goodey, 1964," Nematology, vol. 11, no. 4, pp. 523–540, 2009.

107. C. D. Giorgi, P. Veronico, F. D. Luca, A. Natilla, C. Lanave, and G. Pesole, "Structural and evolutionary analysis of the ribosomal genes of the parasitic nematode Meloidogyne artiellia suggests its ancient origin," Molecular and Biochemical Parasitology, vol. 124, no. 1-2, pp. 91–94, 2002.

108. G. Anthoine and D. Mugniéry, "Variability of the ITS rDNA and identification of Nacobbus aberrans (Thorne, 1935) Thorne & Allen, 1944 (Nematoda: Pratylenchidae) by rDNA amplification," Nematology, vol. 7, no. 4, pp. 503–516, 2005.

109. L. Al-Banna, V. Williamson, and S. L. Gardner, "Phylogenetic analysis of nematodes of the genus Pratylenchus using nuclear 26S rDNA," Molecular Phylogenetics and Evolution, vol. 7, no. 1, pp. 94–102, 1997.

110. X. Li and J. Zheng, "Identification of four Pratylenchus species based on morphology and PCR-RFLP of rDNA-ITS," Acta Phytopathologica Sinica, vol. 43, no. 4, pp. 444–448, 2013.

111. M. T. Britton, C. A. Leslie, G. H. McGranahan, and A. M. Dandekar, "Functional genomic analysis of walnut-nematode interactions," Walnut Research Reports 2009, California Walnut Board, 2010.

112. J. C. Wang, G. M. Huang, Y. D. Wei et al., "Phylogenetic analysis of Pratylenchus (Nematoda: Pratylenchidae based on ribosomal internal transcribed spacers (ITS) and D2/D3 expansion segments of 28S rRNA gene," Acta Zootaxonomica Sinica, vol. 4, pp. 687–693, 2012.

113. D. Y. Chen, H. F. Ni, and T. T. Tsay, "Identification of a new recorded stunt nematode Tylenchorhynchus zeae (Nematoda: Belonolaimidae) in Taiwan," Plant Pathology Bulletin, vol. 16, no. 2, pp. 79–86, 2007.

114. J. Liu, R. E. Berry, and A. F. Moldenke, "Phylogenetic relationships of entomopathogenic nematodes (Heterorhabditidae and Steinernematidae) inferred from partial 18S rRNA gene sequences," Journal of Invertebrate Pathology, vol. 69, no. 3, pp. 246–252, 1997.

115. M. V. Blanco, P. Lax, J. C. Rondan Dueñas, C. N. Gardenal, and M. E. Douc, "Morphological and molecular characterization of the entomoparasitic nematode Hammerschmidtiella diesingi (Nematoda, Oxyurida, Thelastomatidae)," Acta Parasitologica, vol. 57, no. 3, pp. 302–310, 2012.
116. S. A. Nadler, E. Bolotin, and S. P. Stock, "Phylogenetic relationships of Steinernema Travassos, 1927 (Nematoda: Cephalobina: Steinernematidae) based on nuclear, mitochondrial and morphological data," Systematic Parasitology, vol. 63, no. 3, pp. 161–181, 2006.
117. O. Douda, M. Zouhar, E. Nováková, J. Mazáková, and P. Ryšánek, "Variability of D2/D3 segment sequences of several populations and pathotypes of potato cyst nematodes (Globodera rostochiensis, Globodera pallida)," Plant Protection Science, vol. 46, no. 4, pp. 171–180, 2010.
118. S. P. Stock, J. F. Campbell, and S. A. Nadler, "Phylogeny of Steinernema travassos, 1927 (Cephalobina: Steinernematidae) inferred from ribosomal DNA sequences and morphological characters," Journal of Parasitology, vol. 87, no. 4, pp. 877–889, 2001.
119. E. E. Morris, R. M. Kepler, S. J. Long, D. W. Williams, and A. E. Hajek, "Phylogenetic analysis of Deladenus nematodes parasitizing northeastern North American Sirex species," Journal of Invertebrate Pathology, vol. 113, pp. 177–183, 2013.
120. Q. Yu, P. de Groot, I. Leal, C. Davis, W. Ye, and B. Foord, "Characterization of Deladenus siricidicola (Tylenchida: Neotylenchidae) ssociated with Sirex noctilio (Hymenoptera: Siricidae) in Canada," Intemational Joumal of Nematology, vol. 19, no. 1, pp. 23–32, 2009.
121. K. Zhang, H. Liu, J. Sun et al., "Molecular phytogeny of geographical isolates of Bursaphelenchus xylophilus: implications on the origin and spread of this species in China and worldwide," Journal of Nematology, vol. 40, no. 2, pp. 127–137, 2008.
122. R. B. Gasser and H. Hoste, "Genetic markers for closely-related parasitic nematodes," Molecular and Cellular Probes, vol. 9, no. 5, pp. 315–319, 1995.
123. T. O. Powers, T. C. Todd, A. M. Burnell et al., "The rDNA internal transcribed spacer region as a taxonomic marker for nematodes," Journal of Nematology, vol. 29, no. 4, pp. 441–450, 1997.
124. R. Floyd, E. Abebe, A. Papert, and M. Blaxter, "Molecular barcodes for soil nematode identification," Molecular Ecology, vol. 11, no. 4, pp. 839–850, 2002.
125. A. Eyualem and M. Blaxter, "Comparison of biological, molecular, and morphological methods of species identification in a set of cultured Panagrolaimus isolates," Journal of Nematology, vol. 35, no. 1, pp. 119–128, 2003.
126. P. De Ley, I. T. De Ley, K. Morris et al., "An integrated approach to fast and informative morphological vouchering of nematodes for applications in molecular barcoding," Philosophical Transactions of the Royal Society B, vol. 360, no. 1462, pp. 1945–1958, 2005.
127. N. R. R. Da Silva, M. C. Da Silva, V. F. Genevois et al., "Marine nematode taxonomy in the age of DNA: the present and future of molecular tools to assess their biodiversity," Nematology, vol. 12, no. 5, pp. 661–672, 2010.
128. M. A. Smith, D. M. Wood, D. H. Janzen, W. Hallwachs, and P. D. N. Hebert, "DNA barcodes affirm that 16 species of apparently generalist tropical parasitoid flies (Diptera, Tachinidae) are not all generalists," Proceedings of the National Academy of Sciences of the United States of America, vol. 104, no. 12, pp. 4967–4972, 2007.
129. H. van Megen, S. van den Elsen, M. Holterman et al., "A phylogenetic tree of nematodes based on about 1200 full-length small subunit ribosomal DNA sequences," Nematology, vol. 11, no. 6, pp. 927–950, 2009.
130. A. Yu. Ryss, "Phylogeny of the order Tylenchida (Nematoda)," Russian Journal of Nematology, vol. 1, no. 2, pp. 74–95, 1993.
131. L. K. Carta, A. M. Skantar, and Z. A. Handoo, "Molecular rDNA phylogeny of telotylenchidae siddiqi, 1960 and evaluation of tail termini," Journal of Nematology, vol. 42, no. 4, pp. 359–369, 2010.
132. R. A. Bedding, "Controlling the pine-killing woodwasp, Sirex noctilio, with nematodes," in Use of Microbes For Control of Invasive Arthropods, A. E. Hajek, T. R.

Glare, and M. O'Callaghan, Eds., pp. 213–235, Springer, Amsterdam, The Netherlands, 2009.

133. A. R. Maggenti, M. Luc, D. J. Raski, R. Fortuner, and E. Geraert, "A reappraisal of Tylenchina (Nemata). 2. Classification of the suborder Tylenchina (Nemata: Diplogasteria)," Revue De Nématologie, vol. 10, no. 2, pp. 135–142, 1987.
134. V. N. Chizhov and S. N. Kruchina, "Phylogeny of the nematode order Tylenchida (Nematoda)," Zoologichesky Zhurnal, vol. 67, no. 9, pp. 1282–1293, 1988.
135. M. Luc, A. R. Maggenti, R. Fortuner, D. J. Raski, and E. Geraert, "A Reappraisal of Tylenchina (Nemata) 1. For a New Approach to the Taxonomy of Tylenchina," Revue De Nématologie, vol. 10, no. 2, pp. 127–134, 1987.
136. O. V. Slobodyanyuk, "Host specificity in tylenchids of fleas," in Abstracts of Papers Presented at the Russian Society of Nematologists 1st English Language International Symposium, pp. 102–103, Zoological Institute of the Russian Academy of Sciences, 1995.
137. K. D. Elsey, "Parasitism of some economically important species of chrysomelidae by nematodes of the genus howardula," Journal of Invertebrate Pathology, vol. 29, no. 3, pp. 384–385, 1977.
138. G. O. Poinar Jr., J. Jaenike, and D. D. Shoemaker, "Howardula neocosmis sp.n. parasitizing North American Drosophila (Diptera: Drosophilidae) with a listing of the species of Howardula Cobb, 1921 (Tylenchida: Allantonematidae)," Fundamental and Applied Nematology, vol. 21, no. 5, pp. 547–552, 1998.
139. R. L. H. Dennis, L. Dapporto, S. Fattorini, and L. M. Cook, "The generalism-specialism debate: the role of generalists in the life and death of species," Biological Journal of the Linnean Society, vol. 104, pp. 725–737, 2011.
140. C. E. Richmond, D. L. Breitburg, and K. A. Rose, "The role of environmental generalist species in ecosystem function," Ecological Modelling, vol. 188, no. 2-4, pp. 279–295, 2005.
141. M. E. J. Woolhouse, L. H. Taylor, and D. T. Haydon, "Population biology of multihost pathogens," Science, vol. 292, no. 5519, pp. 1109–1112, 2001.
142. H. D. Loxdale, G. Lushai, and J. A. Harvey, "The evolutionary improbability of "generalism" in nature, with special reference to insects," Biological Journal of the Linnean Society, vol. 103, no. 1, pp. 1–18, 2011.
143. J. C. Castillo, S. E. Reynolds, and I. Eleftherianos, "Insect immune responses to nematode parasites," Trends in Parasitology, vol. 27, no. 12, pp. 537–547, 2011.
144. R. Kassen, "The experimental evolution of specialists, generalists, and the maintenance of diversity," Journal of Evolutionary Biology, vol. 15, no. 2, pp. 173–190, 2002.
145. N. I. Samia, K. L. Kausrud, H. Heesterbeek et al., "Dynamics of the plague-wildlife-human system in Central Asia are controlled by two epidemiological thresholds," Proceedings of the National Academy of Sciences of the United States of America, vol. 108, no. 35, pp. 14527–14532, 2011.
146. H. E. Kaiser and A. J. Boucot, "Specialisation and extinction: cope's law revisited," Historical Biology, vol. 11, no. 1–4, pp. 247–265, 1996.
147. C. Darby, A. Chakraborti, S. M. Politz, C. C. Daniels, L. Tan, and K. Drace, "Caenorhabditis elegans mutants resistant to attachment of Yersinia biofilms," Genetics, vol. 176, no. 1, pp. 221–230, 2007.
148. G. A. Eroshenko, N. A. Vidyaeva, and V. V. Kutyrev, "Comparative analysis of biofilm formation by main and nonmain subspecies Yersinia pestis strains," FEMS Immunology and Medical Microbiology, vol. 59, no. 3, pp. 513–520, 2010.
149. G. A. Eroshenko, N. A. Vidyaeva, L. M. Kukleva et al., "Studies of biofilm formation in non-pigmented and plasmid-deprived mutants of Yersinia pestis on biotic surfaces, in vivo and in vitro conditions," Problems of Particularliy Dangerous Infections, vol. 113, pp. 45–49, 2012.

CHAPTER 12

Rogue Sperm Indicate Sexually Antagonistic Coevolution in Nematodes

Ronald E. Ellis[1], Lukas Scha¨ rer[2]

Department of Molecular Biology, Rowan University SOM, Stratford, New Jersey, United States of America, 2 Evolutionary Biology, Zoological Institute, University of Basel, Basel, Switzerland

ABSTRACT

Intense reproductive competition often continues long after animals finish mating. In many species, sperm from one male compete with those from others to find and fertilize oocytes. Since this competition occurs inside the female reproductive tract, she often influences the outcome through physical or chemical factors, leading to cryptic female choice. Finally, traits that help males compete with each other are sometimes harmful to females, and female countermeasures may thwart the interests of males, which can lead to an arms race between the sexes known as sexually antagonistic coevolution. New studies from *Caenorhabditis* nematodes suggest that males compete with each other by producing sperm that migrate aggressively and that these sperm may be more likely to win access to oocytes. However, one byproduct of this competition appears to be an increased probability that these sperm will go astray, invading the ovary, prematurely activating oocytes, and sometimes crossing basement membranes and leaving the gonad altogether. These harmful effects are sometimes observed in crosses between animals of the same species but are most easily detected in interspecies crosses, leading to dramatically lowered fitness, presumably because the competitiveness of the sperm and the associated female countermeasures are not precisely matched. This mismatch is most obvious in crosses involving individuals from androdioecious species (which have both hermaphrodites and males), as predicted by the lower levels of sperm competition these species experience. These results suggest a striking example of sexually antagonistic coevolution and dramatically expand the value of nematodes as a laboratory system for studying postcopulatory interactions.

On the Origin of Species focused almost exclusively on the role of natural selection in evolution [1], but Darwin realized that animals also compete for mates and described the process of sexual selection at length in a later book [2]. The simplest examples involve combat like that between male elephant seals fighting for access to females. However, sexual selection also includes many other types of interactions. For example, some male birds have elaborate plumage because females favor this trait when choosing mates (reviewed in [3]). In their simplest form, these interactions can be thought of as parts of a triangle—competition between two males forming the base and the interactions between each of the males and the female forming the two legs.

MANY TYPES OF SEXUAL SELECTION OCCUR AFTER COPULATION

It took almost a century for scientists to realize that sexual selection often continued after copulation was finished [4]. If several males mate with the same female (which appears to be the rule rather than the exception), their sperm will often compete within her reproductive tract for access to her oocytes. During mating, males therefore try to enhance the success of their own sperm by actively displacing those from previous males, by chemically preventing females from further mating, or by making sperm that are competitively superior in number or phenotype (reviewed in [5],[6]). Because these interactions are played out inside the female, she will often influence and sometimes control the postcopulatory actions of both her mates and their sperm, giving her opportunities to favor the use of one male's sperm over another. This bias is often termed "cryptic female choice" and represents the postcopulatory equivalent of more familiar types of female choice (reviewed in [7],[8]).

Postcopulatory Sexual Selection Can Lead to Conflicts between Males and Females

Since male-male competition often selects for ejaculates that are more abundant or more persistent than desirable from the female's perspective, females often evolve countermeasures that control what males (or their ejaculates) can do to them, which in turn favors male traits that can overcome these female defenses. This sexual conflict between mates following insemination results in ongoing sexually antagonistic coevolution. Striking examples include manipulative seminal fluids in fruit flies [8], sperm with bristles in some free-living flatworms [9], and traumatic insemination that bypasses the normal route of fertilization in a whole range of organisms [10]. Similar phenomena also occur during sperm-egg interactions in free-spawning marine organisms and in many cases have led to striking patterns in the evolution of the morphologies and molecules that mediate these interactions [8],[11].

EVOLUTIONARY BIOLOGY OF *CAENORHABDITIS*

Although the nematode *Caenorhabditis elegans* originally became famous for developmental genetics [12], its genus is now an important model for evolutionary research (reviewed in [13]). Because field studies are still difficult [14], most of that work is done by comparing gene functions in the laboratory, either between different wild isolates or across species. These studies have become easier with the availability of many newly discovered species, nine genome-sequencing projects (Figure 1), and the widespread availability of RNA interference [15] and gene-editing techniques [16]. However, the most valuable feature is hermaphroditic genetics.

SELF-FERTILE HERMAPHRODITES

C. elegans and two related species produce self-fertile, sequential hermaphrodites. Such *XX* individuals are female in most respects but can make their own sperm early in life and use them later for self-fertilization. The same species also make *XO* males, which mate with the hermaphrodites in a sexual system called androdioecy. However, males are infrequent in the wild, and outcrossing is less common than selfing in these androdioecious species [17],[18]. Since most other *Caenorhabditis* species are dioecious (comprised of males and females), self-fertility must have arisen on three independent occasions (Figure 1) [14],[19],[20]. The existence of hermaphrodites has been of great utility because it (1) simplifies genetic research in *C. elegans* and *C. briggsae*, (2) provides an important trait to follow in evolutionary studies, and (3) allows direct tests of many theories about sexual reproduction and behavior.

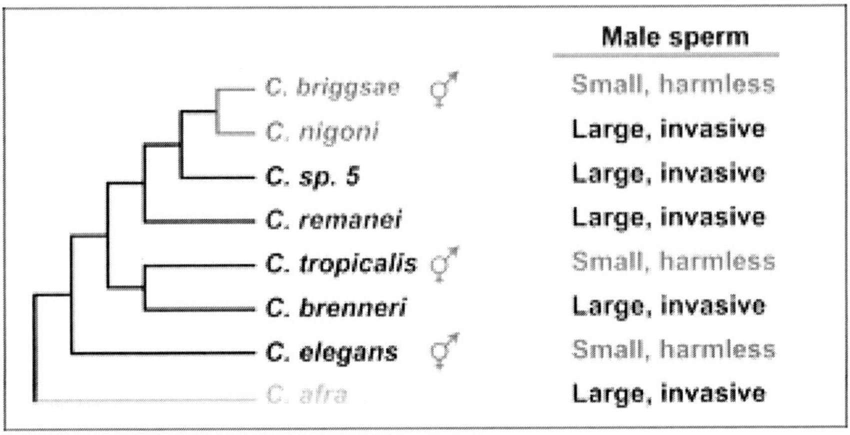

Figure 1. Hermaphrodites have evolved in three independent lineages in *Caenorhabditis.* Only species with sequenced genomes are shown. Androdioecious species (comprised of males and hermaphrodites) are marked with a red symbol, and the others are dioecious (comprised of males and females). The two species in blue are able to interbreed and produce fertile offspring, and the outgroup for the elegans group is orange. Modified from Kiontke et al. [14] and Félix et al. [38]. See main text for details on the types of male sperm.

SPERM COMPETITION IN NEMATODES

One of the best examples of such a test involves sperm competition. When *C. elegans* males mate with hermaphrodites, their sperm take precedence, and the hermaphrodite sperm are largely excluded from the two spermathecae (Figure 2A); this pattern of sperm precedence is due to the superiority of the male sperm themselves, rather than timing factors or seminal fluid [21]. The advantage could be due to size, since male sperm are larger than hermaphrodite sperm in all three androdioecious species of *Caenorhabditis* [22],[23]. Supporting this idea, direct competition between wild isolates with different sperm sizes showed that males with larger sperm outcompete those with smaller sperm [24], and experimental evolution in conditions that favored multiple matings led to a rapid increase in the size of male sperm within a few dozen generations [25]. Thus, classical studies with *C. elegans* provide strong evidence for sperm competition between males (Figure 2B) and for the superiority of male sperm over hermaphrodite sperm. Because *C. elegans* males appear to be rare in the wild, multiple matings by hermaphrodites should be uncommon, and sperm competition is probably weaker than in dioecious species.

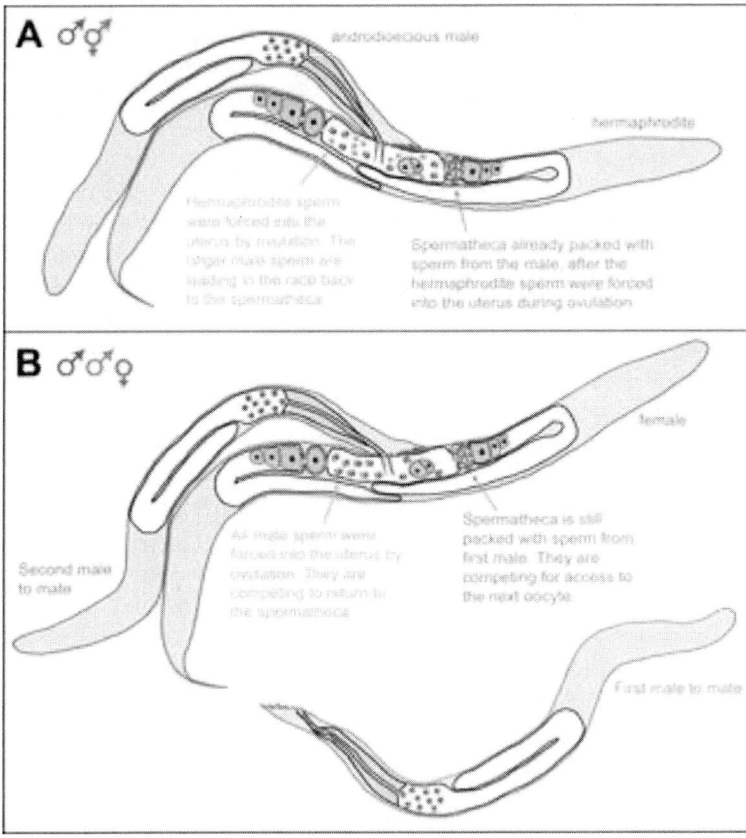

Figure 2. In *Caenorhabditis* nematodes, sperm compete for position in the spermathecae.(A) Diagram of a hermaphrodite (red) mating with a male (blue). The hermaphrodite gonad is bilaterally symmetric with a central uterus. The male is ejaculating larger sperm (blue) into the uterus, and they outcompete the smaller hermaphrodite sperm (pink) in the race to repopulate each spermatheca (purple) after sperm are dislodged during ovulation. (B) Diagram of a female (red) mating sequentially with two males (one green and the other blue). The female's gonad resembles that of the hermaphrodite in (A). The spicules from the blue male have penetrated the vulva, and he is ejaculating (blue) sperm into the uterus. These sperm will compete with those from the first male (green) for positions in the two spermathecae (purple), where they wait for the chance to fertilize oocytes. Although the sperm from the first male have already taken the best positions (shown in the right spermatheca), they will be displaced into the uterus each time an oocyte is ovulated (shown in the left spermatheca) and must compete with those from the second male to reestablish their positions. Although displacement has been directly observed, additional factors that remain unknown might help influence competition among these sperm.

Studies from *C. elegans* revealed additional factors that might be involved in postcopulatory sexual selection, although they have yet to be studied in an evolutionary context (Figure 3). First, oocytes are involved in attracting sperm towards the spermathecae, and their absence causes sperm to wander aimlessly [26]. This attraction depends on a complex mixture of prostaglandins that is secreted by the oocytes [27]. It would be fascinating to know if the spectrum of prostaglandins has changed during *Caenorhabditis* evolution and if any of these molecules play a role in cryptic female choice. Second, sperm release small membrane-bound packets that contain major sperm protein, which stimulates oocytes to mature and the somatic gonad to contract during ovulation [28],[29]. It is possible that these packets contain additional signals too. Thus, male and female components engage in complex physical and molecular interactions after mating, and these interactions might be under strong selective pressure.

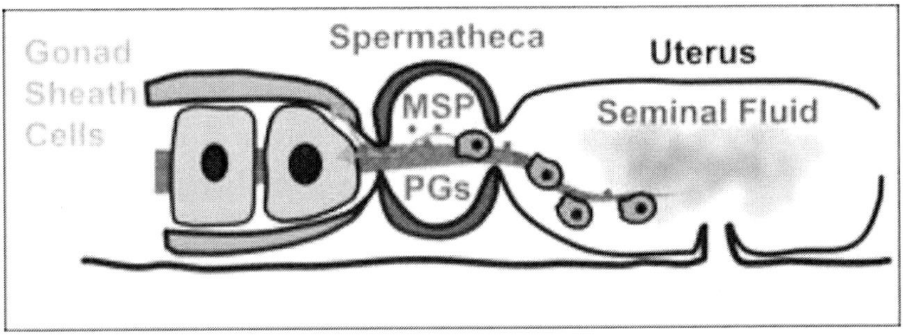

Figure 3. Male and female cues involved in gamete behavior. The male ejaculates sperm (blue) and seminal fluid (green) into the uterus. The seminal fluid contains signals that activate the sperm so that they extend pseudopods and are able to crawl. The seminal fluid is probably complex and might contain additional signals. In addition, active sperm release small membrane-bound packets (blue circles) that contain major sperm protein (MSP), which stimulates oocytes to mature. MSP also causes the gonad sheath cells (orange) to contract and force oocytes through the distal spermathecal valve into the spermatheca (purple). Finally, the oocytes release a complex mixture of prostaglandins (PGs) that guide the sperm. Each of these signals has the potential to mediate sperm competition and/or cryptic female choice.

INTERSPECIES CROSSES AND POSTCOPULATORY SEXUAL SELECTION

Interspecies crosses provide a powerful technique for studying these reproductive interactions [30] and are used with great success in a new study by Ting et al. [31]. Animals from a wide variety of *Caenorhabditis* species can easily be induced to mate with each other in the laboratory; some crosses do not result in fertilization, others lead to inviable offspring, and a few produce hybrids [32]–[34]. Particular attention has been focused on the androdioecious species *C. briggsae* and the dioecious species *C. nigoni*, which are so closely related that they can produce fertile offspring [35].

Now, Ting et al. report a remarkable discovery—in many cases when males from one *Caenorhabditis* species mate with hermaphrodites or females from another, they significantly decrease their fertility [31]. This effect is caused by two distinct interactions. First, these male sperm often physically displace other sperm from the two spermathecae, as expected from previous observations of crosses between *C. elegans* males and hermaphrodites. Second, sperm sometimes invade the ovary, where they induce premature maturation of young oocytes and disrupt the development of new gametes. Occasionally, such ectopic sperm even cross a basement membrane and escape from the gonad altogether. The fact that *C. brenneri* and *C. nigoni* males are equally effective at displacing *C. tropicalis* sperm but *C. brenneri* males more severely affect fertility suggests that both types of interactions matter.

Some of the species in these experiments were androdioecious, whereas others were dioecious. Since sperm competition should be less intense in androdioecious species, observing how they respond in these crosses should provide a strong test for whether postcopulatory selection is involved. Indeed, the authors found that hermaphrodites were much more susceptible to harmful sperm than females. Furthermore, males from androdioecious species made the least harmful sperm detected in any of their crosses. Thus, competition between male sperm is probably at the root of the phenomena they describe.

Finally and perhaps most intriguingly, the authors found that male sperm sometimes go astray even in crosses between males and females of the same species. Thus, the interspecies crosses may simply provide a more sensitive way to measure interactions that are going on within individual species in the wild. A simple model that can explain their results is that dioecious males are under intense selection to produce highly migratory sperm, which will have the best chance to find good positions in the spermathecae for fertilizing oocytes (Figure 4A). However, the aggressiveness of these sperm means that females need to develop appropriate countermeasures, such as changes in their chemical signals or the physical strength of the distal spermathecal valve, to keep the sperm contained and prevent them from entering regions of the female gonad where they could cause harm. If the competitiveness of a male's sperm is not in sync with the countermeasures of his mate, some overzealous sperm could go rogue, causing a significant loss in fitness (Figure 4B). Thus, selection on males could favor highly migratory sperm that outcompete those from other males, even if they occasionally lower female fertility. In addition, selection in females should favor protective countermeasures that restore normal fertility but may decrease male fitness. This sexually antagonistic coevolution is expected to cause rapid changes in both sexes, which are revealed when animals from different species are used in experimental crosses.

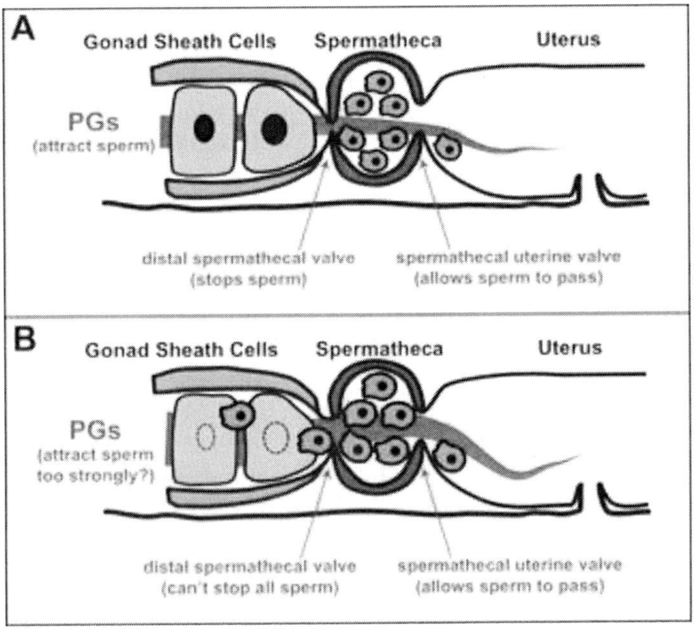

Figure 4. Sperm vigor and female defenses undergo sexually antagonistic coevolution. (A) In a benign interaction between the sexes, male sperm compete for position in the spermatheca, and a combination of female signals and control of the distal spermathecal valve prevent them from entering the ovary. (B) Because of sperm competition, some males develop more competitive sperm that cannot be excluded from the ovary and thus become invasive; resulting fitness costs will favor additional female countermeasures through sexually antagonistic coevolution.

OUTLOOK

These studies show not only that nematodes undergo sperm competition but also that additional postcopulatory interactions are significant and appear to include sexually antagonistic coevolution. Given the broad range of interactions between sperm, oocytes, and chemical and physical cues inside the female reproductive tract (Figure 3) (reviewed in [36]), nematodes in the genus *Caenorhabditis* could provide valuable models for investigating postcopulatory sexual selection and sexual conflicts in the laboratory.

New isolates of different *Caenorhabditis* species are being established at a considerable rate (e.g., [14]), which should also permit study of postcopulatory mechanisms in a variety of wild populations. The detection of rogue sperm in some intraspecies crosses might be explained not only by the greater aggressiveness of sperm from dioecious species but also by the high levels of genetic diversity in these species [37], which should make it easier to detect mismatches between male persistence and female countermeasures. Thus, it would be fascinating to analyze the role that genetic variability in dioecious species plays in the range of sperm phenotypes and interactions and to see if the low levels of diversity in androdioecious species limit this range in the wild. If genetic diversity is important, it might even be possible to find specific strains of androdioecious species that produce ectopic sperm in crosses. Finally, we might be able to learn if some sperm traits and female countermeasures are restricted to isolated populations, which genes underlie these traits, and how fast these types of coevolutionary changes are occurring, at morphological, biochemical, and molecular levels. The combination of experimental and field approaches available to address these questions promises an exciting future.

ACKNOWLEDGMENTS

We thank Gillian Stanfield for comments.

REFERENCES

1. Darwin C (1859) On the Origin of Species by Means of Natural Selection, or the Preservation of Favored Races in the Struggle for Life. London: John Murray.
2. Darwin C (1871) The Descent of Man and Selection in Relation to Sex. London: John Murray.

3. Andersson MB (1994) Sexual selection. Princeton (New Jersey): Princeton University Press.
4. Parker GA (1970) Sperm competition and its evolutionary consequences in the insects. Biol Rev 45: 525–567.
5. Parker GA (1998) Sperm competition and the evolution of ejaculates: towards a theory base. In: Birkhead TR, Møller AP, editors. Sperm competition and sexual selection. London: Academic Press. pp. 3–54.
6. Pizzari T, Parker GA (2009) Sperm competition and sperm phenotype. In: Birkhead TR, Hosken DJ, Pitnick S, editors. Sperm biology: an evolutionary perspective. London: Academic Press. pp. 207–245.
7. Eberhard WG (2009) Postcopulatory sexual selection: Darwin's omission and its consequences. Proc Natl Acad Sci U S A 106 Suppl 1: 10025–10032.
8. Pitnick S, Wolfner MF, Suarez SS (2009) Ejaculate-female and sperm-female interactions. In: Birkhead TR, Hosken DJ, Pitnick S, editors. Sperm biology: an evolutionary perspective. London: Academic Press. pp. 247–304.
9. Schärer L, Littlewood DT, Waeschenbach A, Yoshida W, Vizoso DB (2011) Mating behavior and the evolution of sperm design. Proc Natl Acad Sci U S A 108: 1490–1495.
10. Lange R, Reinhardt K, Michiels NK, Anthes N (2013) Functions, diversity, and evolution of traumatic mating. Biol Rev Camb Philos Soc 88: 585–601.
11. Howard DJ, Palumbi SR, Birge LM, Manier MK (2009) Sperm and speciation. In: Birkhead TR, Hosken DJ, Pitnick S, editors. Sperm biology: an evolutionary perspective. London: Academic Press. pp. 367–403.
12. Brenner S (1974) The genetics of Caenorhabditis elegans. Genetics 77: 71–94.
13. Ellis RE, Lin S-Y (2014) The evolutionary origins and consequences of self-fertility in nematodes. F1000Prime Rep In press.
14. Kiontke KC, Félix MA, Ailion M, Rockman MV, Braendle C, et al. (2011) A phylogeny and molecular barcodes for Caenorhabditis, with numerous new species from rotting fruits. BMC Evol Biol 11: 339.
15. Nuez I, Félix MA (2012) Evolution of Susceptibility to Ingested Double-Stranded RNAs in Caenorhabditis Nematodes. PLoS ONE 7: e29811.
16. Wei Q, Shen Y, Chen X, Shifman Y, Ellis RE (2014) Rapid creation of forward-genetics tools for C. briggsae using TALENs: lessons for nonmodel organisms. Mol Biol Evol 31: 468–473.
17. Barriere A, Félix MA (2005) Natural variation and population genetics of Caenorhabditis elegans. In: The C. elegans Research Community, editor. WormBook. Available: http://www.wormbook.org/chapters/www_naturalvariationgenetics/naturalvariationgenetics.pdf. Accessed 23 June 2014. doi/10.1895/wormbook.1.43.1
18. Sivasundar A, Hey J (2005) Sampling from natural populations with RNAi reveals high outcrossing and population structure in Caenorhabditis elegans. Curr Biol 15: 1598–1602.
19. Cho S, Jin SW, Cohen A, Ellis RE (2004) A phylogeny of Caenorhabditis reveals frequent loss of introns during nematode evolution. Genome Res 14: 1207–1220.
20. Kiontke K, Gavin NP, Raynes Y, Roehrig C, Piano F, et al. (2004) Caenorhabditis phylogeny predicts convergence of hermaphroditism and extensive intron loss. Proc Natl Acad Sci U S A 101: 9003–9008.
21. LaMunyon CW, Ward S (1995) Sperm precedence in a hermaphroditic nematode (Caenorhabditis elegans) is due to competitive superiority of male sperm. Experientia 51: 817–823.
22. LaMunyon CW, Ward S (1999) Evolution of sperm size in nematodes: sperm competition favours larger sperm. Proc Biol Sci 266: 263–267.
23. Baldi C, Viviano J, Ellis RE (2011) A bias caused by ectopic development produces sexually dimorphic sperm in nematodes. Curr Biol 21: 1416–1420.
24. LaMunyon CW, Ward S (1998) Larger sperm outcompete smaller sperm in the nematode Caenorhabditis elegans. Proc Biol Sci 265: 1997–2002.

25. LaMunyon CW, Ward S (2002) Evolution of larger sperm in response to experimentally increased sperm competition in Caenorhabditis elegans. Proc Biol Sci 269: 1125–1128.
26. Kubagawa HM, Watts JL, Corrigan C, Edmonds JW, Sztul E, et al. (2006) Oocyte signals derived from polyunsaturated fatty acids control sperm recruitment in vivo. Nat Cell Biol 8: 1143–1148.
27. Hoang HD, Prasain JK, Dorand D, Miller MA (2013) A heterogeneous mixture of F-series prostaglandins promotes sperm guidance in the Caenorhabditis elegans reproductive tract. PLoS Genet 9: e1003271.
28. Miller MA, Nguyen VQ, Lee MH, Kosinski M, Schedl T, et al. (2001) A sperm cytoskeletal protein that signals oocyte meiotic maturation and ovulation. Science 291: 2144–2147.
29. Kosinski M, McDonald K, Schwartz J, Yamamoto I, Greenstein D (2005) C. elegans sperm bud vesicles to deliver a meiotic maturation signal to distant oocytes. Development 132: 3357–3369.
30. Hill KL, L'Hernault SW (2001) Analyses of reproductive interactions that occur after heterospecific matings within the genus Caenorhabditis. Dev Biol 232: 105–114.
31. Ting JJ, Woodruff GC, Leung G, Shin N-R, Cutter AD, et al. (2014) Intense sperm-mediated sexual conflict promotes reproductive isolation in Caenorhabditis Nematodes. PLoS Biol 7: e1001915.
32. Baird SE, Sutherlin ME, Emmons SW (1992) Reproductive isolation in Rhabditidae (Nematoda: Secernentea); mechanisms that isolate six species of three genera. Evolution 46: 585–594.
33. Baird SE, Yen WC (2000) Reproductive isolation in Caenorhabditis: terminal phenotypes of hybrid embryos. Evol Dev 2: 9–15.
34. Baird SE (2002) Haldane's rule by sexual transformation in Caenorhabditis. Genetics 161: 1349–1353.
35. Woodruff GC, Eke O, Baird SE, Félix MA, Haag ES (2010) Insights into species divergence and the evolution of hermaphroditism from fertile interspecies hybrids of Caenorhabditis nematodes. Genetics 186: 997–1012.
36. Ellis RE, Stanfield GM (2014) The regulation of spermatogenesis and sperm function in nematodes. Semin Cell Dev Biol 29C: 17–30.
37. Dey A, Chan CK, Thomas CG, Cutter AD (2013) Molecular hyperdiversity defines populations of the nematode Caenorhabditis brenneri. Proc Natl Acad Sci U S A 110: 11056–11060.
38. Félix MA, Braendle C, Cutter AD (2014) A streamlined system for species diagnosis in Caenorhabditis (Nematoda: Rhabditidae) with name designations for 15 distinct biological species. PLoS ONE 9: e94723.

CHAPTER 13

Interspecific competition between entomopathogenic nematodes (*Steinernema*) is modified by their bacterial symbionts (*Xenorhabdus*)

Mathieu Sicard*[1,2], Julie Hinsinger[1], Nathalie Le Brun[1], Sylvie Pages[3], Noël Boemare[3] and Catherine Moulia[1]

[1]*Laboratoire Génome, Populations, Interactions, Adaptation UMR 5171 CNRS, Université de Montpellier 2, Place Eugène Bataillon cc. 63, 34095 Montpellier, France,*

[2]*Laboratoire de Génétique et Biologie des Populations de Crustacés, UMR 6556 CNRS, Université de Poitiers, 40 avenue du Recteur Pineau, 86022 Poitiers, France and*

[3]*Laboratoire Ecologie microbienne des insectes et interactions hôte-pathogène UMR 1133 INRA, Université de Montpellier 2 cc. 54, 34095 Montpellier, France*

ABSTRACT

Background

Symbioses between invertebrates and prokaryotes are biological systems of particular interest in order to study the evolution of mutualism. The symbioses between the entomopathogenic nematodes *Steinernema* and their bacterial symbiont *Xenorhabdus* are very tractable model systems. Previous studies demonstrated (i) a highly specialized relationship between each strain of nematodes and its naturally associated bacterial strain and (ii) that mutualism plays a role in several important life history traits of each partner such as access to insect host resources, dispersal and protection against various biotic and abiotic factors. The goal of the present study was to address the question of the impact of *Xenorhabdus* symbionts on the progression and outcome of interspecific competition between individuals belonging to different *Steinernema* species. For this, we monitored experimental interspecific competition between (i) two nematode species: *S. carpocapsae* and *S. scapterisci* and (ii) their respective symbionts: *X. nematophila* and *X. innexi* within an experimental insect-host (*Galleria mellonella*). Three conditions of competition between nematodes were tested: (i) infection of insects with aposymbiotic IJs (i.e. without symbiont) of both species (ii) infection of insects with aposymbiotic IJs of both species in presence of variable proportion of their two *Xenorhabdus* symbionts and (iii) infection of insects with symbiotic IJs (i.e. naturally associated with their symbionts) of both species.

Results

We found that both the progression and the outcome of interspecific competition between entomopathogenic nematodes were influenced by their bacterial symbionts. Thus, the results obtained with aposymbiotic nematodes were totally opposite to those obtained with symbiotic nematodes. Moreover, the experimental introduction of different ratios of *Xenorhabdus* symbionts in the insect-host during competition between *Steinernema* modified the proportion of each species in the adults and in the global offspring.

Conclusion

We showed that *Xenorhabdus* symbionts modified the competition between their *Steinernema* associates. This suggests that *Xenorhabdus* not only provides *Steinernema* with access to food sources but also furnishes new abilities to deal with biotic parameters such as competitors.

BACKGROUND

Symbioses between the entomopathogenic nematodes *Steinernema* spp. and the enterobacteriacae *Xenorhabdus* spp. are associations in which both partners receive benefits from each other [1, 2, 3]. In the soil, the infective juveniles (IJs) of the nematodes act as vectors dispersing the bacteria from insect host to insect-host and in turn, the bacteria increase the nematode's fitness within the insects hosts [3, 4]. Previous studies showed that these symbioses were highly specific and that no *Steinernema* spp. was able to associate with a *Xenorhabdus* spp. genetically distant from its natural one [2, 5, 6]. As the bacterial dispersion is totally dependent upon the fitness of the nematode within the insect-host, it is possible that *Xenorhabdus* spp. might select special traits in order to enhance their vector's fitness. It is known that *Xenorhabdus* spp. are beneficial to their nematodes in providing the latter with a better ability to kill the insect and feed on it [1, 7, 8]. Previous studies that focused on two different *Steinernema* species (*S. carpocapsae* and *S. scapterisci*) have provided us with insights into the association characteristics [2, 5, 6, 9, 10, 11]. Although the two nematode species demonstrated increased fitness when they parasitized insect-hosts with their own native symbiont, *S. scapterisci* appeared less dependent upon its native symbiont (*X. innexi* [12]) than *S. carpocapsae* (associated with *X. nematophila*). Thus, *S. scapterisci*'s symbiont increased the reproductive rate of its naturally associated nematode by a factor of 1.3, whereas *X. nematophila* increased the reproductive rate of its naturally associated nematode by sevenfold (i.e. *S. carpocapsae*) [3]. Moreover, *S. scapterisci*, transported 700-fold fewer cells of its *Xenorhabdus* than *S. carpocapsae* (i.e. ~50 bacteria per nematode for *S. carpocapsae* and ~0.07 bacteria per nematode for *S. scapterisci*) [3]. These two *Steinernema* species also differed in their ability to deal with non-native *Xenorhabdus* strains in case of co-infection in an insect [5, 6]. While *S. scapterisci* reproduced in co-infection situations with all the tested *Xenorhabdus* strains (even if its reproduction was better with its native one than with others), *S. carpocapsae* could not reproduce at all with most of them in the same situation [5, 6]. Despite these specific differences, the global trend emerging from these previous experiments was that non-naturally associated *Xenorhabdus* strains tend to be antagonist against nematodes species which cannot disperse them. One can easily think that this antagonistic effect of *Xenorhabdus* strains on the fitness of nematodes naturally associated with others *Xenorhabdus* strains could be selected in case of frequent interspecific competition. In such an evolutionary context, we can postulate that each *Xenorhabdus* strain should try to provide its own nematode-vector with competitive advantages by producing antagonistic molecules against foreign nematodes. Indeed, a previous study has shown with the association *S. carpocapsae-X. nematophila* as a model-system that, in insects co-infected by antagonistic *Xenorhabdus*, *X. nematophila* partly counteracted their antagonistic effect on the nematodes fitness most probably by the mean of bacteriocins [13, 14].

The goal of the present study was to address the question of the impact of *Xenorhabdus* symbionts on the progression and outcome of interspecific competition between individuals belonging to different *Steinernema* species. For this, we monitored experimental interspecific competition between (i) two nematode species: *S. carpocapsae* and *S. scapterisci* and (ii) their respective symbionts: *X. nematophila* and *X. innexi* within an insect-host (*Galleria mellonella*). In this study, three conditions of competition between nematodes were tested: (i) infection of insects with aposymbiotic IJs (i.e. without symbiont) of both species (ii) infection of insects with aposymbiotic IJs of both species in presence of variable proportion of their two *Xenorhabdus* symbionts and (iii) infection of insects with symbiotic IJs (i.e. naturally associated with their symbionts) of both species.

RESULTS

Proportion of each *Xenorhabdus* in insect's hemolymph 72 h post-infection with IJs

We observed that both bacterial strains were able to multiply and co-exist within the hemolymph of the insect. Nevertheless, *X. nematophila* was clearly less represented within the hemolymph 72 h post-infection with nematodes when an initial injection of a suspension containing 50% of each bacterium in the insect was performed (Fig 1). When 70% and 90% of *X. nematophila* were injected, the two bacteria were found in a very variable proportion and co-existed (Fig 1). In the competition resulting from infection of insects with symbiotic IJs of both nematode species, as well as when 100% of *X. nematophila* were injected into insects infected with aposymbiotic IJs of both nematode species, no *X. innexi* were detected within the hemolymph (Fig 1).

Assessment of nematodes maturation in competition

To know if GFP labelling of *X. nematophila*, employed to discriminate the two nematode species within the global offspring, triggered differences on the progress of competition between nematodes, the data obtained with or without GFP labelling of *X. nematophila* in each competition situation were compared with a Mann Withney test. We showed that GFP labelling had no statistically significant effect on both (i) the ratio of *S. carpocapsae* among all females found in the insect 150 h after infection (N = 4; $4 < U < 14$; $0.072 < P < 0.449$) and (ii) the ratio of *S. carpocapsae* among all males found in the insect 150 h after infection ($21 < N < 33$; $214,000 < U < 398,00$; $0.172 < P < 0.780$). Because of the non-significant differences observed in these comparisons, the results obtained in each situation with and without GFP labelled *X. nematophila* were pooled in further analyses. The ratios of *S.*

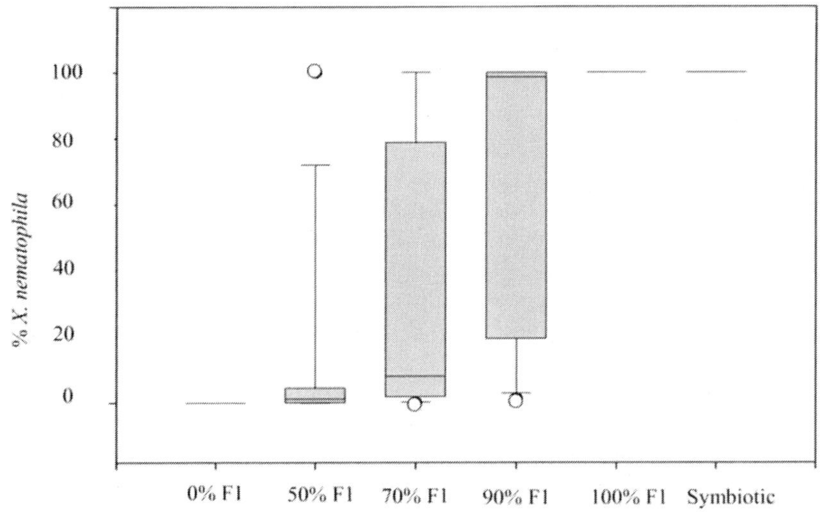

Figure 1. Percentile distribution of the percentages of *X. nematophila* within *Xenorhabdus sp.* found in the hemolymph of the insects 48 h post-infection. Results are given as box plots, where the horizontal line indicate the median (50th of the data), the bottom and the top of the box indicate the first quartiles (25th of the data) and the third quartiles (75th of the data). The whiskers the range of the data (10th of the data and 90th of the data). Others dots are outliers which are under the 10th percentile of the data or over 90th percentile of the data. The abscissa shows the condition of the initial infection.

carpocapsae among all males found in the insects were highly heterogeneous for the different competition conditions (KW: H = 164.052; $P <$ 0.0001). The Noether test showed that no significant difference in the ratio of *S. carpocapsae* males occurred under four competition conditions: (i) competition resulting from infection of insects with aposymbiotic IJs of both nematode species without injection of bacterium, (ii) competition resulting from infection with aposymbiotic IJs of both nematode species in insects injected with 100% of *X. innexi*, (iii) competition resulting from infection with aposymbiotic IJs of both nematode species with 50% of each symbiont injected in the insects and (iiii) competition resulting from infection with aposymbiotic nematodes in insects injected with 70% of *X. nematophila* and 30% of *X. innexi* (24 < N < 63, 0.464 $< z <$ 2.568, 0.441 $< P <$ 1). In these four conditions, the ratios of *S. carpocapsae* among males were significantly lower than in the three other conditions of competition: (i) competition resulting from infection with aposymbiotic IJs in insects injected with 90% of *X. nematophila* and 10% of *X. innexi*, (ii) competition resulting from infection between aposymbiotic IJs in insects injected with 100% of *X. nematophila* and (iii) competition resulting from infection with symbiotic IJs (28 < N < 63; 3.418 <*z* < 9.143; 0.0001 <*P* < 0.0264) (see Figure 2). These results showed that an increase in the proportion of *X. nematophila* within bacteria injected into the insect triggered an increase of *S. carpocapsae* within adult worms found in the insects. The ratios of *S. carpocapsae* among females found in the insects were also highly heterogeneous for the different competition conditions (KW: *H* = 24.946; *P* < 0.001). The Noether test showed that the ratios of *S. carpocapsae* within females were lower in competition resulting from infection with aposymbiotic IJs in insects injected with 100% of *X. innexi* than (i) in competition resulting from infection with aposymbiotic IJs in insects injected with 100% of *X. nematophila* and (ii) in competition resulting from infection with symbiotic IJs (5 < N < 8; 3.260 <*z* < 4.028; 0.002 <*P* < 0.046). In competition resulting from infection with aposymbiotic IJs in insects injected with 100% of *X. innexi*, no *S. carpocapsae* females were observed.

Assessment of the nematode offspring

Total number of infective juveniles (total nematode offspring)

The total number of IJs emerging from insects in the different competition situations for *S. carpocapsae* and *S. scapterisci* was heterogeneous under the different competition conditions (K.W. *H* = 71.041, *P* < 0.0001). The Noether test showed that competition resulting from infection with aposymbiotic IJs without injection of bacteria into the insect resulted in the production of significantly fewer IJs than (i) competition resulting from infection with symbiotic IJs (*z* = 7.296, *P* < 0.001), (ii) competition resulting from infection with aposymbiotic IJs in insects injected with 50% of each symbiont (*z* = 6.108, *P* < 0.0001), (iii) competition resulting from infection with aposymbiotic IJs in insects injected with 70% of *X. nematophila* and 30% of *X. innexi* (*z* = 5.174, *P* < 0.0001) and (iiii) competition resulting from infection between aposymbiotic IJs in insects injected with 90% of *X. nematophila* and 10% of *X. innexi* (*z* = 4.304, *P* < 0.001).

Proportion of each nematode species in the total nematode offspring

In the competition situation resulting from infection with aposymbiotic IJs of both nematode species without injection of bacteria into the insect, the proportion of each nematode species in the total offspring emerging from four insects was assessed with PCR-RFLP made after global extraction of

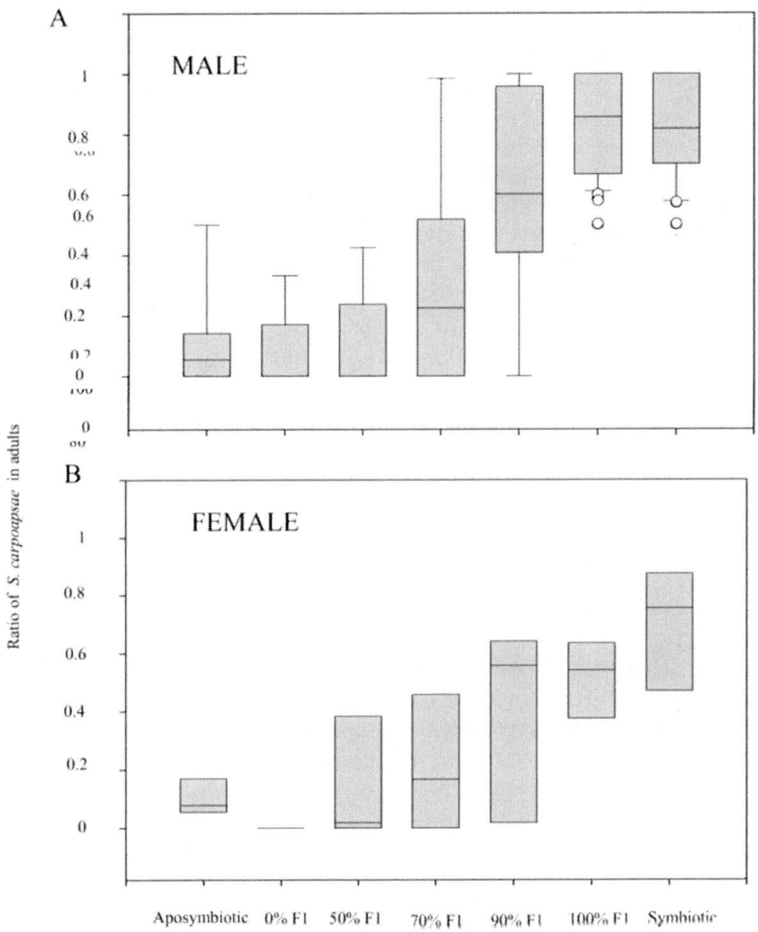

Figure 2. Percentile distribution of the ratio of *S. carpocapsae* within the total number of (A) nematode males and (B) nematode females found in the insect 120 h post-infection. Results are given as box plots, where the horizontal line indicate the median (50th of the data), the bottom and the top of the box indicate the first quartiles (25th of the data) and the third quartiles (75th of the data). The whiskers the range of the data (10th of the data and 90th of the data). Others dots are outliers which are under the 10th percentile of the data or over 90th percentile of the data. The abscissa shows the condition of the initial infection.

DNA from pools of 500 IJs. In control samples, the presence of 25% and 10% of *S. carpocapsae* IJs within pools of 500 IJs containing respectively 75% and 90% of *S. scapterisci* were easily detected (i.e. three bands were observed after incubation of the PCR product with Hind III) (Figure 3). In all the four pools of 500 IJs tested here, only one band was observed after incubation of the PCR product with Hind III (Figure 3). This result suggested that after competition between aposymbiotic nematodes without injection of bacteria, more than 90% of all IJs emerging from the insect belonged to *S. scapterisci*. In other competition situations, the proportion of each nematode species was evaluated by counting the proportion of nematodes harbouring GFP labelled *X. nematophila* within the global offspring. GFP-fluorescence of *X. nematophila* is a good indicator of *S. carpocapsae*'s IJs because pilots experiments showed that GFP labelled bacteria were still observable in IJs that were stored during 6 months at 8°C and that after several generations within different insects, at each generation, 95% of the IJs emerging from the insects habored GFP labelled bacteria. Nevertheless, in order to check that the GFP expression was not lost all along our experiments, we checked GFP expression for all the *X. nematophila* isolated from pools of 500 IJs and confirmed that none of them did not express GFP. This result confirmed the perfect stability of GFP labelled *X. nematophila* all along our experiments. Competition resulting from infection with aposymbiotic IJs in insects injected with (i) 50% of each symbiont, (ii) 70% of *X. nematophila* and 30% of *X. innexi* and (iii) 90% of *X. nematophila* and 10% of *X. innexi* gave offspring in which almost no IJs harboured GFP labelled *X. nematophila* (Figure 4). These observations showed that *S. scapterisci* was highly dominant in the total offspring emerging from insects after such competitions. For competition resulting from infection with aposymbiotic IJs in insects injected with 100% of *X. nematophila*, the proportion of IJs harbouring GFP within the total offspring was very variable from 0 to 100% with almost all the combinations (for the distribution, see Figure 4). For competition resulting from infection with symbiotic IJs, most of the IJs emerging from the insects (90 ± 22%) harboured GFP labelled *X. nematophila* (Figure 4). The latter result suggested that in competition resulting from infection with symbiotic IJs, most of the offspring emerging from the insects belonged to the species *S. carpocapsae*.

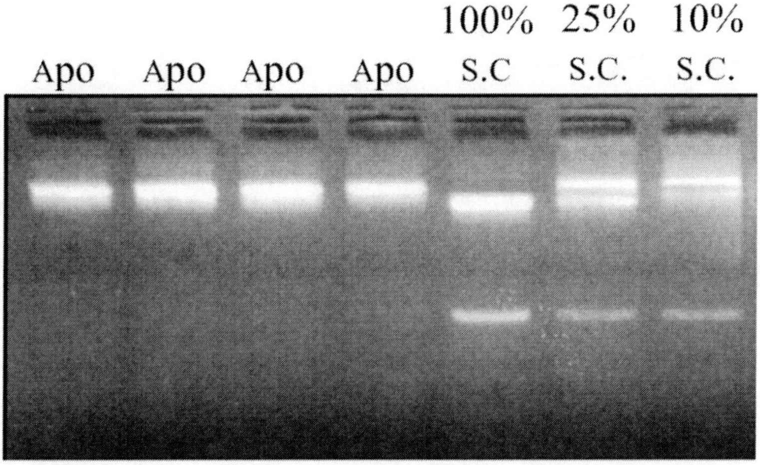

Figure 3. Migration of the PCR-RFLP products on agarose gel. The wells annotated (Apo) contained PCR-RFLP products coming from DNA extraction of pools of 500 IJs having emerged from competitions between aposymbiotic nematodes without the injection of bacteria. The well annotated 100% s.c contained 500 IJs of *S. carpocapsae*. The wells annotated 10% s.c and 25% s.c. respectively contained 10% of *S. carpocapsae* and 90% of *S. scapterisci* and 25% of *S. carpocapsae* and 75% of *S. scapterisci*.

DISCUSSION

Many invertebrates are associated with prokaryotes that provide them with the ability to live in extreme environments [15, 16, 17]. The most documented symbioses are those where symbionts provide their hosts with new abilities to take advantage of poorly nutritive resources [15, 18, 19, 20]. Nevertheless, symbionts can play other important roles on diverse life history traits of their hosts, for example, the bacterium *Vibrio* which provides its sepiolids host with a better discretion towards its predators [21, 22]. We already knew that *Xenorhabdus* symbionts provided *Steinernema* nematodes with diverse adaptations to their environment: (i) they produce virulence factors in order to kill the insect [23, 24, 25], (ii) they lead to a better access to the insect biomass and are themselves food for the nematodes [3, 4], (iii) they produce antibiotics which counteract the multiplication of both distant and

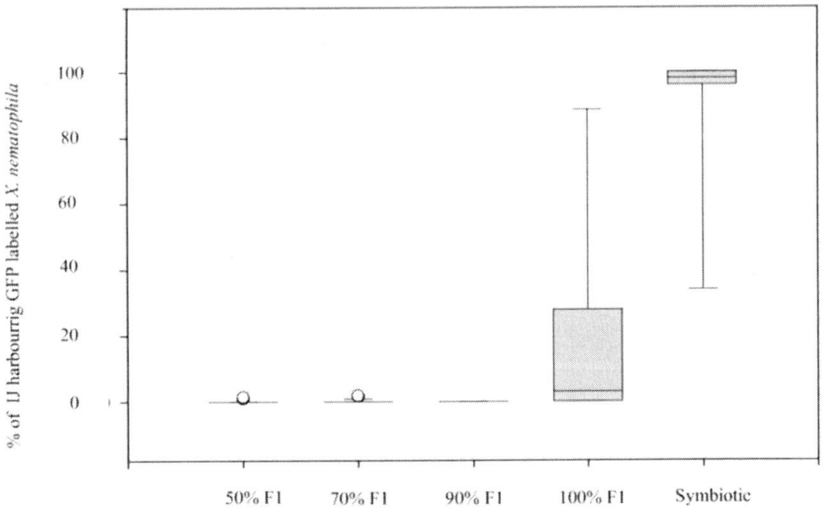

Figure 4. Percentile distribution of the percentages of IJs harbouring GFP labelled *X. nematophila*. Results are given as box plots, where the horizontal line indicate the median (50th of the data), the bottom and the top of the box indicate the first quartiles (25th of the data) and the third quartiles (75th of the data). The whiskers the range of the data (10thof the data and 90th of the data). Others dots are outliers which are under the 10th percentile of the data or over 90th percentile of the data. The abscissa shows the condition of the initial infection.

closely related bacteria [26, 27, 28, 29]. The goal of this study was to infer the abilities of *Xenorhabdus* to modulate the progression and outcome of interspecific competition between *Steinernema* species. For this, we made experimental competitions between *S. carpocapsae* and *S. scapterisci* and their bacterial symbionts. We assessed several parameters of these competitions: (i) the proportion of each symbiont (i.e. *X. nematophila* and *X. innexi*) within the hemolymph of the insect after delay of multiplication, (ii) the ratio of *S. carpocapsae* among the adults found in the insects and (iii) the offspring of each nematode species after competition. The assessment of bacterial multiplication within insect hemolymph showed that *X. innexi* appeared more competitive than *X. nematophila*. Moreover, in insects infected with aposymbiotic IJs and in which 50% of each *Xenorhabdus* strains were initially injected, we found *X. innexi* highly dominant within the hemolymph after 48 h of multiplication (Figure 1). Nevertheless, the ability of *X. innexi* to outcompete *X. nematophila* was not sufficient enough in the symbiotic competition where *S. scapterisci* IJs brought 700-fold less of their symbionts than *S. carpocapsae* in the insects

[3]. In competition resulting from infection of insects with symbiotic IJs, our results showed that such a competition took place within the insects where only *X. nematophila* proliferated.

In this study, we studied both (i) the development of nematodes within the insects and (ii) their ability to reproduce in them. Regarding the first aspect, we showed that the nematode's development into adults differed significantly for (i) competition resulting from infection with aposymbiotic IJs of both nematode species without injection of bacteria and (ii) competition resulting from infection between symbiotic IJs of both nematode species. The competition between aposymbiotic nematodes (without injection of bacteria) allowed evaluation of the intrinsic ability of each nematode species to outcompete the other one without the help of its symbiotic *Xenorhabdus*. In such a situation, we showed that (i) significantly fewer nematodes developed into adults and that (ii) *S. scapterisci* was highly dominant in both males and females found in the insects. These results showed that *S. scapterisci* had a better competitive ability without its *Xenorhabdus* symbiont than did *S. carpocapsae* (Figure 2). On the contrary, in the competition between symbiotic nematodes, *S. carpocapsae* was highly dominant in adults (Figure 2). Experiments in which the amount of each *Xenorhabdus sp.* initially introduced in the insect was experimentally modified showed that the proportion of adults from each nematode species in the insects was directly linked to the proportion of their bacterial symbionts within the insect hemolymph. (Figure 2). The more *X. nematophila* was dominant within the insect hemolymph, the more the proportion of *S. carpocapsae* in adult worms found in the insects increased. The analyses of the offspring emerging from the insects showed that competition resulting from infection with aposymbiotic IJs without injection of bacteria almost only produced IJs belonging to the species *S. scapterisci*. On the contrary, and in accordance with the results obtained on adults development, almost all the IJs emerging from the competition resulting from infection with symbiotic IJs belonged to the species *S. carpocapsae*.

Regarding the offspring production of each nematode species, the experimental modification of the bacterial symbiotic environment within the insects led to surprising results. In all insects where *X. innexi* was injected [even at the lowest dose (10% of the bacteria injected], almost no *S. carpocapsae* were found in the offspring. The inability of *S. carpocapsae* to reproduce in such situations was certainly due to the antagonistic effect of *X. innexi* toward this nematode which has already been demonstrated. Indeed, previous studies had shown that aposymbiotic *S. carpocapsae* were unable to reproduce in insects where only *X. innexi* was injected [5] but that the co-injection of *X. nematophila* with *X. innexi* led to a partially re-established reproduction of *S. carpocapsae* [13]. In the present study, even competition situations where *X. nematophila* co-infected the insects with *X. innexi* gave almost no offspring for *S. carpocapsae.* In such competition situations, it seems that *X. nematophila* was not able to counteract the effect. We can thus postulate that the intrinsic ability of *S. scapterisci* to outcompete by itself *S. carpocapsae* reinforces the antagonistic effect of *X. innexi* on *S. carpocapsae*'s reproduction. In such conditions, *S. scapterisci* and *X. innexi* acted in synergy to outcompete *S. carpocapsae* and *X. nematophila* leading to the absence of reproduction for the latter nematode. On the contrary, in the competition situation which was closest to the natural situation (i.e. symbiotic nematodes), the association *S. carpocapse-X. nematophila* totally outcompeted the association *S. scapterisci-X. innexi.* The outcome of the competition between symbiotic nematodes in favour of *S. carpocapsae* is quite understandable since this competition occurred in a bacterial environment constituted exclusively of *X. nematophila.* This dominance of *X. nematophila* within the insect hemolymph is due to the fact that *S. scapterisci* is poorly associated with its symbiont compared to *S. carpocapsae* (700-fold less cell transported) [3]. Thus, dramatically less cells of *X. innexi* than *X. nematophila* were released by the worms within the insect hemolymph.

CONCLUSION

This study shows that the bacterial symbionts *Xenorhabdus* modify the progress and outcome of interspecific competitions between their *Steinernema* associated nematodes. This suggests that *Xenorhabdus* not only provides *Steinernema* with access to food sources but also furnishes new abilities to deal with biotic parameters such as competitors. Thus, the evolution of the specific association between *Steinernema* and *Xenorhabdus* in environments where interspecific competitions occurred frequently should lead to the selection for an increased number of bacterial cells retained by each nematode in order to outcompete other specific nematodes-bacteria complexes. A previous study has reported that there is dramatically different bacterial retention between some species of *Steinernema* [3]. This bacterial retention variability could be linked to differences in the biotic environments where each symbiotic association evolves. Indeed, *S. carpocapsae* is a worldwide distributed nematode contrary to *S. scapterisci* which is only found in South-America [30, 31, 32]. Moreover, *S. carpocapsae* is supposed to have a wider insect-host range than *S. scapterisci* [33]. These specific ecological traits of each *Steinernema* species suggest that *S. carpocapsae* could have evolved under a higher competition pressure than *S. scapterisci.* This evolutionary constraint could have induced a selection toward a tighter relationship (i.e. more bacterial cells by IJ) in the *S. carpocapsae-X. nematophila* association than in the *S. scapterisci-X. innexi* association.

METHODS

Insects, nematodes and bacteria

The two *Steinernema* species naturally associated with their *Xenorhabdus* (naturally symbiotic nematodes) were established in the laboratory as soon as they were sampled by successive experimental infections of the last instar of the wax moth *Galleria mellonella*. The aposymbiotic and symbiotic IJs used for experiments were always freshly emerged from insects. Insect hosts were reared in the dark in aired plastic boxes at 28°C, 65% RH, on a diet of pollen and wax. Aposymbiotic IJs (i.e. without symbiont) of the species *S. carpocapsae* and *S. scapterisci* were obtained by disinfecting nematode eggs with a bleach solution as described previously [34]. The symbiotic IJs of *S. carpocapsae* associated with GFP (Green Fluorescent Protein) labelled *X. nematophila* (strain F1D3) were produced as described previously [34]. The GFP labeling is based on a plasmid expression which is very stable even if it is not recombined in the bacterial chromosome.

Bacterial suspensions were prepared by transferring a single colony of one bacterial strain to 5 ml of Luria-Bertani broth for liquid culture incubated at 28°C for 15 h. 100 μl of this liquid subculture was used to perform a culture to reach an optical density of 0.7 (600 nm wavelength). Before inoculation into the insect, the number of bacterial cells in each culture was counted with a Thoma cell and diluted in order to obtain a suspension of 100 cells/μl. A control of the actual number of bacteria in the injected suspension was measured by plating it onto three NBTA plates [35]. These plates were incubated at 28°C for 48 h. Then, the colonies that grew on these plates were counted. An experiment was kept only if the number of colonies ranged from 1500 to 2500.

General settings

We performed three different competition experiments: (i) competition resulting from infection with aposymbiotic IJs without injection of bacteria into the insects, (ii) competition resulting from infection with aposymbiotic nematodes in insects followed by injection of variable proportions of their respective bacterial symbiont (*X. nematophila* and *X. innexi*) and (iii) competition resulting from infection with symbiotic IJs (i.e. IJs of nematodes naturally containing *Xenorhabdus* in their guts). In each of these competition experiments, we monitored (i) the proportion of each bacterial symbiont within the hemolymph 72 h post-infection with IJs, (ii) the proportion of males and females from each species within adult nematodes found in the insect and (iii) the proportion of each nematode species within the offspring (IJs). In order to be able to specifically distinguish IJs emerging from the cadaver, we labelled *S. carpocapsae* with its symbiotic bacteria, *X. nematophila*, hosting a plasmid expressing GFP. Previous studies showed that 96% of *S. carpocapsae* IJs were associated with *X. nematophila* and that *S. scapterisci* was unable to associate with this bacterial strain [34, 36, 37]. To be able to evaluate *S. carpocapsae*'s offspring within the total number of IJs emerging from an insect, we made two different sets of insects for each type of competition (except the competition between aposymbiotic nematodes without injection of bacteria): (i) one set of insects was infected with GFP labelled *X. nematophila* and (ii) one set with wild type of *X. nematophila*. (for sample sizes, see Table 1).

Table 1. Sample sizes for each type of experiment

Type of nematodes involved in competition	Proportion of *X. nematophila*/*X. innexi* injected into the insect	Number of insects used for the bacterial assessment	Number of insects used to assess the proportion of males of each species	Number of insects used to assess the proportion of females of each species	Number of insects used to assess the reproductive rate of each species	Total number of insects
Aposymbiotic nematodes	none	10	28	4	37	74
	0/100	10	29	5	15	54
	50/50	10*	46*	8*	33*	97
	70/30	10*	50*	8*	28*	96
	90/10	10*	52*	8*	28*	98
	100	10*	48*	8*	20*	86
Symbiotic nematodes	none	10*	63*	8*	37*	118

* half of these experiments were performed with GFP labelled *X. nematophila*

Experimental infection

Forty IJs of each species (aposymbiotic or symbiotic depending on the type of experiment) were counted and deposited into 1.5-ml Eppendorf tubes containing a filter paper. A last instar of *G. mellonella* was then introduced into each Eppendorf, and the nematodes and the insect-host were incubated together at 24°C during 24 h. For competition resulting from infection with aposymbiotic and symbiotic IJs, no bacteria were experimentally introduced into the insect. For competition resulting from infection with aposymbiotic IJs in insects with a controlled and variable bacterial symbiont environment, 24 h after infection with aposymbiotic IJs, insects were injected with different proportions of the two bacterial symbionts (*X. nematophila* and *X. innexi*) [13]. The different proportions tested were (i) 100% of *X. innexi*, (ii) 50% of each symbiont, (iii) 70% of *X. nematophila* and 30% of *X. innexi*, (iiii) 90% of *X. nematophila* and 10% of *X. innexi* and (iiiii) 100% of *X. nematophila* (for sample sizes, see table 1).

Proportion of each *Xenorhabdus* in insect's hemolymph 72 h post-infection with IJs

For infections made with aposymbiotic IJs followed by the injection of variable proportions of each symbiont, the proportion of each symbiont was assessed 72 h post-infection with aposymbiotic IJs (i.e. 48 h after the direct injection of *Xenorhabdus* cells into the insects). For infections made with symbiotic IJs, the same assessment was made 72 h post-infection with symbiotic IJs. To evaluate the proportion of each symbiotic bacterium within the insect, the hemolymph of each *Galleria mellonella* was collected using a syringe. The sampled hemolymph was diluted by 10^5 and 100 µl of the obtained suspension was plated out onto three NBTA plates. The plates were then incubated at 28°C during 48 h. After incubation, few cells from each colony were sampled with the help of sterile tooth picks and transferred in lines onto two NBTA plates containing control samples of *X. nematophila* and *X. innexi*. One of these plates was incubated at 28°C during 48 h and the other one at 37°C during 48 h. These two different incubation temperatures helped us to discriminate between the two *Xenorhabdus* strains. Indeed, both of these bacteria grew at 28°C but only *X. innexi* was also able to grow at 37°C [38]. The proportion of *X. nematophila* in the insect was then obtained by subtracting the number of colonies grown at 28°C from the number of colonies grown at 37°C and then dividing the result by the number of colonies grown at 28°C.

Assessment of nematode maturation in competition

120 h post-infection by IJs, the insects were dissected and adult nematodes were transferred to separate Eppendorfs containing sterile Ringer. Males were then mounted with a drop of water between slide and coverslip and the species was determined after spicule observation [39]. The proportion of *S. carpocapsae* within all males found in an insect was calculated (for the number of insect analysed, see table 1). For each tested condition, all nematode females (i.e. from 2 to 37 per insect) found in four insects were separately analysed by PCR-RFLP and assigned to one nematode species. In order to do that, DNA from each female was separately extracted as described previously [40]. Then, the ITS region was amplified as described previously [39]. 10 µl of PCR products were then incubated at 37°C with 0.5 µl of the restriction enzyme Hind III, 2 µl of enzyme buffer and 7.5 µl of

sterile water. Previous studies showed that Hind III cuts the amplified ITS fragment for *S. carpocapsae* and not for *S. scapterisci* [39]. After such analyses, the ratios of *S. carpocapsae* within nematode females found in each insect were calculated.

Assessment of the nematode offspring

The total number of offspring produced by each infection was separately harvested in 50-ml Falcon flasks two months after infection and stored at 8°C. The total number of IJs produced was then evaluated under binocular microscope, using 1 ml of the suspension taken from the Falcon flask on a grid drawn on a 6-cm Petri dish. Then, for each IJ emergence coming from the competitions performed with GFP labelled *X. nematophila*, we (i) examined the vesicle of 100 IJs by epifluorescence microscopy and (ii) made isolation of bacteria contained in 500 IJs as previously described by Sicard et al., 2003. The first experiment led to assign each IJ to one nematode species. Indeed, if GFP labelled *X. nematophila* were observed within their vesicles, the IJs were assigned to *S. carpocapsae*, if not to *S. scapterisci*. The second experiment led to check if all the clones of *X. nematophila* isolated from IJs expressed GFP. In case of competition resulting from infection with aposymbiotic IJs without injection of bacteria into the insects, the presence/absence of each nematode species in the offspring was evaluated by performing PCR-RFLP (as described above) with DNA extraction on four pools of 500 IJs coming from four different insects. Preliminary experiments showed that when IJs of *S. carpocapsae* were mixed with IJs of *S. scapterisci* three bands were observed after PCR-RFLP. Such an approach led to detect 10% of IJs of *S. carpocapsae* within pools containing 90% of *S. scapterisci*. We used this latter experiment as a standard in our experiment (see Fig 3).

Statistical analyses

In order to compare two non-parametric distributions, we used the Mann Withney test. To compare more than two non-parametric distributions, we performed a Kruskal-Wallis (KW) test followed by the pair wise comparison test of Noether [41].

ACKNOWLEDGEMENTS

Research in the GPIA and EMIP labs are funded by CNRS, INRA and IFREMER.

We would like to thank Alain Givaudan and Anne Lanois for the construction of the GFP labelled strain of *X. nematophila*, Céline Cavaillé for technical assistance and Marie-Claire Britton for language corrections. We also thank three anonymous reviewers for comments on our manuscript.

AUTHORS' CONTRIBUTIONS

MS designed experiments, carried out some of them, interpreted the data and wrote the first draft of the manuscript. JH carried out main experiments and revised the manuscript, NLB carried out some experiments and revised the manuscript, SP helped to the design of some experiments, NB revised the manuscript and CM participated to the design of the study, the interpretation of the data and revised the manuscript. All authors read and approved the final manuscript.

REFERENCES

1. Poinar GOJ: The presence of Achromobacter nematophilus in the infective stage of a Neoaplectana sp. (Steinernematidae: Nematoda). Nematologica. 1966, 12: 105-108.
2. Akhurst RJ: Neoaplectana species: Specificity of association with bacteria of the genus Xenorhabdus. Exp Parasitol. 1983, 55 (2): 258-263.
3. Sicard M, Le Brun N, Pages S, Godelle B, Boemare N, Moulia C: Effect of native Xenorhabdus on the fitness of their Steinernema hosts: contrasting types of interaction. Parasitol Res. 2003, 91 (6): 520-524.
4. Poinar G, Thomas GM: Significance of Achromobacter nematophilus Poinar and Thomas (Achromobacteriaceae: Eubacteriales) in the development of the nematode, DD-136 (Neoplectana sp. Steinernematidae). Parasitology. 1966, 56: 385-390.
5. Sicard M, Ferdy JB, Pagès S, Le Brun N, Godelle B, Boemare N, Moulia C: When mutualists are pathogens: an experimental study of the symbioses between Steinernema (entomopathogenic nematodes) and Xenorhabdus (bacteria). J Evol Biol. 2004, 17 (5): 985-993.
6. Sicard M, Ramone H, Le Brun N, Pagès S, Moulia C: Specialization of the entomopathogenic nematode Steinernema scapterisci with its mutualistic Xenorhabdus symbiont. Naturwissenschaften. 2005, 92 (10): 472-476.
7. Boemare N, Bonifassi E, Laumond C, Luciani J: Experimental study of the pathogenic action of the nematode Neoaplectana carpocapsae Weiser; gnotobiological research using the insect Galleria mellonella L. Agronomie. 1983, 3 (5): 407-415.
8. Boemare N, Givaudan A, Brehelin M, Laumond C: Symbiosis and pathogenicity of nematode-bacterium complexes. Symbiosis. 1997, 22 (1-2): 21-45.
9. Aguillera MM, Smart GC: Development, reproduction, and pathogenicity of Steinernema scapterisci in monoxenic culture with different species of bacteria. J Invertebr Pathol. 1993, 62 (3): 289-294.
10. Grewal PS, Matsuura M, Converse V: Mechanisms of specificity of association between the nematode Steinernema scapterisci and its symbiotic bacterium. Parasitology. 1997, 114 (5): 483-488. 1
11. Bonifassi E, Saux MFL, Boemare N, Lanois A, Laumond C, Smart G: Gnotobiological study of infective juveniles and symbionts of Steinernema scapterisci: A model to

clarify the concept of the natural occurrence of monoxenic associations in entomopathogenic nematodes. J Invertebr Pathol. 1999, 74 (2): 164-172.
12. Lengyel K, Lang E, Fodor A, Szallas E, Schumann P, Stackebrandt E: Description of four novel species of Xenorhabdus, family Enterobacteriaceae: Xenorhabdus budapestensis sp. nov., Xenorhabdus ehlersii sp. nov., Xenorhabdus innexi sp. nov., and Xenorhabdus szentirmaii sp. nov. Syst Appl Microbiol. 2005, 28 (2): 115-122.
13. Sicard M, Tabart J, Boemare N, Thaler JO, Moulia C: Effect of phenotypic variation in Xenorhabus nematophila on its mutualistic relationships with the entomopathogenic nematode Steinernema carpocapsae. Parasitology. 2005, 131: 687-694.
14. Thaler JO: Purification and characterization of a particulate bacteriocin of Xenorhabdus nematophilus strain F1. Proc Int Colloq Invertebr Pathol Microb Control. 1994, 6: 3-4.
15. Douglas AE: Mycetocyte symbiosis in Insects. Biol rev. 1989, 64: 409-434.
16. Saffo MB: Invertebrates in endosymbiotic associations. Amer Zool. 1992, 32: 557-565.
17. Dimijian GG: Evolving together: the biology of symbiosis, part 1. Proc (Bayl Univ Med Cent). 2000, 13 (3): 217-226.Google Scholar
18. Cavanaugh CM, Gardiner SL, Jones ML, Jannasch HW, Waterbury JB: Prokaryotic cells in the hydrothermal vent tube worm Riftia Pachyptila Jones Possible chemoautotrophic symbionts. Science. 1981, 213: 340-342.
19. Baumann P, Chi-Yung L, Baumann L, Rouhnakhsh D, Moran NA, Clark MA: Mutualistic associations of aphids and prokaryotes: bology of the genus Buchnera. Appl Environ Microbiol. 1995, 61: 1-7.
20. Graf J: Symbiosis of Aeromonas and Hirudo medicinalis, the medicinal leech. ASM news. 2000, 66 (3): 147-153.
21. Nishiguchi M, Ruby E, McFall-Ngai M: Competitive dominance among strains of luminous bacteria provides an unusual form of evidence for parallel evolution in sepiolid squid-Vibrio symbioses. Appl Environ Microbiol. 1998, 64 (9): 3209-3213.
22. Visick KL, McFall-Ngai MJ: An exclusive contract: specificity in the Vibrio fischeri-Euprymna scolopes partnership. J Bacteriol. 2000, 182 (7): 1779-1787.
23. Forst S, Dowds B, Boemare N, Stackebrandt E: Xenorhabdus and Photorhabdus spp.: Bugs that kill bugs. Annu Rev Microbiol. 1997, 51: 47-72. 10.1146/annurev.micro.51.1.47.
24. ffrench-Constant R, Waterfield N, Daborn P, Joyce S, Bennett H, Au C, Dowling A, Boundy S, Reynolds S, Clarke D: Photorhabdus: towards a functional genomic analysis of a symbiont and pathogen. FEMS Microbiol Rev. 2003, 26 (5): 433-456.
25. Duchaud E, Rusniok C, Frangeul L, Buchrieser C, Givaudan A, Taourit S, Bocs S, Boursaux-Eude C, Chandler M, Charles JF, Dassa E, Derose R, Derzelle S, Freyssinet G, Gaudriault S, Medigue C, Lanois A, Powell K, Siguier P, Vincent R, Wingate V, Zouine M, Glaser P, Boemare N, Danchin A, Kunst F: The genome sequence of the entomopathogenic bacterium Photorhabdus luminescens. Nat Biotechnol. 2003, 21 (11): 1307-1313.
26. Akhurst RJ: Antibiotic activity of Xenorhabdus spp., bacteria symbiotically associated with insect pathogenic nematodes of the families Heterorhabditidae and Steinernematidae. J Gen Microbiol. 1982, 128 (12): 3061-3065.
27. Boemare NE, Boyer-Giglio MH, Thaler JO, Akhurst RJ, Brehélin M: Lysogeny and bacteriocinogeny in Xenorhabdus nematophilus, and other Xenorhabdus spp. Appl Environ Microbiol. 1992, 58 (9): 3032-3037.
28. Maxwell PW, Chen GH, Webster JM, Dunphy GB: Stability and activities of antibiotics produced during infection of the insect Galleria mellonella by two isolates of Xenorhabdus nematophilus. Appl Environ Microbiol. 1994, 60 (2): 715-721.
29. Thaler JO, BoyerGiglio MH, Boemare NE: New antimicrobial barriers produced by Xenorhabdus spp. and Photorhabdus spp. to secure the monoxenic development of entomopathogenic nematodes. Symbiosis. 1997, 22 (1-2): 205-215.

30. Hominick WM: Biogeography. Entomopathogenic Nematology. Edited by: Gaugler R. 2002, Wallingford, New York , CABI Publishing UK, 1: 115-143.
31. Stock S: Presence of Steinernema scapterisci Nguyen et Smart parasitizing the mole cricket Scapteriscus borellii in Argentina. Nematol medit. 1992, 20 (2): 163-165.
32. Stock SP, Gardner SL, Wu FF, Kaya HK: Characterization of two Steinernema scapterisci populations (Nemata: Steinernematidae) using morphology and random amplified polymorphic DNA markers. J Helminthol Soc Wash. 1995, 62 (2): 242-249.
33. Laumond C, Mauléon H, Kermarrec A: Données nouvelles sur le spectre d'hôtes et le parasitisme du nématode entomophage Neoaplectana carcpocapsae. Entomophaga. 1979, 24 (1): 13-27.
34. Sicard M, Brugirard-Ricaud K, Pages S, Lanois A, Boemare NE, Brehélin M, Givaudan A: Stages of infection during the tripartite interaction between Xenorhabdus nematophila, its nematode vector, and insect hosts. Appl Environ Microbiol. 2004, 70 (11): 6473-6480.
35. Akhurst RJ, Boemare NE: A numerical taxonomic study of the genus Xenorhabdus (Enterobacteriaceae) and proposed elevation of the subspecies of X. nematophilus to species. J Gen Microbiol. 1988, 134 (7): 1835-1845.
36. Vivas EI, Goodrich-Blair H: Xenorhabdus nematophilus as a model for host-bacterium interactions: rpoS is necessary for mutualism with nematodes. J Bacteriol. 2001, 183 (16): 4687-4693.
37. Martens EC, Goodrich-Blair H: The Steinernema carpocapsae intestinal vesicle contains a subcellular structure with which Xenorhabdus nematophila associates during colonization initiation. Cell Microbiol. 2005, 7 (12): 1723-1735.
38. Tailliez P, Pagès S, Ginibre N, Boemare N: New insight of the diversity in the Xenorhabdus genus including the description of ten novel species. Int J Syst Evol Microbiol. 2006,
39. Hominick WM, Briscoe BR, del Pino FG, Heng JA, Hunt DJ, Kozodoy E, Mracek Z, Nguyen KB, Reid AP, Spiridonov S, Stock P, Sturhan D, Waturu C, Yoshida M: Biosystematics of entomopathogenic nematodes: current status, protocols and definitions. J Helminthol. 1997, 71 (4): 271-298.
40. Sicard M, Desmarais E, Lambert A: Molecular characterisation of Diplozoidae populations on five Cyprinidae species: consequences for host specificity. C r Acad Sci Sér 3. 2001, 324: 709-717.
41. Scherrer B: Biostatistique. Edited by: Morin G. 1984, Paris, France

CHAPTER 14

Parasitic Nematode Immunomodulatory Strategies: Recent Advances and Perspectives

Dustin Cooper and Ioannis Eleftherianos *

Insect Infection and Immunity Lab., Department of Biological Sciences, Institute of Biomedical Sciences, The George Washington University, Washington DC 20052, USA

ABSTRACT

More than half of the described species of the phylum Nematoda are considered parasitic, making them one of the most successful groups of parasites. Nematodes are capable of inhabiting a wide variety of niches. A vast array of vertebrate animals, insects, and plants are all identified as potential hosts for nematode parasitization. To invade these hosts successfully, parasitic nematodes must be able to protect themselves from the efficiency and potency of the host immune system. Innate immunity comprises the first wave of the host immune response, and in vertebrate animals it leads to the induction of the adaptive immune response. Nematodes have evolved elegant strategies that allow them to evade, suppress, or modulate host immune responses in order to persist and spread in the host. Nematode immunomodulation involves the secretion of molecules that are capable of suppressing various aspects of the host immune response in order to promote nematode invasion. Immunomodulatory mechanisms can be identified in parasitic nematodes infecting insects, plants, and mammals and vary greatly in the specific tactics by which the parasites modify the host immune response. Nematode-derived immunomodulatory effects have also been shown to affect, negatively or positively, the outcome of some concurrent diseases suffered by the host. Understanding nematode immunomodulatory actions will potentially reveal novel targets that will in turn lead to the development of effective means for the control of destructive nematode parasites.

Keywords: nematode; parasitism; immunomodulation; immune evasion; host innate immunity; adaptive immunity

1. INTRODUCTION

Nematodes, or roundworms, are a group of parasites that infect a wide range of hosts. Plants, insects and vertebrate animals all serve as hosts to several parasitic nematode species. Successful parasitization of a host requires both the penetration and invasion of host tissues, as well as persistence within the host's body. The continued success of nematodes as parasites is, in large part, due to a multifaceted approach to survival within the host [1,2,3]. During the infection process, nematodes face a powerful innate immune system that regulates a variety of immune effector molecules. Survival and replication of nematode parasites in the host is achieved by suppressing, evading, or modulating the activity of certain host immune signaling pathways, as well as the immune functions that these pathways control [4,5].

Nematode immunomodulation is a strategy that promotes the survival of nematode parasites in the host by altering the activation of the host immune response during infection. This tactic involves the synthesis and release of molecules as excreted-secreted products into the host tissue during the invasion process. Molecules involved in nematode immunomodulation vary greatly depending on their mode of action and the type of host [6,7,8]. They include proteinaceous molecules, such as serpins and proteases, and small RNAs that are capable of mimicking host miRNAs [9,10,11,12,13,14]. While the effects of nematode immunomodulatory molecules are studied with respect to their contribution to nematode survival and killing of the host, they are usually broad-spectrum, and can impact host responses to concurrent diseases [15]. For example, molecules resulting in anti-inflammatory responses have been implicated in low prevalence of inflammatory bowel disease in populations with high incidence of nematode infections [16]. Conversely, nematode parasitism concurrent to malarial infection may modulate host immune responses against the latter [17].

While immunomodulation mechanisms are widespread among nematode parasites, very rarely are the mechanisms used against vastly different hosts taken together to display the equally broad variation observed among them [18]. This review summarizes recent progress on the identification and characterization of mechanisms used by parasitic nematodes to modulate various aspects of the host innate immune response of insects, plants, and vertebrate animals. Such information is invaluable for devising novel approaches to provide efficient protection to harmful parasitic nematode of humans or to determine mechanisms for enhancing the parasitic activity of beneficial nematode species that are used for the biological control of noxious insect pests.

2. NEMATODE IMMUNOMODULATION IN INSECTS

Entomopathogenic nematodes are a group of roundworms that are able to infect and kill insects. To aid in this, they have developed an obligate mutualistic association with certain bacterial symbionts. This relationship is highly specific and has been a useful model to study the evolutionary origins of mutualism [19,20]. Research on entomopathogenic nematodes has focused on two genera: *Steinernema*, which associates with *Xenorhabdus* bacteria, and *Heterorhabditis*, which associates with *Photorhabdus* bacteria. Entomopathogenic nematodes provide shelter and serve as a vector for the bacteria, carrying them from one host to another. In return, once the nematodes have invaded an insect host, the bacteria are regurgitated and released within the insect hemocoel. The bacteria then secrete a cocktail of toxins and virulence factors and replicate rapidly in various tissues of the insect, the carcass of which is finally consumed by the nematode parasites [21,22,23].

Infection by nematode-bacteria complexes does not, however, go unnoticed by the insect host. The insect innate immune system has in place a series of mechanisms that promote the detection of the mutualistic partners in order to restrain their dissemination in the host [3,24]. Humoral and cellular immune responses make up the majority of the insect innate immune response [25,26]. Humoral immune responses are responsible for activating genes necessary for synthesizing and secreting antimicrobial peptides from the fat body (equivalent to the mammalian liver) into the hemolymph (insect blood) [27,28,29]. Cellular immune responses are regulated by the function of hemocytes (insect blood cells) [30]. During infection, hemocytes proliferate and contribute to the formation of cell aggregates, phagocytosis, and encapsulation, among other defense mechanisms [31,32]. The melanization response, which is regulated by the enzyme phenoloxidase, involves the synthesis and deposition of melanin that is directed against the invading microbes [33,34].

During the course of infection within an insect host, nematodes must evade, suppress, or modulate the insect immune response in order to survive, expel their mutualistic bacteria, and complete several life-cycles [35]. Established mechanisms that nematodes employ to modulate the insect humoral and cellular immune reactions involve the production of molecules that are secreted by the parasites during the infection process. Proteases have been identified as a major contributor of host immunomodulation by entomopathogenic nematodes. These molecules can also aid in nematode invasion of the host and digestion of proteins within the infected insect [35]. Previous research has led to the purification and characterization of a trypsin-like serine protease that is secreted by *S. carpocapsae* infective juveniles during infection of *Galleria mellonella* larvae [36] (Figure 1). This protease was shown to possess a function in regulating immunomodulation and host immune inhibition, and thus protecting the nematode from the potent effects of the insect immune system. Following characterization, in vitro analysis of the trypsin-like serine protease revealed its prophenoloxidase inhibitory activity in a dose-dependent manner. In addition, incubation of hemocytes with the purified protease resulted in severe changes to the morphology of the cells. Untreated hemocytes were found to be flat, while treated hemocytes appeared round or spherical. The immunomodulatory activity also involved substantial decrease in the ability of hemocytes to perform cell spreading, a process whereby cells adhere to a foreign microbe and spread along the surface in order to phagocytose or encapsulate their target [37]. Further tests using *G. mellonella*-derived hemocytes co-incubated with three different entomopathogenic nematode species indicated that hemocytes were able to respond to *H. bacteriophora*, but not to *S. carpocapsae* or *S. glaseri*, suggesting that this kind of immunomodulation of insect hemocytes is probably specific to *Steinernema* nematodes.

S. carpocapsae nematodes also produce another molecule with immunomodulatory properties. The function of Sc-SRP-6, a novel serpin-like inhibitor, was investigated because of its elevated expression levels during the invasive stage of the nematode parasite [38], which suggested a potential role in nematode virulence (Figure 1). Functional analysis of Sc-SRP-6 revealed two roles in protecting the invading nematodes from the host immune response. First, Sc-SRP-6 inhibits the hydrolysis of food particles by reducing trypsin activity of the insect digestive enzymes. This finding indicates a role in the protection of *S. carpocapsae* from proteolytic enzyme activity. Second, Sc-SRP-6 was found to interfere with clot formation in infected insects [38,39]. In particular, binding assays revealed that Sc-SRP-6, together with three hemolymph plasma proteins, form complexes that prevent the incorporation of melanin into the clot matrix, an essential step in the encapsulation and nodulation immune processes [32,40]. Thus, this serpin-like inhibitor secreted by *S. carpocapsae* provides protection against insect digestive enzymes and participates in the inhibition of the melanization and encapsulation reactions against the nematodes.

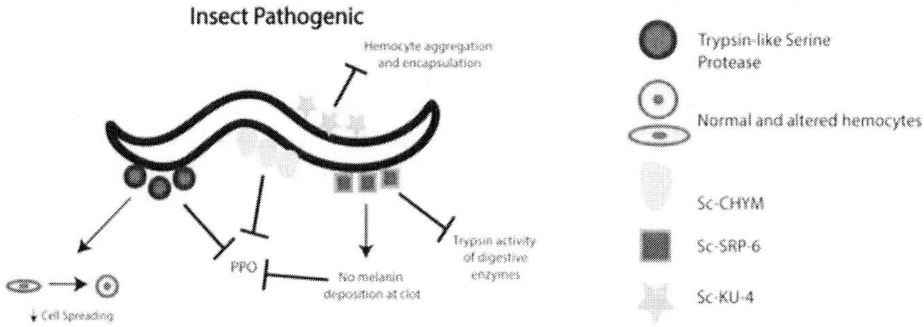

Figure 1. Insect Pathogenic nematodes modulate the phenoloxidase response via the secretion of a trypsin-like serine protease, Sc-CHYM, and Sc-SRP-6. The identified Trypsin-like serine protease has also been shown to limit spreading of insect hemocytes. Sc-SRP-6 modulates the host immune response by altering the activity of digestive enzymes. Sc-KU-4 acts by modulating the ability of hemocytes to aggregate or encapsulate foreign microbes.

Additional compounds with immunomodulatory characteristics have been identified and functionally characterized by examining the mixture of molecules produced by infective stage *S. carpocapsae* nematodes. One such compound, a chymotrypsin serine protease virulence factor designated "Sc-CHYM", disrupts the activation of prophenoloxidase, which interrupts the melanization response and as a consequence prevents nematode encapsulation by hemocytes [41] (Figure 1). Another molecule, Sc-KU-4, has been identified as a Kunitz-family protease inhibitor [42] (Figure 1). The inhibitor was found to target insect immune recognition proteins in order to inhibit hemocyte aggregation, as well as the encapsulation of foreign particles. While these immune processes were functionally inhibited, it was also observed that Sc-KU-4 was not able to prevent their initial activation.

3. NEMATODE IMMUNOMODULATION IN PLANTS

Plant-parasitic nematodes are capable of infecting and eliciting an innate immune response from their host plants [43,44]. Because of their drastic impact on food crops worldwide, a significant amount of research has gone into investigating the interactions between nematode strategies and molecules that are produced by the parasites during infection, and plant immune functions that are activated in response to nematode invasion [45]. Plant parasitic nematodes can be sedentary or migratory, with the latter causing extensive cellular damage to roots. It is the sedentary Heteroderidae family of nematodes, however, that are responsible for the majority of damage caused to food crops. The Heteroderidae family contains the cyst and root-knot nematodes, both of which employ similar strategies to parasitize plants. Motile, infective juveniles penetrate the root and induce permanent feeding sites. Once the nematodes begin feeding near the vascular tissue, they become sedentary and progress to the adult stage, during which they produce eggs [46]. The plant innate anti-nematode immune response relies on pattern recognition receptors to detect invading parasites in the extracellular matrix. Upon nematode detection, receptor-like kinases activate immune signaling pathways, giving rise to a variety of chemical defenses [47]. A common host defense is the programmed cell death of tissues surrounding the site of infection, preventing the spread of invading pathogens [48]. Modulation of host immunity by parasitic nematodes is necessary to prevent the programmed cell death of tissues surrounding the nematode feeding site, which would halt the growth and development of the parasite. The secretion of effectors capable of disrupting the receptor-like kinases or immune signaling pathways is an effective approach used by sedentary nematodes to aid in infection [49].

Globodera rostochiensis, a potato cyst nematode, secretes a venom allergen-like protein, *Gr-VAP1*, concurrent with the secretion of enzymes that are responsible for the breakdown of the plant cell wall [50] (Figure 2). Because venom allergen-like proteins are conserved among a wide range of parasitic nematodes and are produced during the initial stages of parasitism, Gr-VAP1 was considered a potential modulator of the plant innate immune response. To investigate this, the effects of *Gr-VAP1* silencing on *G. rotochiensis* invasion and the heterologous expression of this protein in plants were observed. Knockdown of the venom allergen-like protein significantly impaired the ability of *G. rotochiensis* to successfully invade plant tissue, while expression of the protein in plants resulted in increased susceptibility of the plant to different types of pathogens. These results provided additional proof that Gr-VAP1 not only plays a role in, but also is required for nematode immunomodulation in plants. Further investigation indicated that Gr-VAP1 associates with proteases arising from the destruction of plant cell walls. These proteases are responsible for regulating signaling to pattern recognition receptors to activate immune signaling pathways, leading to programmed cell death of affected tissues. Through the secretion of Gr-VAP1, plant parasitic nematodes are capable of modulating this mechanism such that pattern recognition receptors are unable to recognize fragments of destroyed cell wall, thus preventing the induction of the plant innate immune response [50].

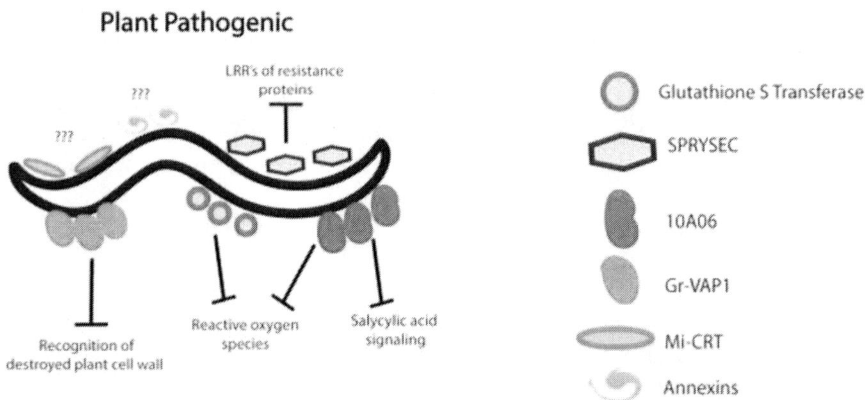

Figure 2. Plant Pathogenic nematodes use a multitude of strategies for host immunomodulation. The venom allergen-like proteins Gr-VAP1 can disrupt the host's initial recognition of nematode invasion. Other proteins such as glutathione S-transferase or 10A06 interfere with reactive oxygen species produced in response to recognition of an invading nematode. 10A06 also interrupts salicylic acid signaling. SPRYSEC proteins interact resistance proteins to alter pathogen recognition. Some molecules, such as MiCRT, or groups of molecules, such as the annexins, are implicated in immunomodulation; however, their mechanism of action remains unknown.

Modern advances in genomics and bioinformatics have allowed researchers to probe into the genomes of plant-parasitic nematodes in order to identify potential effectors produced and secreted into host plant tissues [51]. By investigating gene expression patterns in *G. rostochiensis*, researchers have discovered a novel gene family coding for proteins capable of interacting with plant resistance proteins [52]. One of the identified proteins secreted by these nematodes, called Secreted SPRY Domain-Containing Protein (SPRYSEC), was shown to interact with the leucine-rich repeat region of a coiled-coil nucleotide-binding pathogen resistance protein in tomato plants (Figure 2). It was hypothesized that this interaction promotes nematode virulence by modulating the activity of host immune receptor proteins. Transcriptome analysis of the root-knot nematode *Meloidogyne incognita* has revealed a group of genes that are up-regulated in the parasitic juveniles [53]. RNA-interference mediated silencing of the gene *Mi-gsts-1*, which encodes a glutathione S-transferase, revealed that the protein is produced in, and secreted by, the subventral oesophageal glands (Figure 2). The secreted enzyme is thought to modulate plant defenses in such a way that the nematode is protected from the negative effects of reactive oxygen species produced in response to parasitic attack. Similarly, the secreted effector protein 10A06 from the cyst-nematode *Heterodera schachtii* was found to protect the nematode from antioxidant plant-defenses during the infection process [54] (Figure 2). This protein is also thought to modulate salicylic acid defense signaling, thereby promoting nematode infectivity toward *Arabidopsis thaliana* plants. Although its specific mode of action has yet to be clarified, Mi-CRT, a calreticulin secreted by *M. incognita*, has been found to be important in nematode infectivity because silencing the expression of Mi-CRT reduces *M. incognita* virulence, whereas overexpression of this protein in *A. thaliana* increases sensitivity to the nematodes (Figure 2). It is thought that Mi-CRT is able to control the induction of plant defenses regulated by pathogen-associated molecular patterns [55]. More recent transcriptome analysis of infective juvenile *H. avenae* has shown the up-regulation of genes encoding annexins, a multigene family of well-known calcium-regulated proteins, capable of binding the phospholipids of cell membranes (Figure 2). Annexins are expressed in the oesophageal and sub-ventral secretory gland cells of plant parasitic nematodes and are potentially involved in nematode immunomodulation processes [56,57]. The same study also reports the identification of genes coding for fatty acid and retinol-binding proteins which may inhibit jasmonic acid synthesis and consequently restrict lipid-related plant defenses [56,58].

4. NEMATODE IMMUNOMODULATION IN VERTEBRATES

Parasitic nematodes constitute one of the major threats to human health, causing diseases of major socioeconomic importance worldwide. Recent estimates indicate that more than one billion people are infected with parasitic nematodes, and more than a dozen species routinely parasitize humans [59,60]. The vertebrate innate immune response to parasitic nematodes is complex and varies greatly with the species and severity of nematode infection. Invasion of the host tissues by parasitic nematodes activates the complement system, a collection of proteins responsible for identifying pathogens and directing the innate immune response. Leukocytes are recruited to the site of infection and are responsible for enhancing the inflammatory response and release of cytokines, among a variety of other processes [61]. Mast cells and eosinophils participate in the innate immune response to nematode infection due to their role as potent effectors of a wide variety of cytokines and chemokines. Direct activation of leukocytes occurs via host tissue damage caused by the invading parasites [62]. Recruitment of other types of leukocytes, such as neutrophils, macrophages, basophils, innate lymphoid cells and dendritic cells, leads to the production of toxic free radicals, phagocytosis, and eventually, the development of an adaptive immune response [63,64]. However, similarly to insects and plants, certain vertebrate-parasitic nematodes have evolved immunomodulatory mechanisms that promote nematode infection by interrupting one or more effectors of the innate immune response [65].

Cysteine protease inhibitors, or cystatins, produced by parasitic nematodes have been identified as a major class of molecules with immunomodulatory properties (Figure 3). Cystatins act in a multitude of ways to interfere with host effector mechanisms to prevent nematode clearance. Cystatins secreted by nematodes have been shown to inhibit two classes of cysteine proteases: legumains, which are responsible for antigen processing and presentation, and cathepsin L and S, which are involved in the processing of polypeptides. Inhibition of legumains is especially relevant to nematode infection because it prevents the generation of MHC class II molecules that recognize nematode antigens, thus preventing the induction of an adaptive immune response [66]. Nematode-secreted cystatins are also capable of modulating the host cytokine response in order to create favorable conditions for the invading parasites. More precisely, cystatins strongly enhance the production of the anti-inflammatory cytokine IL-10, which in turn restricts the activity of T cell-mediated responses [67]. A cystatin secreted by *Heligmosomoides polygyrus* modulates the activity of dendritic cells during the anti-nematode immune response in mice [68]. Interestingly, suppression of dendritic cell responses is a common feature of many mammalian parasites [69]. Dendritic cells exposed to recombinant cystatin expressed fewer MHC-II molecules as well as reduced CD40 and CD86, two proteins necessary for T-cell differentiation. The interference with antigen processing generated by the cystatin ultimately results in an immune response modulated in favor of nematode infection. Another cystatin secreted by *Acanthocheilonema vitae* modulates the proinflammatory effects of microglia by altering the expression of key cytokines [70]. Recombinant cystatin induces downregulation of iNOS as well as COX-2, both proinflammatory cytokines. It also induces the upregulation of IL-10, further promoting an anti-inflammatory effect upon microglia stimulated with lipopolysaccharide.

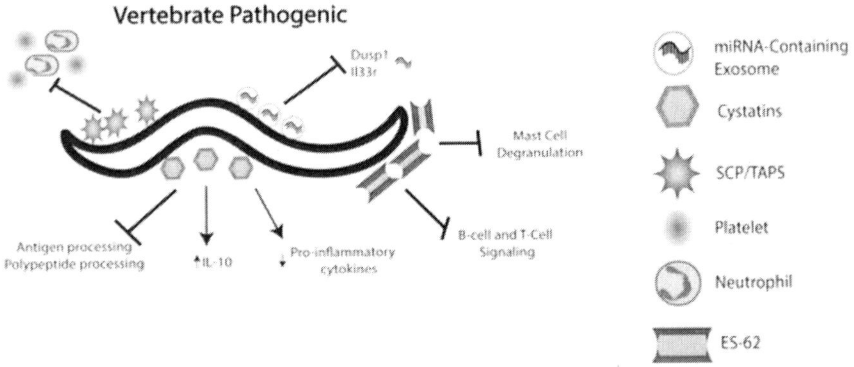

Figure 3. Vertebrate Pathogenic nematodes rely heavily on secreted cystatins for host immunomodulation. Cystatins influence the regulation of host cytokine production, as well as the development of adaptive immune responses by altering antigen and polypeptide processing. SCP/TAPS are believed to assist in inhibiting platelet and neutrophil activity. ES-62 alters a number of immune functions including B and T-Cell activation/proliferation via receptor signaling inhibition and the inhibition of mast cell degranulation through the formation of a complex with TLR-4. Recently, miRNA-containing exosomes secreted by some nematode species have been linked to altered host immune responses.

In addition, transcriptome profiling studies in mammalian parasitic nematodes have revealed molecules with known or predicted host immunomodulatory properties. The identified molecules include homologs of B-cell inhibitors, several serpins and neutrophil inhibitory factors in the barber's pole worm *Haemonchus contortus* [71], as well as proteases that likely act by degrading host intestinal mucins and secretory leukocyte peptidase inhibitor related proteins with anti-inflammatory function in *Trichuris* whipworms [72]. Also revealed in these studies are several genes encoding a "sperm-coating protein-like extracellular domain" (also called SCP/Tpx-1/Ag5/PR-1/Sc7 or SCP/TAPS) in *Strongyloides* and *Trichostrongylus* nematode parasites [73,74] (Figure 3). SCP/TAPS have been proposed to participate in the inhibition of neutrophil and platelet activity as well as other immunomodulatory functions in hookworm infections [75,76].

The rodent filarial nematode *Acanthocheilonema viteae* secretes the glyprotein ES-62, which possesses immunomodulatory properties because it is able to interact with a variety of immune cells and is therefore capable of regulating the host immune system [77] (Figure 3). In particular, ES-62 interferes with molecular events that control B-cell and T-cell receptor signaling, which leads to the substantial inhibition of B cell and T cell activation and proliferation [78]. Also, ES-62 has an impact on antigen-presenting cells, such as macrophages and dendritic cells, and ES-62-mediated modulation requires the presence of TLR4, but not TLR2 and TLR6 [79]. In addition, ES-62 has been shown to inhibit mast cell degranulation by forming a complex with TLR-4 at the plasma membrane [80]. Impairment of the host immune response due to the multiple effects of ES-62 enhances nematode infectivity by promoting the survival and persistence of the parasites. Interestingly, infection by certain nematode species has been shown to slow down or circumvent the outbreak of various autoimmune or allergic related diseases, which is consistent with the theory of the 'hygiene hypothesis' [81]. A complete understanding of the molecular function of immunomodulatory molecules, such as ES-62, will potentially lead to the identification of novel anti-inflammatory drugs with distinct mechanisms of action [82].

There is also growing evidence for the involvement of exosomes in the regulation of several physiological processes including immunological function in response to various stresses [83] (Figure 3). Recent results indicate that certain parasitic nematodes synthesize exosomes that function by delivering small RNA molecules to mammalian cells to modulate the host immune response. It was recently shown that RNA secreted from *Heligmosomoides polygryus*, a nematode that infects the gastrointestinal tract of mice, contains 263 microRNAs (miRNAs), many of which are highly conserved throughout the nematode evolutionary history [14]. Interestingly, most of the identified miRNAs share seed sites with mouse miRNAs, suggesting that nematode-derived miRNAs may be capable of interacting with the mRNA targets of mouse-derived miRNAs. It was further determined that *H. polygyrus* miRNAs are secreted in vesicles also containing homologs for proteins of mammalian exosomes, and that these proteins can protect the vesicles from host degradation. Finally, in vivo analysis showed that nematode-secreted vesicles are internalized by mouse intestinal epithelial cells. This leads to changes in gene expression, most notably in the significant down-regulation of *Dusp1* and *Il33r*, both of which are implicated in mounting the type II innate immune response. Similar methods of immunomodulation have been identified for *Brugia malayi*, the nematode responsible for lymphatic filariasis [84], and species of *Onchocerca* nematodes, which parasitize cattle and humans [85]. Infective stage *B. malayi* release exosome-like vesicles containing miRNAs. These exosome-like vesicles are internalized by host macrophages, which leads to their activation [84]. By analyzing serum and plasma from humans and cattle infected with *O. volvulus* and *O. ochengi*, respectively, researchers were able to identify miRNAs that were potentially secreted by nematodes in a similar manner as described above [85]. These findings demonstrate the direct involvement of nematode-secreted exosomes containing miRNAs in altering several aspects of host immunity in order to promote the persistence of the parasites and the establishment of disease.

5. CONCLUSIONS AND FUTURE DIRECTIONS

To be as prolific a parasite as nematodes are, they must evolve and employ robust mechanisms that promote their survival in the host. Immunomodulatory mechanisms are induced by parasitic nematodes to increase their infectivity once they have invaded their host. While by no means exhaustive, the mechanisms of nematode immunomodulation reported here display the variation of modulatory approaches developed by different types of parasitic nematodes to cripple the host immune response. While the general mechanism of nematode immunomodulation involves the production of molecules capable of altering certain host immune responses, the nature of the secreted compounds used by parasitic nematodes are drastically different. Despite great advances in our understanding of the mechanistic basis of nematode immunomodulatory processes, there exists a large number of parasitic nematodes that are able to infect different types of hosts, and that are capable of altering innate immune responses through the use of yet unidentified immunomodulatory molecules.

Identification and characterization of the molecular and functional basis of nematode immunomodulation will be an important step towards a better understanding of the range of mechanisms used by parasites to evade the host immune system in order to promote their survival, growth, and proliferation in the host. While elucidation of the mechanism(s) by which individual molecules operate is important, understanding how a specific nematode coordinates the activity and timing of different immunomodulatory processes will lead to an appreciation of the entire process of nematode parasitism. Determination of novel nematode immunomodulatory mechanisms may also provide unique and specific targets for developing novel anti-nematode therapies. For instance, eliminating crucial immunomodulatory molecules from human parasitic nematodes could allow the host immune response to function at full capacity, resulting in reduced nematode infectivity levels, and consequently rapid and efficient clearance of the parasites. Finally, a more complete understanding of the nematode immunomodulatory molecules that are involved in creating an anti-inflammatory environment in the host may contribute to the development of treatments for reducing the impact of inflammatory diseases [15].

ACKNOWLEDGMENTS

The authors thank members of the Eleftherianos laboratory and the Department of Biological Sciences at GWU for fruitful discussions. Research in the Eleftherianos laboratory is funded by grants from the National Institutes of Health—National Institute of Allergy and Infectious Diseases (1R01AI110675, 1R56AI110675-01 and 1R21AI109517) and the Columbian College of Arts and Sciences at George Washington University.

AUTHOR CONTRIBUTIONS

Dustin Cooper wrote the paper and Ioannis Eleftherianos revised it.

CONFLICTS OF INTEREST

The authors declare no conflict of interest.

REFERENCES

1. Maizels, R.M.; Blaxter, M.L.; Scott, A.L. Immunological genomics of Brugia malayi: Filarial genes implicated in immune evasion and protective immunity. Parasite Immunol. 2001, 23, 327–344.
2. Friberg, I.M.; Bradley, J.E.; Jackson, J.A. Macroparasites, innate immunity and immunoregulation: Developing natural models. Trends Parasitol. 2010, 26, 540–549.
3. Castillo, J.C.; Reynolds, S.E.; Eleftherianos, I. Insect immune responses to nematode parasites. Trends Parasitol. 2011, 27, 537–547.
4. Maizels, R.M.; Gomez-Escobar, N.; Gregory, W.F.; Murray, J.; Zang, X. Immune evasion genes from filarial nematodes. Int. J. Parasitol. 2001, 31, 889–898.
5. Hewitson, J.P.; Grainger, J.R.; Maizels, R.M. Helminth immunoregulation: The role of parasite secreted proteins in modulating host immunity. Mol. Biochem. Parasitol. 2009, 167, 1–11.

6. Klei, T.R. Immunological control of gastrointestinal nematode infections. Vet. Parasitol. 1997, 72, 507–523.
7. Hartmann, S.; Lucius, R. Modulation of host immune responses by nematode cystatins. Int. J. Parasitol. 2003, 33, 1291–1302.
8. McSorley, H.J.; Hewitson, J.P.; Maizels, R.M. Immunomodulation by helminth parasites: Defining mechanisms and mediators. Int. J. Parasitol. 2013, 43, 301–310.
9. Deehan, M.R.; Frame, M.J.; Parkhouse, R.M.; Seatter, S.D.; Reid, S.D.; Harnett, M.M.; Harnett, W. A phosphorylcholine-containing filarial nematode-secreted product disrupts B lymphocyte activation by targeting key proliferative signaling pathways. J. Immunol. 1998, 160, 2692–2699. [PubMed]
10. Wilson, E.H.; Deehan, M.R.; Katz, E.; Brown, K.S.; Houston, K.M.; O'Grady, J.; Harnett, M.M.; Harnett, W. Hyporesponsiveness of murine B lymphocytes exposed to the filarial nematode secreted product ES-62 in vivo. Immunology 2003, 109, 238–245.
11. Miller, J.E.; Horohov, D.W. Immunological aspects of nematode parasite control in sheep. J. Anim. Sci. 2006, 84, E124–E132.
12. Soares, M.F.; Araújo, C.A. Helminth products as a potential therapeutic strategy for inflammatory diseases. Inflamm. Allergy Drug Targets 2008, 7, 113–118.
13. Ansell, B.R.; Schnyder, M.; Deplazes, P.; Korhonen, P.K.; Young, N.D.; Hall, R.S.; Mangiola, S.; Boag, P.R.; Hofmann, A.; Sternberg, P.W.; et al. Insights into the immuno-molecular biology of Angiostrongylus vasorum through transcriptomics--prospects for new interventions. Biotechnol. Adv. 2013, 31, 1486–1500.
14. Buck, A.H.; Coakley, G.; Simbari, F.; McSorely, H.J.; Quintana, J.F.; Le Bihan, T.; Kumar, S.; Abreu-Goodger, C.; Lear, M.; Harcus, Y.; et al. Exosomes secreted by nematode parasites transfer small RNAs to mammalian cells and modulate innate immunity. Nat. Commun. 2014, 5, 1–10.
15. Jackson, J.A.; Turner, J.D.; Kamal, M.; Wright, V.; Bickle, Q.; Else, K.J.; Ramsan, M.; Bradley, J.E. Gastrointestinal nematode infection is associated with variation in innate immune responsiveness. Microbes Infect. 2006, 8, 487–492.
16. Whelan, R.A.K.; Hartmann, S.; Rausch, S. Nematode modulation of inflammatory bowel disease. Protoplasma 2012, 249, 871–886.
17. Su, Z.; Segura, M.; Morgan, K.; Loredo-Osti, J.C.; Stevenson, M.M. Impairment of protective immunity to blood-stage malaria by concurrent nematode infection. Infect. Immun. 2005, 73, 3531–3539.
18. Harnett, W. Secretory products of helminth parasites as immunomodulators. Mol. Biochem. Parasitol. 2014, 195, 130–136.
19. Kaya, H.K.; Gaugler, R. Entomopathogenic nematodes. Annu. Rev. Entomol. 1993, 38, 181–206.
20. Dillman, A.R.; Sternberg, P.W. Entomopathogenic nematodes. Curr. Biol. 2012, 22, R430–R431.
21. Goodrich-Blair, H.; Clarke, D.J. Mutualism and pathogenesis in Xenorhabdus and Photorhabdus: Two roads to the same destination. Mol. Microbiol. 2007, 64, 260–268.
22. Waterfield, N.R.; Ciche, T.; Clarke, D. Photorhabdus and a host of hosts. Annu. Rev. Microbiol. 2009, 63, 557–574.
23. Rodou, A.; Ankrah, D.O.; Stathopoulos, C. Toxins and secretion systems of Photorhabdus luminescens. Toxins 2010, 2, 1250–1264.
24. Eleftherianos, R.H.; ffrench-Constant, R.H.; Clarke, D.J.; Dowling, A.J.; Reynolds, S.E. Dissecting the immune response to the entomopathogen Photorhabdus. Trends Microbiol. 2010, 18, 552–560.
25. Lemaitre, B.; Hoffmann, J. The host defense of Drosophila melanogaster. Annu. Rev. Immunol. 2007, 25, 697–743.
26. Jiang, H.; Vilcinskas, A.; Kanost, M.R. Immunity in lepidopteran insects. Adv. Exp. Med. Biol. 2010, 708, 181–204. [PubMed]
27. Imler, J.L.; Bulet, P. Antimicrobial peptides in Drosophila: Structures, activities and gene regulation. Chem. Immunol. Allergy 2005, 86, 1–21. [PubMed]

28. Casanova-Torres, Á.M.; Goodrich-Blair, H. Immune signaling and antimicrobial peptide expression in Lepidoptera. Insects 2013, 4, 320–338.
29. Rolff, J.; Schmid-Hempel, P. Perspectives on the evolutionary ecology of arthropod antimicrobial peptides. Philos. Trans. R. Soc. Lond. B Biol. Sci. 2016.
30. Ribeiro, C.; Brehélin, M. Insect haemocytes: What type of cell is that? J. Insect Physiol. 2006, 52, 417–429.
31. Marmaras, V.J.; Lampropoulou, M. Regulators and signalling in insect haemocyte immunity. Cell. Signal. 2009, 21, 186–195.
32. Honti, V.; Csordás, G.; Kurucz, É.; Márku, R.; Andó, I. The cell-mediated immunity of Drosophila melanogaster: Hemocyte lineages, immune compartments, microanatomy and regulation. Dev. Comp. Immunol. 2014, 42, 47–56.
33. Eleftherianos, I.; Revenis, C. Role and importance of phenoloxidase in insect hemostasis. J. Innate Immun. 2011, 3, 28–33.
34. Lu, A.; Zhang, Q.; Zhang, J.; Yang, B.; Wu, K.; Xie, W.; Luan, Y.X.; Ling, E. Insect prophenoloxidase: The view beyond immunity. Front. Physiol. 2014, 5, 252.
35. Brivio, M.F.; Mastore, M.; Pagani, M. Parasite-host relationship: A lesson from a professional killer. Invertebr. Surviv. J.2005, 2, 41–53.
36. Balasubramanian, N.; Toubarro, D.; Simões, N. Biochemical study and in vitro insect immune suppression by a trypsin-like secreted protease from the nematode Steinernema carpocapsae. Parasite Immunol. 2010, 32, 165–175.
37. Stanley, D.; Haas, E.; Miller, J. Eicosanoids: Exploiting insect immunity to improve biological control programs. Insects 2012, 3, 492–510.
38. Toubarro, D.; Avila, M.M.; Hao, Y.; Balasubramanian, M.; Jing, Y.; Montiel, R.; Faria, T.Q.; Brito, R.M.; Simões, M. A serpin released by an entomopathogen impairs clot formation in insect defense system. PLoS ONE 2013, 8, e69161.
39. Theopold, U.; Krautz, R.; Dushay, M.S. The Drosophila clotting system and its messages for mammals. Dev. Comp. Immunol. 2014, 42, 42–46.
40. Satyavathi, V.V.; Minz, A.; Nagaraju, J. Nodulation: An unexplored cellular defense mechanism in insects. Cell. Signal.2014, 26, 1753–1763.
41. Balasubramanian, N.; Hao, Y.J.; Toubarro, D.; Nascimento, G.; Simões, N. Purification, biochemical and molecular analysis of a chymotrypsin protease with prophenoloxidase suppression activity from the entomopathogenic nematode Steinernema carpocapsae. Int. J. Parasitol. 2009, 39, 975–984.
42. Toubarro, D.; Avila, M.M.; Montiel, R.; Simões, N. A pathogenic nematode targets recognition proteins to avoid insect defenses. PLoS ONE 2013, 8, e75691.
43. Williamson, V.M.; Gleason, C.A. Plant-nematode interactions. Cur. Opin. Plant Biol. 2003, 6, 327–333.
44. Kenney, E.; Eleftherianos, I. Entomopathogenic and plant pathogenic nematodes as opposing forces in agriculture. Int. J. Parasitol. 2016, 46, 13–19.
45. Williamson, V.M.; Kumar, A. Nematode resistance in plants: The battle underground. Trends Genet. 2006, 22, 396–403.
46. Phillips, W.S.; Kieran, S.R.; Zasada, I.A. The relationship between temperature and development in Globodera ellingtonae. J. Nematol. 2015, 47, 283–289. [PubMed]
47. Goverse, A.; Smant, G. The activation and suppression of plant innate immunity by parasitic nematodes. Annu. Rev. Phytopathol. 2014, 52, 243–265.
48. Williams, B.; Dickman, M. Plant programmed cell death: Can't live with it; Can't live without it. Mol. Plant Pathol.2008, 9, 531–544.
49. Jaouannet, M.; Rosso, M.N. Effectors of root sedentary nematodes target diverse plant cell compartments to manipulate plant functions and promote infection. Plant Signal. Behav. 2013, 8, e25507.
50. Lozano-Torres, J.L.; Wilbers, R.H.P.; Warmerdam, S.; Finkers-Tomczak, A.; Diaz-Granados, A.; van Schaik, C.C.; Helder, J.; Bakker, J.; Goverse, A.; Schots, A.; et al. Apoplastic venom allergen-like proteins of cyst nematodes modulate the activation of basal plant innate immunity by cell surface receptors. PLoS Pathog. 2014, 10, 1–18.

51. Rehman, S.; Gupta, V.K.; Goyal, A.K. Identification and functional analysis of secreted effectors from phytoparasitic nematodes. BMC Microbiol. 2016, 16, 48.
52. Rehman, S.; Postma, W.; Tytgat, T.; Prins, P.; Qin, L.; Overmars, H.; Vossen, J.; Spiridon, L.N.; Petrescu, A.J.; Goverse, A.; et al. A secreted SPRY domain-containing protein (SPRYSEC) from the plant-parasitic nematode Globodera rostochiensis interacts with a CC-NB-LRR protein from a susceptible tomato. Mol. Plant Microbe Interact. 2009, 22, 330–340.
53. Dubreuil, G.; Magliano, M.; Deleury, E.; Abad, P.; Rosso, M.N. Transcriptome analysis of root-knot nematode functions induced in the early stages of parasitism. New Phytol. 2007, 176, 426–436.
54. Hewezi, T.; Howe, P.J.; Maier, T.R.; Hussey, R.S.; Mitchum, M.G.; Davis, E.L.; Baum, T.J. Arabidopsis spermidine synthase is targeted by an effector protein of the cyst nematode Heterodera schachtii. Plant Physiol. 2010, 152, 968–984.
55. Jaouannet, M.; Magliano, M.; Arguel, M.J.; Gourgues, M.; Evangelisti, E.; Abad, P.; Rosso, M.N. The root-knot nematode calreticulin Mi-CRT is a key effector in plant defense suppression. Mol. Plant Microbe Interact. 2013, 26, 97–105.
56. Kumar, M.; Gantasala, N.P.; Roychowdhury, T.; Thakur, P.K.; Banakar, P.; Shukla, R.N.; Jones, M.G.; Rao, U. De novo transcriptome sequencing and analysis of the cereal cyst nematode, Heterodera avenae. PLoS ONE 2014, 9, e96311.
57. Prior, A.; Jones, J.T.; Blok, V.C.; Beauchamp, J.; McDermott, L.; Cooper, A.; Kennedy, M.W. A surface-associated retinol- and fatty acid-binding protein (Gp-FAR-1) from the potato cyst nematode Globodera pallida: Lipid binding activities, structural analysis and expression pattern. Biochem. J. 2001, 356, 387–394.
58. Gao, B.; Allen, R.; Maier, T.; Davis, E.L.; Baum, T.J.; Hussey, R.S. The parasitome of the phytonematode Heterodera glycines. Mol. Plant Microbe Interact. 2003, 16, 720–726.
59. De Silva, N.R.; Brooker, S.; Hotez, P.J.; Montresor, A.; Engels, D.; Savioli, L. Soil-transmitted helminth infections: Updating the global picture. Trends Parasitol. 2003, 12, 547–551.
60. Hotez, P.J.; Fenwick, A.; Savioli, L.; Molyneux, D.H. Rescuing the bottom billion through control of neglected tropical diseases. Lancet 2009, 373, 1570–1575.
61. Martinez, F.O.; Helming, L.; Gordon, S. Alternative activation of macrophages: An immunologic functional perspective. Annu. Rev. Immunol. 2009, 27, 451–483.
62. Perrigoue, J.G.; Marshall, F.A.; Artis, D. On the hunt for helminths: Innate immune cells in the recognition and response to helminth parasites. Cell. Microbiol. 2008, 10, 1757–1764.
63. De Veer, M.J.; Kemp, J.M.; Meeusen, E.N.T. The innate host defence against nematode parasites. Parasite Immunol.2007, 29, 1–9.
64. Grencis, R.K. Immunity to helminths: Resistance, regulation, and susceptibility to gastrointestinal nematodes. Annu. Rev. Immunol. 2015, 33, 201–225.
65. Maizels, R.M.; Balic, A.; Gomez-Escobar, N.; Nair, M.; Taylor, M.D.; Allen, J.E. Helminth parasites–masters of regulation. Immunol. Rev. 2004, 201, 89–116.
66. Dall, E.; Brandstetter, H. Structure and function of legumain in health and disease. Biochimie 2016, 122, 126–150.
67. Schierack, P.; Lucius, R.; Sonnenburg, B.; Schilling, K.; Hartmann, S. Parasite-specific immunomodulatory functions of filarial cystatin. Infect. Immun. 2003, 71, 2422–2429.
68. Sun, Y.; Liu, G.; Li, Z.; Chen, Y.; Liu, Y.; Liu, B.; Su, Z. Modulation of dendritic cell function and immune response by cysteine protease inhibitor from murine nematode parasite Heligmosomoides polygyrus. Immunology 2012, 138, 370–381.
69. Flávia Nardy, A.; Freire-de-Lima, C.G.; Morrot, A. Immune evasion strategies of Trypanosoma cruzi. J. Immunol. Res.2015, 2015, 178947.
70. Behrendt, P.; Arnold, P.; Brueck, M.; Rickert, U.; Lucius, R.; Hartmann, S.; Klotz, C.; Lucius, R. A helminth protease inhibitor modulates the lipopolysaccharide-induced proinflammatory phenotype of microglia in vitro. Neuroimmunomodulation 2016, 23, 109–121.

71. Schwarz, E.M.; Korhonen, P.K.; Campbell, B.E.; Young, N.D.; Jex, A.R.; Jabbar, A.; Hall, R.S.; Mondal, A.; Howe, A.C.; Pell, J.; et al. The genome and developmental transcriptome of the strongylid nematode Haemonchus contortus. Genome Biol. 2013, 14, R89.
72. Foth, B.J.; Tsai, I.J.; Reid, A.J.; Bancroft, A.J.; Nichol, S.; Tracey, A.; Holroyd, N.; Cotton, J.A.; Stanley, E.J.; Zarowiecki, M.; et al. Whipworm genome and dual-species transcriptome analyses provide molecular insights into an intimate host-parasite interaction. Nat. Genet. 2014, 46, 693–700.
73. Hunt, V.L.; Tsai, I.J.; Coghlan, A.; Reid, A.J.; Holroyd, N.; Foth, B.J.; Tracey, A.; Cotton, J.A.; Stanley, E.J.; Beasley, H.; et al. The genomic basis of parasitism in the Strongyloides clade of nematodes. Nat. Genet. 2016, 48, 299–307.
74. Cantacessi, C.; Mitreva, M.; Campbell, B.E.; Hall, R.S.; Young, N.D.; Jex, A.R.; Ranganathan, S.; Gasser, R.B. First transcriptomic analysis of the economically important parasitic nematode, Trichostrongylus colubriformis, using a next-generation sequencing approach. Infect. Genet. Evol. 2010, 10, 1199–1207.
75. Moyle, M.; Foster, D.L.; McGrath, D.E.; Brown, S.M.; Laroche, Y.; De Meutter, J.; Stanssens, P.; Bogowitz, C.A.; Fried, V.A.; Ely, J.A.; et al. A hookworm glycoprotein that inhibits neutrophil function is a ligand of the integrin CD11b/CD18. J. Biol. Chem. 1994, 269, 10008–10015. [PubMed]
76. Del Valle, A.; Jones, B.F.; Harrison, L.M.; Chadderdon, R.C.; Cappello, M. Isolation and molecular cloning of a secreted hookworm platelet inhibitor from adult Ancylostoma caninum. Mol. Biochem. Parasitol. 2003, 129, 167–177.
77. Pineda, M.A.; Lumb, F.; Harnett, M.M.; Harnett, W. ES-62, a therapeutic anti-inflammatory agent evolved by the filarial nematode Acanthocheilonema viteae. Mol. Biochem. Parasitol. 2014, 194, 1–8.
78. Al-Riyami, L.; Harnett, W. Immunomodulatory properties of ES-62, a phosphorylcholine-containing glycoprotein secreted by Acanthocheilonema viteae. Endocr. Metab. Immune Disord. Drug Targets 2012, 12, 45–52.
79. Goodridge, H.S.; Marshall, F.A.; Else, K.J.; Houston, K.M.; Egan, C.; Al-Riyami, L.; Liew, F.Y.; Harnett, W.; Harnett, M.M. Immunomodulation via novel use of TLR4 by the filarial nematode phosphorylcholine-containing secreted product, ES-62. J. Immunol. 2005, 174, 284–293.
80. Melendez, A.J.; Harnett, M.M.; Pushparaj, P.N.; Wong, W.S.; Tay, H.K.; McSharry, C.P.; Harnett, W. Inhibition of Fc epsilon RI-mediated mast cell responses by ES-62, a product of parasitic filarial nematodes. Nat. Med. 2007, 13, 1375–1381.
81. Harnett, W.; Harnett, M.M. Filarial nematode secreted product ES-62 is an anti-inflammatory agent: Therapeutic potential of small molecule derivatives and ES-62 peptide mimetics. Clin. Exp. Pharmacol. Physiol. 2006, 33, 511–518.
82. Al-Riyami, L.; Pineda, M.A.; Rzepecka, J.; Huggan, J.K.; Khalaf, A.I.; Suckling, C.J.; Scott, F.J.; Rodgers, D.T.; Harnett, M.M.; Harnett, W. Designing anti-inflammatory drugs from parasitic worms: A synthetic small molecule analogue of the Acanthocheilonema viteae product ES-62 prevents development of collagen-induced arthritis. Med. Chem. 2013, 56, 9982–10002.
83. Beninson, L.A.; Fleshner, M. Exosomes: An emerging factor in stress-induced immunomodulation. Semin. Immunol.2014, 26, 394–401.
84. Zamanian, M.; Fraser, L.M.; Agbedanu, P.N.; Harischandra, H.; Moorhead, A.R.; Day, T.A.; Bartholomay, L.C.; Kimber, M.J. Release of small RNA-containing exosome-like vesicles from the human filarial parasite Brugia malayi. PLoS Negl. Trop. Dis. 2015, 9, e0004069.
85. Quintana, J.F.; Makepeace, B.L.; Babayan, S.A.; Ivens, A.; Pfarr, K.M.; Blaxter, M.; Debrah, A.; Wanji, S.; Ngangyung, H.F.; Bah, G.S.; et al. Extracellular Onchocerca-derived small RNAs in host nodules and blood. Parasites Vectors 2015, 8, 58.

CHAPTER 15

Prevalence of Soil Transmitted Nematodes on Nukufetau, a Remote Pacific Island in Tuvalu

Rick Speare*[1], Falatea Fab Latasi[2], Tekaai Nelesone[2], Sonia Harmen[1], Wayne Melrose[1], David Durrheim[1,3] and Jorg Heukelbach[4]

[1]*Anton Breinl Centre for Public Health and Tropical Medicine, WHO Collaborating Centre for Lymphatic Filariasis, James Cook University, Townsville 4811, Queensland, Australia,*

[2]*Ministry of Health, Funafuti, Tuvalu,*

[3]*Health Protection, Hunter New England Population Health, Locked Bag 10, Wallsend 2287, New South Wales, Australia and*

[4]*Department of Community Health, School of Medicine, Federal University of Ceará, Fortaleza CE 60430-140, Brazil*

ABSTRACT

Background

The population of Nukufetau, a remote coral atoll island in Tuvalu in the Western Pacific, received annual mass drug administration (MDA) of diethylcarbamazine and albendazole under the Pacific Elimination of Lymphatic Filariasis program in 2001, 2002 and 2003, with the last MDA occurring six months before a cross-sectional survey of the whole population for soil transmitted helminths (STH).

Methods

A cross-sectional survey in May 2004 recruited 206 residents (35.2% of the population) who provided a single faecal sample that was preserved, concentrated and examined microscopically.

Results

Overall prevalence of STH was 69.9%; only hookworm and *Trichuris trichiura* were diagnosed. *Trichuris* was present in 68.4% with intensity of infection being light in 56.3%, medium in 11.7% and heavy in 0.5%. Hookworm occurred in 11.7% with intensity of infection 11.2% being light and medium in 0.5%. Twenty individuals (9.7%) had dual infections. The prevalence of *Trichuris* was constant across all ages while the prevalence of hookworm was significantly lower in residents below 30 years of age. In the age group 5–12 years comparison of results with a 2001 survey [1] suggested that the prevalence of STH has declined minimally, due to sustained high prevalence of *Trichuris*, while hookworm has declined dramatically from 34.4% to 1.6%.

Conclusion

The results of this survey suggest that although the MDA appears to have reduced hookworm prevalence in residents below 30 years of age, there has been minimal effect on *Trichuris* prevalence. An integrated program to control STH is required.

BACKGROUND

Infections with soil transmitted helminths (STH) are common worldwide, with prevalence and parasite burden being particularly high in developing countries among school-aged children [2], although in many communities the prevalence of hookworm is higher in adults than in children [3]. In children, STH are associated with anaemia, stunting, underweight and poor school performance [4, 5].

A survey of STH in 13 Pacific island countries and territories (PICT) in 2001–2 found a very narrow spectrum of nematode parasites in children with a wide range of prevalences [1]. In this survey, prevalences in children aged 5 to 12 years ranged between 1% and 97% with prevalences lower then 5% in Niue and the Cook Islands, but very high prevalences (>80%) on the Marshall Islands, Kiribati, and Tuvalu (Fig 1). The reasons for this variation are not known.

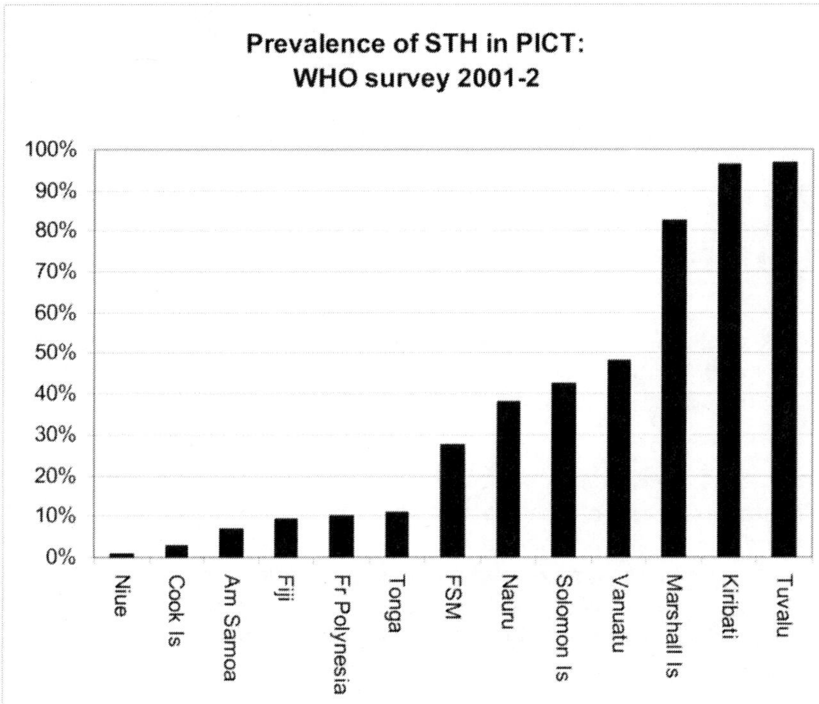

Figure 1. Prevalences of soil transmitted nematodes in children aged 5–12 years in PICT in 2001–2 (data reanalysed from Hughes et al 2004 [1] table 4).

In Tuvalu, samples from 118 children were analyzed, with 32 of these from the island of Nukufetau. Tuvalu had the highest prevalence (97%) of the 13 PICT, and only hookworms and whipworms (*Trichuris trichiura*) were detected [1]. There was not a single case of *Ascaris* infection. The species of hookworm was not identified. Two schools were surveyed, one on the main island of Funafuti with a resident population of 4,452, and another on the remote island of Nukufetau, with a resident population of 585.

Tuvalu had no programs or strategies to control STH before 2001, but in 2001, after the faecal survey, annual mass drug administration (MDA) of diethylcarbamazine (DEC) and albendazole was commenced for lymphatic filariasis elimination [6]. MDA was administered to all residents aged two years and older except for pregnant women and people considered too ill due to other diseases. The coverage (number of residents who received and consumed the drugs/total population) for Tuvalu was 81% in 2001, 47% in 2002 and 83% in 2003 [6].

This paper reports the results of a survey for STH amongst all age groups of Nukufetau residents in 2004, three years after the previous school-based survey [1] and following three rounds of MDA using albendazole and DEC. The survey was an opportunistic one, not linked to the MDA, but initiated to establish the need for a STH control program. The latest MDA had occurred six months prior to this survey.

METHODS

A cross-sectional survey was conducted on the island of Nukufetau, Tuvalu. Nukufetau is located at S8°15' E178°22', approximately 110 km north west of Funafuti, the capital island of Tuvalu. At the time of the last census in 2002, the island had 585 residents. All islands in Tuvalu are coral atolls.

In May 2004, all residents of Nukufetau were provided with containers for collection of faeces. To each specimen returned, SAF (sodium acetate, acetic acid and formaldehyde) solution was added, approximately equal to the volume of the faecal sample, and the faeces macerated to form a slurry. The fixed samples were transported to Princess Margaret Hospital Laboratory (PMHL) on Funafuti for examination. Details recorded for each sample were the resident's name, age and gender.

At the PMHL approximately 0.2 g of faeces was placed in a plastic conical 10 ml centrifuge tube with 8 ml of 10% formalin and 1 ml of petrol, and vigorously shaken by hand. The tubes were then centrifuged for 10 min at roughly 500 G. The ring of petrol and associated debris was loosened in each tube and all fluid in the tube tipped out. Volume was then reconstituted to 1 ml using tap water, mixed and 0.05 ml examined under two cover-slips (24 × 24 mm) placed side by side on a microscope slide at ×100 magnification using a compound light microscope. One slide was prepared from each specimen. Results were reported as number of eggs seen per slide and then calculated as eggs per gram (EPG) by multiplying by 200. The multiplication factor was determined by dilution of original sample at fixation (×2), 0.2 g in test tube (×5), examination of 0.05 ml of 1 ml (×20). Egg results were reported as light, medium and heavy using the criteria of WHO [7]: an egg count of 1–999, 1000–9999 and ≥9999 EPG was considered as light, moderate and heavy infection for *T. trichiura*, and 1–1999, 2000–3999 and ≥3999 EPG as light, moderate and heavy infection for hookworms.

Data was entered into an Excel file, checked for errors that may have occurred during data entry and analysed using SPSS for Windows (version

12.0.1) to generate descriptive statistics. Chi-square tests were used to test for differences in prevalence by gender and age category. Geometric mean intensity of EPG for *Trichuris* was calculated for all participants. The study was performed under James Cook University Ethics Committee Approval number H1423 and with the permission and full involvement of the Ministry of Health, Tuvalu.

RESULTS

Two hundred and six specimens were collected, 35.2% of the population of Nukufetau. Males provided 96 specimens (46.6%) and females 110 (53.4%). Mean age was 33 years with a median of 34 years and a range from seven months to 84 years. All age-groups were represented (Table 1) with 20–29 years under-represented at 11.8% and >49 years over-represented at 60.5%. STH were diagnosed in 69.9%. (Table 1).

Table 1. Prevalence of STH in Nukutefau residents by age-group, 2004. Population data are from 2002 census.

Age group (years)	Total population	Number examined (% pop of age group)	Hookworm positive (%)	*Trichuris* positive (%)	STH positive (%)	*Trichuris* Geometric Mean Intensity EPG*
0–9	171	62 (36.3)	1 (1.6)	43 (69.4)	43 (69.4)	54.2
10–19	95	26 (27.4)	1 (3.8)	19 (73.1)	19 (73.1)	67.6
20–29	51	6 (11.8)	0 (0.0)	5 (83.3)	5 (83.3)	40.9
30–39	70	15 (21.4)	3 (20.0)	10 (66.7)	10 (66.7)	35.3
40–49	79	25 (31.6)	5 (20.0)	16 (64.0)	17 (68.0)	17.6
50–59	52	37 (71.2)	9 (24.3)	27 (73.4)	27 (73.0)	56.0
60–69	47	23 (48.9)	3 (13.0)	13 (56.5)	14 (60.9)	20.3
≥70	20	12 (60.0)	2 (16.7)	8 (66.7)	9 (75.0)	35.5
Total	585	206 (35.2)	24 (11.7)	141 (68.4)	144 (69.9)	41.2
CI 0.95			9.5–13.9%	65.2–71.6%	66.7–73.1%	

*Geometric Mean Intensity EPG calculated for all subjects as per Mani et al (2004).

Only hookworm and *Trichuris* infections were diagnosed. The overall prevalence of hookworm was 11.7% with a prevalence of 2.1% in residents under 30 years of age and 19.6% amongst residents 30 years of age or older (Table 1). There was a significant difference in prevalence by age category (p = .014). The overall prevalence of *Trichuris* was 68.4% with no significant difference in prevalence by age group. Twenty (9.7%) residents had concomitant infections with hookworm and *Trichuris*, and only three were infected by hookworm alone. There was no significant difference in prevalence of *Trichuris* or hookworm infections by gender.

Amongst the 61 children aged 5–12 years, 46 (75.4%) had STH, all with *Trichuris* and one (1.6%) had a concomitant infection with hookworm.

Fifty percent (12/24) of the children less than 5 years had STH including 25% of the 12 children 2 years of age and younger. Only one child under 5 years, a 4 year old, had hookworm infection. This child also had *Trichuris* as did the remaining 12 STH positive children. The youngest child with *Trichuris* was 1 year of age.

For individuals with *Trichuris* infection the mean (±std) and maximum EPG were 673 (±1510) and 15,200 respectively. The proportions with light, medium and heavy infections were 56.3%, 11.7% and 0.5% respectively. For hookworm the mean (±std) and maximum EPG were 242 (±407) and 2,040 respectively. The proportions with light and medium infections were 11.2% and 0.5% respectively. There were no heavy infections. For the 61 children 5–12 years of age surveyed, intensity of infection with *Trichuris* was light in 57.4%, medium in 16.4% and heavy in 1.6%.

DISCUSSION

To the best of our knowledge, this is the first published cross-sectional prevalence survey of STH amongst an entire island population in the Western Pacific Region. The prevalence of STH in this remote atoll island was high (69.9%), mainly due to infection with *T. trichiura*, and the spectrum of geohelminths was narrow. Trichuriasis was equally distributed across all ages while the prevalence of hookworm infection was low (2.1%) under 30 years of age, but 19.6% in residents aged 30 years or older. Since the species of hookworm in Tuvalu is unknown, the high prevalence warrants studies to determine the species and to evaluate the health impact of hookworm infection, particularly the possible occurrence of iron deficiency anaemia in reproductive age and older women. Interestingly, not a single case of *Ascaris* infection was detected, confirming Hughes et al's findings for Tuvalu [1].

The only available baseline data on STH for the population of Nukufetau prior to the MDA campaigns comes from the survey amongst children aged 5–12 years in 2001 [1]. Although the prevalence of STH in this age range has changed only slightly in Nukufetau since 2001, this is due to the high prevalence of *Trichuris*. In the survey in 2001 in the age group 5–12 years, prevalence of *Trichuris* was 84% (n = 32) while in our survey prevalence was 75% (n = 61), a decline in prevalence of 11%. However, hookworm has declined dramatically in this age-group from 34.4% to 1.6%, a 95% reduction in prevalence. Intensity of infection with hookworm has remained constant with all cases in the light category [1]. Intensity of *Trichuris* in this age group has not changed significantly, with light 59.4% and medium 25% in 2001 and light 57.4%, medium 16.4% and heavy 1.6% in this survey. However, direct comparison is difficult since the participants in neither survey were randomly selected [8]. Only one of the 24 children under 5 years, a 4 year old, had hookworm, a prevalence of 4.2% for children less than 5 years of age. Children less than 3 years at the time of this survey had not participated in any MDA round, while those aged 3, 4 and 5 years and older had possibly participated in one, two and three MDA rounds respectively. The transmission of hookworm may have decreased as a result of the annual MDA campaigns.

MDA with DEC and albendazole achieved a reduction in hookworm prevalence in Indonesia from 25.3% to 5.9% [3]. Although on Nukufetau the hookworm prevalence in adults was high, without a pre-MDA baseline amongst adults comments on trends cannot be made. The initial hookworm prevalence was 50% in Indonesian residents older than 50 years and demonstrated a dramatic reduction in this age group after two rounds of MDA [3]. A single dose of 400 mg albendazole achieved an average cure rate of 78% in 68 studies [9]. Hence, for Nukufetau after albendazole had been administered annually for three years, a lower prevalence would have been expected.

The coverage of MDA in Nukufetau, at least in the first and third round, was very good. However, compliance with the MDA may have affected the success of treatment, especially in the middle and older age groups, which showed higher prevalences of hookworm infection. Unfortunately, individual data on compliance was not available. However, at the mass treatment rounds, men were more likely to be unavailable for observed treatment owing to their being absent while fishing or travelling.

Rate of re-infection after treatment can be considerable [10], but the low prevalence of hookworm in children suggests that incidence is low and that re-infection is unlikely to account for the high adult prevalence. *Trichuris* eggs can survive for extended times in the environment as compared to hookworm eggs and larvae, which may have contributed to a higher prevalence of trichuriasis than hookworm infection after mass treatment. Additionally, hookworm may have developed resistance to albendazole. However, resistance of human hookworms to anthelmintics has not been definitely reported. Resistance to pyrantel was suspected in Australia in *Ancylostoma duodenale* with the strain susceptible to albendazole [11]. Reduced efficacy of mebendazole was reported in Mali [12] and after 5 years of intensive therapy in Zanzibar [13]. Although anthelmintic resistance is less likely to account for persistence of hookworm than non-compliance in MDA, standardised resistance testing should be conducted on this population of hookworms in Nukufetau [14, 15].

This study highlights the persistence of *Trichuris* infection after three rounds of MDA using albendazole and DEC. Studies from Haiti, India and Sri

Lanka reported in contrast to our data reduction of the prevalence of trichuriasis after mass treatment with albendazole and DEC [16, 17, 18]. DEC has no significant activity against hookworm and *Ascaris lumbricoides* [19] and is unlikely to be active against *T. trichiura*. Albendazole has limited activity against *Trichuris* with prevalences after the standard single dose of 400 mg declining between 23% to 48% [9, 20, 21] or showing no effect at all [10]. The prevalence of *Trichuris* was reduced between 7% and 48%, the latter from a cohort study, of *Trichuris* in Indonesia after two rounds of MDA using albendazole and DEC [3]. A cure rate of 30% with the same anthelminthic combination against *T. trichiura* was reported in Sri Lanka [22]. Ongoing transmission despite repeated MDA campaigns in Nukufetau is suggested by the high rate (66%) of *Trichuris* in children under 5 years of age. Another anthelminthic, such as mebendazole or ivermectin, which has greater efficacy against *Trichuris* than albendazole should be used in Tuvalu to eliminate *Trichuris*. Mebendazole has been reported to have a high efficacy against *Trichuris* infections [23]. The data on efficacy of single dose ivermectin for *T. trichiura* are inconsistent with cure rates ranging from 0% to 88% [24, 25, 26, 27, 28], and two doses of ivermectin showed cure rates between 84% and 100% [29, 30]. Ivermectin with albendazole had a higher cure rate than albendazole alone (79% versus 30% respectively) [22].

The PICT share particular epidemiologic and ecological features that may have contributed to these different results. For example, the parasite fauna is different from other parts of the world, which is exemplified by the complete absence of ascariasis and the high prevalence of trichuriasis in our study population. Different strains with different susceptibilities to anthelmintic drugs may also prevail.

For Nukufetau chemotherapy should be combined with an integrated control strategy to reduce the contamination of the island with eggs [5]. Behaviours that lead to eggs gaining access to soil and surviving the two weeks required for development in the egg have not been investigated in PICT. Homes on Nukufetau have septic sewage systems, but the quality of systems is not high. Rain is the only available source of potable water and when rainwater tanks become depleted during the dry season, residents may be reluctant to use water for toilets and personal hygiene (TN, personal observation). An association between inadequate water supplies in schools and STH was found in 14 PICT [1]. Defecation on the beach or in the sea is not uncommon in Nukufetau (TN, personal observation). The survival of hookworm eggs and larvae and *T. trichiura* eggs in sea-water appear not to have been tested. Experimentally sodium chloride at 2% concentration and higher had a deleterious effect on development of hookworm eggs and larvae [31]; seawater contains approximately 3.5% sodium chloride. However, the embryo of *Trichuris*, protected by a thick shell, could potentially survive immersion in sea-water. This hypothesis warrants investigation as it has implications for management strategies for *Trichuris* in small island nations.

CONCLUSION

We have shown that on a remote island in the Western Pacific the spectrum of STH present in the general population is quite narrow with only hookworm and *Trichuris* infections present and that the prevalence of STH remains high after three rounds of MDA with albendazole and DEC. It appears that the MDA for lymphatic filariasis elimination has resulted in a significant reduction in hookworm prevalence in residents under 30 years of age. However, we assume based on previous data in children aged 5–12 years [1] and the high prevalence overall that annual community DEC and albendazole administration has had a minimal effect on *T. trichiura* prevalence. Alternative anthelmintics plus measures to reduce environmental contamination with helminth eggs need to be considered as part of an integrated STH control program.

ACKNOWLEDGEMENTS

The study was supported by a grant from Glaxo Smith Kline to the WHO Collaborating Centre for Lymphatic Filariasis at James Cook University. Jorg Heukelbach was supported by an Endeavour Australia Research Scholarship. We wish to thank Vicki Saunders for statistical advice.

COMPETING INTERESTS

The WHO Collaborating Centre for Lymphatic Filariasis received funding for filariasis research on Tuvalu from Glaxo Smith Kline, the manufacturers of albendazole.

AUTHORS' CONTRIBUTIONS

RS: Assisted in project design, conducted the survey, performed laboratory examination, did statistical analysis and wrote the initial draft of the manuscript. FFL: Assisted in project design, conducted the survey, performed laboratory examination and contributed to the manuscript; TN: Assisted in project design and contributed to the manuscript; SH: Contributed to the manuscript; WM: Assisted in project design and contributed to the manuscript; DD: Assisted in project design and contributed to the manuscript; JH: Contributed to the manuscript.

REFERENCES

1. Hughes RG, Sharp DS, Hughes MC, Akau'ola S, Heinsbroek P, Velayudhan R, Schulz D, Palmer K, Cavalli-Sforza T, Galea G: Environmental influences on helminthiasis and nutritional status among Pacific schoolchildren. Int J Envir Hlth Res. 2004, 14: 163-177.
2. Savioli L, Stansfield S, Bundy DA, Mitchell A, Bhatia R, Engels D, Montresor A, Neira M, Shein AM: Schistosomiasis and soil-transmitted helminth infections: forging control efforts. Trans Roy Soc Trop Med Hyg. 2002, 96: 577-579.
3. Oqueka T, Supali T, Ismid IS, Purnomo, Ruckert P, Bradley M, Fischer P: Impact of two rounds of mass drug administration using diethylcarbamazine combined with albendazole on the prevalence of *Brugia timori* and of intestinal helminths on Alor Island, Indonesia. Filaria J. 2005, 4: 5
4. Stephenson LS, Holland CV, Cooper ES: The public health significance of *Trichuris trichiura*. Parasitology. 2000, 121: S73-S95.
5. Urbani C, Palmer K: Drug-based helminth control in Western Pacific countries: a general perspective. Trop Med Int Health. 2001, 6: 935-944.
6. PacELF: PacELF Progress Report January – December 2004. 2005, Suva, PacELF Home Office, [http://www.pacelf.org/reports/]
7. Montresor A, Crompton DWT, Bundy DAP, Hall A, Savioli L: Guidelines for the evaluation of soil-transmitted helminthiasis and schistosomiasis at community level. 1998, WHO/CTD/SIP/98, [http://whqlibdoc.who.int/hq/1998/WHO_CTD_SIP_98.1.pdf]
8. Mngomezulu N, Govere JM, Durrheim DN, Speare R, Viljoen L, Appleton C, Booman M: Burden of schistosomiasis and soil-transmitted helminth infections in primary school children in Mpumalanga, South Africa, and implications for control. S Afr J Sci. 2002, 98: 607-610.
9. Horton J: Albendazole: a review of anthelmintic efficacy and safety in humans. Parasitology. 2000, 121: 113-132.
10. Saathoff E, Olsen A, Kvalsvig JD, Appleton CC: Patterns of geohelminth infection, impact of albendazole treatment and re-infection after treatment in schoolchildren from rural KwaZulu-Natal, South-Africa. BMC Infect Dis. 2004, 4: 27
11. Reynoldson JA, Behnke JM, Pallant LJ, Macnish MG, Gilbert F, Giles S, Spargo RJ, Thompson RCA: Failure of pyrantel in treatment of human hookworm infections (*Ancylostoma duodenale*) in the Kimberley region of North West Australia. Acta Trop. 1997, 68: 301-312.
12. de Clercq D, Sacko M, Behnke J, Gilbert F, Dorny P, Vercruysse J: Failure of mebendazole in treatment of human hookworm infections in the southern region of Mali. Am J Trop Med Hyg. 1997, 57: 25-30.
13. Albonico M, Bickle Q, Ramsan M, Montresor A, Savioli L, Taylor M: Efficacy of mebendazole and levamisole alone or in combination against intestinal nematode infections after repeated targeted mebendazole treatment in Zanzibar. Bull World Health Organ. 2003, 81: 343-352.

14. Albonico M, Engels D, Savioli L: Monitoring drug efficacy and early detection of drug resistance in human soil-transmitted nematodes: a pressing public health agenda for helminth control. Int J Parasitol. 2004, 34: 1205-1210.
15. Albonico M, Wright V, Ramsan M, Haji HJ, Taylor M, Savioli L, Bickle Q: Development of the egg hatch assay for detection of anthelminthic resistance in human hookworms. Int J Parasitol. 2005, 35: 803-811.
16. Mani TR, Rajendran R, Sunish IP, Munirathinam A, Arunachalam N, Satyanarayana , Dash AP: Effectiveness of two annual, single-dose mass drug administrations of diethylcarbamazine alone or in combination with albendazole on soil-transmitted helminthiasis in filariasis elimination programme. Trop Med Int Health. 2004, 9:
17. de Rochars MB, Direny AN, Roberts JM, Addiss DG, Radday J, Beach MJ, Streit TG, Dardith D, Lafontant JG, Lammie PJ: Community-wide reduction in prevalence and intensity of intestinal helminths as a collateral benefit of lymphatic filariasis elimination programs. Am J Trop Med Hyg. 2004, 71: 466-470.
18. de Silva NR, Pathmeswaran A, Fernando SD, Weerasinghe CR, Selvaratnam RR, Padmasiri EA, Montresor A: Impact of mass chemotherapy for the control of filariasis on geohelminth infections in Sri Lanka. Ann Trop Med Parasitol. 2003, 97 (4): 421-425.
19. Meyrowitsch DW, Simonsen PE: Short communication: efficacy of DEC against *Ascaris* and hookworm infection in schoolchildren. Trop Med Int Health. 2001, 6: 739-742.
20. Bennett A, Guyatt H: Reducing intestinal nematode infection: efficacy of albendazole and mebendazole. Parasitol Today. 2000, 16: 71-74.
21. Adams VJ, Lombard CJ, Dhansay MA, Markus MB, Fincham JE: Efficacy of albendazole against the whipworm *Trichuris trichiura* – a randomised, controlled trial. S Afr Med J. 2004, 94: 972-976.
22. Ismail MM, Jayakody RL: Efficacy of albendazole and its combinations with ivermectin or diethylcarbamazine (DEC) in the treatment of *Trichuris trichiura* infections in Sri Lanka. Ann Trop Med Parasitol. 1999, 93: 501-504.
23. Legesse M, Erko B, Medhin G: Comparative efficacy of albendazole and three brands of mebendazole in the treatment of ascariasis and trichuriasis. East Afr Med J. 2004, 81: 134-138.
24. Belizario VY, Amarillo ME, de Leon WU, de los Reyes AE, Bugayong MG, Macatangay BJ: A comparison of the efficacy of single doses of albendazole, ivermectin, and diethylcarbamazine alone or in combinations against *Ascaris* and *Trichuris* spp. Bull World Health Organ. 2003, 81: 35-42.
25. Freedman DO, Zierdt WS, Lujan A, Nutman TB: The efficacy of ivermectin in the chemotherapy of gastrointestinal helminthiasis in humans. J Infect Dis. 1989, 159: 1151-1153.
26. Taticheff S, Kebede A, Bulto T, Werkeneh W, Tilahun D: Effect of ivermectin (Mectizan) on intestinal nematodes. Ethiop Med J. 1994, 32: 7-15.
27. Marti H, Haji HJ, Savioli L, Chwaya HM, Mgeni AF, Ameir JS, Hatz C: A comparative trial of a single-dose ivermectin versus three days of albendazole for treatment of *Strongyloides stercoralis* and other soil-transmitted helminth infections in children. Am J Trop Med Hyg. 1996, 55: 477-481.
28. Ranque S, Chippaux JP, Garcia A, Boussinesq M: Follow-up of *Ascaris lumbricoides* and *Trichuris trichiura* infections in children living in a community treated with ivermectin at 3-monthly intervals. Ann Trop Med Parasitol. 2001, 95: 389-393.
29. Naquira C, Jimenez G, Guerra JG, Bernal R, Nalin DR, Neu D, Aziz M: Ivermectin for human strongyloidiasis and other intestinal helminths. Am J Trop Med Hyg. 1989, 40: 304-309.
30. Heukelbach J, Wilcke T, Winter B, Sales de Oliveira FA, Sabóia Moura RC, Harms G, Liesenfeld O, Feldmeier H: Efficacy of ivermectin in a patient population concomitantly infected with intestinal helminths and ectoparasites. Arzneimittelforschung. 2004, 54: 416-421.

31. Nicholl W: Observations on the influence of salt and other agents in modifying the larval development of the hookworms: *Ankylostoma duodenale* and *Necator americanus*. Parasitology. 1917, IX: 155-189.

CHAPTER 16

The Dual Effects of Root-Cap Exudates on Nematodes: From Quiescence in Plant-Parasitic Nematodes to Frenzy in Entomopathogenic Nematodes

Ivan Hiltpold[1], *, Geoffrey Jaffuel[2] and Ted C. J. Turlings[2]

[1] *Division of Plant Sciences, University of Missouri, 205 Curtis Hall, Columbia, MO 65211-7020, USA*

[2] *FARCE laboratory, University of Neuchâtel, Emile-Argand 11, 2000 Neuchâtel, Switzerland*

ABSTRACT

To defend themselves against herbivores and pathogens, plants produce numerous secondary metabolites, either constitutively or *de novo* in response to attacks. An intriguing constitutive example is the exudate produced by certain root-cap cells that can induce a state of reversible quiescence in plant-parasitic nematodes, thereby providing protection against these antagonists. The effect of such root exudates on beneficial entomopathogenic nematodes (EPNs) remains unclear, but could potentially impair their use in pest management programmes. We therefore tested how the exudates secreted by green pea (*Pisum sativum*) root caps affect four commercial EPN species. The exudates induced reversible quiescence in all EPN species tested. Quiescence levels varied with the green pea cultivars tested. Notably, after storage in root exudate, EPN performance traits were maintained over time, whereas performances of EPNs stored in water rapidly declined. In sharp contrast to high concentrations, lower concentrations of the exudate resulted in a significant increase in EPN activity and infectiousness, but still reduced the activity of two plant-parasitic nematode species. Our study suggests a finely tuned dual bioactivity of the exudate from green pea root caps. Appropriately formulated, it can favour long-term storage of EPNs and boost their infectiousness, while it may also be used to protect plants from plant-parasitic nematodes.

Keywords: Below-ground food web, biological control, entomopathogenic nematode (EPN), *Heterodera glycine*, *Heterorhabditis bacteriophora*, *Meloidogyne incognita*, plant defence, plant-parasitic nematode, root-cap exudate

INTRODUCTION

The estimated production of over 200,000 different compounds (Barber and Martin, 1976) ranks plant's secondary metabolites among the most diverse functional chemicals in the biosphere. Thousands of papers have been devoted to the description of ecological and evolutionary processes mediated by plant secondary metabolites (Dudareva *et al.*, 2013). Stahl (1888) was one of the first to suggest that these metabolites are involved in plant defences and they have since been proposed to contribute as an important selecting force in the interaction between plants and herbivores (Ehrlich and Raven, 1964). Historically, work on chemical plant defences has focused on foliar parts of plants (Hartmann, 2007) and roots have been mostly ignored (Hunter, 2001). Yet up to 20% of the photosynthetic fixed carbon is exuded via belowground tissues as secondary metabolites (Barber and Martin, 1976) and these exudates shape rhizospheric interactions (e.g. Hiltpold and Turlings, 2012; Rasmann *et al.*, 2012*a* and references therein; Erb *et al.*, 2013).

In the context of plant and plant-parasitic nematode interactions, root exudates are often beneficial to the plant antagonists, as they may use plant secondary metabolites to locate host plants (Prot, 1980; Rolfe *et al.*, 2000; Curtis *et al.*, 2009; Reynolds *et al.*, 2011). In addition to attracting nematodes, root exudates also trigger egg hatching in certain plant-parasitic nematode species (e.g. Perry and Clarke, 1981; Dennijs and Lock, 1992; Perry and Gaur, 1996; Gaur *et al.*, 2000; Wesemael *et al.*, 2006; Pudasaini *et al.*, 2008; Khokon *et al.*, 2009; Oka and Mizukubo, 2009).

Yet root exudates can also protect roots against plant-parasitic nematodes. For instance, metabolites exuded from the root-cap cells of legumes and maize (*Zea mays*) slow down movement in plant-parasitic nematodes, sometimes resulting in a state of quiescence, reducing the ability of the nematodes to infect the plant (Zhao *et al.*, 2000). This apparent defence mechanism has been found in several plant species and is usually reversible (Hubbard *et al.*, 2005).

In addition to plant-parasitic nematodes, root exudates have also been reported to affect entomopathogenic nematodes (EPNs), plant protagonists in the rhizosphere. Rasmann *et al.* (2005) found that insect herbivory on maize roots induces the emission of volatile secondary metabolites that can attract the EPN *Heterorhabditis megidis* Poinar, Jackson and Klein. A similar phenomenon has since been shown for belowground interactions between plants and EPNs in other systems involving cultivated plants (Ali *et al.*, 2010; Hiltpold *et al.*, 2010c; Turlings *et al.*, 2012), as well as in wild ecosystems (Rasmann *et al.*, 2011). As EPNs are obligate parasites that kill their insect host in a very short period of time (Dillman *et al.*, 2012), they are considered as very potent biological control agents in several cropping systems (Grewal *et al.*, 2005). Hubbard *et al.* (2005) found that green pea (*Pisum sativum*) root cap exudate induces quiescence in the EPN *Steinernema feltia* Filipjev, implying that this putative plant defence strategy may impair EPN efficacy near root tips, where many herbivores prefer to feed.

With the current study we aimed at assessing the effect of such exudate on four commercial EPN species. As different crop varieties can greatly vary in defensive metabolite patterns (Takabayashi *et al.*, 1991; Loughrin *et al.*, 1995; Wooley *et al.*, 2007; Erb *et al.*, 2011), we tested the exudate of a number of different pea cultivars for their potency in inducing quiescence in EPNs. In addition, to evaluate the impact of the root-cap exudate on the quality of stored EPNs over time, we performed a series of experiments to assess EPN infectiousness, mobility, and lipid content after recovery from induced quiescence. We also examined to what extent diluting the concentration of the exudate affects quiescence induction and whether heat or cold treatment has an effect on exudate activity.

MATERIALS AND METHODS

Plant material

Pisum sativum L. cv. Lancet seeds (Wyss Samen und Planzen AG, Switzerland) were sterilized in 95% ethanol for 5min. Seeds were then rinsed and immersed in pasteurized distilled water for 12h. Imbibed seeds were germinated for 3 days at 25°C in the dark in plastic boxes (15×13.5×5cm) on a 1.5cm layer of 1.0% PhytoAgar (Duchefa Biochemie, The Netherlands).

Root-cap exudate collection

Exudates were collected from 15 pea germinates by arranging them on a Teflon plate at the periphery of a 1ml drop of ultra-pure water, with only the tips (~5mm) submerged. After 2min of immersion, the liquid was gently agitated in order to disperse the border cells and the associated exudate. The solution obtained was then centrifuged at 4000 *g* for 10min in a 1.5ml centrifuge tube (Vaudaux-Eppendorf AG, Switzerland), resulting in a pellet of the suspended border cells. The supernatant, defined as root-cap exudate, was collected. Following this technique, exudates from several hundred of pea germinates were collected, and pooled in one solution. To facilitate the experimentation, this last solution was split into 7ml amber vials (Suppelco, Sigma-Aldrich Chemie GmbH, Switzerland) before being stored at –20°C.

Exudate-induced quiescence and recovery of EPNs

The pea root-cap exudate was tested on the following EPN species: *Heterorhabditis bacteriophora* Poinar (Rhabditida: Heterorhabditidae), *Heterorhabditis megidis* Poinar, Jackson and Klein (Rhabditida: Heterorhabditidae), *Steinernema feltiae* Filipjev (Rhabditida: Steinernematidae), and *S. carpocapsae* Weiser (Rhabditida: Steinernematidae), following the methodology by Hubbard *et al.* (2005). *H. bacteriophora*, *S. feltiae*, and *S. carpocapsae* were provided by Landi-Reba AG (Switzerland), and *H. megidis* by Becker Underwood (UK).

Briefly, ~1000 active infective juveniles of each EPN species were taken from batches freshly hatched from *Galleria mellonella* L. (Lepidoptera: Galleridae). EPNs were suspended in water and aliquots of 30 infective juveniles in 50 µl of water were pipetted into 36 wells of a 96-well plate (BD Biosciences, CA, USA) and 175 µl of root cap exudate was added to each EPN suspension. After 12h, using a dissecting microscope, the number of quiescent EPN-infective juveniles was evaluated by direct counting of individuals exhibiting, or not exhibiting, an active sinusoidal form and movement. Instead of root-cap exudate, 175 µl of water was pipetted into the wells of the control plates. The experiment was repeated 10 times for each EPN species.

To evaluate EPN recovery from exudate-induced quiescence, 125 µl was pipetted out of the wells, leaving settled EPN in the bottom, and replaced with the same volume of water. After 12h, resumption of EPN activity was evaluated using a dissecting microscope. To test if recovered EPNs were still infectious, five times 10 individuals per species per replicate were sampled and applied on top of *G. mellonella* larvae that had been individually placed in wells of a 24-well plate. The baits were then covered with 10% moist sand (white sand, Migros, Switzerland) and stored at 25°C in the dark. EPN infection was assessed after 48h.

Quiescence induction by the root exudates of various green-pea cultivars

Following the procedures described above, eight additional cultivars (cv. Kelvin, cv. Kelvin Sprinter, cv. Merveilles précoces, and cv. Sprint, Wyss Samen und Planzen AG, Switzerland; cv. Exquis, cv. Surab, cv. Picolo, and cv. Carmini, Landi-Reba AG, Switzerland) were tested for their ability to induce quiescence in *H. bacteriophora*. Water was used as a control and the EPNs were kept in the respective solutions for 12h before the number of quiescent nematodes was recorded. Each test was repeated 10 times.

Effect of root-cap exudate on the quality of EPNs over time

Either 41.25ml of root-cap exudate from green pea cv. Lancet or the same amount of water (control) were poured into 50ml culture flasks (BD Biosciences, CA, USA), and 7.5ml of water containing 20 000 *H. bacteriophora*infective juveniles were added. The flasks were then stored at room temperature (25°C) in the dark. Every third day, starting 12h after the induction of quiescence, the infectiousness, mobility, and lipid content of the nematodes from both solutions were assessed for 18 days.

EPN infectiousness

EPN infectiousness was measured using the five-on-one sand well bioassay (modified after Grewal, 2005): In a 24-well tissue plate (BD Biosciences, CA, USA), 10 *G. mellonella* larvae were individually placed in wells and covered with 10% moist sand (white sand, Migros, Switzerland). Five infective juveniles were pipetted from either root-cap exudate or water solution and added to each well. 24-well plates were stored at 25°C for 3 days and *G. mellonella*larvae were dissected. The number of infected insect larvae was recorded.

EPN mobility

EPN mobility was measured using a method modified after Tomalak, 2005. Into a 10-cm diameter Petri dish, we pipetted a 10-μl drop of 0.1% caffeine (99% pure, Alfa Aesar GmbH, Germany) on top of the 10ml of 2% solidified agar. About 100 EPNs from either root-cap exudate or water formulations were slowly pipetted onto a 5-mm diameter filter paper (cut out of bigger filter papers; Schleicher & Schuell GmbH, Germany). To facilitate removal of excessive water, the filter paper was laid on a paper towel. The filter was transferred immediately onto the drop of caffeine, with the EPNs on the upper side of the disc. The Petri dish was covered and stored in the dark at room temperature for 30min. As caffeine strongly repels EPNs, motile EPNs would quickly crawl away from this region onto the agar plate. After incubation, the disc was rinsed in a 1.5-ml centrifuge tube (Vaudaux-Eppendorf AG, Switzerland) containing water and the total number of sessile nematodes, recovered from the disc, was counted and the percentage of mobile EPNs calculated. This experiment was repeated five times per treatment.

Estimation of the neutral lipid content

Estimation of the neutral lipid content was performed using a method modified after Patel *et al.*, 1997). About 50 infective juveniles were sampled from each formulation described above. They were poured into a glass vial (7ml amber glass vial; Suppelco, Sigma-Aldrich Chemie GmbH, Switzerland) and flooded with 70% ethanol saturated with Oil Red O (Sigma-Aldrich Chemie GmbH, Switzerland). The glass vials were incubated at 60°C for 20min. The excess of staining solution was removed before the addition of 5ml of glycerol:water (50:50 v/v) solution. EPNs were left to settle overnight at room temperature. After placing the EPNs on a slide, the lipid content of the stained nematodes was estimated by comparison with a neutral lipid index scale [from 8 (fully stained) to 1 (no staining): Patel *et al.*, 1997].

Heat and cold stability of the green-pea exudate

The stability of green-pea exudate exposed to a temperature gradient was assessed. Aliquots of the exudate were either frozen (−80°C and −20°C) for 12h or heated up (20, 30, 40, 60, 80, and 100°C) for 10min. Controls consisted of water only, subjected to the same temperature treatments. For all treatments, induction of EPN quiescence was tested as described above and repeated seven times with two batches of EPNs.

Dilution of root cap exudate and the effect on bioactivity

Following the procedure described in the previous sections, different concentrations of pea root-cap exudate were tested. Exudate from green pea cv. Lancet was collected and freeze dried. The resulting material was diluted in 1ml (original concentration), 1.25, 1.5, 1.75, 2, 3, and 4ml of water. Quiescence induction in *H. bacteriophora* was tested for each concentration; controls consisted of water only. The number of quiescent nematodes was recorded after 12h of induction. Each test was repeated 15 times.

Increasing activity of EPNs was observed with the reduction of the exudate concentration. Therefore, we assessed EPN activity more precisely by counting the number of oscillations per minute for individual *H. bacteriophora* after 12h exposure to a low concentration of pea root-cap exudate (in 1.5ml of water). Head oscillations of EPNs exposed to the original concentration of the exudate (in 1ml of water) and to water only were also recorded. This test was repeated with 10 different individuals from three different batches of nematodes.

Following the five-on-one methodology (see above and Grewal, 2005), the infectiousness of highly active EPNs in low concentration green pea exudate was compared to those in water only. In 24-well plates, 12 *G. mellonella* larvae were covered with 10% moist sand previously mixed with either 1.5× diluted green-pea exudate or water (9:1, v/v). Five infective juveniles of the EPN *H. bacteriophora* were individually transferred in each well. Plates were wrapped in Parafilm and incubated at 25°C in the dark. The number of EPN-infected larvae was recorded every 12h for 96h, starting 12h after EPN inoculation. This experiment was replicated four times with three different batches of EPNs.

The effect of the exudate concentrations was also tested on plant-parasitic nematodes. Following the procedure described above, the head oscillation counts per minute of the soybean cyst nematode *Heterodera glycine* Ichinohe (Tylenchida: Heteroderinae) and the root-knot nematode *Meloidogyne incognita* Kofoid and White (Tylenchida: Heteroderinae) were recorded in 1.5×-diluted exudate and in water.

Statistical analyses

Statistical tests were conducted in SigmaPlot 12.1.0.15 (Systat Software GmbH Germany).

Quiescence induction by the root exudates of various green pea cultivars

To determine differences in quiescence induced by the different cultivars we analysed the data using a one-way ANOVA procedure. Statistical differences within groups were evaluated using a Tukey post-hoc test.

Effect of root-cap exudate on the quality of EPNs over time

EPN infectiousness, mobility, and neutral lipid content were tested using an RM two-way ANOVA procedure with treatment and time as factors. Statistical differences within groups were evaluated using a Tukey post-hoc test.

Heat and cold stability of the green pea exudate

Differences in induced quiescence after temperature treatments were measured using a one-way ANOVA on ranks procedure. Statistical differences within groups were evaluated using a Tukey post-hoc test.

Dilution of root-cap exudate and the effect on bioactivity

The effect of dilution on the induced quiescence to EPNs was tested using one-way ANOVA on ranks procedure. Statistical differences within groups were evaluated using a Tukey post-hoc test.

Increased activity was tested with a one-way ANOVA comparing the EPN head oscillations when exposed to both concentrations of green pea exudate or water. Infectiousness of EPNs exposed to either water or low-concentration exudate was tested using an RM two-way ANOVA procedure with treatment and time as factors. Statistical differences within groups were evaluated using a Tukey post-hoc test.

Differences between *H. glycine* and *M. incognita* head oscillations per minute recorded in water or in both concentrations of exudate were tested with a one-way ANOVA and a one-way ANOVA on ranks, respectively.

RESULTS

Exudate-induced quiescence and recovery of EPNs

Each EPN species that experienced contact with the green pea root exudate exhibited a state of quiescence (Table 1). Recovery was complete for *S. carpocapsae*, *H. megidis*, and *H. bacteriophora*. A large majority of *S.*

*feltiae*recovered from quiescence. Each EPN species was still infectious after recovering from induced quiescence (Table 1).

Table 1. Exudate-induced quiescence, recovery, and subsequent infectiousness of three EPN species

	Quiescence (%)	Recovery (%)	Infectiousness
Steinernema feltia	91±6	57±12	+[a]
S. carpocapsae	93±4	100	+
Heterorhabditis megidis	89±9	97±2	+
H. bacteriophora	95±2	100	+

[a]Some individuals remained quiescent after removal of the exudate. Nonetheless, larvae treated with recovered *S. feltia* were infected.

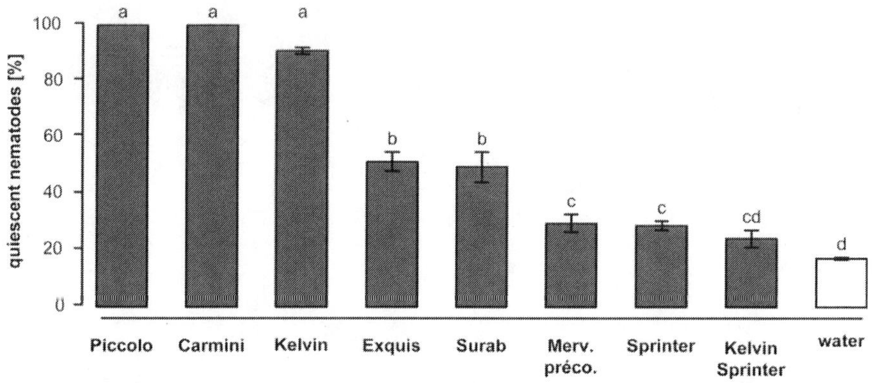

Figure 1. Induced quiescence in *Heterorhabditis bacteriophora* varies among green pea cultivars. Induced EPN quiescence was significantly influenced by the cultivar the root-cap exudates were collected from (grey bars). Three cultivars induced quiescence close or equal to 100% whereas some barely reached 30%, but were still different from the percentage of inactive nematodes recorded in water (white bar). One cultivar (Kelvin Sprinter) was not different from water. Letters indicate statistical differences. Bars represent SEM.

Quiescence induction by the exudates of various green-pea cultivars

The induction of quiescence in *H. bacteriophora* was highly dependent on the green-pea genotypes (Figure 1, ANOVA, $F_{8,93} = 26.686, P \leq 0.001$). Exudate from cv. Kelvin Sprinter had the lowest induction rate of quiescence (~25%), whereas the exudate of some cultivars resulted in almost 100% quiescence (i.e. cv. Picolo) (Figure 1).

Effect of root-cap exudate on the quality of EPNs over time

Induced quiescence in *H. bacteriophora* helped to maintain its performance over time (Figure 2A, B, C). Overall, infectiousness of *H. bacteriophora* that had been stored in exudate was higher than for EPNs that had been stored in water (Figure 2A; two-way RM ANOVA, $F1,139 = 12.05, P \leq 0.001$). After recovering from quiescence, EPNs that had been stored in exudate were significantly more mobile than EPNs that had been stored in water (Figure 2A; two-way RM ANOVA, $F1,139 = 66.839, P \leq 0.001$) and this remained significantly higher over the 18 days of the experiment (Figure 2B; two-way RM ANOVA, $F6,139 = 3.845, P = 0.003$). Lipid reserves in quiescent nematodes were significantly higher than in EPNs stored in water (Figure 2C; two-way RM ANOVA, $F1,139 = 53.129, P \leq 0.001$) and remained so over the course of the experiment (Figure 2C; two-way RM ANOVA, $F6,139 = 2.414, P = 0.03$).

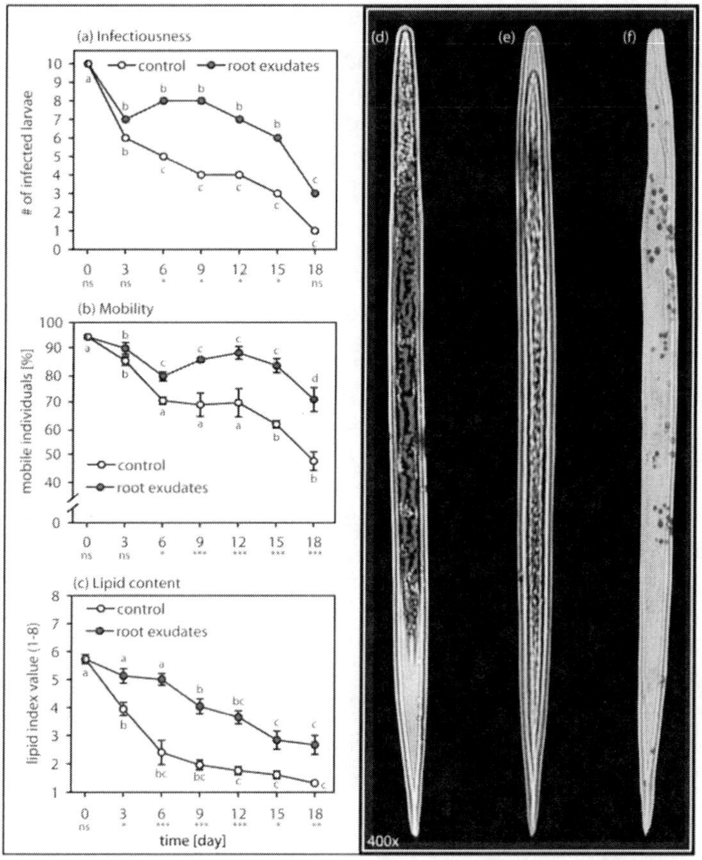

Figure 2. Induced quiescence enhances over time storage of the EPN *Heterorhabditis bacteriophora*. Quality traits were better conserved over time when EPNs were stored in a solution with root cap exudates from green pea cv. Lancet (grey circle) as compared to EPNs stored in water (white circle). (A) EPN infectiousness remained significantly higher in root-cap exudates as compared to water during most of the experiment. (B) When stored in root-cap exudates, EPNs stayed mobile for longer compared to the EPNs in water over most of the duration of the experiment. (C) Because EPNs stored in root-cap exudate were quiescent, their lipid reserves were maintained at a significantly higher level over time compared to lipid reserves measured in EPNs stored in water. (D–F) After being stained with Oil Red O, lipids appear in red inside the nematode body. A visual assessment allows an evaluation of the lipid content according to the 1–8 lipid index (Patel *et al.*, 1997). According to this index, (D) fully stained EPNs were given an 8, (E) intermediately stained EPNs received a 4, and (F) individuals without staining were recorded as 1. Level of significance between treatments on one day are indicated as not significant (ns), $P < 0.05$ (*), $P \leq 0.01$ (**), $P \leq 0.001$ (***). Statistical differences within treatments over time are noted with letters. Bars indicate SEM.

Heat and cold stability of the green-pea exudate

Exposing the exudate to different temperatures did not impair its bioactivity, and quiescence induction levels were similar to those measured in the control exudate. After heating or freezing, the exudate still induced quiescence in exposed *H. bacteriophora*, which was not the case for the water control [Supplementary Data (heat: ANOVA on ranks, $H = 45.419$, $P < 0.001$); Supplementary Data (cold: ANOVA on ranks, $H = 33.289$, $P \leq 0.001$)].

Dilution of root-cap exudate and the effect on bioactivity

Reducing the exudate concentration quickly reached a threshold beyond which induction of quiescence in *H. bacteriophora* failed (Figure 3; ANOVA on ranks, $H7,15 = 52.42$, $P \leq 0.001$), but instead resulted in increased movement (Figure 4A; ANOVA, $F2{-}89 = 491.427$, $P \leq 0.001$). At a low concentration, the exudate had a positive impact on the infectiousness of *H. bacteriophora* (Figure 5; RM two-way ANOVA, $F1,107 = 20.991$, $P = 0.006$), and EPN exposed to low concentration exudates were more effective at killing *G. mellonella* over time as compared to EPN exposed to water only (Figure 5; RM two-way ANOVA, $F8,107 = 5.88$, $P \leq 0.001$). Contrary to the effect on EPNs, the same low concentration of exudate significantly reduced the mobility of the tested plant-parasitic nematode species *H. glycine* (Figure 4B; ANOVA, $F2{-}89 = 225.023$, $P \leq 0.001$) and *M. incognita*(Figure 4C; ANOVA on ranks, $H = 67.876$, $P \leq 0.001$).

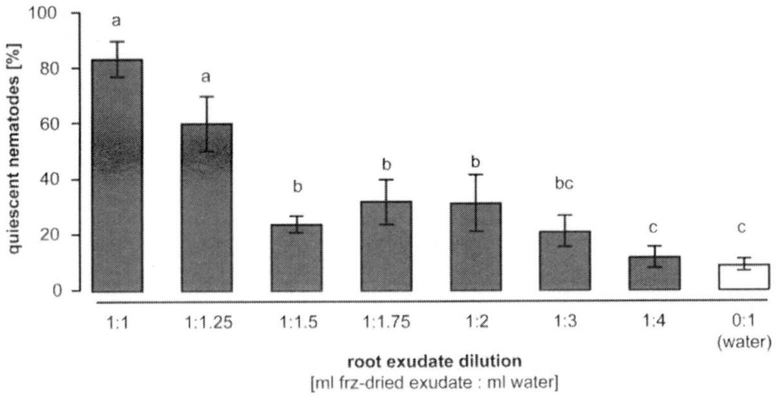

Figure 3. Level of induced quiescence in EPNs depends on root-cap exudate concentration. Lowering the concentration of the root-cap exudate reduced the induction of quiescence in the EPN *Heterorhabditis bacteriophora*. When freeze-dried exudates were diluted in 1.5× more water than its original concentration, the quiescence was reduced 4-fold compared to the original concentration. Low concentrations of exudates did not induce more quiescence than water. Letters indicate statistical differences. Bars indicate SEM.

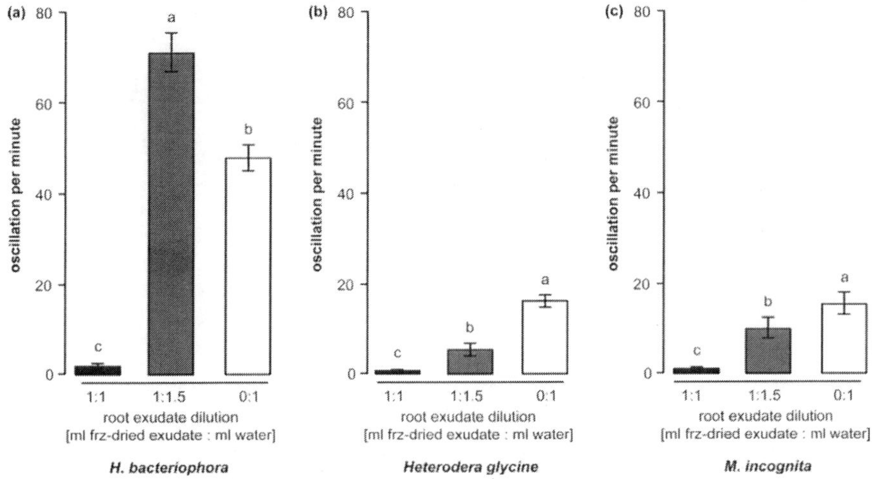

Figure 4. Low root-exudate concentration has a dual effect on nematode activity. (A) The number of oscillations per minute was significantly increased when the EPN *Heterorhabditis bacteriophora* was exposed to 1.5×-diluted exudates (grey bar) as compared to those in water (white bar). The original (1×) concentration of exudate induced almost complete quiescence in the exposed EPN. (B) The plant-parasitic nematode *Heterodera glycine* oscillation counts were higher in diluted exudate as compared to the original concentration. Whereas quiescence induction was incomplete at low concentration, the plant-parasitic nematode oscillation activity was still almost 3-fold lower than observed in water alone. (C) A similar pattern was observed for the plant-parasitic nematode *Meloidogyne incognita*. Letters indicate statistical differences. Bars indicate SEM.

DISCUSSION

Earlier observations had revealed that green pea root-cap exudate causes a loss of mobility and induces a state of quiescence in several plant-parasitic nematode species (Zhao *et al.*, 2000; Hubbard *et al.*, 2005), as well as in the bacteria-feeding nematode *Cenorhabditis elegans* Maupas (Zhao *et al.*, 2000; Hubbard *et al.*, 2005;), and the EPN *S. feltiae* (Hubbard *et al.*, 2005). The present study demonstrates that the activity of other EPN species also drops dramatically when exposed to the green pea root-cap exudate (Table 1). Quiescence is usually a reversible response in nematodes to toxic or unfavourable environmental conditions (Evans and Perry, 1976). As plant-parasitic nematodes, as well as other pathogens, preferably choose the

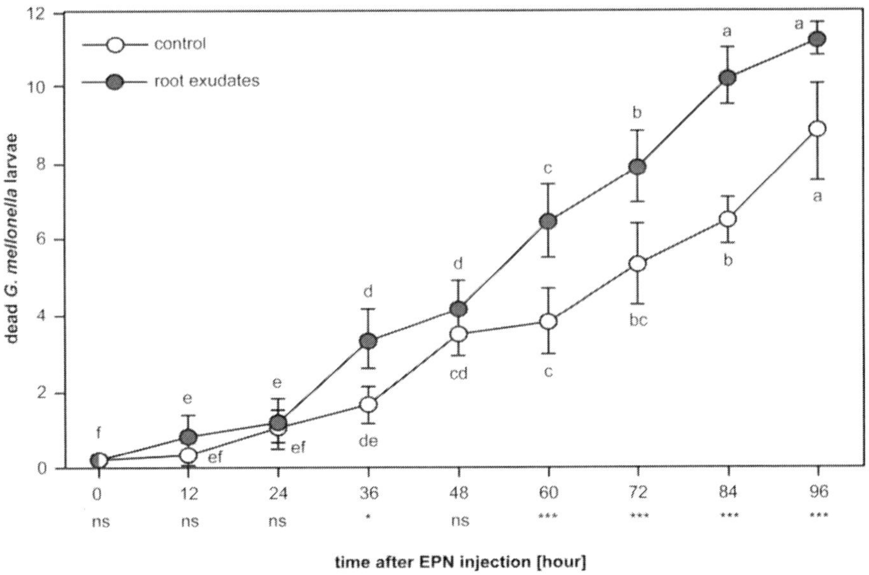

Figure 5. Low concentrations of exudate boosted the infectiousness of the EPN *Heterorhabditis bacteriophora*. EPNs in low concentrations of pea root-cap exudate (grey circles) were quicker and superior in infecting *Galleria mellonella* larvae than those in water alone (white circles). Levels of significance between treatments on one day are indicated as not significant (ns), $P < 0.05$ (*), $P \leq 0.01$ (**), $P \leq 0.001$ (***). Statistical differences within treatment over time are noted with letters. Bars indicate SEM.

elongation zone right behind the root tip to penetrate root tissues (Prot, 1980; Curl and Truelove, 1986; Gunawardena and Hawes, 2002), the induction of quiescence in root pathogens around this susceptible region has been interpreted as a defensive mechanism (Hawes *et al.*, 2000; Zhao *et al.*, 2000; Wuyts *et al.*, 2006). Recent findings showed that plant histone-linked extracellular DNA (exDNA) might be involved in this putative root tip defence (Wen *et al.*, 2009; Hawes *et al.*, 2011; Hawes *et al.*, 2012;). Functioning as an extracellular trap attracting and immobilizing pathogens (Hawes *et al.*, 2012), exDNA has been shown to be a key component in plant resistance to infection by pathogenic fungi (Wen *et al.*, 2009). By reducing or inhibiting the motility of these pathogens, roots can grow away from the threat, avoiding damage to their apical tips, while offering less vulnerable regions to herbivores. Interestingly, this is in contradiction to the common dogma that plants cannot escape a threat by moving away. Indeed, in addition to immobilizing pathogens (Hawes *et al.*, 2000), roots can potentially grow through a cumulative total of 6 m of soil within a day (Lynch and van Beem, 1993). This provides the roots with an escape strategy that is comparable to animals fleeing unfavourable biotic and abiotic conditions. To determine whether this strategy may indeed be effective against the relatively large and mobile nematodes requires additional experiments. The present study shows that the exudates involved in this potential defence mechanism not only affects the roots' 'foes' but can also momentarily impair beneficial microorganisms such as EPNs (Table 1).

As with other plant defence traits, the quiescence potency of root exudates varies a lot among plant families (Hubbard *et al.*, 2005), but we also found dissimilarities among genotypes of a particular species (Figure 1). The results presented here provide a new example of how plant breeding can affect the concentrations of secondary metabolites that are important for plant performance (e.g. Hubbard *et al.*, 2005; Rasmann *et al.*, 2005; Köllner *et al.*, 2008; Hiltpold *et al.*, 2010c; Erb *et al.*, 2011; Robert *et al.*, 2012; Meihls *et al.*, 2013).

Induced quiescence in EPNs may have interesting applications in biological control. Indeed, EPN mass production has been optimized (Ehlers, 2001), but downstream processes such as storage and transport can still be problematic. Inducing quiescence in EPNs with the root exudate significantly maintained EPN performance traits such as infectiousness, mobility, and lipid content (Figure 2A–C). As in insect parasitoids (Casas *et al.*, 2005; Denis *et al.*, 2013), EPNs rely on their lipid reserves while foraging and infecting insect hosts (Patel *et al.*, 1997). Loss of lipids was significantly lower in quiescent *H. bacteriophora* than in EPNs stored in water (Figure 2C), possibly explaining their higher mobility and infectiousness after recovery (Figure 2A and B). Similar to root penetration by plant-parasitic nematodes (Hubbard *et al.*, 2005), infectiousness and mobility of EPNs recovering from quiescence tended to increase (Figure 2A and B), suggesting a general increase of nematode activity after quiescence recovery. Whereas the timescale used in this study is not pertinent for long-term storage, these data clearly imply that the addition of the active compounds to the storage formulation will increase the EPN shelf life.

Induced quiescence in EPNs can be problematic when aiming at controlling insect pests in the rhizosphere. However, this apparent biological conflict might not occur in natural environments. Indeed, at lower concentrations of the exudate, as can be expected in the field, EPN activity was significantly boosted, while plant-parasitic nematodes remained quiescent (Figure 4). Further tests under natural conditions must be performed, but the enhanced infectiousness of 'frenetic' EPNs (Figure 5) suggests a dual effect of the tested exudate: root defences can simultaneously subdue 'foes', and invigorate 'friends'.

The present study reinforces the idea that plant exudates can help shape the dynamics of interactions not only in the rhizosphere, but also at more distant regions in the soil (Rasmann *et al.*, 2012*b*; Erb *et al.*, 2013; Hiltpold *et al.*, 2013;). The strategy of plants to protect their root tips with exudates is in apparent conflict with an alternative strategy, the attraction of natural enemies upon herbivorous attack. Yet, because exudate bioactivity seems tuned to have opposite effects on plant mutualists and antagonists, the two strategies might effectively complement each other. The complexity of biotic and abiotic interactions in the rhizosphere is huge and difficult to fully understand, but each new finding leads to a better and broader understanding of this very particular ecosystem. Therefore the prospect of harnessing root exudation and exploiting it for enhanced pest control (Hiltpold and Turlings, 2012) may indeed be realistic (Degenhardt *et al.*, 2009; Hiltpold *et al.*, 2010a, b; Ali *et al.*, 2012; Hiltpold *et al.*, 2012).

ACKNOWLEDGMENTS

We thank the FARCE lab members for their support and stimulating discussions. We thank Kim Jemmely for his help with the collection of exudate and the conduction of bioassays. We are also grateful to Wyss Samen und Planzen AG (Switzerland), Landi-Reba AG (Switzerland), Becker Underwood (UK), and Prof. M. Mitchum (University of Missouri, USA) for freely providing seeds and nematodes.

REFERENCES

1. Ali JG, Alborn HT, Campos-Herrera R, Kaplan F, Duncan LW, Rodriguez-Saona C, Koppenhöfer AM, Stelinski LL. 2012. Subterranean, herbivore-induced plant volatile increases biological control activity of multiple beneficial nematode species in distinct habitats. PLoS One 7, e38146.
2. Ali JG, Alborn HT, Stelinski LL. 2010. Subterranean herbivore-induced volatiles released by citrus roots upon feeding by Diaprepes abbreviatus recruit entomopathogenic nematodes. Journal of Chemical Ecology 36, 361–368.
3. Barber DA, Martin JK. 1976. The release of organic substances by cereal roots into soil. New Phytologist 76, 60–80.
4. Casas J, Pincebourde S, Mandon N, Vannier F, Poujol R, Giron D. 2005. Lifetime nutrient dynamics reveal simultaneous capital and income breeding in a parasitoid. Ecology 86, 545–554.
5. Curl EA, Truelove B. 1986. The rhizosphere. In: Trolldenier G, ed. Advanced series in agricultural sciences. Berlin: Springer-Verlag, 124–125.
6. Curtis RHC, Robinson AF, Perry RN. 2009. Hatch and host location. In: Perry R, Moens M, Starr J, eds. Root-knot nematodes. Wallingford, UK: CABI Publishing, 139–162.

7. Degenhardt J, Hiltpold I, Köllner TG, Frey M, Gierl A, Gershenzon J, Hibbard BE, Ellersieck MR, Turlings TCJ. 2009. Restoring a maize root signal that attracts insect-killing nematodes to control a major pest. Proceedings of the National Academy of Sciences, USA 106, 13213–13218.
8. Denis D, Van Baaren J, Pierre JS, Wajnberg E. 2013. Evolution of a physiological trade-off in a parasitoid wasp: How best to manage lipid reserves in a warming environment. Entomologia Experimentalis et Applicata 148, 27–38.
9. Dennijs L, Lock CAM. 1992. Differential hatching of the patato cyst nematodes Gobodera rostochiensis and Globodera pallida in root diffusates and water of differing ionic composition. Netherlands Journal of Plant Pathology 98, 117–128.
10. Dillman AR, Chaston JM, Adams BJ, Ciche TA, Goodrich-Blair H, Stock SP, Sternberg PW. 2012. An entomopathogenic nematode by any other name. PLoS Pathogens 8, e1002527.
11. Dudareva N, Klempien A, Muhlemann JK, Kaplan I. 2013. Biosynthesis, function and metabolic engineering of plant volatile organic compounds. New Phytologist 198, 16–32.
12. Ehlers RU. 2001. Mass production of entomopathogenic nematodes for plant protection. Applied Microbiology and Biotechnology 56, 623–633.
13. Ehrlich PR, Raven PH. 1964. Butterflies and plants – a study in coevolution. Evolution 18, 586–608.
14. Erb M, Balmer D, De Lange ES et al. 2011. Synergies and trade-offs between insect and pathogen resistance in maize leaves and roots. Plant, Cell and Environment 34, 1088–1103.
15. Erb M, Huber M, Robert CAM, Ferrieri AP, Machado RAR, Arce CCM. 2013. The role of plant primary and secondary metabolites in root– herbivore behaviour, nutrition and Physiology. In: Johnson SN, Hiltpold I, Turlings TCJ, eds. Advances in Insect Physiology , Vol. 45. Oxford, UK: Academic Press, 53–95.
16. Evans AAF, Perry RN. 1976. Survival strategies in nematodes. In: Croll NA, ed. The organisation of nematodes. London, UK: Academic Press, 383–424.
17. Gaur HS, Beane J, Perry RN. 2000. The influence of root diffusate, host age and water regimes on hatching of the root-knot nematode, Meloidogyne triticoryzae. Nematology 2, 191–199.
18. Grewal P. 2005. One-on-five sand-well bioassay for Heterorhabditis bacteriophora. In: Grunder JM, ed. Quality control of entomopathogenic nematodes. Wädenswil, CH: Agroscope FAW, 64–65.
19. Grewal PS, Ehlers RU, Shapiro-Ilan DI. 2005. Nematodes as biocontrol agents. Wallingford, UK: CABI Publishing. Gunawardena U, Hawes MC. 2002. Tissue specific localization of root infection by fungal pathogens: Role of root border cells Molecular PlantMicrobe Interactions 15, 1128–1136.
20. Hartmann T. 2007. From waste products to ecochemicals: Fifty years research of plant secondary metabolism. Phytochemistry 68, 2831–2846.
21. Hawes MC, Curlango-Rivera G, Wen F, White GJ, VanEtten HD, Xiong Z. 2011. Extracellular DNA: The tip of root defenses? Plant Science 180, 741–745.
22. Hawes MC, Curlango-Rivera G, Xiong Z, Kessler JO. 2012. Roles of root border cells in plant defense and regulation of rhizosphere microbial populations by extracellular DNA 'trapping'. Plant and Soil 355, 1–16.
23. Hawes MC, Gunawardena U, Miyasaka S, Zhao XW. 2000. The role of root border cells in plant defense. Trends in Plant Science 5, 128–133.
24. Hiltpold I, Baroni M, Toepfer S, Kuhlmann U, Turlings TCJ. 2010a. Selection of entomopathogenic nematodes for enhanced responsiveness to a volatile root signal helps to control a major root pest. Journal of Experimental Biology 213, 2417–2423.
25. Hiltpold I, Baroni M, Toepfer S, Kuhlmann U, Turlings TCJ. 2010b. Selective breeding of entomopathogenic nematodes for enhanced attraction to a root signal did not reduce their establishment or persistence after field release. Plant Signaling and Behavior 5, 1450–1452.

26. Hiltpold I, Bernklau E, Bjostad LB, Alvarez N, Miller-Struttmann NE, Lundgren JG, Hibbard BE. 2013. Nature, evolution and characterisation of rhizospheric chemical exudates affecting root herbivores. In: Johnson SN, Hiltpold I, Turlings TCJ, eds. Advances in insect physiology , Vol. 45. Oxford, UK: Academic Press, 97–157.
27. Hiltpold I, Hibbard BE, French DW, Turlings TCJ. 2012. Capsules containing entomopathogenic nematodes as a Trojan horse approach to control the western corn rootworm. Plant and Soil 358, 11–25.
28. Hiltpold I, Toepfer S, Kuhlmann U, Turlings TCJ. 2010c. How maize root volatiles influence the efficacy of entomopathogenic nematodes against the western corn rootworm? Chemoecology 20, 155–162.
29. Hiltpold I, Turlings TCJ. 2012. Manipulation of chemically mediated interactions in agricultural soils to enhance the control of crop pests and to improve crop yield. Journal of Chemical Ecology 38, 641–650.
30. Hubbard JE, Flores-Lara Y, Schmitt M, McClure MA, Stock SP, Hawes MC. 2005. Increased penetration of host roots by nematodes after recovery from quiescence induced by root cap exudate. Nematology 7, 321–331.
31. Hunter MD. 2001. Out of sight, out of mind: the impacts of rootfeeding insects in natural and managed systems. Agricultural and Forest Entomology 3, 3–9.
32. Khokon MAR, Okuma E, Rahman T, Wesemael WML, Murata Y, Moens M. 2009. Quantitative analysis of the effects of diffusates from plant roots on the hatching of Meloidogyne chitwoodi from young and senescing host plants. Bioscience Biotechnology and Biochemistry 73, 2345–2347.
33. Köllner TG, Held M, Lenk C, Hiltpold I, Turlings TCJ, Gershenzon J, Degenhardt J. 2008. A maize (E)-ß-caryophyllene synthase implicated in indirect defense responses against herbivores is not expressed in most American maize varieties. The Plant Cell 20, 482–494.
34. Loughrin JH, Manukian A, Heath RR, Tumlinson JH. 1995. Volatiles emitted by different cotton varieties damaged by feeding beet armyworm larvae. Journal of Chemical Ecology 21, 1217–1227.
35. Lynch J, van Beem JJ. 1993. Growth and architecture of seedling roots of common bean genotypes. Crop Science 33, 1253–1257.
36. Meihls LN, Handrick V, Glauser G et al. 2013. Natural variation in maize aphid resistance is associated with 2,4-dihydroxy-7-methoxy-1,4- benzoxazin-3-one glucoside methyltransferase activity. The Plant Cell 25, 2341–2355.
37. Oka Y, Mizukubo T. 2009. Tomato culture filtrate stimulates hatching and activity of Meloidogyne incognita juveniles. Nematology 11, 51–61.
38. Patel MN, Stolinski M, Wright DJ. 1997. Neutral lipids and the assessment of infectivity in entomopathogenic nematodes: Observations on four Steinernema species. Parasitology 114, 489–496.
39. Perry RN, Clarke AJ. 1981. Hatching mechanisms of nematodes. Parasitology 83, 435–449.
40. Perry RN, Gaur HS. 1996. Host plant influences on the hatching of cyst nematodes. Fundamental and Applied Nematology 19, 505–510.
41. Prot J-C. 1980. Migration of plant-parasitic nematodes towards plant roots. Revue de Nématologie 3, 305–318.
42. Pudasaini MP, Viaene N, Moens M. 2008. Hatching of the root-lesion nematode, Pratylenchus penetrans, under the influence of temperature and host. Nematology 10, 47–54.
43. Rasmann S, Ali JG, Helder J, van der Putten WH. 2012a. Ecology and evolution of soil nematode chemotaxis. Journal of Chemical Ecology 38, 615–628.
44. Rasmann S, Erwin AC, Halitschke R, Agrawal AA. 2011. Direct and indirect root defences of milkweed (Asclepias syriaca): trophic cascades, trade-offs and novel methods for studying subterranean herbivory. Journal of Ecology 99, 16–25.
45. Rasmann S, Hiltpold I, Ali JG. 2012b. The role of root-produced volatile secondary metabolites in mediating soil interactions. In: Montanaro G, Dichio B, eds. Advances in selected plant physiology aspects. Rijeka, Croatia: InTech, 269–290.

46. Rasmann S, Köllner TG, Degenhardt J, Hiltpold I, Toepfer S, Kuhlmann U, Gershenzon J, Turlings TCJ. 2005. Recruitment of entomopathogenic nematodes by insect-damaged maize roots. Nature 434, 732–737.
47. Reynolds AM, Dutta TK, Curtis RHC, Powers SJ, Gaur HS, Kerry BR. 2011. Chemotaxis can take plant-parasitic nematodes to the source of a chemo-attractant via the shortest possible routes. Journal of the Royal Society Interface 8, 568–577.
48. Robert CAM, Veyrat N, Glauser G et al. 2012. A specialist root herbivore takes advantage of defensive metabolites to locate nutritious tissues. Ecology Letters 15, 55–64.
49. Rolfe RN, Barrett J, Perry RN. 2000. Analysis of chemosensory responses of second stage juveniles of Globodera rostochiensis using electrophysiological techniques. Nematology 2, 523–533.
50. Stahl E. 1888. Pflanzen und Schnecken: Eine biologische Studie über die Schutzmittel der Pflanzen gegen Schneckenfrass. Jenaer Zeitschrift Medizin Naturwissenschaften , 557–684.
51. Takabayashi J, Dicke M, Posthumus MA. 1991. Variation in composition of predator-attracting allelochemicals emitted by herbivoreinfested plants: Relative influence of plant and herbivore. Chemoecology 2, 1–6.
52. Tomalak M. 2005. Caffeine test for determination of infective juvenile movement ability. In: Grunder JM, ed. Quality control of entomopathogenic nematodes. Wädenswil, CH: Agroscope FAW. Turlings TCJ, Hiltpold I, Rasmann S. 2012. The importance of rootproduced volatiles as foraging cues for entomopathogenic nematodes. Plant and Soil 359, 51–60.
53. Wen F, White GJ, Vanetten HD, Xiong Z, Hawes MC. 2009. Extracellular DNA is required for root tip resistance to fungal infection. Plant Physiology 151, 820–829.
54. Wesemael WML, Perry RN, Moens M. 2006. The influence of root diffusate and host age on hatching of the root-knot nematodes, Meloidogyne chitwoodi and M. fallax. Nematology 8, 895–902.
55. Wooley SC, Donaldson JR, Gusse AC, Lindroth RL, Stevens MT. 2007. Extrafloral nectaries in aspen (Populus tremuloides): Heritable genetic variation and herbivore-induced expression. Annals of Botany 100, 1337–1346.
56. Wuyts N, Maung ZTZ, Swennen R, De Waele D. 2006. Banana rhizodeposition: characterization of root border cell production and effects on chemotaxis and motility of the parasitic nematode Radopholus similis. Plant and Soil 283, 217–228.
57. Zhao XW, Schmitt M, Hawes MC. 2000. Species-dependent effects of border cell and root tip exudates on nematode behavior. Phytopathology 90, 1239–1245.

CHAPTER 17

Meloidogyne Incognita - Rice (Oryza Sativa) Interaction: A New Model System to Study Plant–Root-Knot Nematode Interactions in Monocotyledons

Phong Vũ Nguyễn[1,7], Stéphane Bellafiore[1,5], Anne-Sophie Petitot[1] , Rana Haidar[1,8], Aurélie Bak[1] , Amina Abed[1,4], Pascal Gantet[3,5], Itamara Mezzalira[1,6], Janice de Almeida Engler[2,6]* and Diana Fernandez

[1] *Institut de Recherche pour le Développement, UMR 186 IRD-Cirad-UM2 Résistance des Plantes aux Bioagresseurs, 911 avenue Agropolis, BP 64501, Montpellier, Cedex 5 34394, France.*

[2] *UMR IBSV INRA/CNRS/UNS, 400, Route de Chappes, BP167, Sophia Antipolis, CEDEX F-06903, France.*

[3] *Université Montpellier 2, UMR IRD-UM2 DIADE, 911 avenue Agropolis, BP 64501, Montpellier, Cedex 5 34394, France.*

[4] *INRAA- CRP, BP 37 Mehdi Boualem, Baraki, Algiers, Algeria.*

[5] *Institut de Recherche pour le Développement, LMI RICE, University of Science and Technology of Hanoi, Agricultural Genetics Institute, Hanoi, Vietnam.*

[6] *Embrapa – Recursos Genéticos e Biotecnologia, Brasília - DF 70849-970, Brazil.*

[7] *Nông Lâm University, Linh Trung, Thủ Đức, Hồ Chí Minh city, Việt Nam.*

[8] *INRA, UMR1065 Santé et Agroécologie du Vignoble (SAVE), ISVV, CS 20032, Villenave d'Ornon 33882, France.*

ABSTRACT

Background

Plant-parasitic nematodes developed strategies to invade and colonize their host plants, including expression of immune suppressors to overcome host defenses. *Meloidogyne graminicola* and *M. incognita* are root-knot nematode (RKN) species reported to damage rice (*Oryza sativa* L.) cultivated in upland and irrigated systems. Despite *M. incognita* wide host range, study of the molecular plant - RKN interaction has been so far limited to a few dicotyledonous model plants. The aim of this study was to investigate if the rice cv. Nipponbare widely used in rice genomic studies could be used as a suitable monocotyledon host plant for studying *M. incognita* pathogenicity mechanisms. Here we compared the ability of *M. graminicola* and *M. incognita* to develop and reproduce in Nipponbare roots. Next, we tested if RKNs modulates rice immunity-related genes expression in galls during infection and express the *Mi-crt* gene encoding an immune suppressor.

Results

Root galling, mature females, eggs and newly formed J2s nematodes were obtained for both species in rice cultivated in hydroponic culture system after 4-5 weeks. *Meloidogyne graminicola* reproduced at higher rates than *M. incognita* on Nipponbare and the timing of infection was shorter. In contrast, the infection characteristics compared by histological analysis were similar for both nematode species. Giant cells formed from 2 days after infection (DAI) with *M. graminicola* and from 6 DAI with *M. incognita*. Real-time PCR (qRT-PCR) data indicated that RKNs are able to suppress transcription of immune regulators genes, such as *OsEDS1, OsPAD4* and *OsWRKY13* in young galls. Four *M. incognita* reference genes (*Mi-eif-3, Mi-GDP-2, Mi-Y45F10D.4,* and *Mi-actin*) were selected for normalizing nematode gene expression studies *in planta* and in pre-parasitic J2s. *Meloidogyne incognita* expressed the immune suppressor calreticulin gene (*Mi-crt*) in rice roots all along its infection cycle.

Conclusion

RKNs repress the transcription of key immune regulators in rice, likely in order to lower basal defence in newly-formed galls. The calreticulin Mi-CRT can be one of the immune-modulator effectors secreted by *M. incognita* in rice root tissues. Together, these data show that rice is a well suited model system to study host- *M. incognita* molecular interactions in monocotyledons.

Keywords: Rice Root-knot nematodes Giant cells Immunity suppression Meloidogyne graminicola Meloidogyne incognita

BACKGROUND

Root-knot nematodes (RKNs, *Meloidogyne* spp.) are one of the most economically damaging genera of plant-parasitic nematodes on horticultural and field crops in all temperate and tropical areas (Trudgill and Blok [2001]). In particular, crops important for tropical countries as coffee, cotton, cowpea, peanut, soybean and rice are highly susceptible to RKNs, including *M. incognita* (Kofoid and White 1919; Chitwood 1949), *M. arenaria* (Neal 1889; Chitwood 1949), *M. javanica* (Treub 1885; Chitwood 1949), and *M. graminicola* (Golden and Birchfield [1968]). *Meloidogyne* spp. are obligate plant parasites that settle in roots and complete their life cycle by feeding from host cells (Williamson and Gleason [2003]). Like other plant and animal parasites, plant-parasitic nematodes developed strategies to invade and colonize their host plants, subvert the host machinery to their own benefit and overcome host defenses (Haegeman et al. 2012; Rosso et al. [2012]; Mitchum et al. [2013]). *Meloidogyne* spp. (juveniles stage J2) usually enter the plant through the apex and the root elongation zone, and then migrate between plant cells to reach the young central cylinder. Recent genomic data showed that *M. incognita* and *M. hapla* (Chitwood, 1949) genomes contain a high number of cell wall degrading enzymes, indicating that the nematode may use a combination of mechanical piercing and cell wall softening to enter and migrate into roots (Abad et al. [2008]; Opperman et al. [2008]; Danchin et al. [2010]). Once going into the differentiating vascular tissues, juveniles become sedentary and initiate nourishing feeding site originated from few parenchyma cells. Concomitantly, neighbouring cells divide causing roots to form knots or swellings. It has been shown that secretions from the nematode are crucial in establishment of the nourishing feeding site within the host root (Bellafiore and Briggs [2010]; Rosso et al. [2012]; Mitchum et al. [2013]). By secreting a number of compounds (including effectors) into root cells, RKNs induce their differentiation into hypertrophied, multinucleate and metabolically active feeding cells, named giant cells (GCs) (Kyndt et al. [2013]). Feeding-site formation enables the parasites to pump large amounts of nutrient solutions from the plant's vascular system. The nematode then goes through two developmental stages (J3, J4) to finally differentiate into an adult female which will lay eggsand new juveniles arising from these eggs will, in turn, start a new reproduction. Depending on the host plant and environmental conditions, the cycle lasts 15-45 days (Triantaphyllou and Hirschmann [1960]; Perry and Moens [2011]). It is critical for the nematode to cope with the host immune responses all along the infection process. One strategy is most likely the release of immune-modulatory effectors that block or interact with the plant basal defense network (Bellafiore and Briggs, [2010]). A series of transcriptome analyses in *Arabidopsis* and tomato have shown that, when a clear induction of the cell primary metabolism is evident, the expression of genes related to the plant immune responses are down-regulated in galls during plant-nematode interactions (Barcala et al. [2010]; Caillaud et al. [2008]b).

Until now, functional analysis of *Meloidogyne* spp. effectors has been essentially limited to *M. incognita* and to a lower extent to *M. javanica*, having *A. thaliana* or tomato as host plants. *Meloidogyne incognita* has a wide host range encompassing several hundreds of wild and cultivated plants. It is thus hypothesized that *M. incognita* pathogenicity mechanisms are conserved across plant genera, and even between dicotyledons and monocotyledons (Bellafiore and Briggs [2010]; Rosso et al. [2012]). However, the functional characterization of *M. incognita* effectors in other plant hosts, including monocot species, has been poorly investigated. Among plant species amenable to high-throughput genetic transformation and analysis, rice (*Oryza sativa*) is well-suited as a model monocot and is a crop species of high agronomic value and a RKN host.

Meloidogyne species damage upland rice in Asia, West Africa and Latin America with a prevalence of up to 50% (Bridge et al. [2005]). Up to now, nearly all grown *O. sativa* varieties tested are susceptible to *Meloidogyne* spp. infection, even when differences in host response to *M. graminicola* or *M. incognita* infection can be observed (Diomandé [1984]; Bridge et al. [2005]; Prasad et al. [2006]; de Araújo-Filho et al., 2010). Specific resistances to *Meloidogyne* spp. were identified in the African relative species like *O. glaberrima* (Diomandé [1984]; Soriano et al. [1999]) and progenies derived from inter-specific crosses are currently being tested for nematode resistance.

The *M. graminicola* life cycle in rice roots was recently investigated by histopathological analysis in several *O. sativa* and *O. glaberrima* rice varieties (Cabasan et al. [2012]). Analysis of the molecular rice responses to *M. graminicola* infection showed that the hormone-mediated resistance signaling pathways controlled by salicylic acid (SA), jasmonic acid (JA) and ethylene (ET) are repressed soon after infection by *M. graminicola* (Nahar et al. [2011]).

In contrast, the rice - *M. incognita* interaction has been poorly investigated until now and only little histological data was published (Ibrahim et al. [1973]). How the rice host plant copes with various RKN species has been poorly investigated. Therefore, the aim of this study was to investigate if rice could be used as a suitable host plant for studying *M. incognita* pathogenicity mechanisms and plant defense responses. Here we tested the *O. sativa* cv. Nipponbare for *M. incognita* susceptibility, and compared the ability of *M. graminicola* and *M. incognita* to develop and reproduce in roots of this rice cultivar. Next, we tested the hypothesis that the nematode modulates host defense responses in rice plants. We show that *M. incognita* expressed the calreticulin gene (*Mi-crt*) in infected rice roots and that several rice defense genes expression are down-regulated at an early stage of infection when the nematode starts feeding from root cells. Together, these data show that rice can provide an excellent model system to study host-*M. incognita* molecular interactions in monocotyledons.

RESULTS

Root galling and reproduction of *M. incognita* in *O. sativa* cv. Nipponbare

Reproduction and life cycle duration

Root swellings resulting from one or multiple galls formed within the rice root system were visible around 4 days after inoculation (DAI), and egg-laying females were observed from approximately 22 DAI, with a majority observed after 28 DAI (Figure 1K-L). Freshly hatched juveniles appeared at 35 DAI and continued until 56 DAI. *M. incognita* cycle (J2 to J2) duration on rice cv. Nipponbare was thus estimated to 35 days. Galls were initially formed at the root tips of young plantlets and some infected roots stopped their growth. After 35 DAI, galls were found either at root tips in roots which stopped expanding or were distributed in roots which took over expansion.

Histological analysis of rice roots nematode infection

Nematodes entered the roots, preferentially via the root elongation zone migrating via the cortex to the root tip and then migrating up into the vascular cylinder where it induces feeding cells (Figure 1). At 2 DAI, the observation of transverse sections of roots stained with toluidine blue showed that the nematodes were protruded out or were inside the stele (Figure 1A). Nematodes migrated inter-cellularly and no sign of broken or necrotic root cells was observed (Figure 1B). Along the path of migrating nematodes cortical cells presented hypertrophy and asymmetrical shapes most likely induced by the presence of the infecting nematode. No clear sign of binucleate initiating giant cells were observed although signs of feeding cell induction like condensed cytoplasm could be infrequently seen.

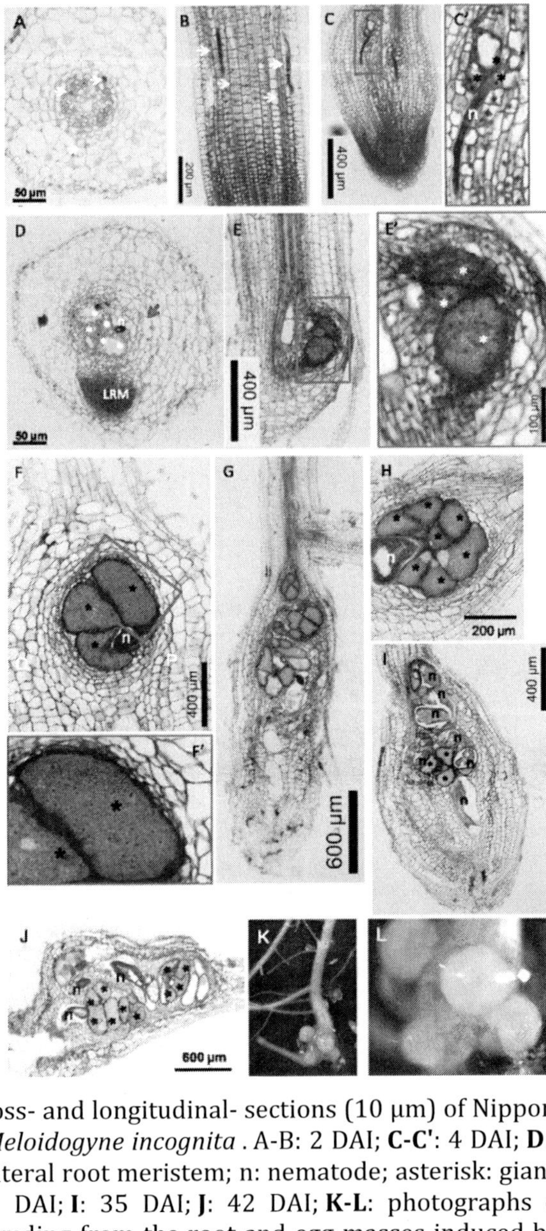

Figure 1. Cross- and longitudinal- sections (10 μm) of Nipponbare rice root infected by *Meloidogyne incognita* . A-B: 2 DAI; **C-C'**: 4 DAI; **D**: 6 DAI; **E-E'**: 7 DAI. (LRM: lateral root meristem; n: nematode; asterisk: giant cell). **F-F'**: 15 DAI; **G-H**: 22 DAI; **I**: 35 DAI; **J**: 42 DAI; **K-L**: photographs of a gall with females protruding from the root and egg masses induced by *M. incognita.* (n: nematode; asterisk: giant cell).

At 4 DAI, a number of nematodes were still migrating and caused root swelling. Within the stele a number of nematodes became sedentary and selected vascular parenchyma tissue containing xylem and phloem to induce their feeding site. Acytokinetic nuclear division was then observed in young feeding cells which presented a dense cytoplasm, identified as newly induced GCs close to the nematode head (Figure 1C,C'). Parenchymatic cells of the vascular cylinder, neighboring giant cells (NCs) divided, lost their typical rectangular shape and presented irregular shapes encircling the giant-feeding cells. New xylem and phloem elements also proliferated in the proximity of giant cells. These NCs showed a denser cytoplasm and nuclei became more prominent when closer to GCs (Figure 1C, and for 6 DAI, Figure 1D).

At 6 to 7 DAI, the formation of several feeding sites within the vascular system was observed and some J2 were still found migrating within the stele. Cross sections of galls showed GCs surrounded by dividing cells apparently originated from the vascular cylinder as well as from the root cortex. A number of cortical cells seemed to divide asymmetrically (red arrow in Figure 1D). The induction of lateral roots originating from pericycle cells was also visible situated at the feeding site (Figure 1D). Multinucleation as well as enlargement of GCs was observed at this stage indicating that intense DNA replication was taking place (Figure 1E, E').

At 15 to 21 DAI (Figure 1F-G), it could be evidenced that most root swellings contained multiple nematode feeding sites (NFS) (Figure 1G). Each NFS contained in average 5 to 7 GCs. GCs appeared more frequently embedded in the vascular cylinder, presented thickened cell walls and contained a dense cytoplasm with a large number of nuclei (per 10 μM section; Figure 1F,F'). Nematodes at this stage (15 to 21 DAI) varied between parasitic J2 and maturing females. Therefore, histological observations at these time points suggest that nematodes underwent molts (J2-J3-J4-young females). GCs at this stage are mature, implying that nuclei division and enlargement ended.

At 32 to 42 DAI, a number of females started egg-laying and due to their large size they were lost during sectioning (Figure 1G-J). GCs devoid of cytoplasm might be due to nematodes which stopped feeding after reproduction or that died.

Root galling and reproduction of *M. graminicola* in *Oryza sativa* cv. Nipponbare

Histological analysis of rice roots nematode infection

Meloidogyne graminicola penetrated rice roots via the root elongation zone, downwards to the root meristem and upwards to the vascular cylinder as observed for *M. incognita*. Examination of roots at 1 DAI confirmed the presence of nematodes in the cortex. Roots showed swellings very early at 2 DAI and galls were obvious at 4 DAI. Histological analysis showed that NFS were visible in swollen root tips from 2 DAI (Figure 2). They were formed of 3-4 young giant cells and located within the vascular cylinder where phloem, xylem and neighboring parenchymatic cells proliferated at the surroundings. At 4 to 7 DAI young GCs contained a dense cytoplasm, presented several nuclei and thicked cell walls compared to walls of vascular parenchyma cells (Figure 2). At 15 DAI giant cells presented multiple enlarged nuclei (Figure 2). Cells in direct contact or very close by GCs contained a dense cytoplasm. Often more than one NFS was observed per root swelling. At 22 DAI mature galls contained egg-laying females which remained embedded in the root tissues exterior to the gall proper (Figure 2). Around 31 DAI GCs were filled with cytoplasm and nuclei of various shapes often clustered. In some galls, some GCs appeared degraded.

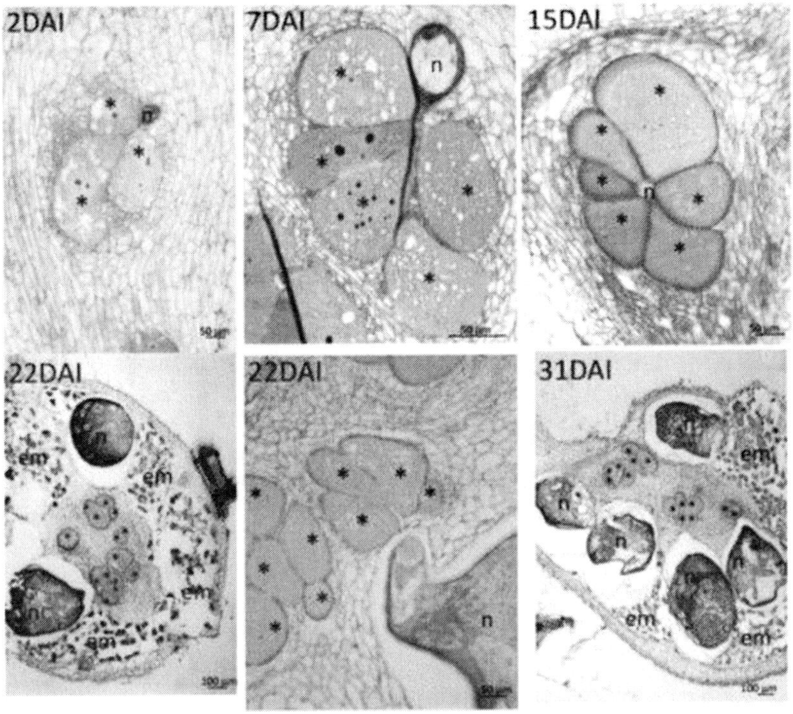

Figure 2. Cross-sections (10 μm) of Nipponbare rice roots infected by *Meloidogyne graminicola* obtained at 2, 7, 15, 22 and 31 days after nematode infection (DAI). (n: nematode; asterisk: giant cell; em: egg masses).

Reproduction and cycle duration

Stage 2 juveniles became parasitic around 4 to 7 DAI and appeared swollen at this stage. At 15 DAI nematode had undergone molts of J2 to J3 and J4 non-feeding finally reaching the female stage (7 to 15 DAI) and egg laying adult females were observed from 18 to 22 DAI. Female bodies and egg masses were embedded within the gall tissues (cortex). At 31 DAI, eggs laid by mature females hatched and most likely infected roots within the same plant.

Nematode genes expression in rice roots

A time course experiment with *O. sativa* cv Nipponbare infected with *M. incognita* was applied to study the *in planta* nematode gene expression. Swollen root tips and galls located outlying from the root apex, were collected at 6, 10 and 20 DAI and RNA was extracted. For each time points, at least 35 plants were inoculated. The experiment was repeated in triplicate.

Transcript accumulation of 5 genes (*Mi-csq-1, Mi-eif-3, Mi-GDP-2, Mi-Y45F10D.4,* and *Mi-actin*) putatively constitutively expressed in *M. incognita* was measured in root samples collected at different times after infection. In all three time-course experiments tested, the nematode genes transcript levels increased in root samples all along infection (Additional file 1: Table S1).

Figure 3 shows the averaged expression levels (as defined by Cq values) of the 5 *M. incognita* genes here analyzed in infected root tissues. On average, a 42-fold (deltaCq = 5.4) difference was found in the transcript accumulation between samples collected at the earliest (6 DAI) and latest (20 DAI) time tested after inoculation with *M. incognita* (Figure 3). *Mi-actin* transcript levels were the most elevated in root samples as compared to other genes tested (Additional file 1: Table S1) and exhibited a 128-fold change accumulation during the time course after infection. The stability of each *M. incognita* gene expression across samples was analyzed using RefFinder (Xie et al. [2012]) in order to select the most reliable genes for using them as *M. incognita* reference genes for qPCR gene expression analysis in rice tissues. Except *Mi-csq-1,* all genes showed good stability across samples (Additional file 2: Figure S1). The *Mi-actin* gene showed the most stable expression pattern (lowest geometric mean) and was used to further normalize *Mi-crt* expression levels.

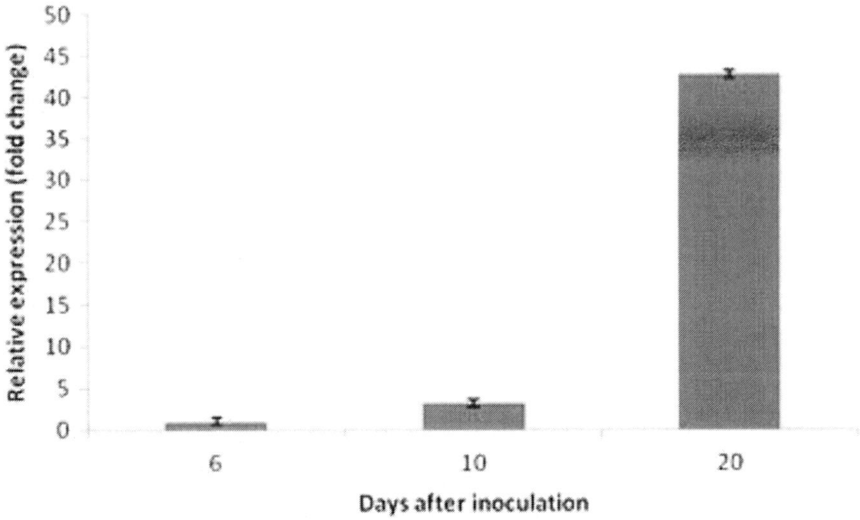

Figure 3. Relative expression average of *M. incognita* constitutive genes in infested rice *O. sativa* cv. Nipponbare roots. Gene expression was measured by reverse transcription-quantitative polymerase chain reaction in plants infested with *M. incognita* at different time points after treatment. Bars are the mean values (± standard error of the mean [SEM]) of *Mi-csq-1, Mi-eif-3, Mi-GDP-2, Mi-Y45F10D.4,* and *Mi-actin* genes expression data (Cq) in three independent biological replicates, with 35 plants per condition. (*n=5; 3, each contained 35 plants pooled*).

In parallel, the *Os-actin* transcripts were quantified to verify the amount of rice transcripts in galls. When compared to other rice reference genes, *Os-actin* showed stable expression patterns in galls and roots of *O. sativa* cv. Nipponbare plants (data not shown). Contrary to the enhanced *Mi-actin* transcript accumulation, *Os-actin* transcripts decreased during the infection course, mainly after 10 DAI (an average difference of 2 cycles between the 6- and 20-DAI samples) (Additional file 2: Figure S2). These data suggest that nematode representation is increasing in the biological material collected from infected roots along the time-course tested. This is in accordance with the histological data showing that the nematode body size dramatically increased inside galls. These data then indicate that the RNA extraction procedure used for the mixed plant-nematode samples was efficient for the purification of the total RNAs from both organisms.

Expression of the *M. incognita* effector gene *Mi-crt* was followed in root samples at 6, 10 and 20 DAI. The *Mi-crt* data were normalized to the *Mi-actin* data and *Mi-crt* gene expression in infected-rice samples was further calculated relative to the 6 DAI sample (Figure 4). *Mi-crt* transcripts were as abundant as *Mi-actin* transcripts, and accumulated at the three infection stages tested. In addition, the *Mi-crt* expression data were not found to significantly differ between 6 DAI, compared to 10 and 20 DAI, as expected from Jaubert et al. ([2005]) in *M. incognita - A. thaliana* interaction.

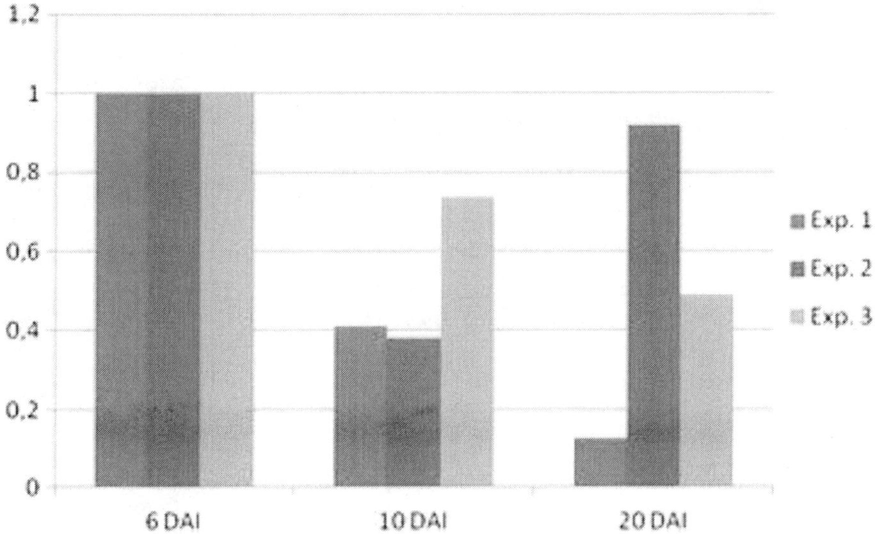

Figure 4. Relative expression of the *Mi-crt* gene in *M. incognita* -infested rice *O. sativa* cv. Nipponbare roots after 6, 10, and 20 dai. Gene expression was measured by reverse transcription-quantitative polymerase chain reaction in plants infested with *M. incognita* at different time points after treatment. The *Mi-crt* data were normalized to the *Mi-actin* data and *Mi-crt* gene expression in infected-rice samples was further calculated relative to the 6 DAI sample. Data presented are means (± standard error of the mean [SEM]) expression in three independent biological replicates, with 35 plants per condition (*n=3, each contained 35 plants pooled*).

Rice defense genes expression

To study the rice defense responses against *Meloidogyne* spp. root samples were collected from Nipponbare plants inoculated with *M. graminicola* or *M. incognita* at 2 and 6 DAI, respectively. These time points correspond to the formation of nutrient feeding sites observed in rice roots for each nematode species, respectively. For each time point chosen, a series of 90 plants were inoculated, and root samples were collected and pooled together. The experiment was repeated in duplicate. Non-inoculated plants served as control at the time of inoculation.

A series of genes involved in the rice immune responses were selected, including those from signaling, salicylic acid (SA)- and jasmonic acid (JA)-dependent resistance signaling pathways (Delteil et al. [2010]). We chose *OsMAPK6*, *OsMAPK5a* and *OsMAPK20* for early signaling (phosphorylation cascades in PTI and ETI), *OsAOS2* (JA pathway), *OsEDS1* and *OsPAD4* (SA-dependent resistance), *OsRAC1* (oxidative burst), *OsNIH1*, and *OsWRKY13* (positive transcriptional regulators of defense genes). Gene expression was normalized to the rice *Os-actin* used in Tao et al. ([2009]) and Delteil et al. ([2011]).

Analysis of gene expression in rice roots challenged with *Meloidogyne* spp. showed that the majority of these genes, except for *NIH1* that was induced two-fold at this infection stage for *M. incognita*, were not induced or down-regulated in the early time points after inoculation (Figure 5) as compared with non-inoculated control plants. This was particularly evident for *OsEDS1*, *OsPAD4* and *OsWRKY13* genes tested that were down-regulated from two- to three-fold at 2 DAI in *M. graminicola* and 6 DAI in *M. incognita*.

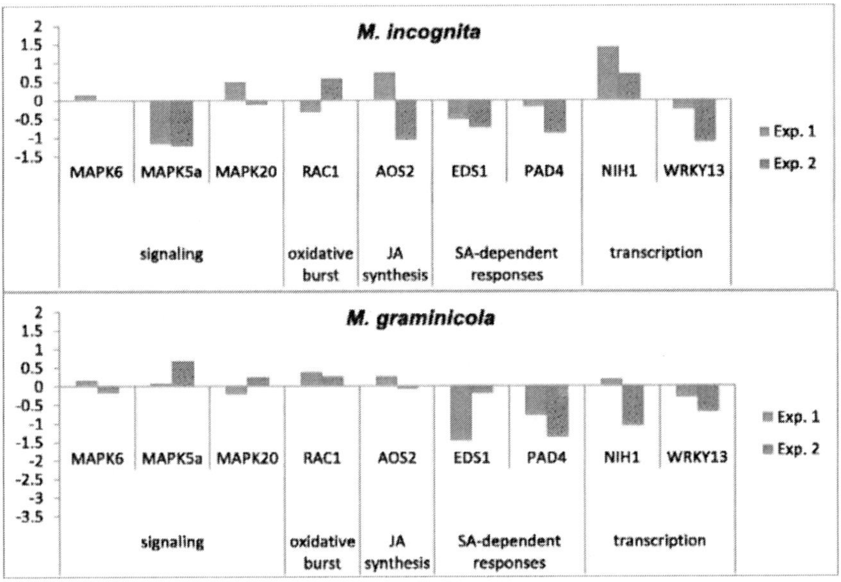

Figure 5. Early regulation of defense-related genes in *O. sativa* cv. Nipponbare roots during *Meloidogyne* infection. Gene expression was measured by reverse transcription-quantitative polymerase chain reaction in plants infested with *M. graminicola* at 2 DAI or with *M. incognita* at 6 DAI. Gene expression was normalized to *Os-actin*. Bars represent the log2 values of the ratio of the mean transcript levels for inoculated vs. non-inoculated plants from two technical replicates. Two independent biological replicates were carried out, with 90 plants per condition.

DISCUSSION

In this study, we showed that *M. incognita* is able to develop and reproduce in *O. sativa* cv. Nipponbare. Root galling, mature females, eggs and freshly hatched J2s were obtained in hydroponic culture system after 4-5 weeks. *Meloidogyne incognita* is naturally found associated to rice in rainfed growing systems but the impact of this RKN species on rice cultivation is lower than *M. graminicola* that is remarkably well adapted to flooded conditions (Fortuner and Merny, [1979]; De Waele and Elsen, [2007]). In this study, we observed that *M. graminicola* reproduced at higher rates than *M. incognita* on Nipponbare and the timing of infection was shorter although the infective behaviour and feeding site morphologies were similar for both nematode species. Juveniles penetrated the root elongation zone and tips, and established in the stele where they induced the formation of nutrient providing feeding sites. Giant cells were already visible at 2 DAI with *M. graminicola* and only at 6 DAI with *M. incognita*. These giant-feeding cells presented walls that thickened during their development exhibiting apparent cell wall ingrowths and interruptions indicative to be plasmodesmata. Development of galls induced by *M. incognita* as well as by *M. graminicola* in *O. sativa* cv. Nipponbare involves hypertrophy of vascular parenchyma cells ending up in a small number of giant cells per gall (up to 7). Concomitantly, hyperplasy is observed in cells neighboring the giant-feeding cells and xylem and phloem cells. A small number of cells around the gall proper also undergo to some extent hypertrophy. In addition, other feeding sites form in the root vicinity meaning that each root swelling may contain multiple galls. Highnumber of feeding sites (or galls) is noteworthy in each root swelling caused by *M. graminicola* as recently reported by Cabasan et al. ([2013]). The presence of multiple galls suggests that during this susceptible interaction hyperplasy might be further induced in the root surrounding area. Therefore, it is tempting to say that multiple gall formation might be facilitated by the presence of neighboring galls. Otherwise, simply multiple galls form close to each other because several juvenile nematodes penetrate concurrently. A gall induced by *M. incognita* in rice roots stays mostly confined to the vascular cylinder whereas apparently galls induced by *M. graminicola* involve a small number of cells outside of the vascular tissue. This is in agreement with galls induced in the dicotyledon model host, *Arabidopsis thaliana* where galls are strictly confined to the vascular tissue (de Almeida Engler et al., [1999]). Galls in *Arabidopsis* do not induce proliferation of endodermal tissue layer, as revealed by lignin stain of galls delimited the feeding site (de Almeida Engler, unpublished data). Number of outer tissue layers of rice galls excluded from the vascular cylinder at different root regions varies from 5 to around 10 cell layers, but this could simply reflect the root region that the gall develops. Therefore, there it is not yet clear that cells outside the vascular cylinder are part of the gall induced by *M. graminicola*. Initial phase of gall development shows giant cells which undergo the first nuclear division accompanied by increase in cytoplasm density. Concomitantly, parenchymatic cells neighboring giant cells (NCs) divide and lose their typical rectangular shape and become more rounded with various shapes encircling the giant-feeding cells. These NCs show a denser cytoplasm and nuclei become more prominent when encircling giant cells, suggesting some kind of cellular communication between giant cells and NCs. The presence of plasmodesmata devoid of callose deposition has been demonstrated for galls induced by root-knot nematodes in *Arabidopsis* roots by Hofmann et al. ([2010]). The presence of lateral root meristems in the gall region was also observed during gall development suggesting local auxin accumulation.

Once galls mature nematode developed into the classical J2-J3-J4-female phases and egg-laying females were observed at 22 DAI with *M. graminicola* and at 32 DAI with *M. incognita*. A conspicuous feature between both *Meloidogyne* species was that *M. incognita* lays eggs outside of the roots whereas *M. graminicola* eggs remain embedded within the root tissues, this latest explaining why *M. graminicola* is a RKN species successful in infecting rice grown in irrigated or flooded systems (Fortuner and Merny, [1979]; Prot and Matias, [1995]; De Waele and Elsen, [2007]). *Meloidogyne graminicola* female body size at 31 DAI was similar to the size of *M. incognita* females. Evidently, egg laying within the root cortex by *M. graminicola* is thus a specific adaptation for the aquatic survival and infection ability. In addition, the fact that *M. graminicola* eggs are kept within the roots while *M. incognita* eggs are extruded out of the root tissue might represent an adaptation to their hosts: to water submersed root systems, or to the soil environment, respectively. Understanding the genetic basis of this adaptation might therefore be a novel clue in nematode parasitism knowledge in plants with submersed roots versus soil grown root systems.

Importantly, data presented here indicate that both RKN species are able to suppress rice basal defence genes expression at early stages after infection. We hypothesize that RKN repress the transcription of key immune regulators in order to lower the basal defence. This is in accordance with Nahar et al. ([2011]) and Ji et al. ([2013]) who reported that rice cv Nipponbare defense genes expression was repressed in roots and giant cells early upon infection with *M. graminicola*. Already at 1 DAI, the mRNA levels of four immune-related genes tested (*OsWRKY45, OsPR1b, OsEin2b,* and *JiOsPR10*) were significantly attenuated in gall tissues (Nahar et al., [2011]). Here we show that the same occurred in rice galls infected by *M. graminicola* at 2 DAI, or by *M. incognita* at 6 DAI, when the nematode had started feeding and that first nutrient feeding sites have been established

(Figures 1 and 2). Eight genes involved in the rice immune responses were tested, including those from cell signaling, SA- and JA-dependent resistance signaling pathways. When overexpressed in rice plants, *OsAOS2, OsRAC1, OsNIH1, OsWRKY13* genes presented enhanced resistance levels to the fungus *Magnaporthe oryzae*, causing rice blast disease and the bacteria *Xanthomonas oryzae* pv. *oryzae* causing bacterial blight (Delteil et al. [2010]). *EDS1* (Enhanced Disease Susceptibility 1) and *PAD4* (Phytoalexin Deficient 4) play a key role in *A. thaliana* R-protein-triggered and basal resistance to invasive biotrophic and hemi-biotrophic pathogens (Falk et al. [1999]; Jirage et al. [1999]). EDS1 and PAD4 are intracellular proteins homolog to lipases that can interact and form different protein complexes. In response to infection, EDS1 and PAD4 activate SA production and signaling, and also mediate antagonism between the JA and ET defense response pathways (Rietz et al. [2011]; Wiermer et al. [2005]). In the same way, *OsWRKY13* serves as a node of the antagonistic crosstalk between SA- and JA-dependent pathways in pathogen-induced defense responses. There are more than 100 *WRKY* transcription factors in the rice genome (Wu et al. [2005]), and many are involved in rice innate immune responses, including *OsWRKY03, OsWRKY71, OsWRKY13, OsWRKY45* (Qiu et al. [2009]; Liu et al. [2013]). *OsWRKY13* mediates disease resistance to bacterial blight and fungal blast through activation of SA-dependent pathways and suppression of JA -dependent pathways (Qiu et al. [2009]). The present study strongly suggests that successful *M. incognita* or *M. graminicola* attack suppresses SA signaling in rice, as revealed by the down-regulation of *OsEDS1, OsPAD4* and *OsWRKY13* genes in newly-formed galls. This result is in accordance with Nahar et al. ([2011]) who observed down-regulation of genes involved in SA/JA/ET signaling in rice cv. Nipponbare challenged with *M. graminicola*. However, rice plants impaired in SA biosynthesis (expressing the *Pseudomonas putida* salicylate hydroxylase *NahG* gene) were only slightly more susceptible to nematode infection, pointing to a positive but minor role for SA in rice defense against *Meloidogyne*. In tomato, *NahG* plants that still produce minimal amounts of SA did not show increased susceptibility to *Meloidogyne* indicating that low levels of SA might be sufficient for basal resistance to root-knot nematodes (Bhattarai et al. [2008]). In addition, successful development of *Meloidogyne* in tomato roots involves the repression of pathogenesis-related (PR) genes associated to SA-dependent systemic acquired resistance (SAR) (Molinari et al. [2013]). Conversely, the JA/ET signalling pathway seems to play a major role in basal defense to nematodes (Bhattarai et al. [2008]; Nahar et al. [2011]). When exogenous ethephon and methyl jasmonate were supplied to rice plants, *M. graminicola* was less effective in counteracting root defense pathways. Here, the *OsAOS2* gene, which codes the first enzyme for JA biosynthesis pathway, was not induced in response to nematode infection at the time studied. This suggests that the plant did not perceive nematode attack or that proper signalisation was impaired.

Mitogen-activated protein kinase (MAPK) cascades are activated in plants during responses to pathogens and mediate intracellular stress responses, including reactive oxygen species (ROS) production, cell death, and activation of PR gene expression. The *OsMAPK6* cascade plays an important role in both pathogen-associated molecular pattern (PAMP)-triggered immunity (PTI) and effector-triggered immunity (ETI) in rice (Liu et al. [2013]). *OsMAPK6, OsMAPK5a* and the rice small GTPase *OsRac1* act together to regulate plant defense responses to pathogens or to pathogen-derived elicitor signaling in rice (Lieberherr et al. [2005]; Liu et al. [2013]). In this study, *OsMAPK5a* was down-regulated in rice galls, and more specifically in response to *M. incognita* infection. Recently, *OsMAPK5* was shown to negatively modulate PR genes expression in rice, such as *PR1* and *PR10* (Xiong and Yang [2003]; Seo et al. [2011]). When *OsMAPK5a* is knocked-down in rice, plants appear more resistant to fungal *(M. oryzea)* and bacterial *(Burkholderia glumae)* pathogens (Xiong and Yang, [2003]). However, *OsMAPK5a* transcription has been reported to be activated in response to *M. oryzae* (for both virulent and avirulent isolates) few hours after inoculation, but it is further down-regulated until 3 DAI (Delteil et al. [2012]). Here it is thus not clear whether *OsMAPK5a* would also act as a negative regulator of rice defense responses to nematodes.

It is now known that the local suppression of host defense signaling observed after nematode infection is the result of virulence effectors secreted into the host plant to facilitate infection (Haegeman et al. [2012]; Mitchum et al. [2013]). *Meloidogyne* spp. potentially secrete hundreds of proteins in its host (Bellafiore et al. [2008]). In this study, we showed that *M. incognita* expressed the calreticulin *Mi-CRT* gene all along its infection cycle in Nipponbare roots. In *A. thaliana*, Mi-CRT is secreted *in planta* throughout parasitism (Jaubert et al. [2005]) and plays a role as immune-modulator in the suppression of plant basal defenses (Jaouannet et al. [2013]). It is thus expected that Mi-CRT play a similar role in interfering with the plant immune system PTI in distantly related hosts. Calreticulins are highly conserved calcium-binding proteins in plants and animals that act as Ca^{2+} - binding chaperones, regulating Ca^{2+} storage and signalling in the cell. But, how Mi-CRT contributes to the infection process of the nematode remains unknown and should be further investigated in particular during rice infection.

CONCLUSIONS

Our data demonstrate that the *M. incognita*-rice pathosystem may be a novel additional model system to dissect the complex cellular and molecular nematode interactions with monocotyledonous host plants. A benefit of the *Meloidogyne*-rice pathosystem over other plant-nematode models is the specific resistance identified in the African relative species *O. glaberrima* (Soriano et al. [1999]). The rice, *O. sativa* and *O. glaberrima* - *Meloidogyne* spp. interactions thus may serve as a model to understand compatible and incompatible plant- nematode interactions respectively and to elucidate the molecular mechanisms developed by these parasites to infect their monocotyledonous host plants.

METHODS

Biological material

Meloidogyne incognita isolate 1 (race 1, USA; Li et al. [2007]) was propagated from greenhouse-grown tomato plants (*Solanum lycopersicum* L. cv. Naine). Second-stage juvenile (J2) nematodes were hatched from sterilized eggs as described (Bellafiore et al. [2008]). Eggs were hatched at room temperature for 5 days, and J2 worms were allowed to crawl through six Kimwipe tissue layers in water with 100 mg/l streptomycin. The population of *M. graminicola* used in all experiments was originally collected from Laurel (Batangas, Philippines) and cultured on *Oryza sativa* cv. IR64. Eggs were extracted from infected roots by shaking *M. graminicola*-infected roots in bleach 0.7% for 5 min and mixing them in a "blender" for 5 times during 1s. Eggs were collected onto a 25 µm mesh and were hatched at room temperature. Only J2 nematode populations collected after a maximum of 96 h were used as inoculums.

Nematode inoculation assays on rice plants

Oryza sativa subsp. *japonica* cv. Nipponbare seeds were germinated on sand wetted with Hoagland ¼ solution for 7 days and then transferred to tubes containing 10 g SAP substrate (Reversat et al. [1999]) wetted with Hoagland ¼ solution (KNO_3 5 mM; KH_2PO_4 1 mM; $Ca(NO_3)_2$ 5 mM; $MgSO_4$ 2 mM; 25 mg iron; trace element). Rice plants were maintained in a growth chamber under controlled conditions at 26°C/24°C day/night temperature, under a 14 h day/ 10 h night light regime (60 µmol m^{-2} s^{-1} illumination) and 78% relative humidity. Three days after transplanting into SAP (Sand and Absorbent Polymer) substrate (Reversat et al. [1999]), the plants were inoculated with 1 ml water containing 400 freshly hatched stage J2 juveniles of *M. graminicola* or *M. incognita*. One day after inoculation (DAI) rice plants were transferred to a 15 ml hydroponic culture system with Hoagland ¼ solution (Reversat et al. [1999]) to synchronize the infection process.

Histopathology study

Infected roots were harvested at 1, 2, 4, 7, 15, 22, 31, 35 and 42 DAI, carefully washed and immediately placed in freshly prepared fixative (2% paraformaldehyde - 1% glutaraldehyde - 1% cafein (Sigma-Aldrich) in 0.5 × phosphate buffer (Sigma-Aldrich). Root tips (1 cm segment) or when visible, galls were excised from each plant, fixed for 15 h in PFA, dehydrated for 1 h in each ethanol dilution (once 50% and twice 70% vol/vol) and embedded in the epoxy resin Technovit 7100 (Kulzer Friedrichsdorf, Germany) according to Pegard et al. ([2005]). Blocks containing galls of different time points were sectioned (10 μm), mounted in 90% glycerol and microscopically observed under UV light (UV filter set A2, Zeiss AXIO Imager). The same sections were subsequently stained (3 min at room temperature) with 0.05% toluidine blue in 0.1 M sodium phosphate buffer, pH 5.5. Images were taken with an Axiocam digital camera (Leica microscope) with standard bright-field optics.

RNA extraction and cDNA synthesis

Root tips were excised from infected and non-infected rice plantlets, immediately frozen in liquid nitrogen and kept at -8°C until use. Total RNA was extracted from rice root samples using the RNeasy Plant kit (Qiagen, France), with addition of an on-column DNase I digestion. For quantification, the absorbance from 1 μL RNA samples were measured using the NanoDrop ND-1000 spectrophotometer (NanoDrop Technologies), and 1% agarose gel was run to visualize the quality of the RNA. First-strand cDNAs were synthesized from 1 μg of total RNA in 20 μl final volume, using Omniscript RT kit (Qiagen) and oligo-dT(18)-MN primer (Eurogentec, Angers, France) following the manufacturer's instructions.

Reference genes selection in *M. incognita*

Candidate reference genes of *M. incognita* to normalize RT-qPCR studies were selected based on a study on the model nematode *Caenorhabditis elegans* (Hoogewijs et al. [2008]). The *M. incognita* ortholog genes were searched in the genome (http://www6.inra.fr/meloidogyne_incognita), and specific primers were designed for RT-qPCR. Only the primers designed for *Mi-csq-1*, *Mi-eif-3C*, *Mi-gpd-2*, *Mi-Y45F10D.4*, and *Mi-actin* genes displayed correct amplification efficiency and specificity. Transcript accumulation of candidate genes was detected in rice roots after infection with *M. incognita*. The stability of each *M. incognita* gene expression during rice infection was analyzed using RefFinder (Xie et al. [2012]; http://www.leonxie.com/referencegene.php) and the most reliable gene was selected to normalise RT-qPCR data on *M. incognita* gene expression analysis in rice tissues.

Primer design and selection

Specific primers were designed from the *O. sativa* or *M. incognita* cDNA sequences using the Beacon Designer 5.0 software (Premier Biosoft International, Palo Alto, CA, USA), with melting temperatures (Tm) of 58 ± 5°C, primers lengths of 18 to 25 bp, and amplicon lengths of 75 to 200 bp. Primers were designed from the 3′ region of the gene to ensure gene specificity. For each primer pair, a preliminary real-time assay was performed on *O. sativa* or *M. incognita* pure and mixed cDNAs samples to evaluate the amplification of non-specific products or primer dimmer artefacts. The primer efficiency was experimentally tested with the LinRegPCR programme developed by Ramakers et al. ([2003]) which uses a linear regression analysis of fluorescence data from the exponential phase of PCR amplification to determine amplification efficiency (E). The specificity of PCR products was checked by melting curve analyses and only primers sets producing a single sequence-specific peak in the dissociation curve were used.

Real-time quantitative PCR assays of gene expression

Primers listed in Table 1 were synthesized by Eurogentec and used at 200 nM final concentration. Quantitative RT-PCR was performed using the Stratagene MX3005P with MxPro v 3.00 software for Comparative Quantitation (Stratagene, La Jolla, CA). Quantitative PCR was carried out on 1.25 ng cDNA in a 15 µL amplification mixture containing MESA BLUE Master Mix Plus for SYBR Assay low ROX (Eurogentec, Belgium). The cycling conditions comprised 10 min polymerase activation at 95°C followed by 40 cycles at 95°C for 15 sec, 55°C for 20 sec and 72°C for 40 sec. Following cycling, the melting curve was determined in the range 55°-°95°C, with a temperature slope of 0.01°C/sec. Each assay was conducted in duplicates and included a no-template control.

Table 1. Primers used for reverse transcription-quantitative polymerase chain reaction gene expression studies

Name	Gene	Forward	Reverse primer
Mi-CRT (calreticulin)	GenBank: AAL40720	GGCTCTGTTGGTATTGACATC	CTTGCGTTCTTCCTCATCTGC
Mi-actin	Minc06769*	GCTTTGCTATGTTGCTTTGG	TGTAAGAAGTCTCGTGAATACC
Mi-csq-1	Minc11275*	TGATTATTTACAGGAAATGTTTGG	TAGGGTCGTCTAAATTTAATTGG
Mi-eif-3C	Minc06181*	AAATTCTTTCGTAGTGCTGTTTC	AAATCTTGTCGTCCTAATGGC
Mi-gpd-2	Minc12412a*	AAGCCGTTCTTTCTTTGTATG	AAGCATAACCTTCATAGATTGG
Mi-Y45F10D.4	Minc08763*	CAAAGATGATCCCACAATAGG	AAAGTTTTGAATTTGGCATCG
Os-actin (actin-1)	RAP-DB: Os03g0718100	CTCTCAGCACATTCCAGCAG	AGGAGGACGGCGATAACAG
OsACS2	RAP-DB: Os03t0225900	GCGAGAGACGGAGAACCC	CGACGAGCAACAGCCTTC
OsMAPK6	RAP-DB: Os06g0154500	GATACATTCGCCAACTTCC	CAGTGATGCCAGGTAAGG
OsMPAK5a	RAP-DB: Os03g17700	GTCTGCTCCGTGATGAAC	TGATGCCTATGATGTTCTCG
OsMPAK20	RAP-DB: Os01t0629900	TCAACTCCAATTCCTGCCAAG	AACAACTCTTCCTGGTCTTGC
OsNH1 (NPR1)	RAP-DB: Os01t0194300	AGAAGTCATTGCCTCCAG	ACATCGTCAGAGTCAAGG
OsRAC1	RAP-DB: Os01t0229400	GCTTCTTCCATAATAACAACG	AGTTTCTTTCTGGTTACATCC
OsEDS1	RAP-DB: Os09t0392100	CAGGAGAGGCAGTGTTAATCAG	GCAAGCGGAGTAAGTGGTATG
OsPAD4	RAP-DB: Os11t0195500	TCAGAGGCAAGGCAGTAGTG	ACCGCTCACGCAGGATAG
OsWRKY13	RAP-DB: Os01t0750100	GCCAGCGGAGAACGAATC	CTCCTCCTGCTTCACAACC

*http://www6.inra.fr/meloidogyne_incognita.

Baseline and threshold values were automatically determined for all plates and genes using the MxPro software ver. 4.1.0.0 (Stratagene). In order to ensure comparability between data obtained from different genes, in each run all samples were in a same plate.

Data analysis

Analysis of RT-qPCR fluorescence data with LinRegPCR determined E values for each reaction. Individual Cq values were considered as valid only if the amplification parameters passed all quality checks. For the relative gene expression in rice and nematode samples, data were analyzed using the MxPro software package to obtain the relative expression levels of rice genes. For each sample, the mean Cq value was calculated based on Cq values of both replicates. *Mi-crt* gene expression was normalized to the *Mi-actin* gene (Table 1). *Os-* genes expression was normalized to the *Os-actin* gene (Table 1). Based on the comparative Ct method, data are either expressed as fold changes to calibrator average or as log (base 2) fold changes to calibrator average.

AUTHORS' CONTRIBUTIONS

Conceived and designed the experiments: SB, DF. Performed the experiments: NVP, SB, ASP, RH, AB, AA, IM, DF. Analyzed the data: NVP, ASP,

RH, JA, IM, DF. Wrote or proofread the paper: DF, JA, PG, ASP, SB, NVP. All authors read and approved the final manuscript.

ACKNOWLEDGEMENTS

We are grateful to Dr. Georges Reversat (IRD) for help in establishing the nematode system in our laboratory and for his constant support and advices. We are grateful to Jamel Aribi (IRD) for plant maintenance and to Myriam Collin (IRD) for technical assistance in histology. Funding for this project was provided by IRD, a Vietnam governmental PhD grant to NVP, a Syrian governmental MSc grant for RH, a Brazilian Federal Agency for Support and Evaluation of Graduate Education (CAPES) PhD grant for IM, and an International Atomic Energy Agency (IAEA) grant for AA. JAE had a 1 year foreign visiting professor grant and DF a 9-months foreign visiting researcher grant (PVE, CSF) to Embrapa-Cenargen from the Brazilian CAPES, and work has also been supported by the Brazilian-French CAPES/COFECUB program. This project is supported by Agropolis Fondation under the reference ID "Identification of nematode (*Meloidogyne* spp.) effectors of pathogenicity in rice (*Oryza sativa*)" AA 1002-003.

COMPETING INTERESTS

The authors declare that they have no competing interests.

REFERENCES

1. Abad P, Gouzy J, Aury J, Castagnone-Sereno P, Danchin E, Deleury E, Perfus-Barbeoch L, Anthouard V, Artiguenave F, Blok V, Caillaud MC, Coutinho PM, Dasilva C, De Luca F, Deau F, Esquibet M, Flutre T, Goldstone JV, Hamamouch N, Hewezi T, Jaillon O, Jubin C, Leonetti P, Magliano M, Maier TR, Markov GV, McVeigh P, Pesole G, Poulain J, Robinson-Rechavi M, *et al.*: Genome sequence of the metazoan plant-parasitic nematode *Meloidogyne incognita* . *Nat Biotechnol* 2008, 26(8):909–915.
2. Barcala M, Garcia A, Cabrera J, Casson S, Lindsey K, Favery B, García-Casado G, Solano R, Fenoll C, Escobar C: Early transcriptomic events in microdissected *Arabidopsis* nematode-induced giant cells. *Plant J* 2010, 61: 698–712.
3. Bellafiore S, Briggs SP: Nematode effectors and plant responses to infection. *Curr Opin Plant Biol* 2010, 13(4):442–448.
4. Bellafiore S, Shen Z, Rosso MN, Abad P, Shih P, Briggs SP (2008) Direct identification of the *Meloidogyne incognita* secretome reveals proteins with host cell reprogramming potential. PLoS Pathog 4(10):p.e1000192 Bellafiore S, Shen Z, Rosso MN, Abad P, Shih P, Briggs SP (2008) Direct identification of the Meloidogyne

incognita secretome reveals proteins with host cell reprogramming potential. PLoS Pathog 4(10):p.e

5. Bhattarai KK, Xie QG, Mantelin S, Bishnoi U, Girke T, Navarre DA, Kaloshian I: Tomato susceptibility to root-knot nematodes requires an intact jasmonic acid signaling pathway. *Mol Plant-Microbe Interact* 2008, 21(9):1205–1214.
6. Bridge J, Plowright RA, Peng D: Nematode Parasites Of Rice. In *Plant-Parasitic Nematodes in Subtropical and Tropical Agriculture*. Edited by: Luc M, Sikora RA, Bridge J. CAB International, Wallingford, U.K; 2005:87–130.
7. Cabasan MTN, Kumar A, De Waele D: Comparison of migration, penetration, development and reproduction of *Meloidogyne graminicola* on susceptible and resistant rice genotypes. *Nematology* 2012, 14(4):405–415.
8. Cabasan MTN, Kumar A, Bellafiore S, De Waele D: Histopathology of the rice root-knot nematode, *Meloidogyne graminicola*, on *Oryza sativa* and *O. glaberrima*. *Nematology* 2013, 00: 1–9.
9. Caillaud M-C, Dubreuil G, Quentin M, Perfus-Barbeoch L, Lecomte P, De A, Engler J, Abad P, Rosso MN, Favery B: Root-knot nematodes manipulate plant cell functions during a compatible interaction. *J Plant Physiol* 2008, 165(1):104–113.
10. Danchin EGJ, Rosso MN, Vieira P, De Almeida-Engler J, Coutinho PM, Henrissat B, Abad P: Multiple lateral gene transfers and duplications have promoted plant parasitism ability in nematodes. *Proc Natl Acad Sci U S A* 2010, 107(41):17651–17656.
11. De Almeida Engler J, De Vleesschauwer V, Burssens S, Celenza JL Jr, Inzé D, Van Montagu M, Engler G, Gheysen G: Molecular markers and cell cycle inhibitors show the importance of cell cycle progression in nematode-induced galls and syncytia. *Plant Cell* 1999, 11: 793–807.
12. De Araújo Filho JV, Machado A, Ferraz L: Host status of some selected Brazilian upland rice cultivars to Meloidogyne incognita race 4 and Rotylenchulus reniformis. *Nematology* 2010, 12(6):929–934.
13. De Waele D, Elsen A: Challenges in tropical plant nematology. *Annu Rev Phytopathol* 2007, 45(1):457–485.
14. Delteil A, Zhang J, Lessard P, Morel JB: Potential candidate genes for improving rice disease resistance. *Rice* 2010, 3(1):56–71.
15. Delteil A, Blein M, Faivre-Rampant O, Guellim A, Estevan J, Hirsch J, Bevitori R, Michel C, Morel JB: Building a mutant resource for the study of disease resistance in rice reveals the pivotal role of several genes involved in defence. *Mol Plant Pathol* 2011, 13(1):72–82.
16. Diomandé M: Response of upland rice cultivars to *Meloidogyne* species. *Revue Nématologie* 1984, 7(1):57–63.
17. Falk A, Feys BJ, Frost LN, Jones JD, Daniels MJ, Parker JE: EDS1, an essential component of R gene-mediated disease resistance in *Arabidopsis* has homology to eukaryotic lipases. *Proc Natl Acad Sci U S A* 1999, 96(6):3292–3297.
18. Fortuner R, Merny G: Root-parasitic nematodes of rice. *Revue Nématologie* 1979, 2(1):79–102.
19. Golden AM, Birchfield W (1968) Rice root-knot nematode *Meloidogyne graminicola* as a new pest of rice. Plant Di Rep 52:423 Golden AM, Birchfield W (1968) Rice root-knot nematode Meloidogyne graminicola as a new pest of rice. Plant Di Rep 52:423
20. Haegeman A, Mantelin S, Jones JT, Gheysen G: Functional roles of effectors of plant-parasitic nematodes. *Gene* 2012, 492(1):19–31.
21. Hofmann J, el Ashry AN, Anwar S, Erban A, Kopka J, Grundler F: Metabolic profiling reveals local and systemic responses of host plants to nematode parasitism. *Plant J* 2010, 62(6):1058–1071.
22. Hoogewijs D, Houthoofd K, Matthijssens F, Vandesompele J, Vanfleteren JR: Selection and validation of a set of reliable reference genes for quantitative sod gene expression analysis in *C. elegans*. *BMC Mol Biol* 2008, 9(1):1–8.

23. Ibrahim IA, Ibrahim IKA, Rezk MA: Host parasite relationship of *Meloidogyne incognita* (Kofoid & White) Chitw. on rice. *Nematol Mediterranea* 1973, 1: 8–14.
24. Jaouannet M, Magliano M, Arguel MJ, Gourgues M, Evangelisti E, Abad P, Rosso MN: The root-knot nematode calreticulin Mi-CRT is a key effector in plant defense suppression. *Mol Plant-Microbe Interact* 2013, 26(1):97–105.
25. Jaubert S, Milac AL, Petrescu AJ, de Almeida-Engler J, Abad P, Rosso MN: *In planta* secretion of a calreticulin by migratory and sedentary stages of root-knot nematode. *Mol Plant-Microbe Interact* 2005, 18(12):1277–1284.
26. Ji H, Gheysen G, Denil S, Lindsey K, Topping JF, Nahar K, Kyndt T: Transcriptional analysis through RNA sequencing of giant cells induced by *Meloidogyne graminicola* in rice roots. *J Exp Bot* 2013, 64(12):3885–3898.
27. Jirage D, Tootle TL, Reuber TL, Frost LN, Feys BJ, Parker JE, Ausubel FM, Glazebrook J: *Arabidopsis thaliana PAD4* encodes a lipase-like gene that is important for salicylic acid signaling. *Proc Natl Acad Sci U S A* 1999, 96(23):13583–13588.
28. Kyndt T, Vieira P, Gheysen G, de Almeida-Engler J (2013) Nematode feeding sites: unique organs in plant roots. Planta, DOI 10.1007/s00425–013–1923-z Kyndt T, Vieira P, Gheysen G, de Almeida-Engler J (2013) Nematode feeding sites: unique organs in plant roots. Planta,
29. Li XQ, Wei JZ, Tan A, Aroian RV: Resistance to root-knot nematode in tomato roots expressing a nematicidal *Bacillus thuringiensis* crystal protein. *Plant Biotechnol J* 2007, 5: 455–464.
30. Lieberherr D, Thao NP, Nakashima A, Umemura K, Kawasaki T, Shimamoto K: A sphingolipid elicitor-inducible mitogen-activated protein kinase is regulated by the small GTPase OsRac1 and heterotrimeric G-Protein in rice. *Plant Physiol* 2005, 138: 1644–1652.
31. Liu W, Liu J, Ning Y, Ding B, Wang X, Wang Z, Wang GL: Recent progress in understanding PAMP- and effector-triggered immunity against the rice blast fungus *Magnaporthe oryzae* . *Mol Plant* 2013, 6(3):605–620.
32. Mitchum MG, Hussey RS, Baum TJ, Wang X, Elling A, Wubben M, Davis EL: Nematode effector proteins: an emerging paradigm of parasitism. *New Phytol* 2013, 199: 879–894.
33. Molinari S, Fanelli E, Leonetti P (2013) Expression of tomato SA-responsive pathogenesis-related genes in Mi-1-mediated and SA-induced resistance to root-knot nematodes. Mol Plant Pathol, DOI: 10.1111/mpp.12085 Molinari S, Fanelli E, Leonetti P (2013) Expression of tomato SA-responsive pathogenesis-related genes in Mi-1-mediated and SA-induced resistance to root-knot nematodes. Mol Plant Pathol, DOI:
34. Nahar K, Kyndt T, De Vleesschauwer D, Höfte M, Gheysen G: The jasmonate pathway is a key player in systemically induced defense against root knot nematodes in rice. *Plant Physiol* 2011, 157(1):305–316.
35. Opperman CH, Bird DM, Williamson VM, Rokhsar DS, Burke M, Cohn J, Cromer J, Diener S, Gajan J, Graham S, Houfek TD, Liu Q, Mitros T, Schaff J, Schaffer R, Scholl E, Sosinski BR, Thomas VP, Windham E: Sequence and genetic map of *Meloidogyne hapla* : A compact nematode genome for plant parasitism. *Proc Natl Acad Sci U S A* 2008, 105(39):14802–14807.
36. Pegard A, Brizzard G, Fazari A, Soucaze O, Abad P, Djian-Caporalino C: Histological characterization of resistance to different root-knot nematode species related to phenolics accumulation in *Capsicum annuum* . *Phytopathology* 2005, 95: 158–165.
37. Perry RN, Moens N: Introduction To Plant-Parasitic Nematodes: Modes Of Parasitism. In *Genomics and Molecular Genetics of Plant-Nematode Interactions*. Edited by: Jones J, Gheysen G, Fenoll C. Springer, Dordrecht, the Netherlands; 2011:3–20.
38. Prasad JS, Vijayakumar CHM, Sankar M, Varaprasad KS, Prasad MS, Rao YK: Root-knot nematode resistance in advanced back cross populations of rice developed for water stress conditions. *Nematol Mediterr* 2006, 34: 3–8.

39. Prot JC, Matias D: Effects of water regime on the distribution of *Meloidogyne graminicola* and other root-parasitic nematodes in a rice field toposequence and pathogenicity of *M. graminicola* on rice cultivar UPL R15. *Nematologica* 1995, 41: 219–228.
40. Qiu D, Xiao J, Xie W, Cheng H, Li X, Wang S (2009) Exploring transcriptional signalling mediated by *OsWRKY13*, a potential regulator of multiple physiological processes in rice. BMC Plant Biol 9:74 Qiu D, Xiao J, Xie W, Cheng H, Li X, Wang S (2009) Exploring transcriptional signalling mediated by OsWRKY13, a potential regulator of multiple physiological processes in rice. BMC Plant Biol 9:74
41. Ramakers C, Ruijter JM, Deprez RH, Moorman AF: Assumption-free analysis of quantitative real-time polymerase chain reaction (PCR) data. *Neurosci Lett* 2003, 339(1):62–66.
42. Reversat G, Boyer J: Use of a mixture of sand and water-absorbent synthetic polymer as substrate for the xenic culturing of plant-parasitic nematodes in the laboratory. *Nematology* 1999, 1(2):209–212.
43. Rietz S, Stamm A, Malonek S, Wagner S, Becker D, Medina-Escobar N, Vlot AC, Feys BJ, Niefind K, Parker JE: Different roles of Enhanced Disease Susceptibility1 (EDS1) bound to and dissociated from Phytoalexin Deficient4 (PAD4) in *Arabidopsis* immunity. *New Phytol* 2011, 191(1):107–119.
44. Rosso M-N, Hussey R: Nematode Effectors Protein: Targets And Functions In Plant Parasitism. In *Effectors in Plant-Microbe Interactions*. Edited by: Martin F, Kamoun S. Wiley-Blackwell, Oxford, UK; 2012:327–354
45. Seo Y-S, Chern M, Bartley LE, Han M, Jung KH, Lee I, Walia H, Richter T, Xu X, Cao P, Bai W, Ramanan R, Amonpant F, Arul L, Canlas PE, Ruan R, Park CJ, Chen X, Hwang S, Jeon JS, Ronald PC (2011) Towards establishment of a rice stress response interactome. PLoS Genet 7(4):e1002020
46. Soriano IR, Schmidt V, Brar D, Prot JC, Reversat G: Resistance to rice root-knot nematode *Meloidogyne graminicola* identified in *Oryza longistaminata* and *O. glaberrima* . *Nematology* 1999, 1(4):395–398.
47. Tao Z, Liu H, Qiu D, Zhou Y, Li X, Xu C, Wang S: A pair of allelic *WRKY* genes play opposite role in rice-bacteria interactions. *Plant Physiol* 2009, 151(2):936–948.
48. Triantaphyllou AC, Hirschmann H: Post-infection development of *Meloidogyne incognita* Chitwood, 1949 (Nematoda: Heteroderidae). *Ann Inst Phytopathologique Benaki* 1960, 3(1):1–11.
49. Trudgill DL, Blok VC: Apomictic, polyphagous root-knot nematodes: exceptionally successful and damaging biotrophic root pathogens. *Annu Rev Phytopathol* 2001, 39: 53–77.
50. Wiermer M, Feys BJ, Parker JE: Plant immunity: the *EDS1* regulatory node. *Curr Opin Plant Biol* 2005, 8(4):383–389.
51. Williamson VM, Gleason CA: Plant-nematode interactions. *Curr Opin Plant Biol* 2003, 6(4):327–333.
52. Wu KL, Guo ZJ, Wang HH, Li J: The WRKY family of transcription factors in rice and *Arabidopsis* and their origins. *DNA Res* 2005, 12(1):9–26.
53. Xie F, Xiao P, Chen D, Xu L, Zhang B: miRDeepFinder: a miRNA analysis tool for deep sequencing of plant small RNAs. *Plant Mol Biol* 2012, 80: 75–84.
54. Xiong L, Yang Y: Disease resistance and abiotic stress tolerance in rice are inversely modulated by an abscisic acid-inducible Mitogen-Activated Protein Kinase. *Plant Cell* 2003, 15(3):745–759.

CHAPTER 18

Genetic Identification of Anisakid Nematodes Isolated From Largehead Hairtail (Trichiurus Japonicus) in Korea

Jeong-Ho Kim[1]*, Woo-Hwa Nam[1] and Chan-Hyeok Jeon[2]

[1] *Department of Marine Bioscience, Gangneung-Wonju National University, Gangneung, Gangwon 210-702, South Korea.*

[2] *East Coast Life Science Institute, Gangneung-Wonju National University, Gangneung 210-702, South Korea*

ABSTRACT

Background

The nematode species belonging to genus *Anisakis* occur at their third larval stage in numerous marine teleost fish species worldwide and known to cause accidental human infection through the ingestion of raw or undercooked fish or squids. They may also draw the attention of consumers because of the visual impact of both alive and dead worms. Therefore, the information on their geographical distribution and clear species identification is important for epidemiological survey and further prevention of human infection.

Results

For identification of anisakid nematodes species isolated from largehead hairtail (*Trichiurus japonicus*), polymerase chain reaction-restriction fragment length polymorphism (PCR-RFLP) analysis of internal transcribed spacers of ribosomal DNA were conducted. Mitochondrial cytochrome c oxidase subunit 2 gene was also sequenced, and phylogenetic analysis was conducted. From the largehead hairtail ($n = 9$), 1259 nematodes were isolated in total. Most of the nematodes were found encapsulated throughout the viscera (56.2 %, 708/1259) or moving freely in the body cavity (41.5 %, 523/1259), and only 0.3 % (4/1259) was found in the muscles. By PCR-RFLP, three different nematode species were identified. *Anisakis pegreffii* was the most dominantly found (98.7 %, 1243/1259) from the largehead hairtail, occupying 98.7 % (699/708) of the nematodes in the mesenteries and 98.1 % (513/523) in the body cavity. Hybrid genotype (*Anisakis simplex* × *A. pegreffii*) occupied 0.5 %, and *Hysterothylacium* sp. occupied 0.2 % of the nematodes isolated in this study.

Conclusions

The largehead hairtail may not significantly contribute accidental human infection of anisakid nematode third stage larvae because most of the nematodes were found from the viscera or body cavity, which are not consumed raw. But, a high prevalence of anisakid nematode larvae in the largehead hairtail is still in concern because they may raise food safety problems to consumers. Immediate evisceration or freezing of fish after catch will be necessary before consumption.

Keywords: *Anisakis pegreffii Hysterothylacium* sp. Largehead hairtail *Trichiurus japonicus* Cutlass fish

BACKGROUND

Anisakid nematodes (Nematoda: Anisakidae) infect animals belonging to almost all phyla and commonly found in aquatic vertebrates. Their life cycles involve various hosts at different levels in marine food webs. Crustaceans work as the first intermediate hosts; fish and cephalopods work as intermediate or paratenic hosts and marine mammals and fish-eating birds as definitive hosts (Mattiucci and Nascetti 2008). Since Van Thiel et al. (1960) reported that the third stage larvae (L3) of *Anisakis* from the Atlantic herring *(Clupea harengus)* can infect human, the number of clinical cases has been increasing worldwide, with the increase of knowledge about this parasite and improvement of diagnostics, as well as more globalized customs of eating raw marine fish (Cipriani et al. 2015).

The consumption of raw or undercooked fish harboring anisakid nematode L3 can lead to human infection (Audicana and Kennedy 2008). According to Fagerholm (1991), 10 genera exist in the family Anisakidae. However, the human infection is known to be caused frequently by the L3 of the genera *Anisakis* Dujardin, 1845 and less frequently by *Pseudoterranova* Krabbe, 1878 (Sakanari and McKerrow 1989). Of the genus *Anisakis, A. simplex* sensu stricto (s.s.) and *A. pegreffii* are recognized as zoonotic species causing anisakiasis in Asia and Europe, respectively (Mattiucci et al. 2013; Umehara et al. 2007). Thus, not all the nematode larvae found in fish are infective to human, and a clear diagnosis is important for proper epidemiological survey and further prevention of human infection.

Identification of anisakid nematodes has been traditionally conducted by morphological observation, but it is not always sufficient to make a clear identification at the species level because morphological characters of taxonomic significance in this group are very few, particularly for the larval stages lacking reliable diagnostic features at the species level (Mattiucci and Nascetti 2008). Moreover, morphological identification of nematodes is not always efficient when a huge number of the specimens have to be identified or the specimens are physically damaged and the morphological keys are missed. DNA-based diagnostic techniques such as polymerase chain reaction-restriction fragment length polymorphism (PCR-RFLP) of internal transcribed spacer (ITS) region, direct sequencing of ribosomal DNA (rDNA), and mitochondrial DNA cytochrome c oxidase subunit 2 (mtDNA cox2) gene analysis have solved these constraints and have been widely used (Mattiucci and Nascetti 2008 and the references therein). Current taxonomy of the genus *Anisakis* based on these techniques revealed that there are two clades including nine species in this genus, and *A. simplex*, the most frequently encountered *Anisakis* species in human infection belongs to clade I (Mattiucci and Nascetti 2008).

The largehead hairtail or cutlassfish (Trichiuridae: *Trichiurus* species) distributes in circumtropical and temperate waters of the world and has been considered as one of the economically important fish species in the western North Pacific including Korea (Nakamura and Parin 1993). The annual catch of the largehead hairtail in Korea has been fluctuating between 50,000 and 90,000 t, after a sharp decrease in mid-1990 (Kim et al. 2011). Thus, importation of frozen fish from several countries has been increasing to meet the domestic demand (Cha and Kim 2009). Juveniles feed mostly on euphausiids, small pelagic planktonic crustaceans and small fishes, and adults feed mainly on fishes and occasionally on squids and crustaceans

(Nakamura and Parin 1993). All of these food items are known to be the intermediate hosts of anisakid nematodes (Mattiucci and Nascetti 2008).

There have been many papers regarding anisakid nematodes isolated from *Trichiurus* species (Borges et al. 2012; Kong et al. 2015; Lee et al. 2009; Shih 2004; Umehara et al. 2010; Zhang et al. 2013). But, the identification of host species was not clearly conducted, and several scientific names (e.g., *Trichiurus lepturus*, *Trichiurus haumela*, *Trichiurus* sp.) have been indiscriminately used for the largehead hairtail in their studies. However, clear identification of host species, together with the information on their geographical distribution and feeding behavior are important to understand the occurrence of anisakid nematodes and predict the likelihood of infection in a given area. In particular, the largehead hairtails are often sold as fresh or frozen fillets, which are difficult to identify since most of the morphological features are no longer available. In this case, other techniques are necessary to identify these morphologically similar and closely related species in commercial products.

Although the largehead hairtail is commercially and economically important fish species in Korea, its anisakid nematode fauna has not been clearly investigated and only fragmentary information exists. Lee et al. (2009) reported *A. pegreffii* and *Anisakis typica* from several marine fish species including *T. lepturus*, but they did neither mention the origin of the fish nor the prevalence of infection in each host fish species. Moreover, there was no information in the microhabitat of anisakid nematodes in host fish. In this study, we investigated anisakid nematode fauna from the largehead hairtail caught from Korean waters. We collected nematodes in the largehead hairtail freshly caught from coastal waters around Jeju, Korea. The host species were identified by mitochondrial DNA (mtDNA) cytochrome oxidase I (COI) gene sequencing described by Hsu et al. (2009). The collected nematodes were identified by PCR-RFLP with subsequent sequencing and mtDNA cox2 gene sequencing, to provide information on the epidemiology of *Anisakis* infection in Korea.

METHODS

Nematodes collection

Fresh largehead hairtail ($n = 9$) caught by lines were purchased from local fisheries market located in Hallim, Jeju Island, off the southern coast of the Korean Peninsula, 2015. The freshly caught fish were placed on flake ice and immediately transported to the laboratory. All of the individual fish were measured and weighed. Then, they were dissected to collect nematodes. The body cavity was longitudinally opened, and the viscera were carefully examined for collecting nematodes. The muscles were sliced, placed, and pressed between transparent glass plates, then inspected with the naked eyes under the light. All of the collected live nematodes were washed repeatedly with 0.9 % NaCl solution and preserved individually in absolute alcohol for molecular analysis. Prevalence (the number of host infected with parasites divided by the number of hosts examined) and mean intensity (the average of parasite infection among the infected hosts) were used as quantitative descriptors of the parasite population as previously described by Bush et al. (1997).

DNA extraction and PCR

Before genomic DNA extraction, the nematodes were washed three times with PBS, then placed in 1.5 ml Eppendorf tube containing 400 µL TNES lysis buffer and crushed with a sterile pestle. Genomic DNA was extracted using the phenol-chloroform method with slight modification (Wasko et al. 2003). The extracted DNA was resuspended with 50 µL TE buffer (10 mM Tris-HCl, pH 7.5, 1 mM EDTA). DNA concentration and purity were measured using Nanodrop 1000 (Thermo Scientific, USA). The ITS region (ITS1-5.8S-ITS2) of rDNA was amplified using primers A (forward: 5′-GTCGAATTCGTAGGTGAACCTGCGGAAGGATCA-3′) and B (reverse: 5′-GCCGGATCCGAATCCTGGTTAGTTTCTTTTCCT-3′) and B (D'Amelio et al. 2000). Amplification was conducted using MyCyclerTM (BioRad, USA), with the following conditions. Denaturation at 94 °C for 5 min, then 35 cycles at 94 °C for 40 s, 54 °C for 40 s, 72 °C for 90 s, and post-amplification at 72 °C for 7 min.

Digestion of PCR products with restriction enzymes

Restriction enzymes *Hinf* I, *Hha* I and *Rsa* I were used in the RFLP analysis for the amplified PCR products (D'Amelio et al. 2000). Restriction endonuclease digestion was performed using 3 µL of PCR products, 0.5 µL of restriction enzymes, 1 µL of CutSmart® buffer (NEB, USA) and distilled water up to a final volume of 10 µL. The digestion with these restriction enzymes was performed at 37 °C for 90 min. The digested products were analyzed by electrophoresis in 1.5 % agarose gel containing ethidium bromide and photographed using Gel Logic 100 Imaging System (Biostep, Germany).

Mitochondrial DNA cox2 gene amplification and sequencing

For the amplification of mtDNA cox2 gene, the primers 210 (reverse: 5′-CACCAACTCTTAAAATTATC-3′) and 211 (forward: 5′-TTTCTAGTTATATA GATTGRTTYAT-3′) were used (Nadler and Hudspeth 2000). After initial denaturation at 94 °C for 5 min, 35 cycles were run at 94 °C for 40 s, 48 °C for 40 s, and 72 °C for 60 s followed by post-amplification step at 72 °C for 7 min. PCR products were purified by AccuPrep® Gel Purification Kit (Bioneer, Korea) according to the manufacturer's instructions. 10 ng/μL of purified PCR products were directly sequenced by ABI Prism 3730 XL DNA Analyzer (PE Applied Biosystems, USA).

Phylogenetic analysis

The sequence data of mtDNA cox2 gene was aligned with the published sequences in GenBank database (NCBI) using Clustal W method (Thompson et al. 1994). The phylogenetic tree was constructed by MEGA version 6 for comparing the genetic relationship among other published sequences by the neighbor-joining criteria (Tamura et al. 2013). The nucleotide sequences were registered in GenBank.

Host identification

Genomic DNA was extracted from the muscle tissue of largehead hairtail individuals according to the methods described by Wasko et al. (2003). PCR amplification of mtDNA COI gene was conducted with the primers and conditions described by Hsu et al. (2009). PCR products were purified and sequenced by the same protocols mentioned above. The obtained sequences were aligned with other published sequences including *T. lepturus*, *T. japonicus*, and *Trichiurus* sp. registered in NCBI using Clustal W method (Thompson et al. 1994). The phylogenetic tree was constructed by MEGA version 6 for comparing the genetic relationship among other published sequences of *Trichiurus* species by the neighbor-joining criteria (Tamura et al. 2013). The nucleotide sequences were registered in GenBank.

RESULTS

We examined nine individual largehead hairtail (mean whole body length = 90.7 ± 5.2 cm, mean body weight = 462.0 ± 47.3 g). All the examined individual largehead hairtail fish harbored nematodes (100 %, 9/9). In total, 1259 nematodes were found and mean intensity was 139.9 larvae/fish (1259/9). Most of the nematodes were found embedded in the surfaces of viscera including mesenteries (56.2 %, 708/1259) or existed freely in the body cavity (41.5 %, 523/1259). Two percent of nematodes were found in the intestines (0.6 %, 7/1259) or stomach (1.4 %, 17/1259). Only 0.3 % (4/1259) of nematodes was found in the muscles. All the information is summarized in Table 1.

Table 1. Infection levels of nematodes in the largehead hairtail in this study

P[a] (%)	MI[b]	% of nematodes (number of nematodes in each microhabitat/total number of nematodes)					
		Viscera			Body cavity	Muscles	Total
		Intestines	Stomach	Mesenteries			
100.0 (9/9)	139.9 (1259/9)	0.6 (7/1259)	1.4 (17/1259)	56.2 (708/1259)	41.5 (523/1259)	0.3 (4/1259)	100 (1259/1259)

[a]Prevalence of infection
[b]Mean intensity

The amplification of the rDNA region produced approximately 1 kb fragment (data not shown). PCR-RFLP analysis using *Hinf*I, *Hha*I, and *Rsa*I restriction enzymes revealed different banding patterns depending on the nematodes species. Digestion of the PCR product with *Hinf*I produced three different patterns, i.e., 350–300–250 bp, 700–350 bp, and 620–350–300–250 bp, respectively (Figure 2a). Digestion using *Hha*I produced two different patterns, i.e., 550–430 bp and 350–250–200–150 bp (Figure 2b), and digestion using *Rsa*I produced two different patterns, i.e., 550–300 bp and 650–220 bp, respectively (Figure 2c). These banding patterns corresponded to the known patterns of *A. pegreffii* (lane 1, 2 in Figure 1a–c), *Hysterothylacium* sp. (lane 3, 4 in Figure 1a–c), and hybrid genotype (*A. simplex* × *A. pegreffii*) (lane 5, 6 in Figure 1a–c), respectively.

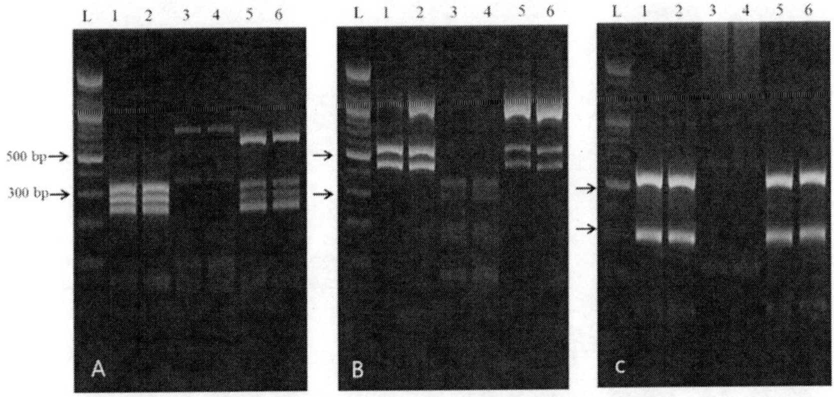

Figure 1. PCR-RFLP profiles of the nematode species obtained by digestion of PCR products with *Hinf*I (**A**), *Hha*I (**B**), and *Rsa*I (**C**) restriction enzymes, respectively (*L* ladder, *1–2*: *A. pegreffii*, *3–4*: *Hysterothylacium* sp., *5–6*: hybrid genotype)

Of 1259 nematodes, most of the nematodes (98.7 %, 1243/1259) were identified as *A. pegreffii*. The rest of them were identified as hybrid genotype, occupying 0.5 % (6/1259) and *Hysterothylacium* sp., occupying 0.2 % (2/1259) of the nematodes. PCR amplification was failed for eight individuals (0.6 %, 8/1259), probably due to the failure of DNA extraction. The results are summarized in Table 2. For *A. pegreffii*, 20 individuals were randomly selected and sequenced for mtDNA cox2 gene because a huge number of samples were collected. For other species, all the individuals were sequenced. All of the samples generated 629 bp size products (data not shown). When the phylogenetic tree was constructed with the mtDNA cox2 sequences of these samples and those previously registered in NCBI, taxonomic position of the *Anisakis* species identified in this study corresponded with the results obtained by PCR-RFLP; The mtDNA cox2 sequences of *A. pegreffii* (Genbank accession number: KU921231-921250) were clustered with those isolated from chub mackerel in Korea and Japan (Figure 2), with 97.9–100.0 % homology. The mtDNA cox2 sequences of *Hysterothylacium* sp. showed the highest homology (92.0 %) with *Hysterothylacium* sp. isolated from chub mackerel (Genbank accession number: KC633443). The mtDNA cox2 sequences of all the hybrid genotype specimens were found to be *A. pegreffii* (data not shown).

Table 2. Anisakid nematodes assemblage of the largehead hairtail examined in this study

Microhabitat/ species name		*A. pegreffii*	Hybrid	*Hysterothylacium* sp.	Fail	Total
Viscera	Intestines	7	0	0	0	7
	Stomach	17	0	0	0	17
	Mesenteries	701	3	0	4	708
Body cavity		514	3	2	4	523
Muscle		4	0	0	0	4
	Total	1243	6	2	8	1259

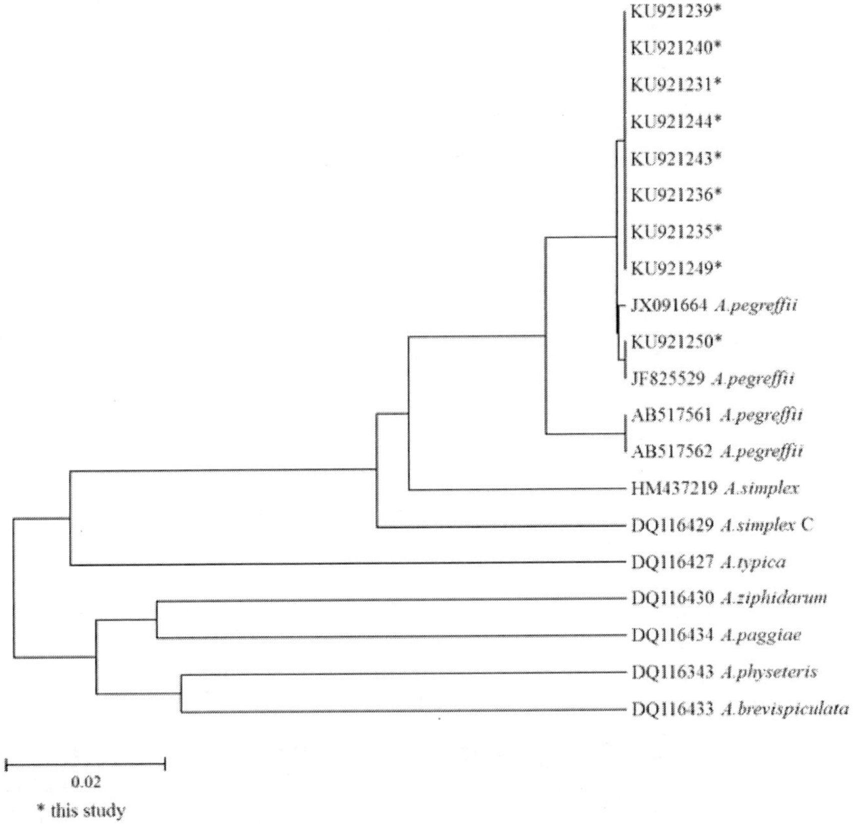

Figure 2. Molecular phylogenetic trees showing the genetic relationships among *Anisakis* spp. based on mtDNA cox2 gene sequences. Analysis was performed using the MEGA 6 program. The *scale bar* indicates evolutionary distance

PCR amplification and sequencing of mtDNA COI sequences from fish samples revealed that all of the nine samples produced one single band of 630 bp (data not shown). The obtained sequences (Genbank accession number: KU963588, 963589) were clustered with those of *T. japonicus* previously registered in GenBank, with 98.4–99.8 % homology (data not shown).

DISCUSSION

Human infection with *A. simplex* L3 accidentally occur when raw or undercooked fish contaminated with the L3 is ingested. Although live L3 can rarely develop to L4 or adults in the human gastrointestinal tracts, they can cause mild to severe epigastric pain, nausea, vomiting, and diarrhea (Audicana and Kennedy 2008). In addition, allergic reactions occur in some patients, which may occur even by the ingestion of dead *A. simplex* larvae (Audicana and Kennedy 2008). In fish, the larvae are mainly located freely in the visceral cavity or encapsulated as flat, tight spirals in and on the visceral organs; however, a minor proportion may migrate into the muscles, which may draw attention of consumers (Levsen and Berland 2012). Thus, it is necessary to have knowledge about the host range of each *Anisakis* species, together with the clear identification and their geographical distribution to avoid accidental human infection and warrant food safety issue.

In this study, most of the nematodes were found throughout the viscera (56.2 %) or from the body cavity of the largehead hairtail (41.5 %) (Table 1). Those from the viscera were embedded in or stick to the surfaces of the mesenteries and intestines, whereas those from the body cavity were freely moving in the body cavity. On the other hand, only 0.4 % (6/1,420) of the nematodes was found in the muscles.

Of 1259 nematodes, most of the nematodes (98.7 %, 1243/1259) were identified as *A. pegreffii* by PCR-RFLP and direct sequencing. Other species, *Hysterothylacium* sp. and hybrid genotype (*A. simplex* × *A. pegreffii*) were rarely found.

A. pegreffii Campana-Rouget and Biocca, 1955 is the dominant species of *Anisakis* in the Mediterranean Sea and also widely distributed in the austral region between 30° N and 55° S. Many pelagic and demersal fish species are known to be the intermediate/paratenic hosts caught in this area, and toothed whales (Delphinidae, Ziiphidae, Physeteridae) and baleen whales (Neobalaenidae) are known to be the definitive hosts (Mattiucci and Nascetti 2008). *A. pegreffii* has been also reported frequently at the larval stage in many fish species in East Asian waters (Bak et al. 2014 and the references therein). Recent reports in Italy suggest that *A. pegreffii* can be the etiological agent of human infection (Mattiucci et al. 2013).

Genus *Hysterothylacium* Ward and Magath, 1917 is one of the common genera in the family Raphidascarididae (Fageholm 1991), in which there are approximately 70 species (Shamsi et al. 2013). The adult stages are usually found in the alimentary canal of predatory teleost, while the L3 are found encapsulated in the mesentery and viscera of various species of marine teleosts and also in various marine invertebrates (Keskin et al. 2015). Thus, it is not host specific in the larval stages, exhibiting a global distribution (Shamsi et al. 2013). Many species of the genus *Hysterothylacium* have been poorly described; thus, their genetic characteristics have been frequently deposited in Genbank without clear identification at the species level. Although the RFLP patterns in this study was in accordance with those of *Hysterothylacium* sp. described in Bak et al. (2014), we could not identify these specimens at the species level because the highest homology of the mtDNA cox2 sequences showed 92 % with that of *Hysterothylacium* sp. registered in GenBank. It would be necessary to conduct morphological observation and further molecular analysis, for clear identification of these specimens.

The prevalence of infection and mean intensity for anisakid nematodes L3 in *T. japonicus* calculated in this study were extremely higher than those described by Lee et al. (2009). There can be several explanations for this difference. In our study, we collected freshly caught *T. japonicus* from Korean waters, whereas they collected *T. lepturus* in a fisheries market, without information on their geographical origin.

In marine food webs, the abundance of phytoplankton is strongly related to temperature and changes in phytoplankton biomass can affect all subsequent stages, first impacting on marine crustaceans, the predominant group in zooplanktons abundance and biomass, and thereafter migratory fish, cephalopods, and predatory marine mammals further up the food chain (Utaaker and Robertson 2015). The life cycles of *Anisakis* species involve various hosts at different levels in marine food webs. The high prevalence of *Anisakis* species in fish hosts in this study is therefore thought to reflect the richness of micro/macroinvertebrates and possibly subsequently have relations with the biomass of other predators such as pinnipeds and cetaceans in waters investigated. Several papers also reported relatively a high prevalence of anisakid nematodes larvae in the largehead hairtail (Borges et al 2012; Kong et al. 2015; Umehara et al. 2010), as in this study. The low number of the nematodes larvae in the previous study (Lee et al. 2009) may just reflect the low abundance of micro/macroinvertebrates in the areas where the largehead hairtail were caught.

Several papers described anisakid nematodes L3 from the largehead hairtail (Borges et al. 2012; Kong et al. 2015; Lee et al. 2009; Shih 2004; Umehara et al. 2010; Zhang et al., 2013), but there was no report in which *T. japonicus* was mentioned as host species, except for our study. It is not clear if the host species mentioned was in fact *T. japonicus* or other *Trichiurus* species in their studies because the identification of host species was not clearly conducted. But except for the paper by Borges et al. (2012), they probably sampled *T. japonicus* because the known geographical distribution of *T. japonicus*, the northwestern Pacific (Hsu et al. 2009; Tzeng et al. 2007) includes their sampling areas.

After *Anisakis* L3 in the invertebrate or primary fish hosts are ingested by predatory fish, they penetrate in the intestinal wall and then may remain within the body cavity or migrate into the musculature or internal organs of the fish. The different relative abundance of these L3 among these microhabitats in fish can be affected by many factors such as the parasites species, the fish species, the fish age, and the environmental conditions (Lymbery and Cheah 2007). Although the mechanisms of larval penetration within fish hosts are still unknown, some studies revealed that L3 migrate from the visceral organs to the muscle after the death of the fish hosts and this postmortem migration is thought to be also affected by interactions among parasites, host, and external environments (Cipriani et al. 2016; Lymbery and Cheah 2007). Thus, the risk of human infection can be enhanced if the fish are ingested whole or if the fish have been kept without evisceration for a period of time.

Several methods such as postharvest handling, freezing, and adequate cooking are suggested for prevention or control measures of the human infection by ingestion of raw or undercooked fish harboring *Anisakis* L3 (Audicana and Kennedy 2008). Deep freezing and/or adequate cooking remain the most effective measures to prevent human infection (Audicana and Kennedy 2008; Lymbery and Cheah 2007). In addition, immediate evisceration or freezing of fish after catch will be necessary for the prevention of human infection (Lymbery and Cheah 2007). In case of our study, most of the *Anisakis* L3 on the largehead hairtail was found from the body cavity and viscera, which are generally not consumed raw and removed before cooking. In addition, *A. simplex* (s.s.), the most commonly involved species in human infection, was not found in this study. Thus, the largehead hairtail may not be the common source of human infection in Korea. But care should be still taken because *A. pegreffii* is also considered as a possible human pathogen (Mattiucci et al. 2013), and a high prevalence of anisakid nematode larvae in the largehead hairtail may raise food safety problems to the consumers. Immediate evisceration or freezing of fish after catch will be necessary before consumption.

CONCLUSIONS

Information on the anisakid nematodes fauna of commercially important fish species is important because the L3 of anisakid nematodes, mostly *A. simplex*, can cause accidental human infection in countries where the customs of eating raw or undercooked fish or cephalopods exist. Given the growing popularity of these customs worldwide, the risk of human infection is widespread, together with the increased number of new host records and expansion of the geographical ranges of *Anisakis* spp. The largehead hairtail, one of the commercially important fish species in Korea, was investigated for anisakid nematodes fauna. Most of the nematodes were found throughout the viscera (58.1 %) and the body cavity (41.5 %), which is not consumed raw. Only 0.3 % was found in the muscles. *A. pegreffii* was the most frequently encountered nematode species (98.7 %) from the largehead hairtail in this study. Detailed information on the distribution of different *Anisakis* species in the largehead hairtail in this study has significant implications for the biology, ecology, and epidemiology of anisakid nematodes.

ABBREVIATIONS

COI, cytochrome c oxidase subunit 1; ITS, internal transcribed spacer; L3, third stage larvae; mtDNA cox2, mitochondrial DNA cytochrome c oxidase subunit 2; RFLP, restriction fragment length polymorphism

ACKNOWLEDGEMENTS

We thank Jeong-Ho Jang, Min-Woo Kim, Chan-Goo Kim, and Hae-Young Yun for their help during the experiments.

FUNDING

This research was supported by the Basic Science Research Program through the National Research Foundation of Korea (NRF) funded by the Ministry of Science, ICT, and future planning (2015R1A2A2A01004406).

AUTHORS' CONTRIBUTIONS

JHK conceived, designed the experiments, and wrote the manuscript. WHN and CHJ carried out sample collections and analyses. JHK, WHN, and CHJ interpreted results. All authors read and approved the final manuscript.

COMPETING INTERESTS

The authors declare that they have no competing interests.

REFERENCES

1. Audicana MT, Kennedy MW. *Anisakis simplex*: from obscure infectious worms to inducer of immune hypersensitivity. Clin Microbiol Rev. 2008;21:360–79.
2. Bak TJ, Jeon CH, Kim JH. Occurrence of anisakid nematode larvae in chub mackerel (*Scomber japonicus*) caught off Korea. Int J Food Microbiol. 2014;191:149–56.
3. Borges JN, Cunha LFG, Santos HLC, Monteiro-Nato C, Santos CP. Morphological and molecular diagnosis of anisakid nematode larvae from cutlassfish (*Trichiurus lepturus*) off the coast of Rio de Janeiro, Brazil. PLOS One. 2012;7:e40447.
4. Bush AO, Lafferty KD, Lotz JM, Shostak A. Parasitology meets ecology on its own terms: Margolis et al., revisited. J Parasitol. 1997;83:575–83.
5. Cha YG, Kim KS. A causality analysis of the prices between imported fisheries and domestic fisheries in distribution channel. J Fish Business Admin. 2009;40:105–26 (in Korean with English summary).
6. Cipriani P, Smaldone G, Acerra V, D'Angelo L, Anastasio A, Bellisario B, et al. Genetic identification and distribution of the parasitic larvae of *Anisakis pegreffii* and *Anisakis simplex* (s. s.) in European hake *Merluccius merluccius* from the Tyrrhenian Sea and Spanish Atlantic coast: implications for food safety. Int J Food Microbiol. 2015;198:1–8.
7. Cipriani P, Acerra V, Bellisario B, Sbaraglia GL, Cheleschi R, Nascetti G, et al. Larval migration of the zoonotic parasite *Anisakis pegreffii* (Nematoda: Anisakidae) in European anchovy, *Engraulis encrasicolus*: implications to seafood safety. Food Control. 2016;59:148–57.
8. D'Amelio S, Mathiopoulos KD, Santos CP, Pugachev ON, Webb SC, Picanço M, et al. Genetic markers in ribosomal DNA for the identification of members of the genus *Anisakis* (Nematoda: ascaridoidea) defined by polymerase-chain-reaction-based restriction fragment length polymorphism. Int J Parasitol. 2000;30:223–6.
9. Fageholm HP. Systematic implications of male caudal morphology in ascaridoid nematode parasites. Syst Parasitol. 1991;19:215–28.
10. Hsu KC, Shih NT, Ni IH, Shao KT. Speciation and population structure of three *Trichiurus* species based on mitochondrial DNA. Zool Studies. 2009;48:835–49.
11. Keskin E, Koyuncu CE, Genc E. Molecular identification of *Hysterothylacium aduncum* specimens isolated from commercially important fish species of Eastern

Mediterranean Sea using mtDNA cox1 and ITS rDNA gene sequences. Parasitol Int. 2015;64:222–8.
12. Kim YH, Yoo JT, Lee EH, Oh TY, Lee DW. Age and growth of largehead hairtail *Trichiurus lepturus* in the East China Sea. Kor J Fish Aquat Sci. 2011;44:695–700.
13. Kong QM, Fan LF, Zhang JH, Akao N, Dong KW, Lou D, et al. Molecular identification of *Anisakis* and *Hysterothylacium* larvae in marine fishes from the East China Sea and the Pacific coast of central Japan. Int J Food Microbiol. 2015;199:1–7.
14. Lee MW, Cheon DS, Choi CS. Molecular genotyping of *Anisakis* species from Korean sea fish by polymerase chain reaction restriction fragment length polymorphism (PCR-RFLP). Food Control. 2009;20:623–6.
15. Levsen A, Berland B. *Anisakis* species. In: Woo PTK, Buchmann K, editors. Fish parasites: pathobiology and protection. CAB International: USA; 2012. p. 298–309
16. Lymbery AJ, Cheah FY. Anisakid nematodes and anisakiasis. In: Murrell KD, Fried B, editors. Food-borne parasitic zoonoses: fish and plant-borne parasites. New York: Springer Science; 2007. p. 185–207.
17. Mattiucci M, Nascetti G. Advances and trends in the molecular systematics of anisakid nematodes, with implications for their evolutionary ecology and host-parasite co-evolutionary process. Adv Parasitol. 2008;66:47–148.
18. Mattiucci M, Fazii P, De Rosa A, Paoletti M, Megna AS, Glielmo A, et al. Anisakiasis and gastroallergic reactions associated with *Anisakis pegreffii* infection, Italy. Emer Inf Dis. 2013;19:196–9.
19. Nadler SA, Hudspeth DSS. Phylogeny of the Ascaridoidea (Nematoda: Ascaridida) based on three genes and morphology: hypotheses of structural and sequence evolution. J Parasitol. 2000;86:380–93.
20. Nakamura I, Parin NV. FAO Species Catalogue. Vol. 15. Snake mackerels and cutlassfishes of the world (Families Gempylidae and Trichiuridae). An annotated and illustrated catalogue of the snake mackerels, snoeks, escolars, gemfishes, sackfishes, domine, oilfish, cutlassfishes, scabbardfishes, hairtails, and frostfishes known to date. FAO Fish Synop. 1993;125:99–106.
21. Sakanari JA, McKerrow JH. Anisakiasis. Clin Microbiol Rev. 1989;2:278–84.
22. Shamsi S, Gasser R, Beveridge I. Description and genetic characterisation of *Hysterothylacium* (Nematoda: Raphidascarididae) larvae parasitic in Australian marine fishes. Parasitol Int. 2013;62:320–8.
23. Shih HH. Parasitic helminth fauna of the cutlass fish, *Trichiurus lepturus* L., and the differentiation of four anisakid nematode third-stage larvae by nuclear ribosomal DNA sequences. Parasitol Res. 2004;93:188–95.
24. Tamura K, Stecher G, Peterson D, Filipski A, Kumar S. MEGA6: Molecular Evolutionary Genetics Analysis Version 6.0. Mol Biol Evol. 2013;30:2725–9.
25. Thompson JD, Higgins DG, Gibson TJ. CLUSTAL W: improving the sensitivity of progressive multiple sequence alignment through sequence weighting, position-specific gap penalties and weight matrix choice. Nucleic Acids Res. 1994;22:4673–80.
26. Tzeng CH, Chen CS, Chiu TS. Analysis of morphometry and mitochondrial DNA sequences from two *Trichiurus* species in waters of the western North Pacific: taxonomic assessment and population structure. J Fish Biol. 2007;70(B):165–76.
27. Umehara A, Kawakami Y, Araki J, Uchida A. Molecular identification of the etiological agent of the human anisakiasis in Japan. Parasitol Int. 2007;56:211–5.
28. Umehara A, Kawakami Y, Ooi HK, Uchida A, Ohmae H, Sugiyama H. Molecular identification of *Anisakis* type I larvae isolated from hairtail fish off the coasts of Taiwan and Japan. Int J Food Microbiol. 2010;143:161–5.
29. Utaaker KS, Robertson LJ. Climate changes and food borne transmission of parasites: a consideration of possible interactions and impacts for selected parasites. Food Res Int. 2015;68:16–23.
30. Van Thiel PH, Kuipers FC, Roskam RT. A nematode parasitic to herring, causing atue abdominal syndromes in man. Trop Geogr Med. 1960;2:97–113

31. Wasko AP, Martins C, Oliveira C, Foresti F. Non-destructive genetic sampling in fish. An improved method for DNA extraction from fish fins and scales. Hereditas. 2003;138:161–5.
32. Zhang LP, Du XJ, An RY, Li L, Gasser RB. Identification and genetic characterization of *Anisakis* larvae from marine fishes in the South China Sea using electrophoretic-guided approach. Electrophoresis. 2013;34:888–94.

CHAPTER 19

Engineering Crop Plants for Nematode Resistance through Host-Derived RNA Interference

Selvaraj Tamilarasan and Manchikatla Venkat Rajam*

Plant Polyamine, Transgenic and RNAi Laboratory, Department of Genetics, University of Delhi, India

ABSTRACT

Plant parasitic nematodes are major threat for crop plants and cause severe economic losses worldwide annually. Various strategies deployed for the control of these notorious parasites has resulted either in limited success, or having huge negative impact on environment. RNA interference (RNAi) is a gene-silencing phenomenon that is conserved in various eukaryotes. Experimentally induced RNAi is highly specific and potent, leading to its wide utilization in functional studies for exploring gene functions. Crops engineered through RNAi have proven to be successful in protection against pest and parasites, including nematodes. Engineering nematode resistance in crop plants through host-derived RNAi is largely based on the selection of target gene. The expression of nematode specific dsRNA in plants generates siRNAs and taken up by nematodes on feeding. Depending upon the function, level of expression and silencing efficacy of the target gene, resistance was determined. RNAi seems to be promising in many aspects, such as providing durable resistance to crops against plant parasitic nematodes in the near future. In the present article, we have reviewed the published work on the host-derived RNAi for developing nematode resistance in plants.

Keywords: RNAi; dsRNA; siRNA; Host-derived RNAi; Plantparasitic nematodes; Nematode resistance

INTRODUCTION

Plant parasitic nematodes (PPN) are major agricultural pests, which infest almost all cultivated crop plants causing drastic economic loss estimated as ~ 157 billion US$ [1]. Clear symptoms of infections are often deficient, leading the nematodes unnoticed by farmers, until the infestation is rigorous. Moreover, nematode infection often results in secondary diseases to the host plant, leading to the underestimation of the exact crop losses attributed by PPN [2]. Among plant parasitic nematodes, most of the damage is often contributed by sedentary endoparasitic genera Meloidogyne spp (root knot nematode (RKN)), Globodera and Heterodera spp (cyst nematode). Root knot nematodes are able to attack wide range of crops (~ 2000), while each species of cyst nematodes has a narrow host range. These parasites invade plant roots as mobile infective second juvenile stage (J2) to reach plant vascular cylinder, where they migrate intercellularly (root-knot nematode), or intracellularly (Cyst nematode), to become sedentary and induce the formation of feeding cells that supports the development and reproduction of females over a period of few weeks. Depending on the nematode species, the feeding cells either develop into specialized multinucleate giant cells (root-knot nematode) or syncytia (cyst nematode), by inducing mitosis without cytokinesis, leading to the formation of galls and subsequently, plant dies after several cycles of nematode infection [3-5].

Nematode Control Strategies and their Limitations

There are several methods widely used for controlling these notorious parasites, such as chemical nematicides, cultural practices, organic means and use of resistant cultivars. Although chemical nematicides are effective means of nematode control in recent decades, issues such as mammalian toxicity, ground water contamination, environmental toxicity and residues in food have caused much stringent rule for their use in agriculture, and in many countries, effective nematicides have been banned for commercial use [2]. Utilization of green manures for nematode control was also carried out in field conditions, but assessing their effectiveness remains challenging. Cultural practices such as crop rotation are mostly used for controlling nematode pests in field conditions, but certain nematodes remain dormant for many years, e.g. cyst nematodes. The cover crops, particularly marigold, have been recommended for nematode suppression. However, marigold as an intercrop has its own limitations, i.e. it is sensitive to short-day photoperiod, and also serves as host for some other pests. For limited range crops, resistant cultivars have been a commercial success, but invariably this approach is unable to control many nematode problems. Few naturally occurring resistant genes have been identified and cloned for developing nematode resistance [6]. However, attempts to transfer resistance from one crop to others have so far been unsuccessful, for reasons remain unclear. Further other approaches, including traditional breeding to develop nematode resistant plants, have resulted in limited success [7]. Therefore, there is an urgent need for developing a novel strategy for successful mitigation of these plant parasitic nematodes. In this regard, RNA interference (RNAi) technology has proven to be novel and successful for the control of various pests and parasites, including nematodes.

RNAi Mechanism

RNAi is an apparently ancient defense mechanism against invading viruses, prevents deleterious effect of transposon movement, regulation of gene expression and chromatin modification. It is a powerful tool to suppress gene expression and analyze gene function. RNAi operates at transcriptional (called as transcriptional gene silencing-TGS), as well as post-transcriptional (called as post-transcriptional gene silencing- PTGS) for gene silencing, which has been previously reported as cosuppression in plants [8], and quelling in fungus [9]. In fact, these three techniques appeared to be remarkably well conserved in several eukaryotes, which is initiated by the action of dicer enzyme on dsRNA, leading to the production of small interfering RNAs (siRNAs), and these siRNAs along with the RNA induced silencing complex (RISC) are involved in sequence specific silencing of the target mRNA (Figure 1) [10-12]. Stoichiometric considerations revealed that a few dsRNA molecules are enough to cause RNAi phenotypes, leading to the discovery that the silencing response can be amplified through the action of RNA-dependent RNA polymerases (RdRPs) on the targeted mRNA, which results in the synthesis of secondary siRNAs [11,13]. In fact, complementary RNA (cRNA) synthesis allows siRNA formation and homology-dependent silencing to spread outside of the initial inducer sequence, a process called transitivity. However, the precise mechanism involved in RdRP-mediated synthesis of secondary siRNAs appears to vary in different organisms. Secondary siRNAs have 5'-triphosphate, are antisense to the target transcript, and are associated with specific Argonaute proteins [14]. In case of C. elegans, transmembrane proteins SID-1 and SID-2 are independently for dsRNA ingestion mediated RNAi. SID-1 is required for the dsRNA uptake into all cells, while SID-2 is required for silencing initiated by dsRNA [15].

RNAi as a Powerful Functional Genomics Tool

RNAi has been successfully utilized for exploring gene functions in many eukaryotes, including nematodes. Induction of RNAi by delivering dsRNA has been successful in non-parasitic nematode, Caenorhabditis elegans, and majority of its genes has been characterized using this technique [16]. Delivering dsRNA to plant parasitic nematodes has been technically challenging, because these obligatory parasites requires living host for feeding and the pre-parasitic nematodes are of non feeding and do not ingest fluid at this stage. Several house keeping and parasitism genes were silenced in vitro for functional studies using this approach are well documented, has been reviewed extensively earlier [7,17].

Initial reports highlighted the utility of RNAi for functional studies in plant parasitic nematodes was achieved by treating dsRNA/siRNAs to the infective stage juveniles, and flurescine isothionate (FITC) was commonly utilized to monitor the up take of dsRNA/siRNAs. In most of the cases, octopomaine, resorcinol, spermidine and serotonin were used to stimulate the uptake of dsRNA from solution [17]. Alternative route for dsRNA up take may also exist in plant parasitic nematodes. Silencing of Meloidogyne artiellia chitin synthase gene was achieved by soaking the intact eggs in dsRNA solution, resulted in depletion of transcripts and delayed hatching of juveniles from treated eggs [18]. Profound silencing in terms of reduced transcript level and abnormal phenotypic behavior could be induced by targeting neuronally expressed genes using various sizes of dsRNA (88 bp, 222 bp, 227 bp, 316 bp) in water, without any stimulated ingestion. This has been suggested that alternative dsRNA up take route were also possible through sensory organs (amphids), secretory/excretory pore, or even the cuticle [19].

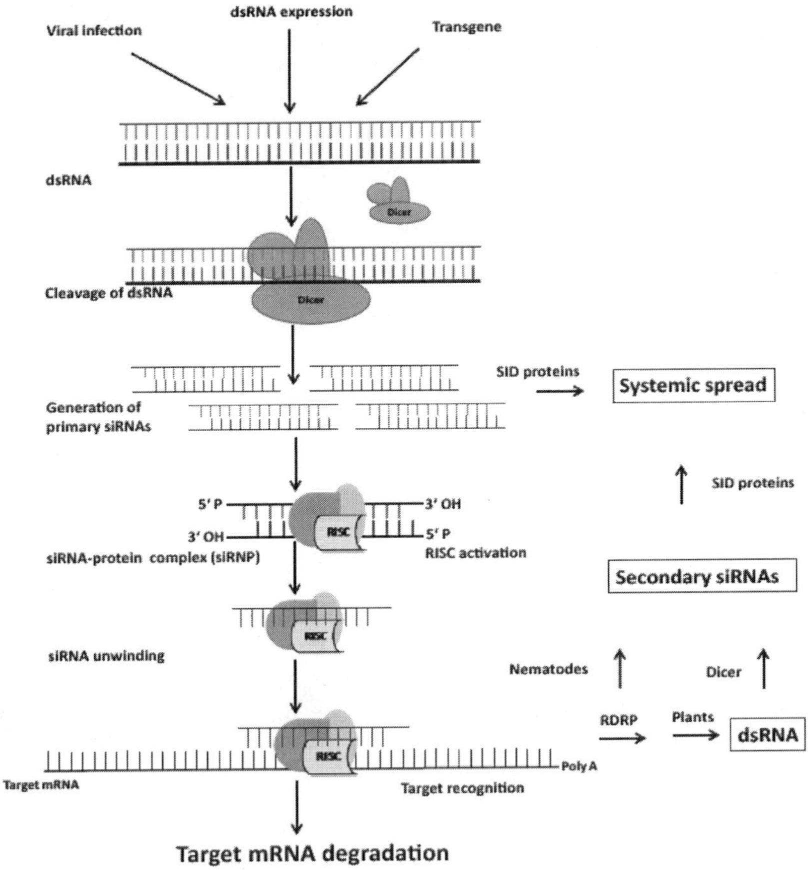

Figure 1: Schematic representation of general RNAi mechanism. Double-stranded RNA (dsRNA) originating from diverse sources, including viral infection, transgenes, hair-pin dsRNA expression, etc. is processed by a Dicer-complex to generate primary siRNAs. The primary siRNAs are loaded to multi-protein RNAinduced silencing complex (RISC) complex, mediates RISC activation, which involves siRNA unwinding, target mRNA recognition and cleavage. Hence, cleaved target mRNA degradation occurs by the action of other nucleases. In *C. elegans*, Argounaute protein loaded with primary siRNA and target mRNA generates secondary siRNAs by the action of RNA-dependent RNA polymerase (RdRP). In case of plants, RdRP recruitment is optional and depends on synthesis of dsRNA and secondary siRNAs are generated by cleavage action of dicer. SID proteins and secondary siRNAs are mainly involved in systemic RNAi [11-14,20].

Due to grand success in functional genomics, RNAi has emerged as versatile tool with wide range of applications in plant biology. Recent developments in RNAi led to the identification of core RNAi components in many eukaryotes, and have envisioned researchers to adopt this tool for wider applications in crop improvement, including engineering crop plants against plant parasitic nematode menace. In the past decade, various studies indicated key developments in the RNAi machinery, and also whole genome sequence data available for few of the plant-parasitic nematodes [3,20], provides a new horizon for the successful exploitation of RNAi for nematode control.

Host-Derived RNAi for the Control of Plant-Parasitic Nematodes

Expression of dsRNA, corresponding to the nematode target gene through host plant, mediates the production of siRNAs. Based on the sequence similarity between the siRNAs and the corresponding target mRNA, silencing is achieved upon nematode feeding. Although most of the reports successfully utilized RNAi for studying the gene function in vitro, however, they were not technically feasible to use as nematode control under field conditions. And major hurdle for potential application of plant-mediated RNAi was whether nematodes ingest dsRNAm when feeding on plant cells expressing dsRNA constructs. These plant parasitic nematodes produces specialized feeding structures inside the plant cell on which they feed through out the development called feeding tube, act as a molecular sieves, and may helps in restricting nematodes from ingesting plant defense related macromolecules [21,22]. And the size exclusion limit is accurately unknown, but it has been reported that up take of GFP (28 kDa) and different other dyes indicated that a size exclusion for these plant parasitic nematodes are between 20 and 40 kDa [17,23]. One recent report indicated that RKN species (M. hapla) infective stage juveniles could ingest a range of Bacillus thuringiensis proteins sizes from 45 kDa to 140 kDa, directly through the stylet under in vitro condition [24]. Moreover, it has been proven that linear dsRNA/siRNA was possibly taken up by plant parasitic nematodes [25-37]. And also, feeding tube would be possibly excludes both siRNA incorporated into the RISC, as well as non-linear dsRNA molecule [17-19,21]. The delivery of parasites gene-specific dsRNA through the host may be more effective and appropriate method for nematode control.

Host-delivered RNAi appears to be most promising strategy for nematode control. Based on the choice of the target gene selected, it can be broadly categorized as dsRNA against the targets, which are constitutively expressed throughout nematode life cycle, genes involved in nematode development, neurotransmission, nematode effectors/ parasitism genes (Table 1). Some of these nematode target genes have been validated experimentally through in vitro RNAi experiments in C. elegans and other plant parasitic nematodes [15,17,25-38]. On the other hand, soon after the nematode infection, there are number of genes differentially regulated in host cells, which are responsible for feeding site establishment, and also helps in successful parasitism [39]. Genes involved in the feeding site formation and other plant genes including those involved in cell cycle, miRNAs which are differentially regulated during nematode infections are of primary interest. There is no direct evidence reported till now for the utilization of candidate microRNAs for nematode control.

Nematode Target Genes

House-keeping genes

These are constitutively expressed throughout the nematode development and implicated in major cellular processes. Targeting these genes through host-generated dsRNA ultimately hampers the nematode development, and hence, impedes parasitic success. The first demonstrated success of host-delivered RNAi for the nematode control was achieved by targeting splicing factor and integrase. Tobacco plants expressing dsRNA of these two genes under the control of constitutive promoter resulted in target gene depletion and also mitigate the nematode (Meloidogyne incognita) establishment [40]. Also, soybean roots expressing RNAi constructs targeting three essential genes involved in mRNA metabolism Heterodera glycines RPS-3a, RPS- 4 and SPK-1 displayed variations in nematode infection, i.e. 80-88% reduction in nematode numbers [41]. In other report, transgenic soybean expressing PRP 17 gene resulted in significant reduction in nematode (H. glycines) infection (53%) and (79%) reproduction. In the same study, two of the other genes Cpn 1 were found to be embryonic lethal, in case of C. elegans was selected, and dsRNA corresponding to these genes expressed through composite soybean plants displayed up to 95% reduction in number of egg masses developed to H. glycines infection was observed [42]. Based on these reports, it may be enticing to choose such genes; some precautions should be administered before selecting such kind of targets. Most of the houses keeping genes are highly conserved across different species, which may have off-target effects on host plant/other beneficial organisms/humans. Transgenic tobacco plants expressing different dsRNA hairpin structures targeting RKN (Meloidogyne javanica) putative transcription factor MjTIS-11, leads to the down-regulation of transcripts in nematodes feeding on the transgenic RNAi lines. Even though the presence of siRNAs in transgenic plants was detected, but it does not lead to any lethal phenotypes [43]. This result highlighting the most important fact, that not all genes are amenable for developing nematode resistance through host-derived RNAi. Hence it is imperative to look for novel and potent targets for successful exploitation of RNAi-mediated nematode control.

Nematode-specific genes

These are most attractive targets for engineering crop plants for nematode resistance through host-derived RNAi. The parasitism/ effector genes which are unique to nematodes have been already identified, available in public databases, and targeting such genes could provide durable resistance with less safety related issues. In fact, utility of such nematode specific targets for developing nematode resistance through plant-mediated RNAi has already been explored. Transgenic Arabidopsis plants expressing dsRNA of sub-ventral esophageal gland cell secreted protein 16D10 conserved region of RKN displayed 65-90% reduction in number of galls, with reduced gall size and corresponding reduction total egg production [30]. Based on their observation, either full length dsRNA or processed siRNAs were sufficient to induce an RNA silencing phenotype. Same strategy has been proven to be successful in Arabidopsis; although no complete nematode resistance was observed when unique and conserved regions of parasitism genes were targeted through in planta RNAi. Reduction of developing nematodes ranged from 23-64% in different RNAi lines. In fact, in planta dsRNA/siRNA up take in cyst nematodes appears to be functional because of the specific down-regulation of target mRNA, and nematodes feeding on RNAi plants were less successful in further development, and also led to a reduction in number of mature females [44]. Silencing of parasitism gene 8D05 of M. incognita through host derived RNAi in Arabidopsis resulted in drastic reduction in gall number [45]. Also, targeting M. incognita calreticulin (MiCRT) through transgenic expression of dsRNA in Arabidopsis resulted in target gene reduction and decreased gall number [46].

Genes involved in development and reproduction

There is increasing evidence based on the sequence data, that it is now possible to utilize genes which are involved in nematode development and reproduction. However, there are few examples available, as they found to be involved in reproduction, and are also specific to nematodes. One such target is major sperm protein (MSP) and silencing soybean cyst nematode (H. glycines) MSP through transgenic soybean derived dsRNA displayed 68% reduction in egg production. Moreover, the progeny of the hatched

Table 1: Host-derived RNAi for the control of plant parasitic nematodes.

Species	Target Gene	Promoter-the control of	Host Plant	Target mRNA depletion	Observed phenotype	Time point	Reference
M. incognita	Integrase Splicing factor	35S	Tobacco	Yes	Decreased number of established nematode	6-7 wpi	Yadav et al. [40]
M. incognita *M. javanica* *M. hapia* *M. arnena*	Secreted peptide 16 D 10	35S	Arabidopsis	Not detected	Decreased number of galls & gall size	4 wpi	Huang et al. [30]
H. glycines	Major sperm protein	ACT2	Soybean	Not detected	Decreased number of eggs	8 wpi	Steeves et al. [47]
M. javanica	Putative transcription factor	35S, TobRB7	Tobacco	Yes	-	6 wpi	Fairbairn et al. [43]
H. schacbii	4G06, ubiquitin-like	35S	Arabidopsis	Yes	Decreased number of females	14 dpi	Sindhu et al. [44]
H. schacbii	3B05, cellulose binding protein	35S	Arabidopsis	Yes	-	14 dpi	Sindhu et al. [44]
H. schacbii	10A06, Zinc finger protein	35S	Arabidopsis	Not detected	Low number of developing females	14 dpi	Sindhu et al. [44]
H. schacbii	8H07, SKP1-like	35S	Arabidopsis	Yes	Decreased number of developing females	14 dpi	Sindhu et al. [44]
H. glycines	Ribosomal protein 3a,4. Spliceosomal SR protein	FMV-sgt	Soybean roots	Not detected	Reduction in number of females	8 dpi	Klink et al. [41]
M. incognita	Tyrosine Phosphate, itochondrial stress -70 protein precursors MSP		Soybean roots	Yes	Decreased number of galls, 5-6 fold reductions in mature female diameter	14 dpi	Sindhu et al. [44]
H. glycines	Fib 1,Y 25	35S	Chimeric soybean root system	Yes	Suppression of nematode reproduction & development	5 wpi	Li et al. [42]
H. glycines	Cpn 1 Y25,Prp 17	35S	Composite soybean plants	Yes	Reduction in reproduction & development	5 wpi	Li et al. [48]
M. incognita	Rpn7	35S	Tomato hairy roots	Yes	Reduction in reproduction & motility	40 dpi	Niu et al. [2012]
M. incognita	Parasitism gene 8D05	35S	Arabidopsis	Not detected	Reduction in gall number	8 wpi	Xue et al. [45]
M. incognita	Calreticulin - MiCRT	35S	Arabidopsis	Yes	Reduction in gall number	8 wpi	Jaouannet et al. [46]

*dpi-days post infection; wpi-weeks post infection

from the eggs of those nematodes feeding on transgenic RNAi lines apparently displayed 75% reduction of egg production, when allowed to infect susceptible wildtype soybean plants. These results strengthen the knowledge about the in planta RNAi can be successfully used to control cyst nematodes as well [47]. Moreover, transgenic soybean expressing dsRNA fragments specific to genes related to H. glycines reproduction or fitness (CPN- 1, Y25, PRP-17), significantly suppressed the nematode reproduction and development [48]. RNAi constructs targeting four different genes (lactate dehydrogenase LD, tyrosine phosphatase TP, mitochondrial stress 70 protein precursor MSP and ATP synthase beta chain mitochondrial protein), of M. incognita, which has high homology with cyst nematodes, and C. elegans genes were transformed with soybean roots and investigated for nematode infection. Out of the four RNAi constructs tested, constructs targeting MSP and TP were showed up to 94% reduction in gall formation [49]. In another report, four RNAi lethal homologues of C. elegans were selected, encoding ribosomal protein 3a and 4, a spliceosomal SR protein and synaptobrevin of H. glycines, and expressed as tandem inverted repeats in transgenic soybean plants inhibited the female development [41]. Similar strategy was deployed by another group, to silence the genes encoding a beta subunit of coatomer (COPI) complex, pre mRNA splicing factor and protein with unknown function, which has led to hampering nematode infection and female development of H. glycines [38]. Although, all the reported experiments on plant-mediated nematode resistance has utilized transgenic plant expressing empty RNAi vector as control in evaluating the resistant phenotype, but most care should be taken in terms of considering many more factors to reduce the unintended noise of off-target effects.

Nematode responsive host plant genes

Based on the existing literature, it is now clear that plant-nematodes have evolved different strategies to facilitate compatible interaction with the host plant and mediates their infection successfully [4,21]. Nematodes secretes a variety of effectors into host cell performing numerous functions, including cell wall degrading, suppression of host defenses and interact with host signaling pathways to facilitate feeding site formation [21]. Soon after plant parasitic nematode infect the roots, many number of host genes becomes up-regulated. This, in turn, facilitates the feeding site establishment and also helps in successful parasitism [50]. Down tuning the expression of those nematode responsive plant genes may be a vital strategy for the control of plant parasitic nematodes. Recently, Cappelle et al. [51] have reported that silencing of nematode induced Arabidopsis thaliana cyclin dependant kinase (AtCDKA; 1) gene under the control of nematode responsive promoter WRKY23 resulted in reduced transcript abundance, and also led to retardation in reproduction of cyst and RKN. Although silencing was observed, significantly fewer galls developed relative to wild-type control plants, but it has not provided complete resistance. Limited success for providing complete resistance could be because of the weakness of the promoter or multigene nature. Targeting multiple host genes involved in the feeding site establishment, under the control of highly specific nematode responsive promoters for stringent control of dsRNA expression, could be appropriate for complete resistance. Because those genes involved in feeding site establishment were detrimental for plant growth, silencing multiple genes may result in deleterious effect on plant development [17]. Another possible strategy for successful nematode control in near future may be through the microRNAs and other small RNAs [39]. A recent report indicated the cyst nematode induced small RNAs and their targets in model plant Arabidopsis [39]. Interestingly, Dicer (DCR) like and RNA-dependent RNA polymerase (RDR) mutants of Arabidopsis exhibited decreased susceptibility to H. schachtii infection, suggesting the possible role of small RNAs in plant nematode interaction. Also, nematode induced small RNAs, including microRNAs and their corresponding targets have been identified. The authors showed that there was a negative correlation between the smallRNA accumulation and their corresponding target transcript abundance. Moreover, it has been suggested that nematode induced microRNAs and small RNAs likely to be involved in feeding site establishment and parasitism, respectively [39]. Over-expression of those nematode induced candidate microRNAs, and/or silencing of their corresponding targets, may provide clear insight about plant-nematode parasitism, and also may provide nematode resistance to crop plants.

Advantages of Using RNAi Technology for PPN Control

High specificity degradation of target mRNA results in effective silencing, which eventually leads to lethal for the pests. Even single mismatch between the target mRNA and corresponding siRNA inhibits gene silencing [52]. In general, terminal bases are tolerant to mismatches, whereas the central bases are not. 9-11 bases are least tolerant to mismatches [53]. Silencing can be induced by the 30-50 nt similarity, with the off-target gene results in silencing in C. elegans [54]. In order to paramount such off-target effects, choice of target gene should be specific to the pathogen/pest, and could have minimal/no sequence homology with the sequences of plants, animals, beneficial insects, as well as free living nematodes [38].

- It is possible to silence gene families with single sequence, in case the selected target sequence in conserved among gene families [55,56]. By targeting several genes with the single gene construct should correlate with greater silencing effect.
- In general, genetic resistance is based on multigenic complex, difficult to transfer it to susceptible cultivars. RNAi strategy could possibly create a single resistant locus that will facilitate the transfer of nematode/pest resistance to other cultivars [38].
- Long range stability of pest resistance may also be possible by deploying RNAi. Based on the choice of target gene, as well as sequences used for the RNAi construct, stability can be maintained, i.e. especially mutation in the selected target gene for developing RNAi construct would considered to be lethal. Also, it is possible to maximize the effectiveness for durable resistance by selecting target gene sequences that show high homology among the nematode population. However, longterm field and greenhouse studies will be needed to verify stability. In order to tackle the RNAi tolerant nematode population caused by random mutation, RNAi locus can be engineered to target multiple genes [38].

Factors Limiting the Exploitation of Host-Derived RNAi for PPN Control

Although RNAi has emerged as a potent and successful technology for crop protection, there are few limitations which need to be properly addressed before adopting this technology. The important factors contributing to the effectiveness and possible solutions to overcome those major limitations were reviewed extensively earlier [57]. Major concern regarding the wider employment of RNAi-based crop protection is off-target effect; host-derived dsRNAs/siRNAs might interfere with the expression of the other non-intended target genes of the host itself, beneficial organisms and humans, etc. based on sequence homology. In order to circumvent the off-target effects, in-depth in silico homology searches were employed, and the region which is having no/minimal homology with the available sequences of the host, as well as other organisms were widely utilized for developing hairpin RNAi constructs [43,55,56]. Some online tools are also available for such analysis, including siRNA scan, pssRNAit, etc. However, the experimental approaches using qPCR and microarray can be deployed to evaluate the off-target effects.

Future Prospects of RNAi-Mediated Nematode Resistance in Crop Plants

Although RNAi-mediated nematode control in crop plants appears to be the most promising strategy, most of the above discussed methodologies did not provide complete resistance against plant parasitic nematodes. One of the possible ways to achieve complete resistance is through the use of chimeric RNAi constructs targeting multiple genes involved in single or multiple pathways of plant parasitic nematodes/nematode responsible plant genes, under the control of tissue-specific or wound-inducible or nematode-inducible promoter. In fact, the simultaneous silencing of multiple genes has been successfully demonstrated in Drosophila [58], C. elegans [59], and having strong therapeutic potential against viral infections and cancers [60]. Further, the enhanced resistance to plant parasitic nematodes by combining the expression of dsRNAs by crossing appropriate Arabidopsis thaliana lines resulted in an additive effect that further reduced nematode numbers and developmental capacity [61]. It is also notable that silencing miRNA pathway genes led to nematode resistance under in vitro conditions [62]. Targeting microRNA pathway, and/or miRNA genes through plants may provide durable and complete resistance against these notorious plant parasitic pests. And also complete nematode resistance may also be achieved through plant expressing artificial miRNA (amiRNA), containing known miRNA genes with the replaced seed region of the vital gene of plant parasitic nematodes. The primary advantage of amiRNA technology over dsRNA mediated silencing is that specificity, wherein long hairpin constructs were cleaved into multiple fragments, which can cause unintended silencing of non-target genes [38]. Interestingly, in a recent report, amiRNA-mediated silencing of endogenous soybean gene (Rhg1 locus of LRR-K) caused detectable reduction in target mRNA abundance, but as expected, there was no significant impact on soybean cyst nematode infectivity were detected [63]. Recent development in molecular techniques utilized for studying gene expression analysis, including next generation analysis, including microarray, lacer capture microdissection and next generation sequencing opened new era in studying plant-nematode interactions, which generated enormous data. Proper mining of these available resources may provide new insights in developing RNAi-mediated nematode resistance in crop plants. On the other hand, genomic and EST sequences are available for few of the important plant-parasitic nematodes, which provide valuable resource for researchers to explore possible potent and specific RNAi targets for successful control of plant-parasitic nematodes. RNAi-mediated crop protection against nematode pests appears to be most promising than other available methods, in terms of efficiency, stability and its ability to suppress gene expression in a regulated manner. Moreover, it appears to be most promising strategy with lesser environmental impact. Further, substantial research on feasibility studies are necessary for predicting ecological risk assessments before transforming this technology to field conditions.

ACKNOWLEDGEMENTS

Manchikatla Venkat Rajam is grateful to the Department of Biotechnology and Department of Science and Technology (DST-Govt. of India), New Delhi, University of Delhi, Delhi, Bejo Sheetal Seeds Pvt. Ltd., Jalna, Maharashtra, India and Sri Biotech Laboratories India Ltd., Hyderabad, India, for generously funding his research programmes on RNAi. The authors thank University Grants Commission for Special Assistance Programme and DST for FIST programme.

REFERENCES

1. Chitwood DJ (2003) Research on plant-parasitic nematode biology conducted by the United States Department of Agriculture-Agricultural Research Service. Pest Manag Sci 59: 748-753.
2. Thomason IA (1987) Challenges facing nematology: Environmental risks with nematicides and the need for new approaches. In: Vistas on Nematology, Society of Nematologists.
3. Abad P, Gouzy J, Aury JM, Castagnone-Sereno P, Danchin EG, et al. (2008) Genome sequence of the metazoan plant-parasitic nematode Meloidogyne incognita. Nat Biotechnol 26: 909-915.
4. Bird DM, Kaloshian I (2003) Are roots special? Nematodes have their say. Physiol Mol Plant Pathol 62: 115-123.
5. Caillaud MC, Dubreuil G, Quentin M, Perfus-Barbeoch L, Lecomte P, et al. (2008) Root-knot nematodes manipulate plant cell functions during a compatible interaction. J Plant Physiol 165: 104-113.
6. Williamson VM (1999) Plant nematode resistance genes. Curr Opin Plant Biol 2: 327-331.

7. Fuller VL, Lilley CJ, Urwin PE (2008) Nematode resistance. New Phytol 180: 27-44.
8. Napoli C, Lemieux C, Jorgensen R (1990) Introduction of a chimeric chalcone synthase gene into petunia results in reversible co-suppression of homologous genes in trans. Plant Cell 2: 279-289.
9. Romano N, Macino G (1992) Quelling: Transient inactivation of gene expression in Neurospora crassa by transformation with homologous sequences. Mol Microbiol 6: 3343-3353.
10. Fire A, Xu S, Montgomery MK, Kostas SA, Driver SE, et al. (1998) Potent and specific genetic interference by double-stranded RNA in Caenorhabditis elegans. Nature 391: 806-811.
11. Pak J, Fire A (2007) Distinct populations of primary and secondary effectors during RNAi in C. elegans. Science 315: 241-244.
12. Hannon GJ (2002) RNA interference. Nature 418: 244-251.
13. Winston WM, Molodowitch C, Hunter CP (2002) Systemic RNAi in C. elegans requires the putative transmembrane protein SID-1. Science 295: 2456-2459.
14. Miska EA, Ahringer J (2007) RNA interference has second helpings. Nat Biotechnol 25: 302-303.
15. Zhuang JJ, Hunter CP (2012) RNA interference in Caenorhabditis elegans: Uptake, mechanism, and regulation. Parasitology 139: 560-573.
16. Kamath RS, Fraser AG, Dong Y, Poulin G, Durbin R, et al. (2003) Systematic functional analysis of the Caenorhabditis elegans genome using RNAi. Nature 421: 231-237.
17. Rosso MN, Jones JT, Abad P (2009) RNAi and functional genomics in plant parasitic nematodes. Annu Rev Phytopathol 47: 207-232.
18. Fanelli E, Di Vito M, Jones JT, De Giorgi C (2005) Analysis of chitin synthase function in a plant parasitic nematode, Meloidogyne artiellia, using RNAi. Gene 349: 87-95.
19. Kimber MJ, McKinney S, McMaster S, Day TA, Fleming CC, et al. (2007) flp gene disruption in a parasitic nematode reveals motor dysfunction and unusual neuronal sensitivity to RNA interference. Faseb J 21: 1233-1243.
20. Opperman CH, Bird DM, Williamson VM, Rokhsar DS, Burke M, et al. (2008) Sequence and genetic map of Meloidogyne hapla: A compact nematode genome for plant parasitism. Proc Natl Acad Sci U S A 105: 14802-14807.
21. Gheysen G, Vanholme B (2007) RNAi from plants to nematodes. Trends Biotechnol 25: 89-92.
22. Hussey RS, Mims CW (1991) Ultrastructure of feeding tubes formed in giant-cells induced in plants by the root-knot nematode Meloidogyne incognita. Protoplasma 162: 99-107.
23. Urwin PE, Moller SG, Lilley CJ, McPherson MJ, Atkinson HJ (1997) Continual green-fluorescent protein monitoring of cauliflower mosaic virus 35S promoter activity in nematode-induced feeding cells in Arabidopsis thaliana. Mol Plant Microbe Interact 10: 394-400.
24. Zhang F, Peng D, Ye X, Yu Z, Hu Z, et al. (2012) In vitro uptake of 140 kDa Bacillus thuringiensis nematicidal crystal proteins by the second stage juvenile of Meloidogyne hapla. PLoS One 7: e38534.
25. Urwin PE, Lilley CJ, Atkinson HJ (2002) Ingestion of double-stranded RNA by preparasitic juvenile cyst nematodes leads to RNA interference. Mol Plant Microbe Interact 15: 747-752.
26. Bakhetia M, Charlton W, Atkinson HJ, McPherson MJ (2005) RNA interference of dual oxidase in the plant nematode Meloidogyne incognita. Mol Plant Microbe Interact 18: 1099-1106.
27. Rosso MN, Dubrana MP, Cimbolini N, Jaubert S, Abad P (2005) Application of RNA interference to root-knot nematode genes encoding esophageal gland proteins. Mol Plant Microbe Interact 18: 615-620.
28. Lilley CJ, Davies LJ, Urwin PE (2012) RNA interference in plant parasitic nematodes: A summary of the current status. Parasitology 139: 630-640.

29. Chen Q, Rehman S, Smant G, Jones JT (2005) Functional analysis of pathogenicity proteins of the potato cyst nematode Globodera rostochiensis using RNAi. Mol Plant Microbe Interact 18: 621-625.
30. Huang G, Allen R, Davis EL, Baum TJ, Hussey RS (2006) Engineering broad root-knot resistance in transgenic plants by RNAi silencing of a conserved and essential root-knot nematode parasitism gene. Proc Natl Acad Sci U S A 103: 14302-14306.
31. Bakhetia M, Urwin PE, Atkinson HJ (2007) QPCR analysis and RNAi define pharyngeal gland cell-expressed genes of Heterodera glycines required for initial interactions with the host. Mol Plant Microbe Interact 20: 306-312.
32. Li J, Todd TC, Lee J, Trick HN (2011) Biotechnological application of functional genomics towards plant-parasitic nematode control. Plant Biotechnol J 9: 936-944.
33. Dalzell JJ, McMaster S, Fleming CC, Maule AG (2010) Short interfering RNA-mediated gene silencing in Globodera pallida and Meloidogyne incognita infective stage juveniles. Int J Parasitol 40: 91-100.
34. Ma HB, Lu Q, Liang J, Zhang XY (2011) Functional analysis of the cellulose gene of the pine wood nematode, Bursaphelenchus xylophilus, using RNA interference. Genet Mol Res 10: 1931-1941.
35. Gaeta RT, Bahaji A, Polack GW, Calderon-Urrea A (2011) Effects of ced-9 dsRNA on Caenorhabditis elegans and Meloidogyne incognita. American Journal Of Agricultural And Biological Sciences 6: 19-28.
36. Matsunaga Y, Kawano K, Iwasaki T, Kawano T (2012) RNA interference-mediated growth control of the southern root-knot nematode Meloidogyne incognita. Biosci Biotechnol Biochem 76: 378-380.
37. Niu J H, Jian H, Xu J, Chen C, Guo Q (2012) RNAi silencing of the Meloidogyne incognita Rpn7 gene reduces nematode parasitic success. European J Plant Pathol 134: 131-144.
38. Li J, Todd TC, Lee J, Trick HN (2011) Biotechnological application of functional genomics towards plant-parasitic nematode control. Plant Biotechnol J 9: 936-944.
39. Hewezi T, Howe P, Maier TR, Baum TJ (2008) Arabidopsis small RNAs and their targets during cyst nematode parasitism. Mol Plant Microbe Interact 21: 1622-1634.
40. Yadav BC, Veluthambi K, Subramaniam K (2006) Host-generated double stranded RNA induces RNAi in plant-parasitic nematodes and protects the host from infection. Mol Biochem Parasitol 148: 219-222.
41. Klink VP, Kim KH, Martins V, Macdonald MH, Beard HS, et al. (2009) A correlation between host-mediated expression of parasite genes as tandem inverted repeats and abrogation of development of female Heterodera glycines cyst formation during infection of Glycine max. Planta 230: 53-71.
42. Li J, Todd TC, Trick HN (2010) Rapid in planta evaluation of root expressed transgenes in chimeric soybean plants. Plant Cell Rep 29: 113-123.
43. Fairbairn DJ, Cavallaro AS, Bernard M, Mahalinga-Iyer J, Graham MW, et al. (2007) Host-delivered RNAi: an effective strategy to silence genes in plant parasitic nematodes. Planta 226: 1525-1533.
44. Sindhu AS, Maier TR, Mitchum MG, Hussey RS, Davis EL, et al. (2009) Effective and specific in planta RNAi in cyst nematodes: expression interference of four parasitism genes reduces parasitic success. J Exp Bot 60: 315-324.
45. Xue B, Hamamouch N, Li C, Huang G, Hussey RS, et al. (2013) The 8D05 parasitism gene of Meloidogyne incognita is required for successful infection of host roots. Phytopathology 103: 175-181.
46. Jaouannet M, Magliano M, Arguel MJ, Gourgues M, Evangelisti E, et al. (2013) The root-knot nematode calreticulin Mi-CRT is a key effector in plant defense suppression. Mol Plant Microbe Interact 26: 97-105.
47. Steeves RM, Todd TC, Essig JS, Trick HN (2006) Transgenic soybeans expressing siRNAs specific to a major sperm protein gene suppress Heterodera glycines reproduction. Func Plant Biol 33: 991-999.

48. Li J, Todd TC, Oakley TR, Lee J, Trick HN (2010) Host-derived suppression of nematode reproductive and fitness genes decreases fecundity of Heterodera glycines Ichinohe. Planta 232: 775-785.
49. Ibrahim HM, Alkharouf NW, Meyer SL, Aly MA, Gamal El-Din Ael K, et al. (2011) Post-transcriptional gene silencing of root-knot nematode in transformed soybean roots. Exp Parasitol 127: 90-99.
50. Gheysen G, Fenoll C (2002) Gene expression in nematode feeding sites. Annu Rev Phytopathol 40: 191-219.
51. Van de Cappelle E, Plovie E, Kyndt T, Grunewald W, Cannoot B, et al. (2008) AtCDKA;1 silencing in Arabidopsis thaliana reduces reproduction of sedentary plant-parasitic nematodes. Plant Biotechnol J 6: 749-757.
52. Elbashir SM, Harborth J, Lendeckel W, Yalcin A, Weber K, et al. (2001) Duplexes of 21-nucleotide RNAs mediate RNA interference in cultured mammalian cells. Nature 411: 494-498.
53. Du Q, Thonberg H, Wang J, Wahlestedt C, Liang Z (2005) A systematic analysis of the silencing effects of an active siRNA at all single-nucleotide mismatched target sites. Nucleic Acids Res 33: 1671-1677.
54. Rual JF, Klitgord N, Achaz G (2007) Novel insights into RNAi off-target effects using C. elegans paralogs. BMC Genomics 8: 106.
55. Gupta A, Pal RK, Rajam MV (2013) Delayed ripening and improved fruit processing quality in tomato by RNAi-mediated silencing of three homologs of 1-aminopropane-1-carboxylate synthase gene. J Plant Physiol 170: 987-995.
56. Sinha R, Rajam MV (2013) RNAi silencing of three homologues of S-adenosylmethionine decarboxylase gene in tapetal tissue of tomato results in male sterility. Plant Mol Biol 82: 169-180.
57. Auer C, Frederick R (2009) Crop improvement using small RNAs: Applications and predictive ecological risk assessments. Trends Biotechnol 27: 644-651.
58. Schmid A, Schindelholz B, Zinn K (2002) Combinatorial RNAi: A method for evaluating the functions of gene families in Drosophila. Trends Neurosci 25: 71-74.
59. Tischler J, Lehner B, Chen N, Fraser AG (2006) Combinatorial RNA interference in Caenorhabditis elegans reveals that redundancy between gene duplicates can be maintained for more than 80 million years of evolution. Genome Biol 7: R69.
60. Grimm D, Kay MA (2007) Therapeutic application of RNAi: is mRNA targeting finally ready for prime time? J Clin Invest 117: 3633-3641.
61. Charlton WL, Harel HY, Bakhetia M, Hibbard JK, Atkinson HJ, et al. (2010) Additive effects of plant expressed double-stranded RNAs on root-knot nematode development. Int J Parasitol 40: 855-864.
62. Dalzell JJ, Warnock ND, Stevenson MA, Mousley A, Fleming CC, et al. (2010) Short interfering RNA-mediated knockdown of drosha and pasha in undifferentiated Meloidogyne incognita eggs leads to irregular growth and embryonic lethality. Int J Parasitol 40: 1303-1310.
63. Melito S, Heuberger AL, Cook D, Diers BW, MacGuidwin AE, et al. (2010) A nematode demographics assay in transgenic roots reveals no significant impacts of the Rhg1 locus LRR-Kinase on soybean cyst nematode resistance. BMC Plant Biol 10: 104.

CHAPTER 20

Damage Thresholds and Population Dynamics of Pratylenchus Penetrans on Carrot (Daucus Carota L. Cv. Nerac) at Three Different Seed Densities

Misghina G[1]. Teklu & Beira[2] H. Meressa[3] & Esther Radtke[4] & Thomas H. Been[1] & Johannes Hallmann[3]

[1]*Agro-Systems ResearchWageningen University and Research CentreWageningenThe Netherlands*

[2].*Laboratory of NematologyWageningen UniversityWageningenThe Netherlands*

[3].*Federal Research Centre for Cultivated PlantsJulius Kühn-Institut, Institute for Epidemiology and Pathogen DiagnosticsMünsterGermany*

[4].*Landwirtschaftskammer Nordrhein-WestfalenBonnGermany*

ABSTRACT

Yield and quality loss of carrot (*Daucus carota* L. cv. Nerac) caused by *Pratylenchus penetrans* and the population dynamics of this nematode were studied in a climate controlled glasshouse. A range of 12 nematode densities was used at three different seed densities of carrot; 2, 4 and 18 seeds pot^{-1}. Seinhorst's yield loss model; $y = m + (1 - m)\ 0.95^{Pi/T-1}$ for $Pi > T$; $y = 1$ for $Pi \leq T$ for Tylenchina was fitted to the yield and quality loss data. Seinhorst's model for population dynamics of migratory nematodes with multiple generations; Pf=M∗Pi/(Pi+M/a) Pf=M∗Pi/(Pi+M/a) was fitted to the data of the final population densities (*Pf*). *P. penetrans* had a significant impact on carrot taproot yield and its quality. The tolerance limits for the relative carrot taproot yield (T_y) were 1.51, 1.88, and 1.37 and those of quality yields (T_q) were 0.67, 0.18, and 0.40 *P. penetrans* (g dry soil)$^{-1}$ at 2, 4 and 18 seeds pot^{-1}, respectively. Both the minimum yield (0.20, 0.29, and 0.60) and the minimum quality yield (0.05, 0.07, and 0.20), expressed as a proportion, increased with seed density at 2, 4 and 18 seeds pot^{-1}, respectively. The model for population dynamics fitted well to the *Pf* data obtained. The maximum multiplication rates (a) were 19.58, 9.99, and 17.54, while the maximum population densities (M) were 49.86, 43.21, and 60.37 *P. penetrans* (g dry soil)$^{-1}$ at 2, 4, and 18 seeds pot^{-1}, respectively. Carrot cv. Nerac can be considered a good host for *P. penetrans*.

Keywords: Lesion nematodes Modelling Population density Quality loss Tolerance limit and yield loss

INTRODUCTION

Carrot (*Daucus carota* L.) is a major vegetable grown in the temperate regions and cooler highlands of the tropics. Its production is affected, among others, by root-knot nematodes (*Meloidogyne* spp.), root-lesion nematodes (*Pratylenchus* spp.), pin nematodes (*Paratylenchus* spp.) and stubby root nematodes (*Paratrichodorus* spp.) (Noling 2012). In particular, infestation by *Meloidogyne hapla*, *M. fallax* and *M. chitwoodi* can lead to severe quality losses due to forking, stunting, excessive formation of lateral roots and galls on the carrot taproots (Wesemael and Moens 2008; Walker 2004; Heve et al. 2015).

In Germany, *M. hapla* was considered the most damaging plant-parasitic nematode on carrots, particularly on light sandy soils (Hahn 1958; Stein 1963; Gugino et al. 2006). Recently root-lesion nematodes (*Pratylenchus* spp.) seem to become one of the constraints of carrot production. A survey of cereal and vegetable production systems under organic farming in Germany revealed more than 90 % of the samples being infested with *Pratylenchus* spp. (Hallmann et al. 2007). A similar survey in northern Tasmania, Australia revealed that 28 out of 33 carrot crops (*ca.* 85 %) were found infested by *Pratylenchus* spp. as compared to only five out of 33 (*ca.* 15 %) by root-knot nematodes (Hay and Pethybridge 2005). *Pratylenchus* spp. are considered the third most economically important plant-parasitic nematodes in crop production (Castillo and Volvas 2007; Jones et al. 2013). Diverse adaptation has enabled *Pratylenchus* spp. to colonize plant species under different environmental conditions within the temperate and tropical zone (Potter and Olthof 1993; Castillo and Volvas 2007).

Generally, yield loss is experienced when population densities at planting are above the tolerance limit (Viaene and Abawi 1996; Schomaker and Been 2013). Yield reductions by *P. penetrans* on carrots and *P. crenatus* on Kuroda type carrots is reported by (Vrain and Belair 1981; Hay and Pethybridge 2005) respectively. Also, the quality of the carrot taproot is affected by *P. penetrans*. Quality losses due to forking of carrot taproots, stunted growth and excessive formation of lateral roots are reported (Coosemans 1975; Vrain and Belair 1981). Similar quality defects on carrot taproots are described for *P. crenatus* (Potter and Olthof 1993). Therefore, knowledge concerning the relation between *Pi* and expected yield and quality loss is of major importance to plan justifiable control strategies (Ferris 1978; Barker and Olthof 1976) to keep population densities below the 'economical' tolerance limit where unacceptable yield or quality losses are suffered (Seinhorst 1967, 1970).

Both in Germany and The Netherlands, carrots for industrial processing are grown only on fields that have been surveyed for *P. penetrans* infestations. Based on soil samples, a contract can be obtained when the nematode density detected is below a threshold of 100 and 75 *P. penetrans* (100 ml soil)$^{-1}$ for the two countries, respectively. These thresholds are based on rough estimations, but generally provide for the absence of serious quality problems caused by *P. penetrans*. In the ProGemüse project (INTERREG IV-A II-3-04-100), a prototype sampling method has been developed to improve the accuracy of the population density estimate used for farmers advice, but information on the tolerance limit, yield and quality loss of carrots, and the population dynamics of *P. penetrans* at different seed densities is still lacking.

In carrot production, seed densities range from 1 to 20 million seeds ha^{-1}, depending on their use either for processing or the fresh market. The seed density is used to determine the size of the carrot taproots produced. The objectives of this study were to determine both the damage thresholds for yield and quality loss of carrot cv. Nerac attacked by *P. penetrans* at different seed densities and its population dynamics. This information is required to help farmers, industrial grower and extension services in their decision making process as to whether a land is suitable for carrot production and if crop protection is needed based on pre-crop sampling results.

MATERIALS AND METHODS

This experiment was carried out at the Julius Kühn-Institute, in Münster, Germany, while at the same time a similar experiment was conducted at Wageningen UR, Wageningen, The Netherlands with carrot cv. Nerac and *M. chitwoodi* (Heve et al. 2015).

Multiplication of *P. penetrans*

Three seeds pot^{-1} of maize (*Zea mays* cv. Husar) were planted in 6-l pots containing a bottom layer of 390 g of non-crushed hydro-grains and 5610 g of pre-glacial silver sand. The bottom of each pot was covered with Ederol filter No. 261, 15 cm diam. (J.H. Ritmeester B.V, Utrecht) to prevent the leaching of the growing medium. A *Pi* of 4 *P. penetrans* (g dry soil)$^{-1}$ was inoculated by adding the required suspension in a shallow groove around the established plants after 2 weeks. The groove was then covered with soil. Pots were watered and fertilized every week for 90 days. Roots were then collected by sieving the soil through 3 mm mesh sieves. The collected roots were gently rinsed, cut into pieces of 1 cm and placed on extraction sieves (150 µm mesh size, diam. 20 cm), and were placed on top of extraction dishes (diam. 25 cm) and incubated in a Seinhorst mist-chamber (Seinhorst 1988). Nematodes were collected every 3 days from the roots of 45 maize plants until a stock suspension, containing sufficient *P. penetrans* (adults and juveniles) for the experiment (11.1 million) was obtained. The stock suspension was stored at 4 °C till the time of inoculation.

Growth medium and pot preparation

For this experiment a steam sterilized field soil was used. After steaming, the soil was exposed to the air for at least 30 days to enable evaporation and dissipation of toxic compounds. The pots used were slightly tapered with a square base of 9.8 by 9.8 cm, a square top of 13.7 by 13.7 cm and 23 cm height, with a capacity of 3.5 l, and eight drainage holes at the bottom. The chosen pot volume and height represented the rooting space of carrots grown on dams in the field. The moisture content of the soil was determined by oven-drying (105 °C) three samples of 3700 g for 24 h. An average moisture content of 10.4 % was measured and 2.9 kg of steam-sterilized soil was then filled into the pots and gently compressed using a wooden stamp. Soil moisture was then adjusted to 17 % before nematode inoculation.

Nematode density preparation and inoculation

The volume of the stock suspension of nematodes was adjusted to the highest density required, after which a dilution series was made according to a log series of 2^x where x is an integer ranging from −3 to 7. To distribute the nematodes more or less uniformly through the soil column of the pot, a 29.7 cm closed bottom needle with side holes was used. Each needle was inserted into the soil down to 2.5 cm above the bottom of the pot and the inoculum was released while gently withdrawing the needle. Controls were inoculated with water. Nine, 3 ml injections pot^{-1}, totalling 27 ml, were used to apply the required number of nematodes to each pot, obtaining a series of densities ranging from 0.125 to 128 *P. penetrans* (g dry soil)$^{-1}$.

Glasshouse conditions and planting

Daily temperature of the glasshouse was maintained at 23 °C and 16 h of day light provided using 600 W, 58,500 lumen lamps (Norka-Lighting®, Hamburg, Germany). In order to synchronise the germination of the carrot seeds with the inoculation of the nematodes into the soil, seeds were soaked in water for 4 days prior to planting. The detailed procedure is described in Heve et al. (2015). After inoculation, each pot was sown with either 2, 4 or 18 seeds pot^{-1} to obtain an equivalent seed density of 1, 2 and 10 million ha^{-1}, respectively, which complies with normal practice. Carrot seeds were planted at 5 mm depth. To make sure that the required numbers of plants per pot of 2, 4 and 18 seeds were obtained, 2, 4 and 4 extra seeds per seed density were added, respectively, based on a 94 % germination rate reported by Heve et al. (2015) for cv. Nerac. Extra seedlings were thinned immediately after germination by removing the aboveground part. Each nematode and seed density combination was replicated five times. In total, 180 pots were used for this experiment.

Watering and fertilizing

To minimize evaporation and avoid splashing during watering, each pot was covered with perforated black polythene-sheet. Throughout the growing period, pots were weighed and watered weekly to restore the soil moisture content to 17 %. In between, five randomly selected pots were weighed and a fixed volume of water was given to each pot, if necessary. The pots were arranged in a complete randomised design on raised benches in a glasshouse and during watering pots were rotated to minimize positional effects. Moreover, the increase in weight due to plant growth was taken into consideration during watering (1 ml extra water for each g of growth based on prior studies for cv. Nerac (Heve et al. 2015). Plants were fertilized weekly with 0.3 % WUXAL® Super liquid foliar fertilizer (Agrarverstand Oberland, Schongau, Germany). On account of the growth pattern described in Heve et al. (2015), carrots were harvested 83 days after germination when they reached the industry desired diameter of the carrot taproot.

Observations and measurements

Seed germination date of carrots was recorded. At harvest, fresh and dry shoot, root and taproot weights were measured for each plant. Dry weights were measured after 24 h oven drying at 105 °C. The diameter and length of each carrot was measured using an electronic Vernier-calliper. The commercial or industrial market quality of the carrot taproots was visually graded with a score of either 0 (absent) or 1 (present) for several quality features, which include carrot taproots being forked or branched, hairy-, stubby, misshapen or rotten. Carrot taproots with growth cracks, pale green coloured heads (>1 cm) and taproot head diameter outside a range of 12–25 mm (all not nematode related) were also rejected according to industrial standards. Carrot taproots were considered as acceptable when all negative quality features were absent.

Estimation of final population densities at harvest

Nematode densities from the organic fraction of each pot were determined from the entire root system and from the first 0.5 mm peel of all carrot taproots, independent of their industrial quality. The roots of each pot were removed from the soil by hand; the soil was then sieved gently through a 3-mm mesh sieve to collect the root hairs. Roots were rinsed with water to remove any adhering soil and then cut into pieces of 1 cm. With a multi-functional carrot peeler with twin swivelling blades (BONET) the carrot peel was collected, cut in pieces of 1 cm^2 and added to the roots (http://bonet.en.alibaba.com/). Both were placed in a 9 cm diam. extraction sieve (150 µm) on top of an extraction dish of 12 cm diam. and placed in a Seinhorst mist-chamber (Seinhorst 1988). In the mist-chamber the sieves were exposed to a 45 min cycle providing a fine mist of 20 °C during 15 min and a rest period of 30 min. Hatched nematodes were tapped every week for 28 days.

Nematodes from the mineral fraction (the soil) were extracted from aliquots of 250 ml soil using the centrifugal flotation ($MgSO_4$) method (Hooper et al. 2005). The nematodes from both fractions were collected into glass bottles and stored in a refrigerator at 4 °C until counted. The final nematode population density (*Pf*), combined number of adults and juveniles was then estimated by adding the nematode densities of both fractions (g dry soil)$^{-1}$.

Statistical analysis and modelling

All data sets were analysed using scripts written in Tinn-R, version 3.0.3.6 and run in R-version 3.0.1 using non-linear regression (Dalgaard 2002). The goodness of fit of models was expressed as the coefficient of determination (R^2), adjusted to degrees of freedom. The Least significant difference (LSD) at $p = 0.05$ was calculated for all estimated parameters between seed density if necessary.

Diameter, length, yield and quality of carrot taproots

Seinhorst's Eq. (1) as described by Schomaker and Been (2013) was fitted to the data of carrot length, –diameter, relative yield (weight of all carrot taproots), and quality yield (weight of all acceptable carrot taproots), based on the standards acceptable for the industry. Before fitting the model to the data, each measurement was averaged over the replicates for each nematode density.

$$\begin{aligned} y &= m + (1-m)\,0.95^{Pi/T-1} && \text{for } Pi > T \\ y &= 1 && \text{for } Pi \leq T \end{aligned} \tag{1}$$

The parameters tolerance limits (T) and relative minimum yield (m) were estimated for all measurements.

Population dynamics

The Pf obtained from each density was first log transformed and then averaged over replications and back transformed. The relation between Pi and Pf was modelled to estimate the maximum multiplication rate (a) and the maximum population density (M) for each seed density, using Eq. (2) for migratory nematodes with more than one generation per growing period (Seinhorst 1966; Teklu et al. 2014).

$$Pf = M^{*}Pi/(Pi + M/a) \tag{2}$$

RESULTS

Carrot establishment

Almost all seedlings at 2 and 4 seeds pot^{-1}, at all population densities, developed into carrot plants. Only at 18 seeds pot^{-1} a negative correlation between *Pi* and plant establishment became noticeable. Carrot plants without carrot taproots occurred only at the highest density of 128 *P. penetrans* (g dry soil)$^{-1}$ at all seed densities. No correlation between taproot presence and *Pi* could be established, except for 18 seeds pot^{-1} (Figure 1). A few averages lie slightly above the intended seed density of the pot which probably resulted from surplus seedlings that escaped the thinning.

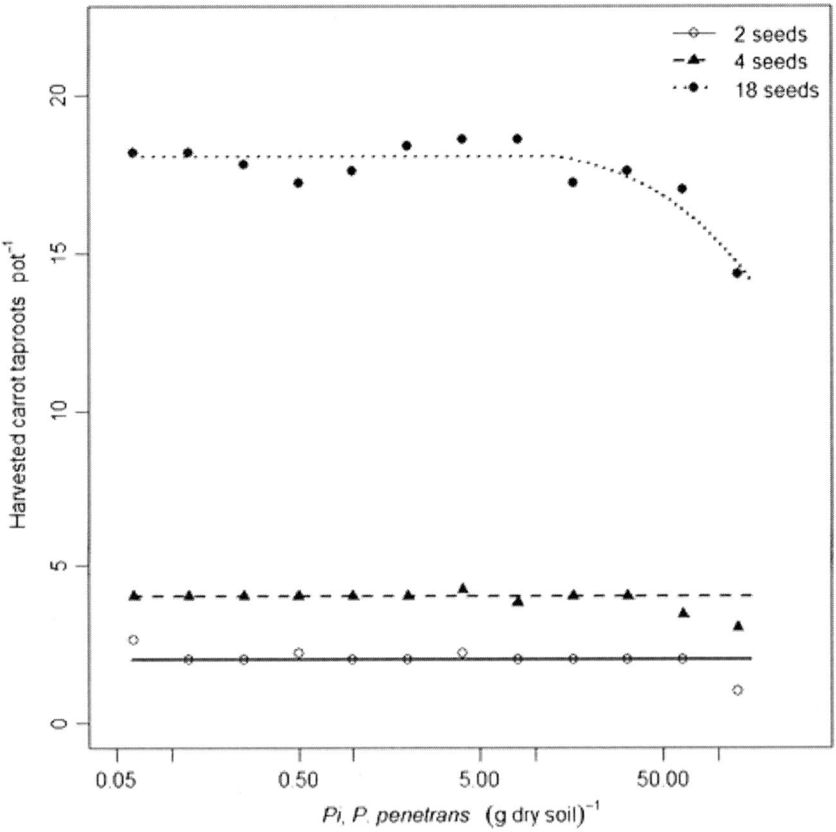

Figure 1. The relation between the initial population density (*Pi*) of *P. penetrans*, plotted on a log scale, and the number of carrot taproots harvested at 2, 4 and 18 seeds pot^{-1}

Carrot taproot growth

Carrot taproots at all seed densities attained a diameter of 12–25 mm, which is required for the canning industry, 83 days after sowing. Both carrot taproot diameter and, especially, carrot taproot length were negatively correlated with increasing *Pi*. No differences were detected for the tolerance limits (T_d) of 0.79, 2.71, and 0.23 *P. penetrans* (g dry soil)$^{-1}$ and the relative minimum diameter (m_d) of 0.90, 0.83, and 0.95 at 2, 4, and 18 seeds pot^{-1}, respectively. The tolerance limits (T_L) for length were 1.33, 1.55, and 1.21 *P. penetrans* (g dry soil)$^{-1}$ for 2, 4, and 18 seeds pot^{-1}, respectively. No statistical difference was found. The relative minimum length (m_L) was 0.47, 0.48, and 0.76 at 2, 4, and 18 seeds pot^{-1}, respectively. The m_L value at 18 seeds pot^{-1} was significantly higher than those at 2 and 4 seeds pot^{-1}. Both maximum diameter and length of carrot taproots were inversely related to seed density. Parameter values are summarized in Table 1.

Taproot yield

Equation 1 was fitted to the data of carrot taproot yield to investigate the relation between *Pi* and yield for each seed density (Figure 2). Carrot taproot yield started to be affected at *Pi* values larger than (T_y) of 1.51, 1.88, and 1.37 *P. penetrans* (g dry soil)$^{-1}$ at respective seed densities. Minimum yields (m_y) of 0.20, 0.29, and 0.60 at 2, 4, and 18 seeds pot^{-1} were estimated, respectively. No significant difference in T_y value could be established between the seed densities. However, the m_y value at 18 seeds pot^{-1} was significantly higher than that of 2 and 4 seeds pot^{-1}. Generally, taproot weight increased with increasing seed densities. Maximum carrot taproot weight $Y_{max(y)}$, of 59.90, 76.14, and 100.32 g were measured at 2, 4, and 18 seeds pot^{-1}, respectively. Estimated model parameter values for carrot taproot yield and LSD are listed in Table 1.

Taproots quality

The relation between *Pi* and quality yield is displayed in Figure 3a and a summary of parameter values is provided in Table 1. Tolerance limits (T_q) of 0.67, 0.18, and 0.40 *P. penetrans* (g dry soil)$^{-1}$ at 2, 4, and 18 seeds pot^{-1} were estimated, respectively. The minimum quality yields (m_q) were 0.05, 0.07, and 0.20 at 2, 4, and 18 seeds pot^{-1}, respectively. No statistical significant difference could be recognized for the parameters T_q and m_q between seed densities. Maximum quality yields $Y_{max(q)}$, of 40.39,

70.06, and 91.25 g were measured at 2, 4, and 18 seeds pot^{-1}, respectively. As $Y_{max\ (q)}$ is seed density dependent, the same data are plotted relatively for a better comparison in Figure 3b.

Table 1. Parameter values for Seinhorst Eq. (1) for the relation between initial population density (*Pi*) of *P. penetrans* and measured variables of carrots. *Pi* and *T* are expressed in *P. penetrans* (g dry soil)$^{-1}$ while, *y* and *m* are proportions and Y_{max} is expressed in grams. Comparison of the parameter values at 2 seeds pot^{-1} was done with that of 4 and 18 seeds pot^{-1}

Parameter values; diameter of carrot taproots										
According to the equation: $y_d = m_d + (1 - m_d)\ 0.95^{Pi/T_d - 1}$ for $Pi > T_d$ and $y_d = 1$ for $Pi \leq T_d$										
Seed density	T_d	m_d	$Y_{max\ (d)}$	$SE_{T\ (d)}$	$SE_{m(d)}$	$SE_{Ymax\ (d)}$	R^2	df	$LSD_{T(d)}$	$LSD_{m\ (d)}$
2	0.79	0.90	22.65	1.09	0.04	0.40	0.52	8	–	–
4	2.71	0.83	20.39	2.82	0.11	0.22	0.71	8	6.41	0.26
18	0.23	0.95	14.96	0.11	0.01	0.10	0.79	8	2.32	0.10
Parameter values; length of the carrot taproots										
According to the equation: $y_L = m_L + (1 - m_L)\ 0.95^{Pi/T_L - 1}$ for $Pi > T_L$ and $y_L = 1$ for $Pi \leq T_L$										
Seed density	T_L	m_L	$Y_{max\ (L)}$	$SE_{T\ (L)}$	$SE_{m\ (L)}$	$SE_{Ymax\ (L)}$	R^2	df	$LSD_{T\ (L)}$	$LSD_{m\ (L)}$
2	1.33	0.47	120.50	1.09	0.13	6.00	0.82	9	–	–
4	1.55	0.48	101.64	1.18	0.12	4.35	0.85	9	3.37	0.38
18	1.21	0.76	54.74	0.35	0.02	0.62	0.94	9	2.41	*0.27
Parameter values; carrot taproot yield										
According to the equation: $y = m_y + (1 - m_y)\ 0.95^{Pi/T_y - 1}$ for $Pi > T_y$ and $y = 1$ for $Pi \leq T_y$										
Seed density	T_y	m_y	$Y_{max\ (y)}$	$SE_{T\ (y)}$	$SE_{m\ (y)}$	$SE_{Ymax\ (y)}$	R^2	df	$LSD_{T\ (y)}$	$LSD_{m\ (y)}$
2	1.51	0.20	59.90	0.71	0.13	2.78	0.91	9	–	–
4	1.88	0.29	76.14	0.44	0.07	1.65	0.97	9	1.75	0.30
18	1.37	0.60	100.32	0.91	0.08	3.16	0.86	9	2.42	*0.32
Parameter values: carrot taproot quality yield										
According to the equation: $y_q = m_q + (1 - m_q)\ 0.95^{Pi/T_q - 1}$ for $Pi > T_q$ and $y_q = 1$ for $Pi \leq T_q$										
Seed density	T_q	m_q	$Y_{max\ (q)}$	$SE_{T\ (m)}$	$SE_{m\ (q)}$	$SE_{Ymax\ (q)}$	R^2	df	$LSD_{T\ (d)}$	$LSD_{m\ (q)}$
2	0.67	0.05	43.39	0.26	0.13	3.64	0.85	9	–	–
4	0.18	0.07	70.06	0.09	0.08	6.15	0.92	9	0.58	0.32
18	0.40	0.20	91.25	0.13	0.07	5.01	0.93	9	0.62	0.31

*Significantly different at $p = 0.05$

Quality losses, associated with *P. penetrans*, included forked, misshapen, constricted, rotted and hairy carrots. Non-nematode related quality damage, like greening of the hypocotyl caused by exposure to sunlight, was also noticed. However, a major part of the quality losses could be attributed to forked and misshapen carrot taproots; both symptoms were positively correlated with *Pi*. Only a small number of carrot taproots suffering from rot were observed, mainly at 2 and 4 seed pot^{-1}. This quality problem was *Pi* independent. Little quality loss was also observed at *Pi* = 0.

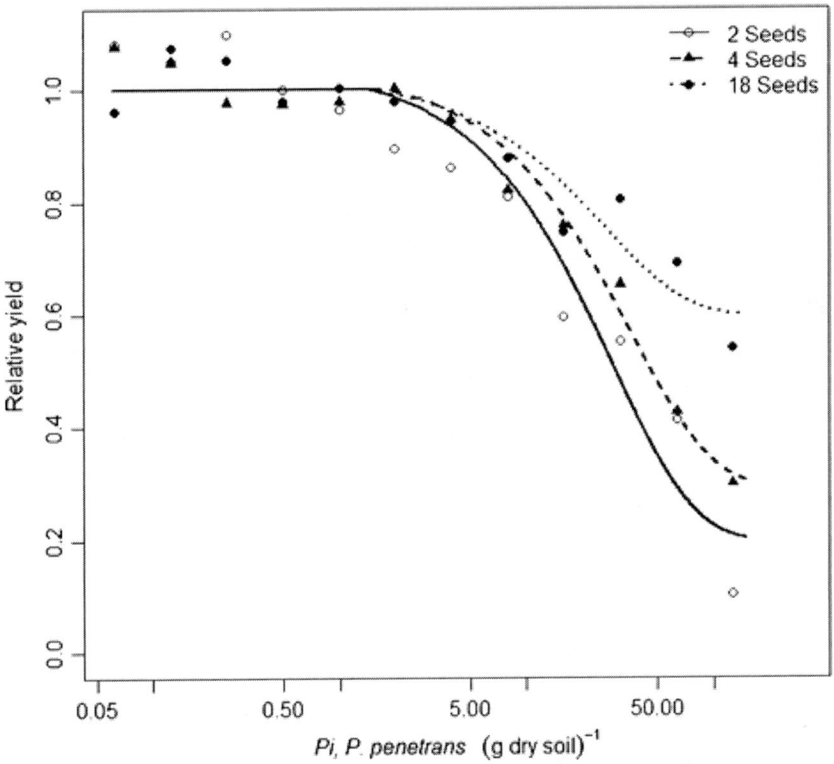

Figure 2. The relation between the initial population density, (*Pi*) of *P. penetrans*, on a log scale, and relative carrot taproot yield at 2, 4 and 18 seeds pot^{-1}. Fitted line according to the Seinhorst yield loss equation: $0.95^{Pi/T_y - 1}$ for $Pi > T_y$ and $y = 1$ for $Pi \leq T_y$

Population dynamics

The model for population dynamics (Eq. 2) fitted well to the relation between *Pi* and *Pf*. All the fitted lines are above the equilibrium density line (*Pf* = *Pi*). The relation between *Pi* and *Pf* at all seed densities according to Eq. (2) is shown in Figure 4. Maximum multiplication rates (*a*) of 19.58, 9.99, and 17.54 and maximum population densities (*M*) of 49.86, 43.21, and 60.37 *P. penetrans* (g dry soil)$^{-1}$ at 2, 4, and 18 seeds pot^{-1} were estimated, respectively. Both values for *a* and *M* at 2 and 18 seeds pot^{-1} were significantly higher than those at 4 seeds pot^{-1} (Table 2).

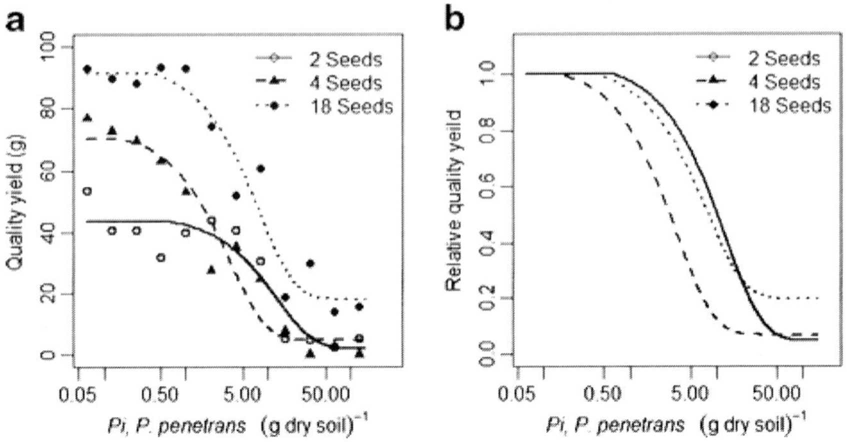

Figure 3. The relation between the initial population density (*Pi*) of *P. penetrans*, on a log scale and **A**: Carrot taproot quality yield (g) and **B**: Relative carrot taproot quality yield at 2, 4, and 18 seeds pot^{-1}. Fitted lines according to the Seinhorst yield loss equation:

$$y_q = m_q + (1 - m_q)\, 0.95^{Pi/T_q - 1} \text{ for } Pi > T_q \text{ and } y_q = 1 \text{ for } Pi \leq T_q$$

Table 2. Parameter values of the population dynamics model for the relation between initial (*Pi*) and final population density (*Pf*) both measured as *P. penetrans* (g dry soil)$^{-1}$ at 2, 4 and 18 seeds pot^{-1}. The model for migratory nematodes with more than one generation: Pf=M∗Pi/(Pi+M/a)Pf=M∗Pi/(Pi+M/a) was used. Multiplication rate (*a*) is dimensionless while, maximum population (*M*) is measured as *P. penetrans* (g dry soil)$^{-1}$. Comparison of the parameter values of 2 seeds pot^{-1} was done with that of 4 and 18 seeds pot^{-1}

Parameter values (Population dynamics)								
Seed density	a	M	SE_a	SE_M	R^2	df	$LSD_{(a)}$	$LSD_{(M)}$
2	19.58	49.86	3.33	7.26	0.93	9	–	–
4	9.99	43.21	1.79	7.95	0.93	9	*0.52	*0.49
18	17.54	60.37	2.79	9.14	0.94	9	*0.49	*0.44

*Significantly different at $p = 0.05$

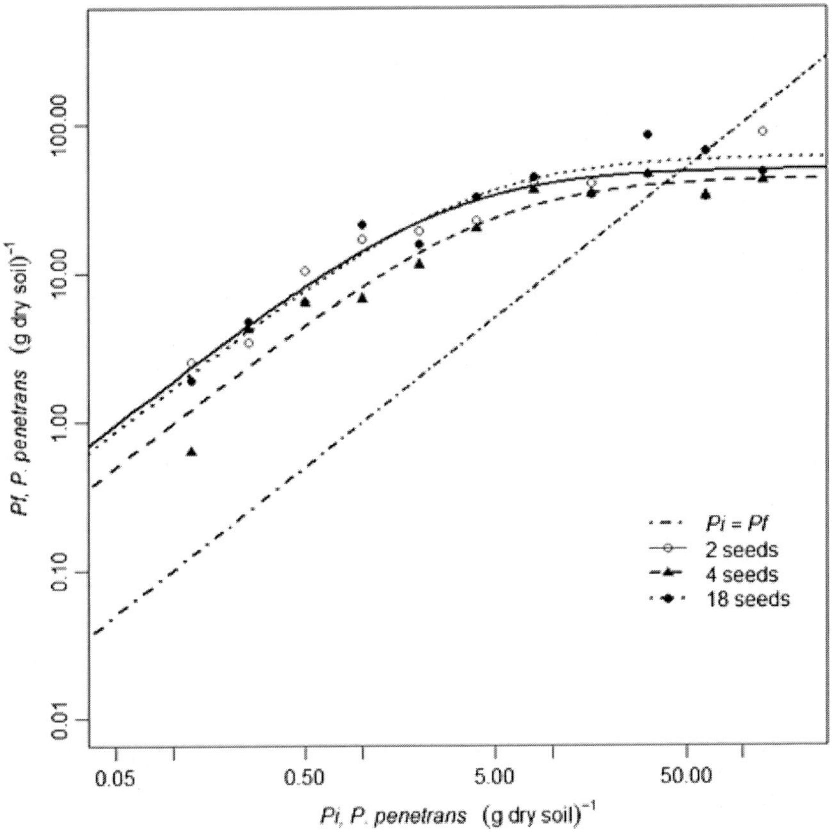

Figure 4. The relation between initial (*Pi*) and final (*Pf*) population densities of *P. penetrans*, on a log scale at 2, 4 and 18 seeds pot^{-1}. Fitted lines according to the equation: Pf=M∗Pi/(Pi+M/aPf=M∗Pi/(Pi+M/a) for population dynamics (Seinhorst 1966). Straight diagonal dashed line: population equilibrium line: *Pf* = *Pi*

DISCUSSION

Establishment and growth

Establishment of seeds into carrot plant was independent of the *Pi*, except at 18 seeds pot^{-1}. This might indicate that at the highest seed density a combined effect of competition between seedlings and attack by high nematode densities occurred. The additional reduction of taproot development by carrot plants only occurred at the highest *Pi* of 128 *P. penetrans* (g dry soil)$^{-1}$at all seed densities. A more pronounced correlation might be obtained when even higher nematode densities would be added to the *Pi* series. However, these densities would be surpassing the *Pi* encountered in the field in spring, which are seldom higher than 50 *P. penetrans* (g dry soil)$^{-1}$ or *ca*. 5000 *P. penetrans* (100 ml soil)$^{-1}$ in The Netherlands (L.P.G. Molendijk, pers.com.). Therefore, little practical value is to be expected in exploring higher *Pi* on the establishment of carrots. Establishment of cv. Nerac into carrot plant was reported to be independent of seed density and *Pi* of *M. chitwoodi*, but carrot plants that developed carrot taproots were negatively correlated with *Pi* at the same seed densities (Heve et al. 2015).

The negative correlation between carrot taproot length and diameter with the *Pi*, confirms previous results from Coosemans (1975) and Vrain and Belair (1981). Also, both taproot diameter and length were negatively correlated with increasing seed density, which reflects the purpose of varying seed densities to provide for different carrot taproot sizes required by the canning industry. These results comply with those of Heve et al. (2015) at identical seeding rates of cv. Nerac exposed to *M. chitwoodi*.

Yield and quality loss

The tolerance limits for carrot yield (T_y) were not affected by increasing seed densities in this experiment. Similar threshold levels for yield loss of carrot were reported by Oostenbrink (1972), Seinhorst (1998) and Pudasaini (2006) with T_y values of 0.3–1.8, 1.4, and 1.03 *P. penetrans* (g dry soil)$^{-1}$, respectively. A T_y value of *ca*. 0.5 *P. crenatus* (g soil)$^{-1}$, was reported for other species of *Pratylenchus* by Weischer and Brown (2000). Seinhorst (1998) and Pudasaini (2006) reported m_y values for carrots of 0.49 and 0.36, respectively, which do not deviate from the range found in this experiment. Values of 0.76–0.78 for m_y of *P. crenatus* on Kuroda-type carrot were reported by Hay and Pethybridge (2005). In that case, differences of host suitability within species and carrot cultivars can be expected. With increasing seed density a higher total marketable carrot taproot was obtained and, more important, an increasing minimum yield (m_y), similar to observations by Heve et al. (2015) with *M. chitwoodi*. This indicates the distribution of the initial population density over a larger root volume per unit soil, when the seed density is increased.

The tolerance limits for quality yield varied between seed densities, but were not statistically significant. Coosemans (1975) reported about 75 % of the carrots being forked at 10 *P. penetrans* (100 ml soil)$^{-1}$ which is equivalent to *ca*. 0.08 *P. penetrans* (g soil)$^{-1}$, a density even lower than the T_q in this experiment. Sato et al. (2009) reports a T_q value of 0.11 *P. penetrans* (g soil)$^{-1}$ for radish. Different to *P. penetrans*, Heve et al. (2015) observed increasing T_q values with an increasing seed density for *M. chitwoodi* on the same carrot cultivar. In general, results confirm that quality loss by *P. penetrans* was more severe than yield loss; both for the parameters *T* and *m* as was also reported by Coosemans (1975) and for *M. chitwoodi* on carrot (Heve et al. 2015) and potato (Norshie et al. 2011).

Population dynamics and management

The relatively higher *a* and *M* values at 2 seeds pot^{-1} compared to those at 4 seeds pot^{-1} and, for *a*, even at 18 seeds pot^{-1} is difficult to elucidate. A possible explanation could be a large variation in the nematode density estimation of *P. penetrans*. Both experience and literature indicate that 50 % of the population of *P. penetrans* is retrieved from the fine root hairs; a fraction which can easily be lost when retrieving the roots from the soil is handled too rough (Pudasaini et al. 2006a). However, extreme care was taken, soil was sieved two to three times to ensure retrieval of the fine roots and any variation would not have been limited to a single seed density. Pudasaini (2006) reported a multiplication rate (a = 5.9) and maximum population density, M = 37.3 *P. penetrans* (g dry soil)$^{-1}$) on carrots, respectively, which is more or less similar to what was found at 2 and 4 seeds pot^{-1} in this study, despite some differences in methodology applied. Heve et al. (2015) observed increasing *a* and *M* values for *M. chitwoodi* with increasing seed density as a result of an increasing root volume per unit soil and, consequently, more space for the nematode to feed.

P. penetrans multiplied well, indicating that cv. Nerac is a good host, unlike *M. chitwoodi* (Heve et al. 2015). This implies, population densities after harvest will be increased and management might be required before the next crop is grown. The use of *Tagetes patula* (marigold) is commercially applied in The Netherlands and proved to be even more effective than soil fumigation to control *P. penetrans* (Evenhuis et al. 2004). Currently, *Tagetes* is grown on 5000 to 6000 ha arable land, mainly aimed to control this nematode in strawberry, flower bulbs, roses, apple orchards, and vegetables (L.P.G. Molendijk, pers.com.). When five crop rotations were compared, population densities of *P. penetrans* were lowest after cropping *Tagetes patula* compared to barley, timothy, annual ryegrass and pearl millet (Kimpinski and Sanderson 2004). According to Pudasaini et al. (2006b), using better sampling methodology than Kimpinski and Sanderson (2004), *Tagetes patula* reduced population densities by more than 90 % in Belgium field trials. An additional 64 % natural population decline in the following fallow period after *Tagetes patula* had been grown and strict weed control was applied (Pudasaini et al. 2006b). The combined use of non-hosts (*Tagetes patula*) and poor hosts (perennial rye-grass, *Avena strigosa*,) and winter fallow provides the possibility to decrease population densities and thus increase both carrot yield and quality. As chemical control becomes more difficult as a result of legislation, there is a need to establish the host status of more crops, or even obtain resistant crops, which could be used in rotations with carrots to manage *P. penetrans*.

CONCLUSION

A correlation between *Pi* of *P. penetrans*, yield, quality and *Pf* was demonstrated for each seed density tested in this experiment. This indicates the possibility to predict yield losses and population development based on pre-plant population densities of *P. penetrans* and, therefore, to manage this nematode using an integrated pest management strategy. The average quality threshold level T_q of 0.4 *P. penetrans* (g dry soil)$^{-1}$ in this experiment equates to about 50 *P. penetrans* (100 ml)$^{-1}$ and deviates little from the currently used standard in Germany and the Netherlands. Hardly any quality yield loss will be suffered at a *Pi* of 0.8 *P. penetrans* (g dry soil)$^{-1}$ or *ca.* 100 *P. penetrans* (100 ml soil)$^{-1}$. At relatively higher *Pi* the possibility of cropping at 10 million seeds ha^{-1} can be considered, which should provide higher quality yields compared to lower seed densities. Actually, higher quality losses up to 25 % of the carrots produced are accepted by the canning industry, e.g. in The Netherlands when fields with favourable low infestations are scarce.

As tolerance limit generally does not deviate between pot and field experiments and is not related to host status the threshold values reported can be considered to be applicable for practical use. Contrary, the minimum yield is much more variable and varies from field to field and year to year due to weather and field conditions. Therefore, validation in the field is needed to ascertain the variation of the minimum yield and quality loss (m_y and m_q) to be encountered. Parallel to this, field tests have to be carried out to parameterize the model of nematode population dynamics as, generally, under glasshouse conditions, due to the completely optimized growing conditions for the host and, therefore, for the tested plant-parasitic nematode, these are higher than under field conditions. This would also provide information concerning future translation of glasshouse data to the field, as has been done for *Globodera pallida* and potato (Been et al. 1994).

ACKNOWLEDGMENTS

The authors would like to acknowledge the EU for funding this research through the ProGemüse project (INTERREG IV-A II-3-04-100). Special thanks to Corrie H. Schomaker for the modelling and statistical analysis. J.E. Beniers, Falko Lange, Mechthild Neuhaus, Agnes Windt, Dorothee Suttorp, W.K. Heve, Ulrike Hakl, Claudia Ergezinger, Sabine Schöbel, Andrea Nicot and Monika Breuch are thanked for their unreserved technical support.

REFERENCES

1. Barker, K. R., & Olthof, T. H. A. (1976). Relationships between nematode population densities and crop responses. Annual Review of Phytopathology, 14, 327–353.
2. Been, T. H., Schomaker, C. H., & Seinhorst, J. W. (1994). An advisory system for the management of potato cyst nematodes (Globodera spp.). In A. J. Haverkort & D. K. L. McKerron (Eds.), Ecology and modelling of potato crops under conditions limiting growth (pp. 305–321). Dordrecht: Kluwer Academic Publishers.
3. Castillo, P. & Volvas, N. (2007). Pratylenchus (Nematoda: Pratylenchidae); Diagnosis, biology, pathogenicity and management. Nematology Monographs and Perspectives 6. Leiden, The Netherlands, Brill, 529 pp.
4. Coosemans, J. (1975). The influence of Pratylenchus penetrans on growth of Impatiens balsamina L., Daucus carota L., Linen usitatissimum L., and Chrysanthemum indicum L. Mededelingen van de Faculteit Landbou wwetenschappen: Gent University, 40, 465–471.
5. Dalgaard, P. (2002). Introductory statistics with R. New York: Springer. 267 pp.
6. Evenhuis, A., Korthals, G. W., & Molendijk, L. P. G. (2004). Tagetes patula as an effective catch crop for long-term control of Pratylenchus penetrans. Nematology, 6, 877–881.
7. Ferris, H. (1978). Nematode economic thresholds: derivation, requirements, and theoretical considerations. Journal of Nematology, 10, 341–350.
8. Gugino, B. K., Abawi, G. S., & Ludwig, J. W. (2006). Damage and management of Meloidogyne hapla using oxamyl on carrot in New York. Journal of Nematology, 38, 483–490.
9. Hahn, S. (1958). Wurzelgallenälchen (Meloidogyne hapla Chitwood) als Freilandschädlinge an Salat und Möhren. Nachrichtenblatt Dtsch Pflanzenschutzdienst Braunschweig, 10, 123–126.
10. Hallmann, J., Frankenberg, A., Paffrath, A., & Schmidt, H. (2007). Occurrence and importance of plant-parasitic nematodes in organic farming in Germany. Nematology, 9, 869–879.
11. Hay, F. S., & Pethybridge, S. J. (2005). Nematodes associated with carrot production in Tasmania, Australia, and the effect of Pratylenchus crenatus on yield and quality of Kuroda-type carrot. Plant Disease, 89, 1175–1180.
12. Heve, W. K., Been, T. H., Schomaker, C. H., & Teklu, M. G. (2015). Damage thresholds and population dynamics of Meloidogyne chitwoodi on carrot (Daucus carota L., cv. Nerac) at different seed densities. Nematology, 17, 501–514.
13. Hooper, D. J., Hallmann, J., Subbotin, S., & Luc, M. (2005). Methods for extracting and detection of plant and soil nematodes. In R. A. Sikora & J. Bridge (Eds.), Plant

parasitic nematodes in subtropical and tropical agriculture (2nd ed., pp. 53–86). Wallingford: CABI Publishing.

14. Jones, J., Haegeman, A., Danchin, E. G. J., Gaur, H. S., Helder, J., Jones, M. G. K., Kikuchi, T., Manzanilla- López, R., Palomares-Rius, J., Wesemael, W. M. L., & Perry, R. N. (2013). Top 10 plant-parasitic nematodes in molecular plant pathology. Molecular Plant Pathology, 14, 946–961.
15. Kimpinski, J., & Sanderson, K. (2004). Effects of crop rotations on carrot yield and on the nematodes Pratylenchus penetrans and Meloidogyne hapla. Phytoprotection, 85, 13–17.
16. Noling, J. W. (2012). Nematode management in carrots. ENY-021 Florida Cooperative Extension Service. http://edis.ifas.ufl.edu/.
17. Norshie, P. M., Been, T. H., & Schomaker, C. H. (2011). Estimation of partial resistance in potato genotypes against Meloidogyne chitwoodi. Nematology, 13, 477–489.
18. Oostenbrink, M. (1972). Evaluation and integration of nematode control methods. In J. M. Webster (Ed.), Economic nematology (pp. 497–514). San Diego: Academic Press.
19. Potter, J. W., & Olthof, T. H. A. (1993). Nematode pests of vegetable crops. In K. Evans, D. L. Trudgill, & J. M. Webster (Eds.), Plant parasitic nematodes in temperate agriculture (pp. 171–207). Wallingford: CAB International.
20. Pudasaini, M. P. (2006). Interaction between the root lesion nematode, Pratylenchus penetrans, and field crops. PhD thesis, Ghent University, Ghent, Belgium.
21. Pudasaini, M. P., Schomaker, C. H., Been, T. H., & Moens, M. (2006a). Vertical distribution of the plant-parasitic nematode, Pratylenchus penetrans, under four field crops. Phytopathology, 96, 226–233.
22. Pudasaini, M. P., Viaene, N., & Moens, M. (2006b). Effect of marigold (Tagetes patula) on population dynamics of Pratylenchus penetrans in a field. Nematology, 8, 477–484.
23. Sato, E., Min, Y. Y., Toyota, K., & Takada, A. (2009). Relationships between the damage to radish caused by the root-lesion nematode Pratylenchus penetrans, its density prior to cultivation and the soil nematode community structure evaluated by polymerase chain reaction-denaturing gradient gel electrophoresis. Soil Science and Plant Nutrition, 55, 478–484.
24. Schomaker, C. H., & Been, T. H. (2013). Plant growth and population dynamics. In R. N. Perry & M. Moens (Eds.), Plant nematology (2nd ed., pp. 301–330). Wallingford: CAB International.
25. Seinhorst, J. W. (1966). The relationships between population increase and population density in plant-parasitic nematodes I. Introduction and migratory nematodes. Nematologica, 12, 157–169.
26. Seinhorst, J. W. (1967). Review of methods for measuring damage caused by nematodes. FAO, Symposium on crop losses, Rome, Italy.
27. Seinhorst, J. W. (1970). Dynamics of populations of plant parasitic nematodes. Annual Review of Phytopathology, 8, 131–156.
28. Seinhorst, J. W. (1988). The estimation of densities of nematode populations in soil and plants. Vaxtskyddsrapporter jordbruk, pp. 43–68.
29. Seinhorst, J. W. (1998). The common relation between population density and plant weight in pot and micro plot experiments with various nematode plant combinations. Fundamental and Applied Nematology, 21, 459–468.
30. Stein, W. (1963). Untersuchungen über den Befall von Möhren durch Meloidogyne hapla Chitwood in zwei aufeinanderfolgenden Anbaujahren. Zeitschrift Pflanzenpathologie und Pflanzenschutz, 70, 217–223.
31. Teklu, M. G., Schomaker, C. H., & Been, T. H. (2014). Relative susceptibilities of five fodder radish varieties (Raphanus sativus var. oleiformis) to Meloidogyne chitwoodi. Nematology, 16, 577–590.
32. Viaene, N. M., & Abawi, G. S. (1996). Damage threshold of Meloidogyne hapla to lettuce in organic soil. Journal of Nematology, 28, 537–545.PubMedPubMedCentral

33. Vrain, T. C., & Belair, G. (1981). Symptoms induced by the lesion nematode, Pratylenchus penetrans on carrot taproots in organic soil. Phytoprotection, 62, 79–81.
34. Walker, G. E. (2004). Associations between carrot defects and nematodes in South Australia. Australasian Plant Pathology, 33, 579–584.
35. Weischer, B., & Brown, D. J. F. (2000). An introduction to nematodes: General nematology. A students' textbook. Pensoft,Sofia, Bulgaria, pp. 187.
36. Wesemael, W. M. L., & Moens, M. (2008). Quality damage on carrots (Daucus carota L.) caused by the root-knot nematode Meloidogyne chitwoodi. Nematology, 10, 261–270.

Index

N

P

S

T